Springer Theses

Recognizing Outstanding Ph.D. Research

Aims and Scope

The series "Springer Theses" brings together a selection of the very best Ph.D. theses from around the world and across the physical sciences. Nominated and endorsed by two recognized specialists, each published volume has been selected for its scientific excellence and the high impact of its contents for the pertinent field of research. For greater accessibility to non-specialists, the published versions include an extended introduction, as well as a foreword by the student's supervisor explaining the special relevance of the work for the field. As a whole, the series will provide a valuable resource both for newcomers to the research fields described, and for other scientists seeking detailed background information on special questions. Finally, it provides an accredited documentation of the valuable contributions made by today's younger generation of scientists.

Theses are accepted into the series by invited nomination only and must fulfill all of the following criteria

- They must be written in good English.
- The topic should fall within the confines of Chemistry, Physics, Earth Sciences, Engineering and related interdisciplinary fields such as Materials, Nanoscience, Chemical Engineering, Complex Systems and Biophysics.
- The work reported in the thesis must represent a significant scientific advance.
- If the thesis includes previously published material, permission to reproduce this must be gained from the respective copyright holder.
- They must have been examined and passed during the 12 months prior to nomination.
- Each thesis should include a foreword by the supervisor outlining the significance of its content.
- The theses should have a clearly defined structure including an introduction accessible to scientists not expert in that particular field.

More information about this series at http://www.springer.com/series/8790

Florian de Nanteuil

Synthesis and Reactivity of Donor-Acceptor Substituted Aminocyclopropanes and Aminocyclobutanes

Doctoral Thesis accepted by
EPFL, the Swiss Federal Institute of Technology
in Lausanne, Switzerland

Springer

Author
Dr. Florian de Nanteuil
Institut des Sciences et Ingénierie Chimiques
Swiss Federal Institute of Technology in Lausanne
Lausanne
Switzerland

Supervisor
Prof. Jérôme Waser
Institut des Sciences et Ingénierie Chimiques
Swiss Federal Institute of Technology in Lausanne
Lausanne
Switzerland

ISSN 2190-5053 ISSN 2190-5061 (electronic)
Springer Theses
ISBN 978-3-319-23005-4 ISBN 978-3-319-23006-1 (eBook)
DOI 10.1007/978-3-319-23006-1

Library of Congress Control Number: 2015946992

Springer Cham Heidelberg New York Dordrecht London

Printed on acid-free paper

Springer International Publishing AG Switzerland is part of Springer Science+Business Media (www.springer.com)

Parts of this thesis have been published in:

1. *[4 + 2] Annulations of Aminocyclobutanes*
   Perrotta, D.; Racine, S.; Vuilleumier, J.; de Nanteuil, F.; Waser J. *Org. Lett.* **2015**, ASAP, DOI: 10.1021/acs.orglett.5b00149.
2. *Diester-Substituted Aminocyclopropanes: Synthesis and Use in [3 + 2] Annulation Reactions.*
   Serrano E.; de Nanteuil F.; Waser J. *Synlett,* **2014**, 25, 2285.
3. *Synthesis of (Carbo)nucleoside Analogues via [3 + 2] Annulation of Aminocyclopropanes.*
   Racine S.; de Nanteuil, F.; Serrano E.; Waser J. *Angew. Chem., Int. Ed.* **2014**, *53*, 8484-8487.
4. *Dynamic Kinetic Asymmetric [3 + 2] Annulation Reactions of Aminocyclopropanes.*
   de Nanteuil, F.; Serrano E.; Perrotta D.; Waser J. *J. Am. Chem. Soc.* **2014**, *136,* 6239.
5. *Synthesis of Aminocyclobutanes via Iron-Catalyzed [2 + 2] Cycloaddition.*
   de Nanteuil, F.; Waser J. *Angew. Chem. Int. Ed.* **2013**, *52,* 9009.
6. *Catalytic Friedel-Crafts Reaction of Aminocyclopropanes.*
   de Nanteuil, F.; Loup, J; Waser J. *Org. Lett.* **2013**, *doi10.1021/ol401616a.*
7. *Catalytic Enantiospecific [3 + 2] Annulation of Aminocyclopropanes with Ketones*
   Benfatti F.; de Nanteuil, F.; Waser J. *Chem. Eur. J.* ***2012**, 18,* 4844.
8. *Iron-Catalyzed [3 + 2] Annulation of Aminocyclopropanes with Aldehydes: Stereoselective Synthesis of Aminotetrahydrofurans*
   Benfatti F.; de Nanteuil, F.; Waser J. *Org. Lett.* **2012**, *14*, 386.
9. *Catalytic [3 + 2] Annulation of Aminocyclopropanes for the Enantiospecific Synthesis of Cyclopentylamines*
   de Nanteuil, F.; Waser J. *Angew. Chem. Int. Ed.* **2011**, *50*, 12075.

Supervisor's Foreword

Many natural products display a saturated polycyclic core allowing a well-defined arrangement of functional groups in space. They can therefore interact with biological targets with high affinity and selectivity, surpassing many synthetic drugs. Nitrogen-containing functional groups are especially important, as the higher electron-density of nitrogen allows stronger hydrogen bonding in biological systems. Nevertheless, the efficient synthesis of such complex ring systems represents a challenge for organic chemistry.

In this thesis, a general approach for accessing nitrogen-substituted hetero- and carbocycles is presented. Through careful tuning of the electronic properties of a nitrogen donor group and a diester acceptor group, the first [3 + 2] annulation reaction between aminocyclopropanes and enol ethers or carbonyl compounds became possible. The reaction proceeded under mild catalytic conditions. The obtained building blocks can be found at the core of bioactive alkaloids, drugs such as Ramipril and biomolecules such as DNA and RNA. Access to enantioenriched compounds became possible through the dynamic kinetic asymmetric annulation of aminocyclopropanes with enol ethers and aldehydes. This impressive transformation process was mediated by a copper catalyst bearing a commercially available ligand and gave both cyclopentylamines and tetrahydrofurylamines in high yield and enantioselectivity. Finally, a synthesis of donor–acceptor aminocyclobutanes via [2 + 2] cycloaddition using a cheap iron catalyst was developed, and they could be used in [4 + 2] annulations to access cyclohexylamines.

In short, the annulation reactions developed in this thesis give access to nitrogen-substituted four-, five- and six-membered rings, all essential building blocks for the synthesis of bioactive compounds. These important fundamental results can now be used to synthesize libraries of molecules for the discovery of new bioactive chemical entities. In fact, the methodology could already be applied in our group to the synthesis of new nucleoside analogues bearing thimine, uracil and fluorouracil nucleobases. When considering that about 45 nucleoside analogues are FDA approved drugs and that the synthesized compounds have unprecedented structures, the obtained molecules could well lead to exciting results in currently ongoing biological tests. From the fundamental point of view, the synthetic potential of nitrogen-substituted small rings has just begun to be investigated, and I am certain that the work described in this thesis will serve as basis for the discovery of new modes of C-C bond activation, catalytic methods or unprecedented chemical structures, and will ultimately result in the efficient synthesis of bioactive natural products and synthetic drugs.

Lausanne
February 2015

Prof. Jérôme Waser

Abstract

The development of new methodologies in organic synthesis can greatly impact related fields such as medicinal chemistry, material science, molecular biology or environmental science. The complexity of the molecular scaffolds required is increasing, as is the demand for selective and efficient synthetic processes. Additionally, the quest for environmentally benign protocols operating at the lowest possible cost are parameters which should be taken into consideration when developing new reactions.

Molecules containing cyclic scaffolds substituted by nitrogen are ubiquitous in natural products such as indole alkaloids or DNA nucleotides as well as in highly potent synthetic pharmaceuticals. There is a high demand for efficient access to these structures in order to discover molecules with new fields of applications.

The annulation of a formal dipole with a dipolarophile provides a very convergent way to access carbo or heterocyclic structures via the construction of multiple carbon-carbon bonds in one step. Donor–acceptor substituted cyclopropanes are known, when activated by a catalyst, to generate reactive 1,3 formal dipoles. Therefore, the use of nitrogen-substituted donor–acceptor cyclopropanes in annulation reactions would provide an efficient access to nitrogen-substituted cyclic structures. The goal of this thesis was to investigate, for the first time, the intermolecular reactivity of aminocyclopropanes as 1,3 formal dipoles.

In this context, we developed the first catalytic [3 + 2] annulation of aminocyclopropanes with enol ethers. The reaction required phthalimide-substituted donor–acceptor cyclopropanes that were easily accessed in one step. The use of a tin catalyst afforded a wide range of polysubstituted cyclopentylamines with high diastereoselectivity and yields up to 99 %. The method occurred with full transfer of stereogenic information affording the products in enantioenriched form.

The first annulation of aminocyclopropanes with aldehydes was then reported. An easy-to-handle, innocuous and inexpensive iron-based catalyst was used for this purpose. The use of the same phthalimide-substituted cyclopropane as before allowed the isolation of substituted heterocycles in excellent diastereoselectivities and yields up to 99 %. The 2-aminotetrahydrofuran scaffolds are found in the core of DNA and RNA molecules. In order to increase the potential of our methods, we adapted the reaction to less reactive ketones. In this case, the use of the same catalytic system than for enol ethers gave aminotetrahydrofurans with a quaternary C5 atom in high yields. The reaction turned out to be diastereoselective as well as enantiospecific, allowing the isolation of the valuable analogues with an enantiomeric ratio of 98:2.

Again, phthalimide-substituted donor–acceptor cyclopropanes were shown to be efficient electrophilic acceptors in the scandium triflate catalyzed Friedel-Crafts alkylation of aromatic nucleophiles. Indoles substituted with electron-rich as well as electron-poor groups were efficiently alkylated in the C3 position. When C3-substituted indoles were employed, a cationic rearrangement afforded the C2-alkylated products. The reaction tolerated a broad range of aromatic nucleophiles and afforded gamma amino acid derivatives present in important pharmacophores with high yields.

In order to access enantioenriched five-membered carbo- and heterocycles without requiring an enantiopure starting material, we developed the first dynamic kinetic asymmetric [3 + 2] annulation reaction of aminocyclopropanes with enol ethers and aldehydes. The donating functionality had to be optimized and it was found that succinimide substituted donor–acceptor cyclopropanes gave optimal results. The reaction was catalyzed by a copper complex in combination with a commercially available bisoxazoline ligand and tolerated both enol ethers and aldehydes as partners. The cyclopentylamines and aminotetrahydrofurans were obtained in up to 99 % yield and a 98:2 enantiomeric ratio. In order to expand the range of accessible nitrogen-substituted cyclic structures, the reactivity of aminocyclobutanes as 1,4 dipoles was also investigated. First, an efficient access to donor–acceptor substituted aminocyclobutanes was developed. Using the same iron catalyst as for the annulation of aldehydes with cyclopropanes, [2 + 2] cycloaddition between enimides and alkylidene malonates afforded aminocyclobutanes with a broad range of substituents. The products were obtained with yields up to 96 % yields and diastereoselectivities superior to 20:1. The reaction was optimized in order to be conducted easily on a multigram scale. The products were converted to peptide surrogates in a three-step protocol.

Finally, these donor-acceptor substituted aminocyclobutanes were successfully used as formal 1,4 dipoles in annulations with silyl enol ethers. Using tin tetrachloride at -40 °C afforded the six-membered ring analogues in yields up to 98 % and with diastereoselectivities up to 20:1.

Keywords Donor-acceptor substituted cyclopropanes · Aminocyclopropanes · [3 + 2] annulations · Dynamic kinetic asymmetric transformation · Donor-acceptor substituted cyclobutanes · Aminocyclobutanes · [4 + 2] annulations

Contents

Abbreviations, Acronyms and Symbols

$[\alpha]_D^{25}$	Specific rotation at 25 °C at the sodium D line
°C	Degrees centigrade
A	Acceptor
Å	Angstrom
aq	Aqueous
Ar	Aryl
atm	Atmosphere
BINAP	2,2'-bis(diphenylphosphino)-1,1'-binaphthyl
BINOL	1,1'-Bi-2-naphthol
Bn	Benzyl
boc	*Tert*butyloxycarbonyl
BOX	Bisoxazoline
br	Broad
brsm	Based on recovered starting material
Bu	Butyl
ca	Circa
calcd	Calculated
cat	Catalytic
COD	Cyclooctadiene
conv	Conversion
cy	Cyclohexane
cyn	Cynnamyl
D	Donor
d	Doublet
DA	Donor-acceptor substituted
DABCO	1,4-diazabicyclo[2.2.2]octane
DBU	1,8-diazabicycloundec-7-ene
DCE	Dichloroethane
DCM	Dichloromethane
DIPEA	Di*iso*propylethylamine

DKR	Dynamic kinetic resolution
DMAP	4-dimethylamino pyridine
DMF	*N,N*-dimethylformamide
DMSO	Dimethyl sulfoxide
DNA	Deoxyribonucleic acid
dr	Diastereomeric ratio
DyKAT	Dynamic kinetic asymmetric transformation
E^+ or El	Electrophile
EDG	Electron donating group
ee	Enantiomeric excess
EI	Electron impact ionization
eq	Equivalent
er	Enantiomeric ratio
ESI	Electrospray ionization
Et	Ethyl
EtOAc	Ethyl acetate
EWG	Electron-withdrawing group
FC	Friedel-Crafts
g	Gram
GC	Gas chromatography
h	Hour(s)
HIV	Human immunodeficiency virus
HPLC	High pressure liquid chromatography
HRMS	High Resolution Mass Spectroscopy
Hz	Hertz
J	Coupling constant
Kcal	Kilo calories
KHMDS	Potassium hexamethyldisilazane
KR	Kinetic resolution
L	Liter
M	Molarity
m	Multiplet
m/z	Mass per electronic charge
Me	Methyl
mg	Milligram
min	Minute(s)
mixt	Mixture
mL	Milliliter
mmol	Millimol
Mp	Melting point
MS	Molecular Sieves
MW	Molecular weight
NGF	Nerve growth factor
NHC	N-heterocyclic carbene
NMR	Nuclear magnetic resonance

Nu	Nucleophile
OTf	Triflate
p	Para
Ph	Phenyl
Phth	Phthalimide
PMP	Paramethoxyphenyl
ppm	Parts per million
Pr	Propyl
Py-BOX	pyridine-bisoxazoline
q	Quartet
quant	Quantitative
QUINOX	2-(4,5-dihydro-2-oxazolyl)quinoline
quint	Quintet
R_f	Retention factor
RNA	Ribonucleic acid
rt	Room temperature
s	Singlet
SALEN	Salicylaldehyde ethylenediamine
sat	Saturated
S_E	Electrophilic Substitution
S_N	Nucleophilic Substitution
sol	Solution
Succ	Succinimide
T	Temperature
t	Triplet
TADDOL	(α,α,α,α-tetraaryl-1,3-dioxolane-4,5-dimethanols)
TBAF	tetra-*n*-butylammonium fluoride
TBDPS	*tert*butyldiphenylsilyl
TBS	*tert*butyldimethylsilyl
^{t}Bu	*tert*butyl
TCNE	tetracyanoethene
TEMPO	(2,2,6,6-tetramethylpiperidin-1-yl)oxy
TES	triethylsilyl
Tf	trifluoromethanesulfonyl
TFA	trifluoroacetic acid
THF	tetrahydrofuran
TIPS	tri*iso*propylsilyl
TLC	Thin layer chromatography
TMS	trimethylsilyl
Tos	Tosyl
δ	NMR chemical shift in ppm
μL	Microliter
ν	Frequency (cm^{-1})

Chapter 1
Introduction

1.1 Introduction

Since the beginning of the last century, organic chemistry has been playing an important role in the development of modern society. It has led to innovation in major disciplines such as drug development, crop protection, material and environmental sciences. However, the demand for molecular scaffolds that are more and more complex challenges the efficiency of current synthetic methods and highlights the urgent need for new strategies. The development of new reactions is a very important process by which progress in the disciplines requiring molecule engineering can be achieved. A new methodology has to bring a solution to recognized problems. For this purpose, it can improve an existing process by, for example, reducing its cost, increasing its field of application or its efficiency. It can also be based on a new strategy to build molecules in a way that did not exist before. This process can be very rewarding as it can also lead to the exploration of a chemical space that was unexploited previously.

In the field of bioactive compounds, carbo- and heterocycles substituted with a nitrogen atom are predominant scaffolds. On one side, the presence of a cyclic structure in a molecule allows the control of the orientation of its substituents in space. It is very important in drug discovery for example, as the interactions of the molecule with a target are directed in the three dimensions of a confined space. A small variation of the orientation of one substituent can have a dramatic effect on the bioactivity of the molecule. On the other side, nitrogen-containing functionalities are essential for life and can be found in numerous biomolecules, from simple amino acids or DNA bases to incredibly complex alkaloids. There, the nitrogen atom can play a key role in the structural organization of macromolecules via non-bonding interactions or be involved in enzymatic transfer of reactive intermediates via charged-relay network.

There is a vast amount of reactions that are aiming to access cyclic structures. In this context, cycloadditions and annulations are powerful synthetic methods to access

F. de Nanteuil, *Synthesis and Reactivity of Donor-Acceptor Substituted Aminocyclopropanes and Aminocyclobutanes*, Springer Theses,
DOI 10.1007/978-3-319-23006-1_1

these structures because multiple bonds are created in one step. However, controlling the stereoselective construction of cyclic molecules in the presence of diverse functionalities is a goal that is difficult to achieve. Additionally, nitrogen is an element which is not trivial to introduce in molecules due to its basicity. Many catalytic systems and reagents are not compatible with molecules containing a nitrogen atom as they can act as strong deactivating ligands or promote side reactions.

Therefore, developing a synthetic method to access these privileged scaffolds containing both the cycle and the nitrogen substitution would be very valuable. In the past, donor-acceptor substituted cyclopropanes and cyclobutanes have proven their synthetic utility as formal 1,3 and 1,4 dipoles. However none of these studies reported the use of nitrogen-substituted cyclopropanes or cyclobutanes to access nitrogen-containing cyclic structures via formal cycloadditions. In the hands of synthetic chemists, such molecules would have a dual interest as the nitrogen atom can be used as a donating group to control the ring-opening and provide the nitrogen functionality present in the targeted products.

In this thesis I will describe the development of new methodologies involving the synthesis and reactivity of small nitrogen-substituted cycloalkanes. The introduction will give a description of the importance of nitrogen-containing cyclic structures that can be access using the annulation methods developed during my thesis. I will then present the structural properties and the reactivity of the cycloalkanes used for the project. In the last part of the introduction, I will focus on the intermolecular reactivity of aminocyclopropanes and aminocyclobutanes in annulation reactions. After introducing the different objectives of the project, the main part of the thesis reporting my results will be split into two parts: first, the development of intermolecular ring-opening reactions of aminocyclopropanes and second the synthesis of aminocyclobutanes and their applications in [4 + 2] annulations. Lastly, after a short conclusion and outlook section, the final chapter of this thesis will include the experimental and spectral data supporting the results.

1.2 Importance of Nitrogen-Substituted Four, Five and Six-Membered Rings

As stated previously in the introduction, the occurrence of nitrogen-substituted cyclic structures is very high in bioactive molecules. Nevertheless, there are other fields of research such as glycomimetics or peptidomimetics where such amino-substituted cyclic structures are also highly relevant [1, 2].

1.2.1 Four-Membered Rings

The aminocyclobutane unit can be found in natural as well as synthetic bioactive structures (Fig. 1.1) [3]. For example, lannotidine F (**1**) enhances nerve growth

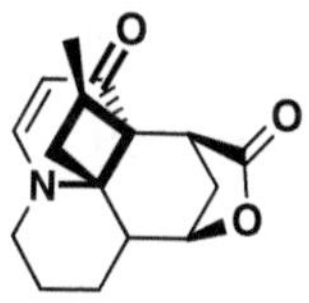

Lannotinidine F (**1**)
NGF stimulation

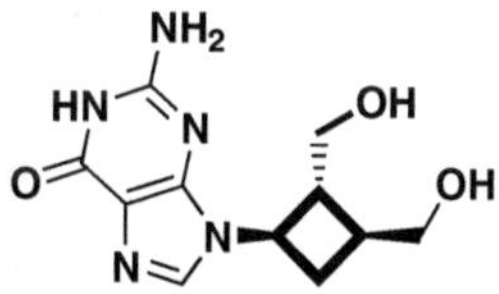

Cyclobut-G (**2**)
Anti-HIV

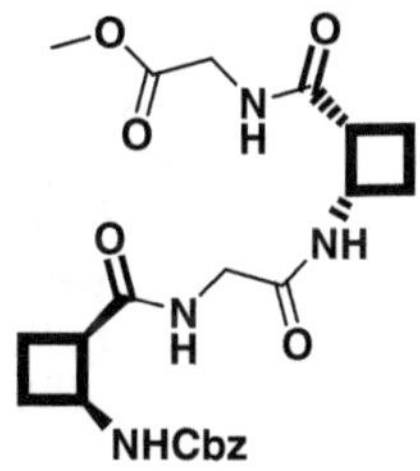

3
Low MW gelator

Fig. 1.1 Molecules containing aminocyclobutanes

factor mRNA expression and is the most potent metabolite isolated from the club mosses *Lannotinum* and *Lannotinum varacrifolium*. Cyclobut-G (**2**) is a synthetic carbocyclic analogue of the powerful anti-HIV natural product, Oxetanocin A [4].

Aminocyclobutanes can also be used as a platform to lock the spatial configuration of their substituents in space. For example, Ortuño and co-workers synthesized aminocyclobutane-based peptide **3** which can self-assemble at a macroscopic level and behaves as a low molecular weight gelator [5, 6]. The β substitution of aminocyclobutanes with a carboxylic acid provides β-amino-acid derivatives that can be used to build protease-resistant peptides [7].

1.2.2 Five-Membered Rings

The five-membered cycles are far more widespread than their four-membered counterparts and in this part we will only focus on aminocyclopentanes and aminotetrahydrofurans as they are relevant to the structures obtained by the methods developed in this thesis. The carbocyclic core can be found in complex molecules such as palau'amine [8] (**4**) or pactamycin [9–11] (**5**) which are two important bioactive natural products. Synthetic and natural carbonucleoside analogues such as abacavir (**6**) and aristeromycin (**7**) are organized around a poly-substituted aminocyclopentane core and possess valuable bioactivities against viral and cancer diseases. With a similar aminocyclopentyl structure, ramipril (**8**) represents billions of dollars in sales for the treatment of heart failure.

The aminotetrahydrofuran structure can be found in the main core of DNA nucleosides such as thymidine (**9**). Since the approval of cytarabin (**10**) in 1969 for the treatment of white blood cell cancers, nucleosides have been a scaffold of choice for the development of new drugs, especially against viral infections (**11** and **12**, Fig. 1.2) and various types of cancer [12].

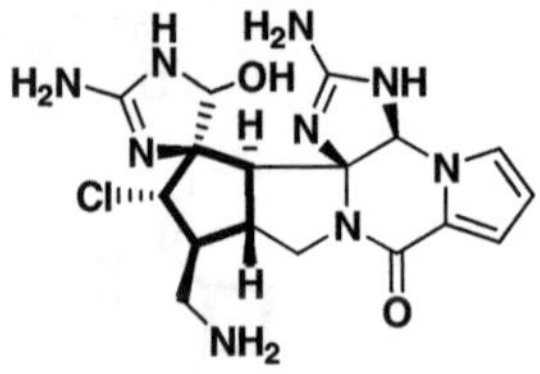

Palau'amine (**4**)
Bioactive natural product

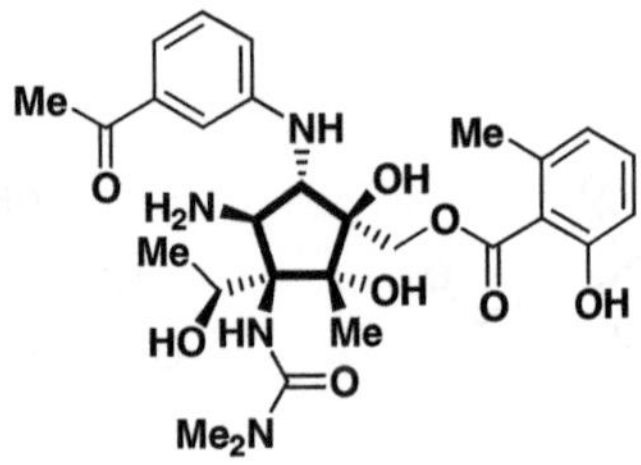

Pactamycin (**5**)
Bioactive natural product

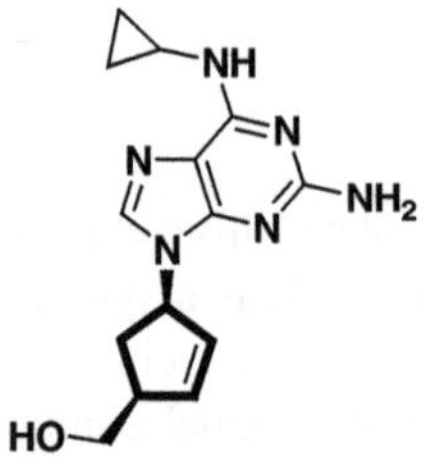

Abacavir (**6**)
Anti-HIV

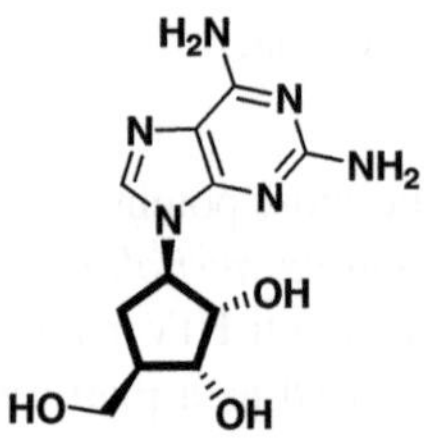

Aristeromycin (**7**)
Antiviral

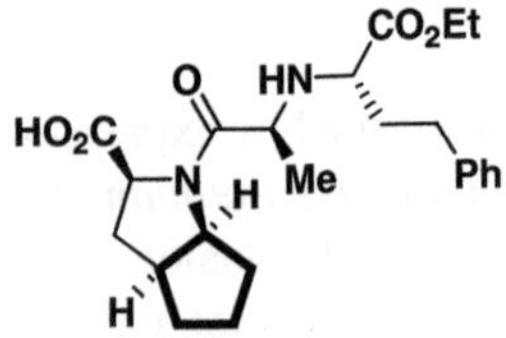

Ramipril (**8**)
Heart diseases

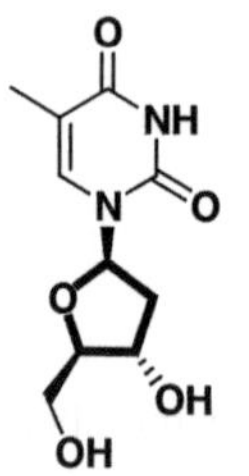

Thymidine (**9**)
DNA

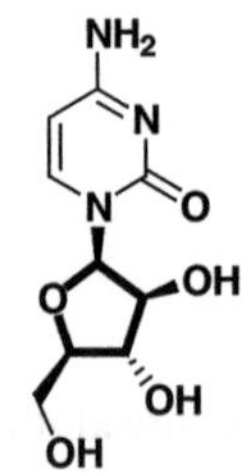

Cytarabine (**10**)
Anti-cancer

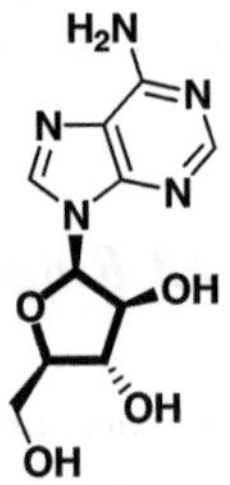

Vidarabine (**11**)
Anti-viral

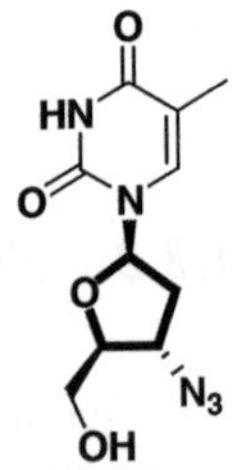

AZT (**12**)
Anti-HIV

Fig. 1.2 Bioactive molecules containing five-membered carbo- and heterocycles

1.2.3 Six-Membered Rings

The six-membered carbocycle is the most stable and therefore probably the most recurrent cyclic structure in organic chemistry. Countless methods have been developed to access this scaffold, the most famous being the Diels-Alder cycloaddition. As expected the number of these cyclic compounds substituted with a nitrogen atom is very large. For examples, pharmaceutical blockbusters representing billions of dollars in sales such as tamiflu (**13**) or oxycodon (**14**) both incorporate this motif. Furthermore, entire families of alkaloids are organized

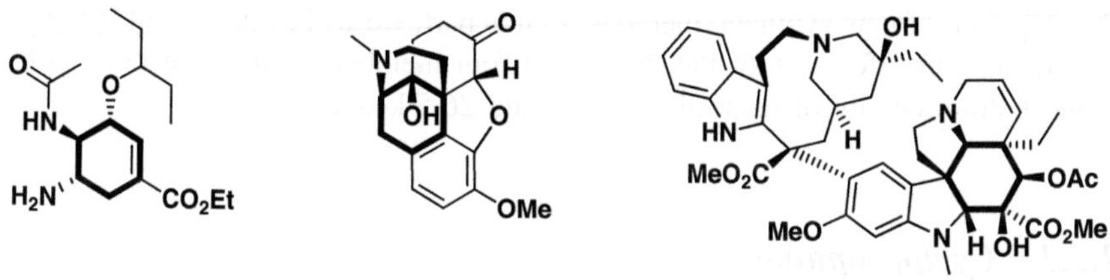

Tamiflu (**13**)
Antiviral

Oxycodon (**14**)
Analgesics

Vinblastine (**15**)
Antitumor

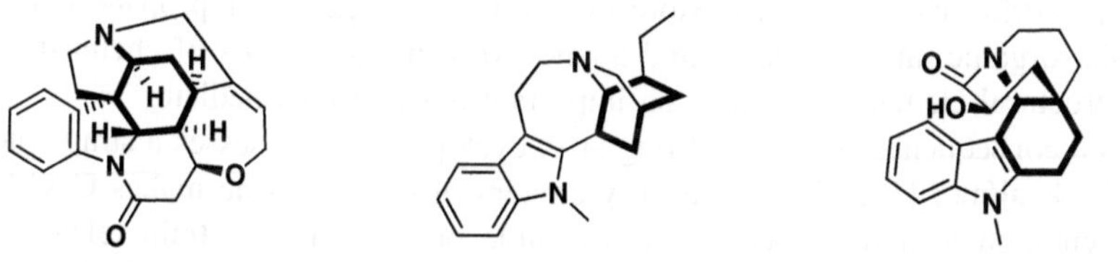

Strychnine (**16**)
Poison

Ibogaine (**17**)
Psychotrope

Kopsihainanine A (**18**)
Antitumor

Fig. 1.3 Bioactive molecules containing aminocyclohexane

Table 1.1 Comparison of selected properties of cycloalkanes

	C–C–C (°)	"Bent" bond (°)	C–C (Å)	Strain energy (kcal/mol)
Cyclopropane	60	18.8	1.512	27.5
Cyclobutane	88	6.7	1.556	26.5
Cyclopentane	109	–	1.546	6.2
Cyclohexane	109.5	–	1.536	0

around a central aminocyclohexyl core and often exhibit important bioactivity (**15**–**18**, Fig. 1.3) [13, 14].

In conclusion, four-, five- and six-membered cyclic structures substituted with nitrogen are omnipresent in bioactive molecules. Methods to selectively access these privileged scaffolds are highly desirable as they can efficiently generate libraries of compounds with potentially high bioactivity.

1.3 Cyclopropanes and Cyclobutanes: General Structure and Reactivity

As we can see in Table 1.1, cycloalkanes possess different structural properties depending on their size. Establishing the relationship between bond lengths, angles, strain energies on one side and reactivity on the other side is very important for the development of new synthetic methods. In this part, we will compare and comment on the structural features of the three- and four-membered rings that possess the highest strain energy of respectively 27.5 and 26.5 kcal/mol.

1.3.1 Cyclopropanes

In 1882, August Freund discovered cyclopropane and assigned it the correct C_3H_6 formula. Its industrial production began around 1930 as it gained strong interest as a very powerful anesthetic [15]. More than 130 years later, cyclopropane is a very popular organic unit that has found applications in many fields of chemistry as a key structural element as well as an important reactive intermediate.

As a consequence of its small ring size, cyclopropane possesses a strain energy of 27.5 kcal/mol [16, 17]. The energy content of cyclopropane makes C–C bond fragmentation a thermodynamically favorable process due to strain release. The extra energy can be decomposed into two main contributions, namely the Bayer and the Pitzer strain [18]. The former results from the angular strain arising from the bond angle of 60° instead of 109.5°. The latter, also called torsional strain, is due to the unfavorable eclipsed conformation of cyclopropyl substituents.

In order to explain the high reactivity of cyclopropane, thermodynamics alone is not sufficient. Cyclobutane possesses similar strain and homolytic cleavage energy (61 kcal/mol for cyclopropane vs. 62.5 kcal/mol for cyclobutane). However, if the reactivity of cyclopropane can be compared to the one of a double bond, it is not the case for cyclobutane. To explain this, there have been many studies and discussions concerning the electronic configuration of cyclopropanes. Three different models can be used to describe these unusual properties.

- The Coulson and Moffit model (Fig. 1.4a) describes cyclopropane as having sp^3 hybridized carbons but with a greater *p*-character. This is required for the bond to accommodate with a smaller angle [19]. C–C bonds are described as banana bonds or bent bonds due to the greater density of electron lying outside the bond axis. This phenomenon has been observed by X-RAY diffraction [20, 21].
- The Walsh model (Fig. 1.4b) considers that cyclopropane is composed of three methylene sp^2 units [22, 23]. In this model, the molecular orbitals are built by combining three sp^2 hybridized orbitals with three *p* orbitals. The poor overlap of the CH_2 orbitals is at the origin of the angular strain and the sp^2-like reactivity of cyclopropane.
- The Dewar model implies that the 6 σ electrons of the three C–C bonds are aromatic as stated by the 4n + 2 rule [24]. A set of physical and chemical properties can therefore be explained using this model, for example, the upfield shift of the protons of cyclopropane in NMR or its reactivity toward electrophiles.

Each of these models has its strengths and weaknesses and some studies have tried to unify this bonding concept in order to picture a clear representation of the electronic structure of the cyclopropane [25]. These theoretical considerations are

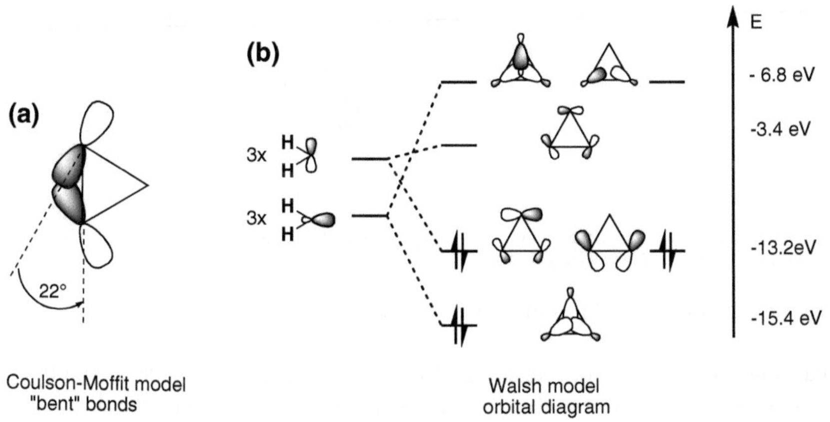

Fig. 1.4 Coulson-Moffit and Walsh model representations

essential for organic chemists as it can help them to discover new concepts of reactivity.

1.3.2 Cyclobutanes

Cyclobutane has a very interesting structure and reactivity profile that lies between the very special cyclopropane and "normal" cycloalkanes. Although it only has one extra carbon atom compared to cyclopropane, this small difference has a large impact on the spatial organization of the atoms as well as its electronic structure [26, 27].

The bond angles in a cyclobutane have a value of 88° instead of the 90° expected in a square planar structure [28]. It implies that the four carbon atoms are not in the same plan but instead adopt a puckered conformation with an angle of *ca* 27° (Fig. 1.5a) that increases bond angle strain. On the other hand, it strongly reduces the torsional strain with a global gain of around 1.5 kcal/mol. The same model of "bent" bonds can be applied to cyclobutane. However, the deviation angle is inferior to the one of cyclopropane with a value of *ca* 7° (Fig. 1.5b).

The methylene units are oriented inward the concave face in order to favor the overlap of the "bent" bonds. As cyclopropanes have shorter bond lengths by *ca* 0.040 Å compared to linear alkanes, it is unexpected to find longer bond lengths in cyclobutane than in cyclohexane or cyclopentane (1.556 vs. 1.536 and 1.546 Å respectively). It can be credited to the presence of C–C transannular interactions (Fig. 1.5c). These interactions are not present in cyclopropane and represent a major contribution to the total strain of cyclobutanes.

For a long time, cyclobutanes have been less exploited than cyclopropanes due to their decreased reactivity. Their use has been mostly limited to intramolecular

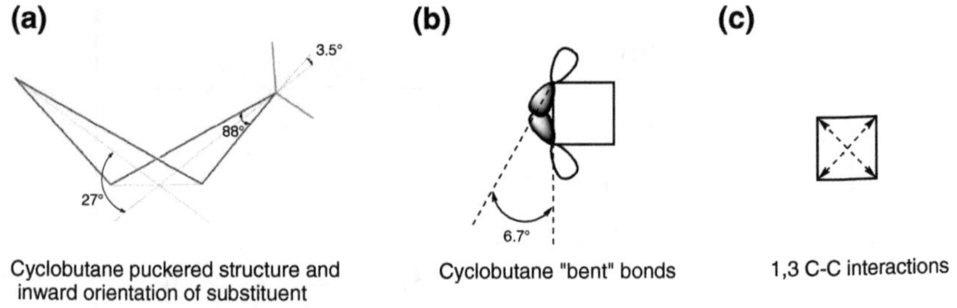

Fig. 1.5 Physico-chemical properties of cyclobutane

reactions such as electrocyclizations. However, since a few decades, cyclobutanes have gain strong interest in the organic chemistry community with the development of new methodologies involving Lewis acid or transition metal activation [29].

1.4 Donor-Acceptor Substituted Cycloalkanes

1.4.1 Structure and Reactivity

The strain energy inherent to small rings is a very powerful driving force for ring-opening reactions. However, under standard conditions of temperature and pressure, these molecules stay relatively inert. Thermal activation is limited for such system as it is very difficult to obtain a regio- and stereoselective pathway to a desired product. One way to control selective ring fragmentation consists in installing activating substituents on selected positions of small cycloalkanes (Fig. 1.6).

Two types of substituents can be used in order to promote ring activation:

- Donor substituents (D) are electron-releasing and therefore stabilize formal positive charges. In this case the ring is nucleophilic and react faster with electrophiles. Such donating groups are aromatic rings, alkoxy, thioesters or alkysilanes. Strong donating heteroatom such as nitrogen have been less investigated due to the difficulty to obtain stable yet reactive cyclopropanes or cyclobutanes.

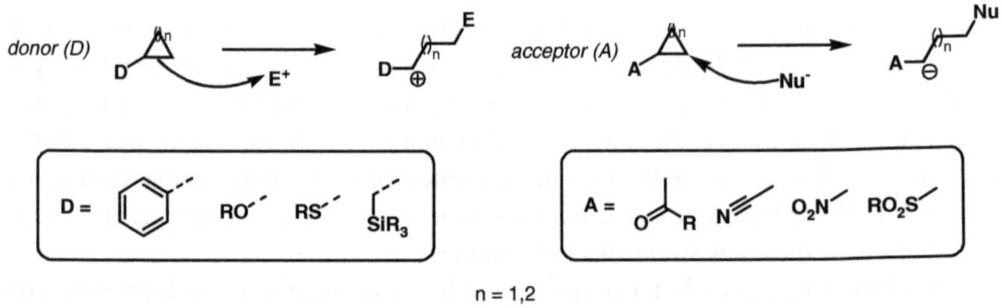

Fig. 1.6 Donor and acceptor substituted small cycloalkanes

- Acceptor (A) substituents are electron-withdrawing and stabilize formal negative charges. The ring will be electron-poor and will react faster with nucleophiles. Typical groups that are used for this purpose are carbonyls, nitro, nitriles or sulfones.

The combination of both electron-donating and withdrawing substituents positioned in a vicinal fashion (*push-pull system*, a, Scheme 1.1) has been most often used. The term donor-acceptor substituted cyclopropanes (DA cyclopropanes) appeared in the 1980s and is attributed to Doyle and Van Leusen [30]. The use of HOMO-raising or LUMO-lowering catalysts, depending on which group is activated, brings a kinetic trigger to promote ring fragmentation.

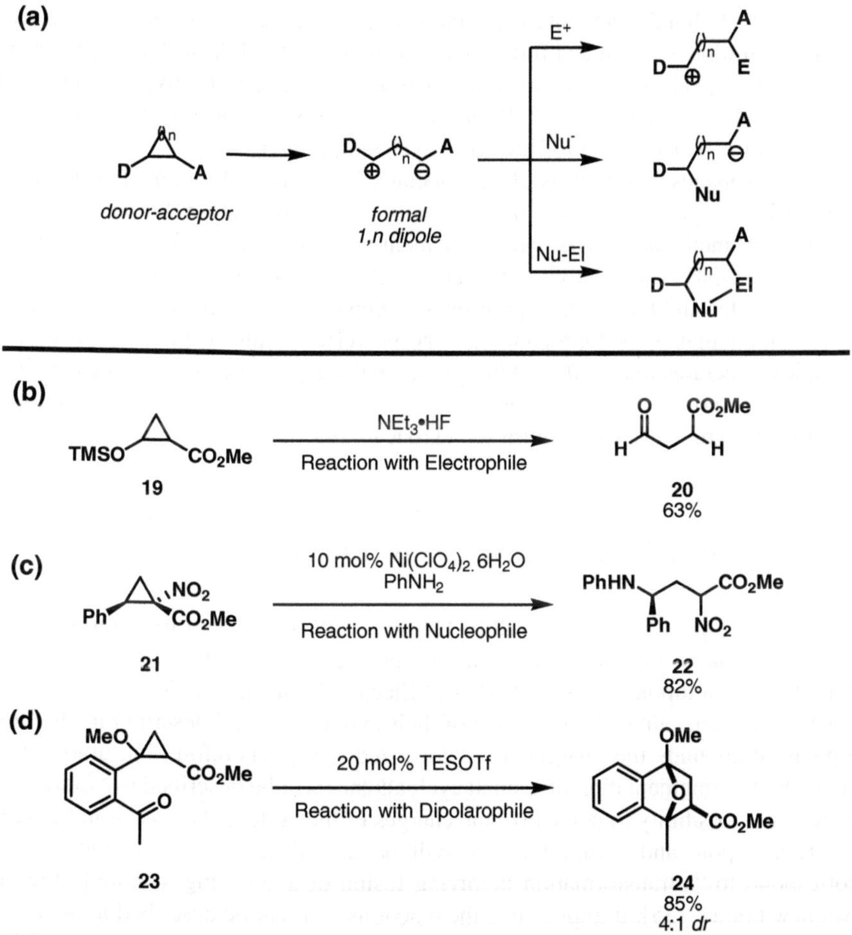

Scheme 1.1 Donor-acceptor substituted small cycloalkanes and examples of their reactivity

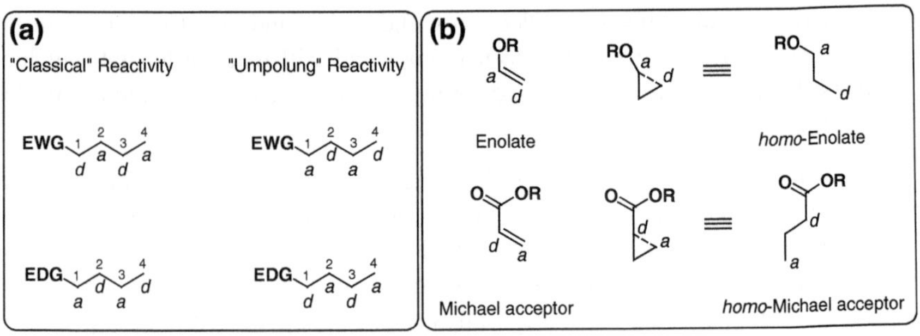

Fig. 1.7 Umpolung reactivity of cyclopropanes

When C–C bond fragmentation occurs, a *push-pull* system generates a reactive formal 1, *n* dipole which can further react with an electrophile such as a proton (b, Scheme 1.1) [31], with a nucleophile such as a free amine to give γ-amino acid analogues **22** (c, Scheme 1.1) [32], or with a dipolarophile such as the π system of carbonyls (d, Scheme 1.1) [33] to form cyclized product **24**.

Cyclopropanes reactivity is often associated to the one of olefins. Furthermore, due to their cyclic structure, the reactivity occurs one atom further from the polarizing functionality. They are consequently considered as *Umpolung* synthons (a, Fig. 1.7) [34]. The reactivity of electron rich cyclopropanes can be associated to *homo* enolates and the electron poor ones to *homo* Michael acceptors (b, Fig. 1.7). This feature makes cyclopropanes a very powerful synthetic platform to access complex structures that could not be synthesized using conventional reactivity. It is important to notice that in the case of cyclobutanes, the presence of an additional atom in the ring causes the return to "classical" reactivity.

1.4.2 Formal Cycloaddition versus Annulation

As the main topic of this thesis will be focused on the creation of new ring systems, it is important to have a look at the terminology for such transformations. The definition for a dipole by the IUPAC is: "Electrically neutral molecules carrying a positive and a negative charge in one of their major canonical descriptions. In most dipolar compounds the charges are delocalized" [35]. Therefore, the dipole generated by the fragmentation of a small cycloalkanes will be described as "formal" as there is no possibility to delocalize the charges of the system. The reaction between a formal dipole and a dipolarophile will be described as an annulation which correspond to "a transformation involving fusion of a new ring to a molecule via two new bonds" [35]. It implies that the reactions that will be described in the result part are not concerted but instead stepwise addition/ring closure events. Other authors would describe these reactions as formal cycloadditions or even cycloadditions.

1.4.3 Synthesis and Reactivity of Donor-Acceptor Substituted Cyclopropanes

DA cyclopropanes have been the focus of many investigations concerning their synthesis or their reactivity[1] and have been crucial building blocks for the total synthesis of important natural products [46–48]. This part will describe in a first time the most common methods developed to access these particular scaffolds and secondly will present their reactivity in intermolecular ring-opening reactions and annulations.

1.4.3.1 Synthesis of Donor-Acceptor Substituted Cyclopropanes

The synthesis of donor-acceptor substituted cyclopropanes is a well-established chemistry and we will not enter deep into details in this chapter [49]. The most efficient methods to access substituted three-membered rings, including DA ones, are represented in Scheme 1.2. The reaction of carbenoid precursors such as diazo or iodonium ylides in presence of a metal complex is the most popular method for the cyclopropanation of electron-rich olefins (a, Scheme 1.2). This method presents two main advantages. First, metallocarbenes are electron-deficient and can react with electron-rich olefins. Second, electron-poor diazo compounds, such as diazocarbonyls, are stable under standard conditions and therefore easy-to-handle. The reaction between organometallic intermediates and olefins has been widely used under the conditions developed by Simmons and Smith [50] and then later modified by Furukawa (b, Scheme 1.2) [51]. The reaction is highly stereospecific and its outcome is strongly improved when allylic alcohols are used as substrates for the reaction.

Finally, the Michael-Initiated Ring Closure (MIRC) is compatible only with αβ-conjugated systems (c, Scheme 1.2). The most famous reaction in this family is the Corey-Chaykovsky cyclopropanation that provides a very simple protocol to cyclize electron-poor olefins with unsubstituted sulfur or phosphorus ylides. The use of diazo compound and MIRC have been the most employed methods for the synthesis of DA cyclopropanes as the donating and the accepting part are directly introduced during the cyclopropanation reaction.

1.4.3.2 Friedel-Crafts Alkylation

There are many reported reactions involving nucleophilic ring opening of DA cyclopropanes. As they involve three-membered rings, these reaction can be described as *homo*-Michael additions. The use of aromatic nucleophiles (Friedel-Crafts alkylations) for the ring opening of DA cyclopropanes gives access to molecules that

[1] For reviews concerning activated cyclopropanes: [36–45].

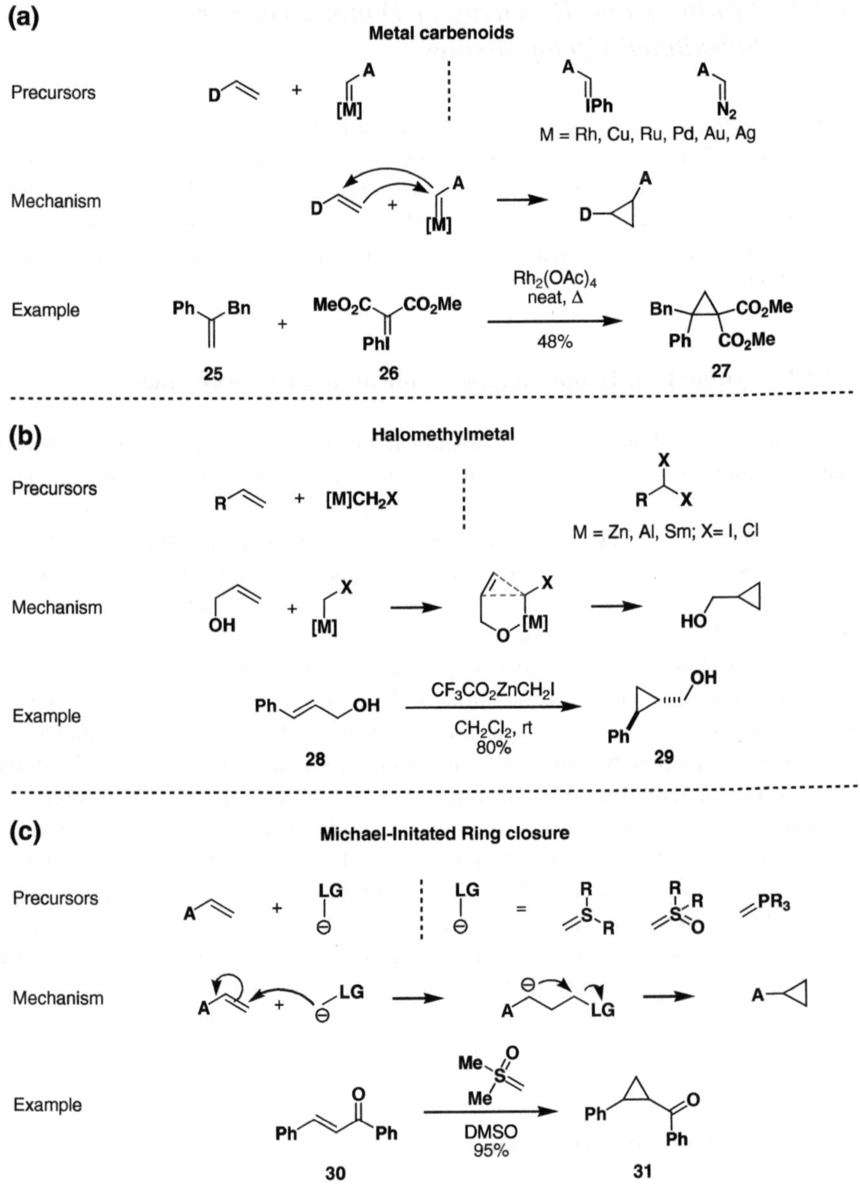

Scheme 1.2 Representative methods for cyclopropane synthesis

would be difficult to construct via classical disconnections. However, the study of such reactions has been limited to few examples.

FC alkylations using DA cyclopropanes as electrophiles were first developed by Kerr and co-workers in 1997 [52]. He discovered that using ytterbium triflate as a catalyst, it is possible to obtain the addition product of N-protected indoles in good

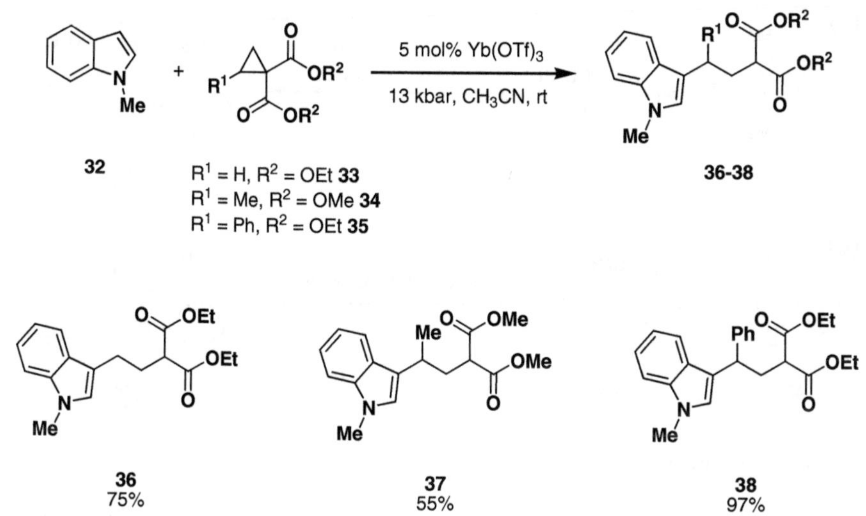

Scheme 1.3 FC-alkylation of N-Methyl indoles

yields (Scheme 1.3). He applied this methodology to acceptor **33** as well as donor-acceptor substituted cyclopropanes **34** and **35**. The donating substituents were methyl for cyclopropane **34** and phenyl for cyclopropane **35**. If unprotected indole was used, cyclization to indoline occurred with cyclopropane **34**. The authors noticed that the substitution took place at the substituted center, indicating a cationic character in the transition state. Kerr and co-workers reported an extension of this method in 2010 with the use of 2-alkynylindoles to access tetrahydrocarbazoles via a tandem FC/Conia-ene reaction [53].

Interestingly, Kerr showed 14 years later that hydrolysis of one ester in the starting cyclopropane was enough to activate it with no need of a catalyst (Scheme 1.4) [54]. The reaction occurred under high pressure (13 kbar) and probably proceeded via intramolecular H-Bonding in cyclopropane **39**. The high pressure was required as conventional and microwave heating failed to deliver the

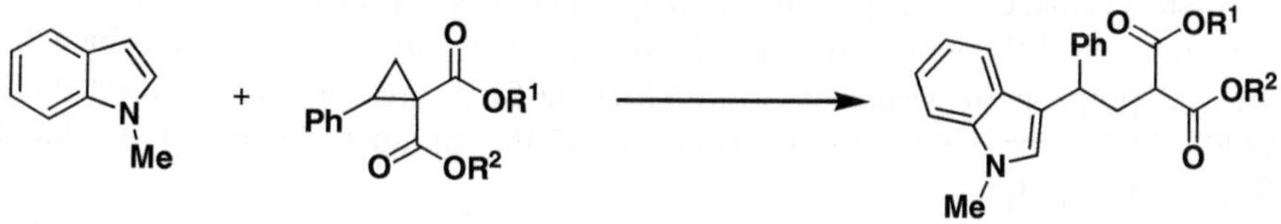

	R_1	R_2	Conditions	Yield
35	Et	Et	5 mol% $Yb(OTf)_3$, rt, 13 kbar, CH_3CN	97%, **38**
39	Me	H	50°C, 13 kbar, CH_3CN	70%, **40**

Scheme 1.4 Lewis acid versus metal-free reactivity

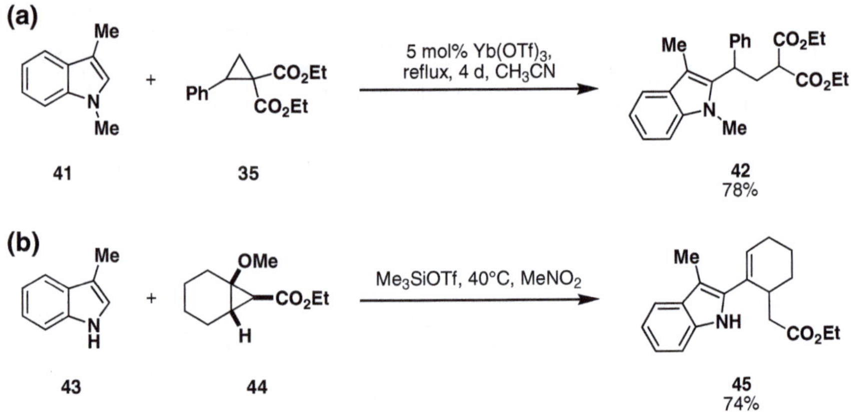

Scheme 1.5 C2 FC alkylations

product. The authors expected the volume of activation for this reaction to be negative and therefore applying high pressure would serve as a driving force. It also allows to suppress spontaneous decarboxylation. This useful metal-free methodology was extended to a range of aromatic-substituted cyclopropanes.

In 2002, the same group studied the reactivity of C3-substituted alkylindoles (Scheme 1.5a) [55]. Compared to the first methodologies, higher temperatures were required. It was shown in the study that the reaction started with a nucleophilic attack on C3 and subsequently 1,2 migration/rearomatization occurred. Another example of C2-FC alkylation was reported by Pagenkopf and co-workers in 2007, this time using bicyclic methoxy-substituted DA cyclopropane **44** (Scheme 1.5b) [56]. Elimination of methanol afforded the unsaturated cyclohexene derivative **45** in 74 % overall yield.

Ivanova and co-workers studied the reactivity between DA cyclopropanes and substituted anthracenes in 2008 [57]. If in the majority of the conditions tested they accessed cyclic products, they were still able to obtain the FC adducts with substituted anthracenes such as **46** (a, Scheme 1.6). The same group also reported later a single example of FC addition of furan **48** to DA cyclopropane **35** (b, Scheme 1.6) [58]. Again the methodology was more focused on accessing cyclized adducts.

Another isolated example of FC alkylation was reported by Gharpure and co-workers in 2012 while studying the reactivity of nitrogen-substituted bicyclic DA-cyclopropane **50**. Using electron-rich arene **51** in presence of TMSOTf, they could obtain one single diastereoisomer of the product **52** in 91 % yield (Scheme 1.7) [59].

As we can see, Friedel-Crafts alkylations of DA cyclopropanes have been reported using different systems. However, only Kerr, Pagenkopf and co-workers developed methodologies that were compatible with a wide range of substrates, the other reaction being single examples of reactivity. Compared to the other types of ring-opening reactions involving DA cyclopropanes, the research concerning FC alkylation has been less intense. It is therefore expected to see the development of

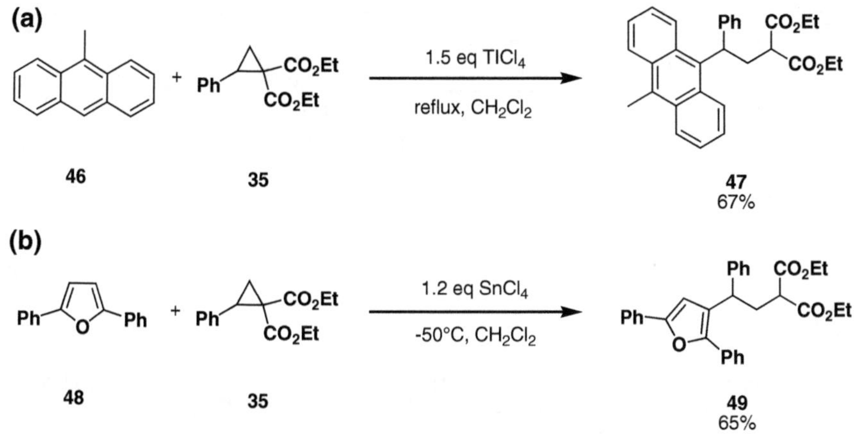

Scheme 1.6 FC alkylations of methylanthracene **46** and diphenylfuran **48**

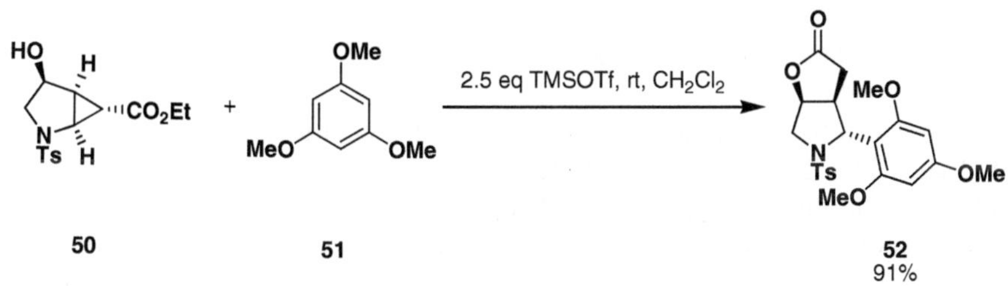

Scheme 1.7 FC alkylation with bicyclic aminocyclopropane **50**

new methodologies involving FC in the future, especially concerning the alkylations with new DA cyclopropanes.

1.4.3.3 Lewis Acid Mediated [3 + 2] Annulation of Olefins and Alkynes

[3 + 2] annulations between DA cyclopropanes and π-systems received large attention during the last decades as they enable the efficient access to five-membered hetero- and carbocycles. The first examples of such reactivity were reported by Stork and Grieco (a, Scheme 1.8) as well as Corey and Balanson in the late 1960s and early 1970s (b, Scheme 1.8) [60, 61]. The reaction involved intramolecular cyclization of unactivated olefins onto DA cyclopropanes. Nucleophilic attack of the olefin occurs via formation of the most stable tertiary cation in **54**. Subsequent ring closure allows the formation of the tricyclic product **55**. Although there is a large proportion of intramolecular version of [3 + 2] annulations in the literature, especially in the field of total synthesis, we will focus on the intermolecular processes that are related to the work presented in this thesis.

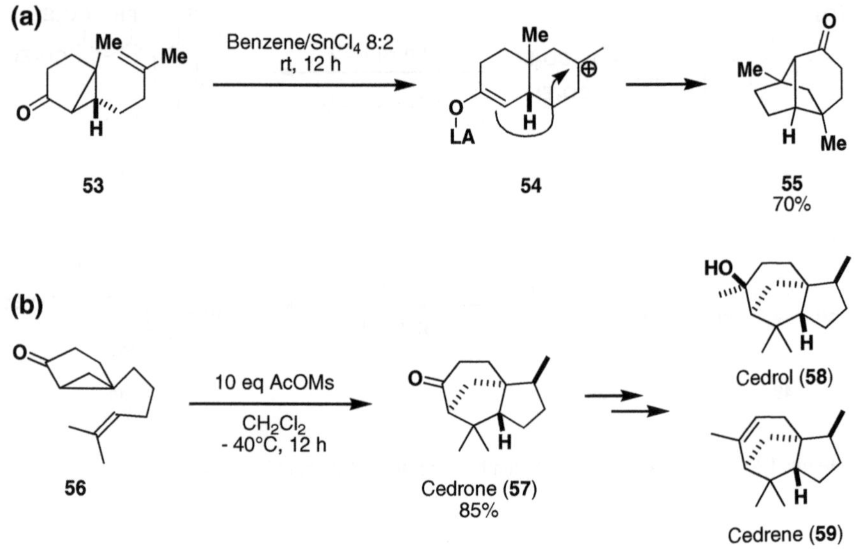

Scheme 1.8 Intramolecular [3 + 2] annulations

In order to have a broad overview of the different structures obtained by intermolecular [3 + 2] annulation of DA cyclopropanes, this section will be organized following the different types of dipolarophiles: electron-poor olefins, unactivated olefins, enol ethers, heteroaromatics, allyl-, allenylsilanes and alkynes. In each case, pioneering results, relevant advances and if reported applications to total synthesis will be presented.

Electron-Poor Olefins

The use of a Lewis acid to catalyze the ring opening of DA cyclopropanes implies that the energetic level of the LUMO is lowered, making the system electrophilic. Consequently, annulations involving electron-poor olefins are very scarce.[2] There is in fact only one example of such reactivity that was reported by Christie, Jones and co-workers. Their initial plan was to perform a cycloaddition between DA cyclopropane **60** and an electron-poor acrylate (Scheme 1.9) [62]. The cobalt-protected alkyne would act as the donor substituent and stabilizes a carbocationic Nicholas intermediate [63]. However, no reaction was observed with this combination of starting materials.

After switching to the more reactive propenal, the reaction gave a 50/50 mixture of cyclopentane **61** and tetrahydrofuran **62**. Considering that the heterocyclic product was also interesting, they did not go further with the annulation of electron-poor olefins and focused on the synthesis of tetrahydrofurans.

---

[2]The electron-poor olefins are more suited for reaction with DA cyclopropanes activated by transition metals. See Sect 1.3.3.6

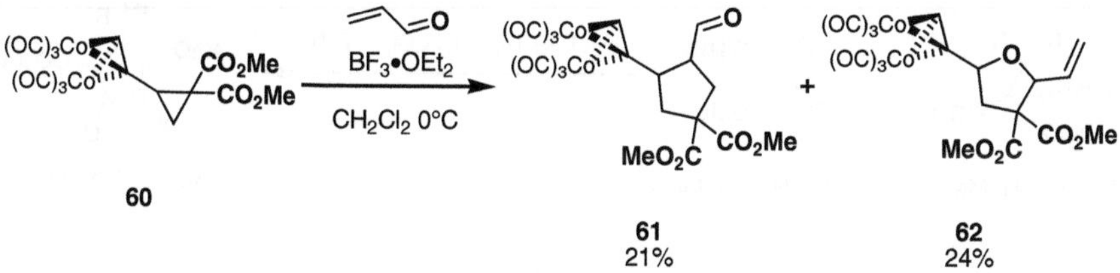

Scheme 1.9 [3 + 2] Annulation using Cobalt protected alkyne-cyclopropane

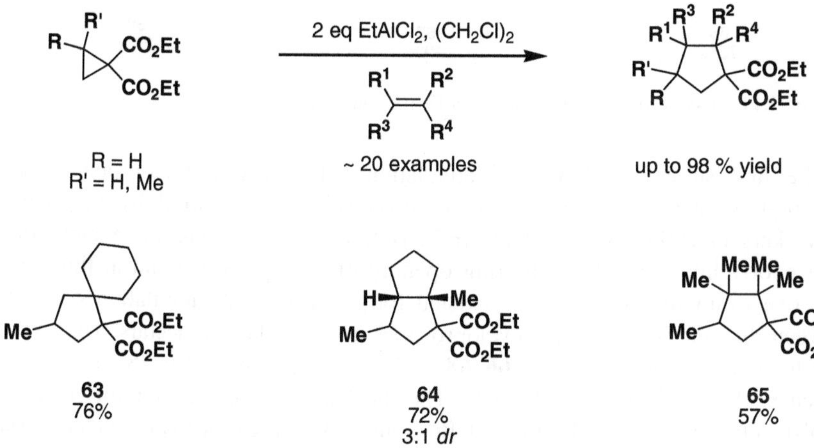

Scheme 1.10 [3 + 2] annulations using unactivated olefins reported by Snider and co-workers

Unactivated Olefins

Probably for the same reasons than for electron-poor olefins, the use of unactivated olefins in combination with an electrophilic activation has been poorly developed. A single study by Snider and co-workers in 1986 reported the use of ethylaluminium-dichloride to catalyze the annulation between alkenes and DA cyclopropanes (Scheme 1.10) [64]. The scope of the reaction was broad. However, stereo- and chemoselectivities were often low. Side-products resulting from hydride shifts, rearrangements, fragmentations or over-reduction by the catalyst were competing with product formation.

Enol Ethers

Most of the dipolarophiles that have been used in [3 + 2] annulations with DA cyclopropanes belong to the electron-rich π-systems. Due to the matching combination of the electrophilic cyclopropanes and the electron-rich olefins, the reaction occurs under mild conditions, displays better regio- and stereoselectivities and tolerates various types of partners.

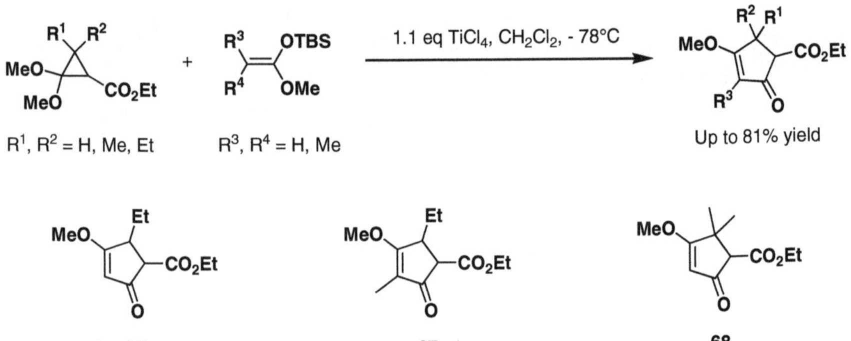

Scheme 1.11 Access to cyclopentenones via [3 + 2] annulation

The first study of a [3 + 2] annulation between DA cyclopropanes and electron-rich olefins in presence of a Lewis acid was reported by Saigo and co-workers in 1990 (Scheme 1.11) [65]. Before this study, carbon nucleophiles were poorly investigated for the ring opening of cyclopropanes and at this time, only one example using cuprates existed [66]. The [3 + 2] annulation occurred between dimethoxy substituted cyclopropanes and silyl ketene acetals. They were able to obtain cyclopentenones **66–68** with yields ranging from 28 to 81 % in presence of a slight excess of $TiCl_4$. The mechanism proposed for this reaction involves nucleophilic attack of the olefin, followed by a Dieckmann condensation. Elimination of one of the two methoxy groups affords the unsaturated products.

In 1991, Kuwajima reported the annulation between TMS-silyl enol ethers and monocarbonyl-butoxy-cyclopropanes (Scheme 1.12) [67]. During this study, they noticed the trans to cis isomerization of the cyclopropane in presence of $SnCl_4$ at −55 °C. They rationalized this result by the formation of a transient zwitterionic intermediate and chelation by tin to both the ester and the methoxy groups. Ethyl ester-substituted cyclopropanes reacted with a stoichiometric amount of tin, but only 3 mol% of catalyst were required for the cyclopropyl ketones. The reaction tolerated di- and trisubstituted olefins and all the products were obtained as unidentified mixtures of diastereoisomers. No product was obtained when tetra-substituted olefins were employed.

The same group extended this methodology in 1993 to thio-substituted DA cyclopropane **75** (Scheme 1.13) [68]. The reaction required 1.1 equivalents of Me_2AlCl at –45 °C and provided cyclopentanes in yields superior than 60 % and good diastereoselectivities for di- to tetrasubstituted olefins. Silyl enol ethers derived from alkyl ketones were used as the ones derived from acetophenone gave mixtures of at least two diastereoisomers. The bulky TBDPS group was selected for this reaction as it gave the best yields compared to TBS or TIPS.

In 2006, Ihara and co-workers exploited the ability of triflic imide to generate a powerful Lewis acid [69] in presence of silyl enol ethers (Scheme 1.14) [70]. The group used this catalyst to promote the [3 + 2] annulation of DA cyclopropanes with

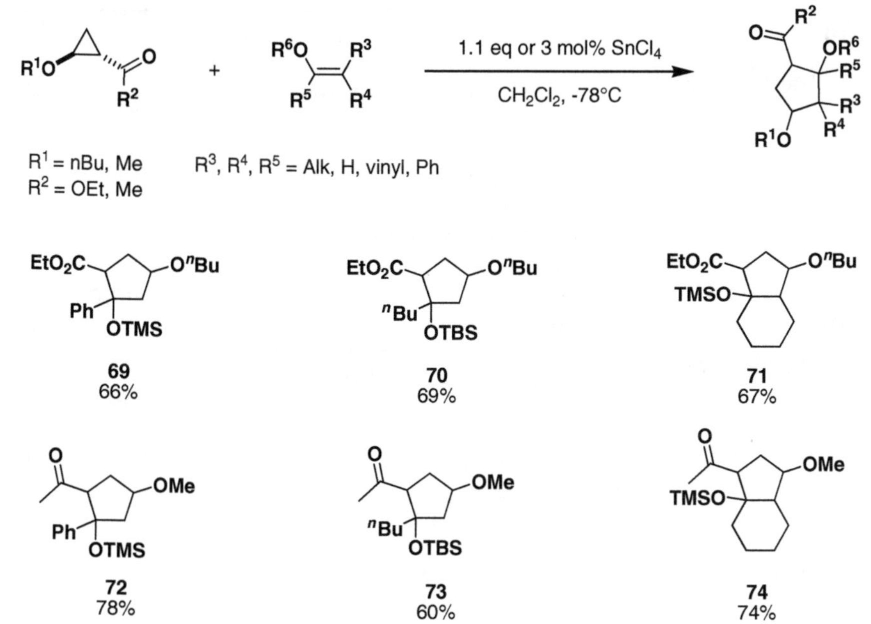

Scheme 1.12 [3 + 2] annulation using silyl enol ethers

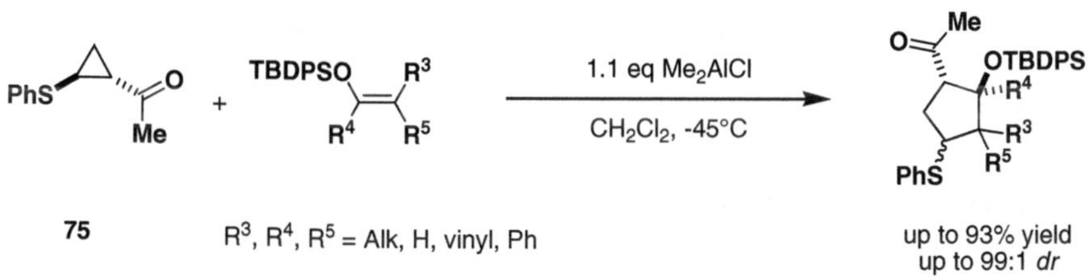

Scheme 1.13 [3 + 2] annulations using sulfur substituted cyclopropane **75**

silyl enol ethers. This strategy was applied in a first time to butyloxy mono-ester cyclopropane **76** using 1 mol% of triflic imide. Except for one derivative of acetophenone, the reaction has been tested only on cyclic TBDPS enol ethers. The yields are ranging from 65 to 70 % but the products were isolated as undetermined mixtures of diastereoisomers. The reaction was then applied to *para*-methoxyphenyl (PMP) mono-ketone cyclopropane **77** under slightly modified conditions. Overall yields are higher in this case but the diastereoselectivity is still poor. The authors ruled out the possibility of catalysis from triflic acid formed in situ and highlighted the versatility of this catalyst by performing a one-pot [4 + 2]/[3 + 2] sequence.

One of the most common problems encountered when studying annulations between silyl enol ethers and malonate-activated cyclopropanes is the retro-aldol reaction that can produce linear side products (Scheme 1.15). The products are still

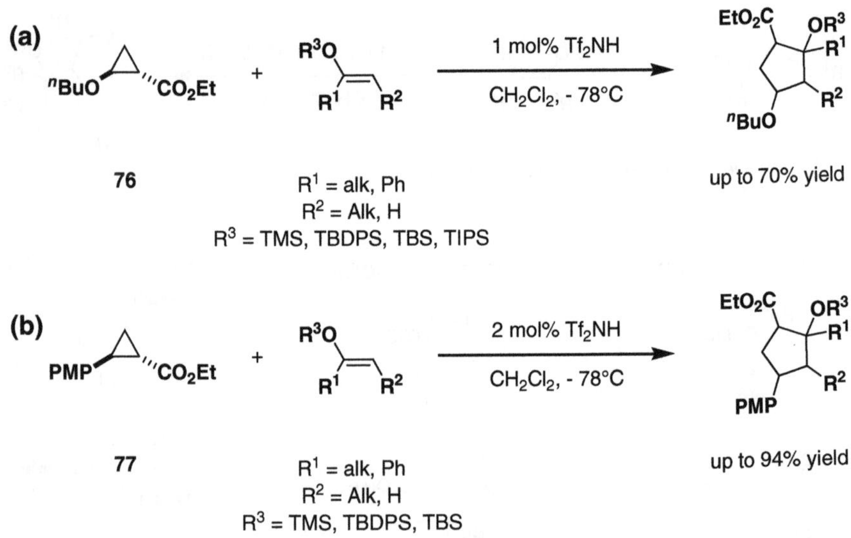

Scheme 1.14 [3 + 2] annulations catalyzed by triflic imide

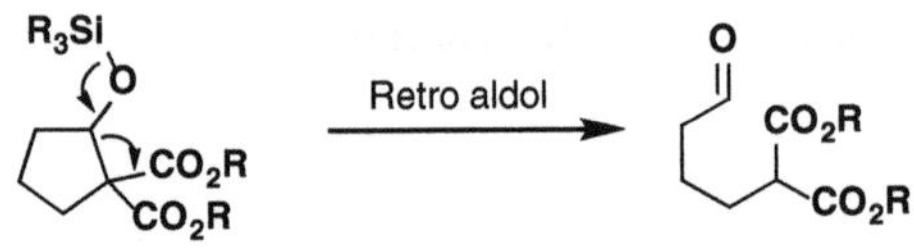

Scheme 1.15 Retro-aldol fragmentation

interesting as they correspond to the homolog version of the Mukaiyama-aldol reaction [71]. Wang and co-workers, for example, developed a method to access exclusively the linear adducts [72].

In 2009 Tang and co-workers studied intensively the selectivity of [3 + 2] annulations between aromatic substituted DA cyclopropanes and silyl enol ethers (Scheme 1.16) [73]. They managed to identify a copper-bisoxazoline ligand (**L**$_1$) complex giving full selectivity for annulation whereas a "naked" copper complex afforded only the homo Mukaiyama-aldol adduct. Compared to other tested Lewis acids such as scandium, tin, indium or iron, this system offered a great improvement in term of yields and stereoselectivities. The authors submitted a large variety of TMS and TBS silyl enol ethers to the reaction conditions in combination with aryl-substituted cyclopropanes to give cyclopentanes such as **79** and **80**. The reaction also afforded cyclopentanes **81** and **82** in good yield and diastereoselectivity with styryl and vinyl substituted DA cyclopropanes.

In 2012, they modified the ester substituents with adamantyl groups, allowing the construction of fused bicycles in good yield and selectivity [74]. They also

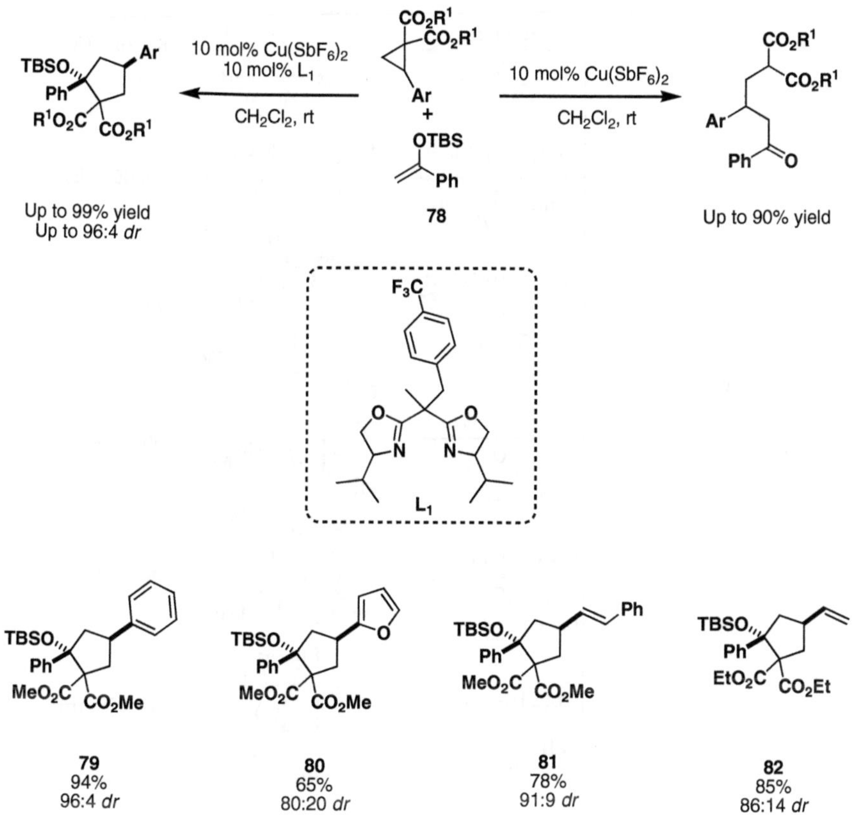

Scheme 1.16 [3 + 2] annulation reported by Wang and co-workers

performed a series of calculations in order to explain the diastereoselectivity observed in the products.

Heteroaromatics

Furans have been poorly exploited in [3 + 2] annulations. They can react as carbon nucleophiles in Friedel-Crafts reaction or as a two or a four carbon dipolarophiles in annulations. Additionally, the products formed can be more reactive than the initial starting materials. In 2009, Budynina and co-workers developed a set of reactions that fully exploit the reactivity of furans (Scheme 1.17) [58]. They studied their behavior under diverse conditions in order to access different products. By changing the substrates, their molar ratios or the Lewis acid, they could optimize [3 + 2] annulations (A), Friedel-Crafts (B), tandem [3 + 2] annulation/intramolecular Friedel-Crafts (C), tandem [3 + 2] annulation/intermolecular Friedel-Crafts (D), tandem [3 + 2]/[3 + 2] annulation (E) and [3 + 3] annulation (F).

Indoles were described in the previous section as efficient nucleophiles in Friedel-Crafts alkylations. However, by modifying the reaction conditions or the

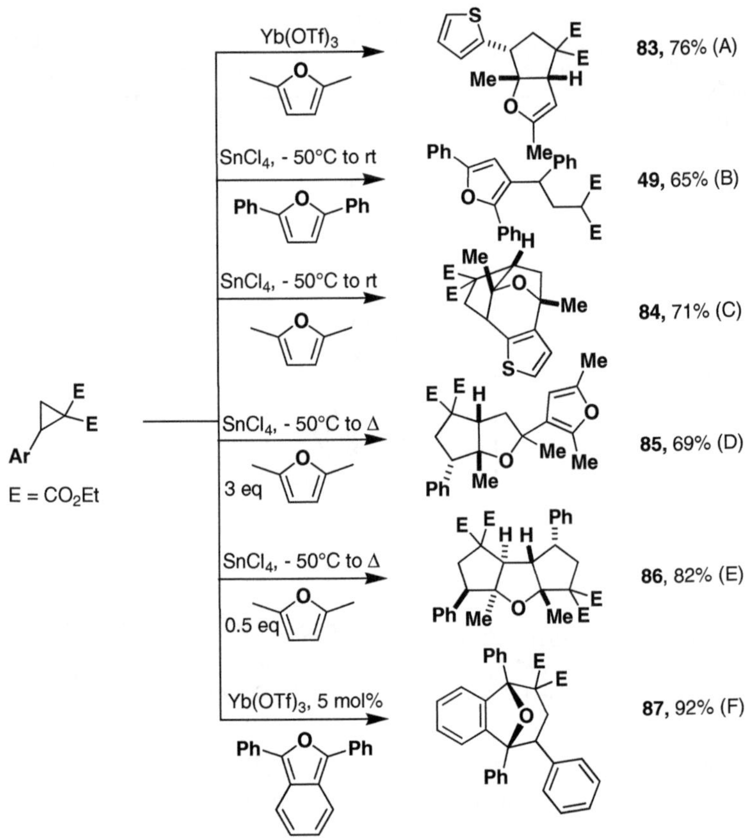

Scheme 1.17 Reactivity between DA cyclopropanes and furans

substrates, they can behave as dipolarophile in [3 + 2] annulations. This reactivity has been disclosed by Kerr in 1999, 2 years after reporting the Friedel-Crafts study [75]. The reaction occurred in presence of $Yb(OTf)_3$ between malonic substituted DA cyclopropanes and protected skatole derivatives, affording the products in yields varying between 27 and 94 % and diastereoselectivities up to 8:1 (Scheme 1.18). The methyl substituent in C3 position prevents rearomatization of the indole and allows cyclization. If the reaction is performed at higher temperature, it can also induce rearrangement to the C2 FC adduct [55].

In 2006, Ila and co-workers studied the annulation between indoles and aryl-substituted DA cyclopropanes [76]. They could use unprotected and unsubstituted indoles that were not compatible with the method published by Kerr and co-workers. Using an excess of $BF_3 \cdot OEt_2$ or $TiCl_4$, they managed to access a wide range of polycyclic structures such as **88–90** in up to 93 % yield (Scheme 1.19).

In 2007, Pagenkopf and co-workers developed the [3 + 2] annulation between alkoxy DA cyclopropanes and C3 unsubstituted indoles [56]. Using one equivalent

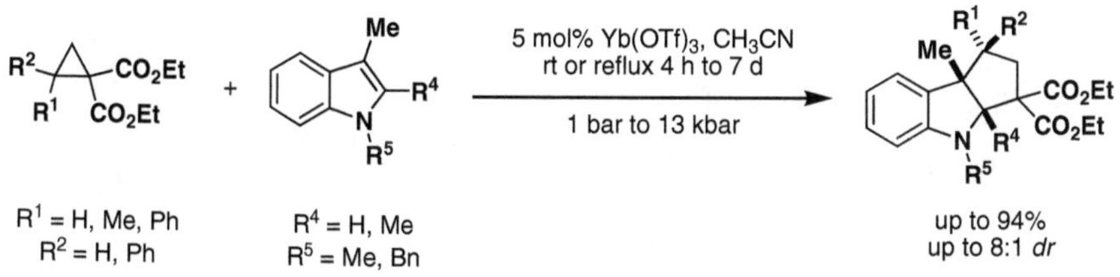

Scheme 1.18 [3 + 2] annulation between skatole derivatives and DA cyclopropanes

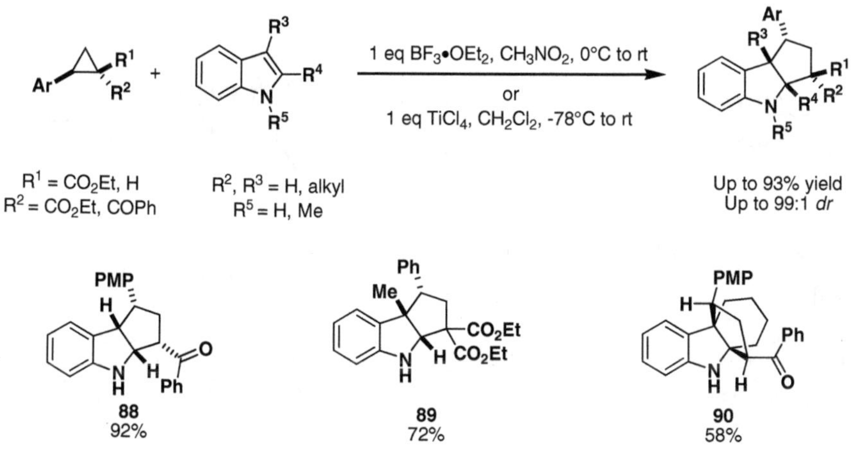

Scheme 1.19 [3 + 2] annulation between unprotected indoles and DA cyclopropanes

of TMSOTf in nitromethane, the reaction afforded a broad range of cyclized products in up to 90 % yield and diastereoselectivities from 1:1 to 99:1.

Allyl- and allenyl silanes

Allyl silanes or allenyl silanes belongs to the family of electron-rich olefins due to the hyperconjugation between the silyl group and the π–system [77]. In 2001, Sugita and co-workers reported the reaction between alkoxy DA-cyclopropanes **91** and allyl silanes (a, Scheme 1.20) [78]. In presence of 1.1 equivalents of $TiCl_4$, the reaction afforded cyclopentanes in yields from 23 to 70 % and diastereoselectivities from 6:4 to 9:1. The main reason accounting for the low yield is the formation of a byproduct **92** resulting from elimination of the silyl group during the reaction. The more stable the silyl, the better the yield. The reaction was applied to methane-chromanones and TIPS allyl silane **93** affording the [4.2.1] bicycles **94** and **96** in up to 83 % yield (b, Scheme 1.21).

Using allenyl silanes as dipolarophile, Yadav and co-workers reported a [3 + 2] annulation involving TBDPS substituted cyclopropyl ketones (Scheme 1.22) [79]. Using 1.3 equivalents of $TiCl_4$, the scope of the reaction included variations on both the starting cyclopropyls and the allenyl substrates. The starting material was used

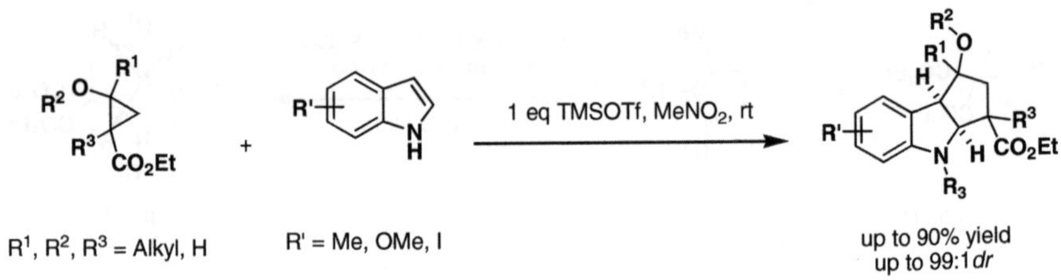

Scheme 1.20 [3 + 2] annulation with indoles reported by Pagenkopf and co-workers

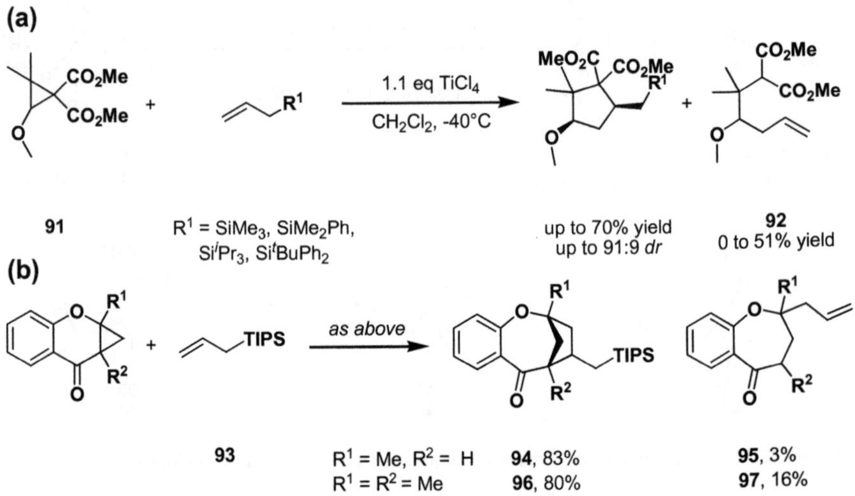

Scheme 1.21 Annulations using allyl silanes

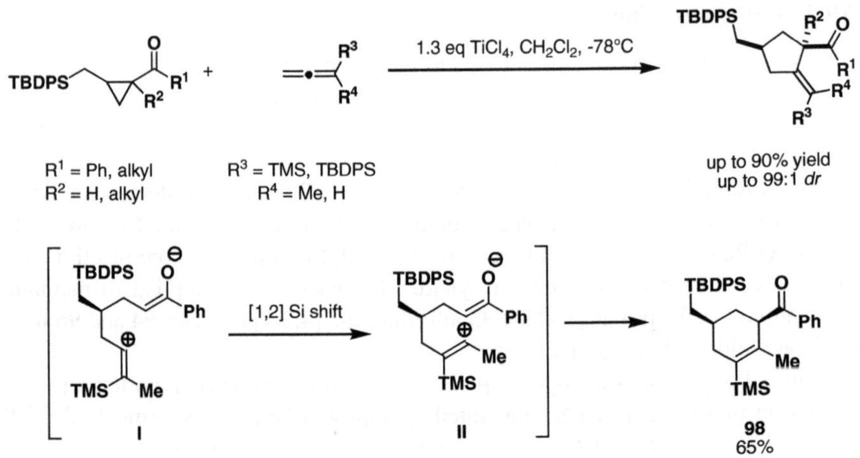

Scheme 1.22 [3 + 2] and [3 + 3] annulations using allenyl silanes

as a mixture of *cis* and *trans* isomers and the products were obtained in yields up to 90 % with good diastereoselectivities. By changing the Lewis acid to Et_2AlCl, they were able to access the [3 + 3] annulation product **98** resulting from a 1,2-silicon shift via intermediates **I** and **II**.

Alkynes

The use of alkynes for the synthesis of cyclopentenes via intermolecular [3 + 2] annulation has been limited to three examples. The first one is attributed to Yadav and co-workers in 2004, who were able to cyclize aryl alkynes with cyclopropyl phenyl ketones such as **99** activated by a silicon group (Scheme 1.23) [80]. In presence of 1.3 equivalents of $TiCl_4$, the reaction delivered cyclopentenes in yields ranging from 55 to 85 % and with variable diastereoselectivities. *Para*-methoxyphenyl substituted alkynes were better substrates in term of yields but afforded the products with a migration of the double bond in conjugation with the carbonyl. Running the reaction in presence of anhydrous K_2CO_3 prevented the formation of the side product. It is also interesting to notice that using *cis* or the *trans* isomer of the starting cyclopropane had no influence on the diastereoselectivity and the yield of the product.

In 2008, Ready and co-workers reported the use of sensitive and very reactive ynol ethers for the synthesis of cyclopentenones (Scheme 1.24) [81]. Mono-ester ethoxycyclopropanes and silyl ynol ethers were reacted in presence of 1 equivalent of aged Me_2AlCl. In fact, a closer look to old bottles of the catalyst revealed the presence of (MeO)AlMeCl which is strong enough to catalyze the reaction without decomposing the sensitive ynol ether. This method was presented by the authors as an alternative to the Pauson-Khand reaction.

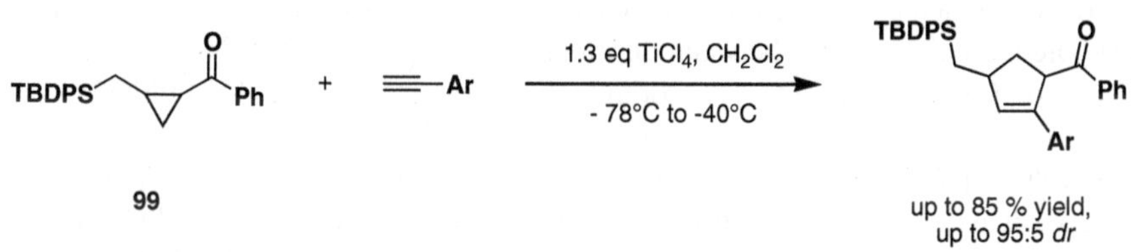

Scheme 1.23 [3 + 2] annulation using alkynes

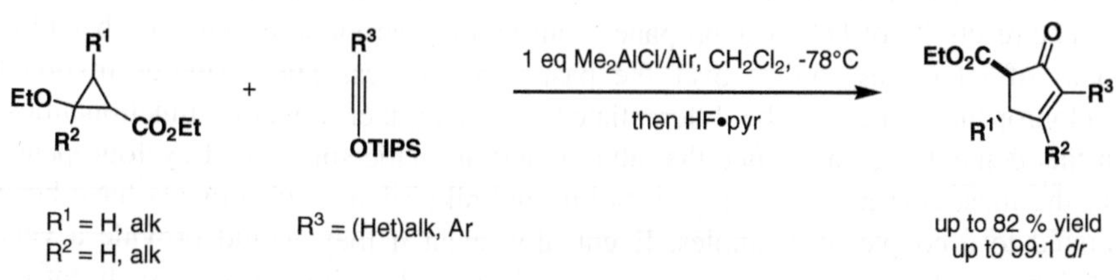

Scheme 1.24 [3 + 2] annulation using silyl ynol ethers

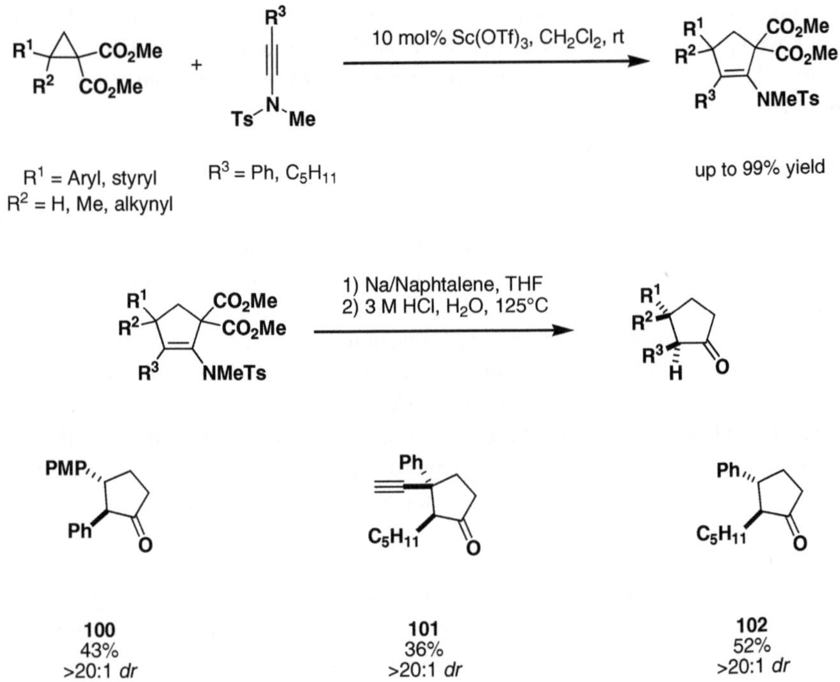

Scheme 1.25 Annulation using ynamines reported by Johnson and co-workers

In 2014, Johnson and co-workers reported for the first time the combination of DA cyclopropanes and ynamines for the synthesis of cyclopentene sulfonamide products (Scheme 1.25) [82]. The ynamines were easily accessed from copper mediated cross-coupling between terminal alkynes and TsNHMe. The reaction required 10 mol% of $Sc(OTf)_3$ at room temperature and afforded the products up to quantitatively. The reaction was compatible with all-carbon quaternary centers on the cyclopropanes, but presented limitation when using extremely rich or poor donor groups. It was shown that the annulation was enantiospecific under the reported conditions. However, the system was not compatible with chiral catalysts as the substrates did not react even after prolonged time. Deprotection of the nitrogen followed by hydrolysis afforded poly-substituted cyclopentanones **100–101** with all-carbon quaternary centers in very good yields and diastereoselectivities.

The reactivity of DA cyclopropanes with various carbon π-systems has been the focus of intensives studies over the past two decades. These studies involved cyclopropanes which can be differentiated mostly by their donating functionalities. In this context, we can notice that alkoxy and aromatic substituted cyclopropanes are the most commonly employed. Sulfur and alkylsilanes substituents have been limited to a couple of examples. Eventually, even if they would provide a very efficient way to access cyclopentylamines, nitrogen donating groups are absent for this type of transformation.

1.4.3.4 Lewis Acid Mediated [3 + 2] Annulation of Carbonyl Compounds

[3 + 2] Annulation between carbonyls and DA cyclopropanes is a very powerful tool to access efficiently tetrahydrofurans [83]. In general, aldehydes were chosen as dipolarophiles due to the lack of reactivity and facial selectivity when ketones were employed. The first efforts towards annulation of DA cyclopropanes and carbonyls were reported in 1981 by Reissig [84] and Brueckner (a, Scheme 1.26) and during the next 24 years, the efficiency of such methods has been limited by the requirement of stoichiometric amounts of Lewis acids. Contributions have been done by Saigo and co-workers in this field (b, Scheme 1.26) [85, 86]. Sugita and co-workers reported the use of catalytic amounts of Lewis acids but only in presence of more strained bicyclic methanochromanone **106** (c, Scheme 1.26) [87, 88].

In 2005, Johnson and co-workers set up a milestone in this field by developing enantiospecific annulations and studying the mechanism of the reaction [89]. They optimized the Lewis acid and found that hafnium triflate afforded a broad range of tetrahydrofurans **108–110** in a highly stereoselective way and in very good yields (Scheme 1.27) [90–92].

In order to study the mechanism of the reaction, the group performed labelled studies, comparing the absolute and relative stereochemistry of starting cyclopropanes and products (Scheme 1.28). The zwitterionic intermediate being ruled out by the enantiospecificity of the reaction, four different mechanisms could be expected

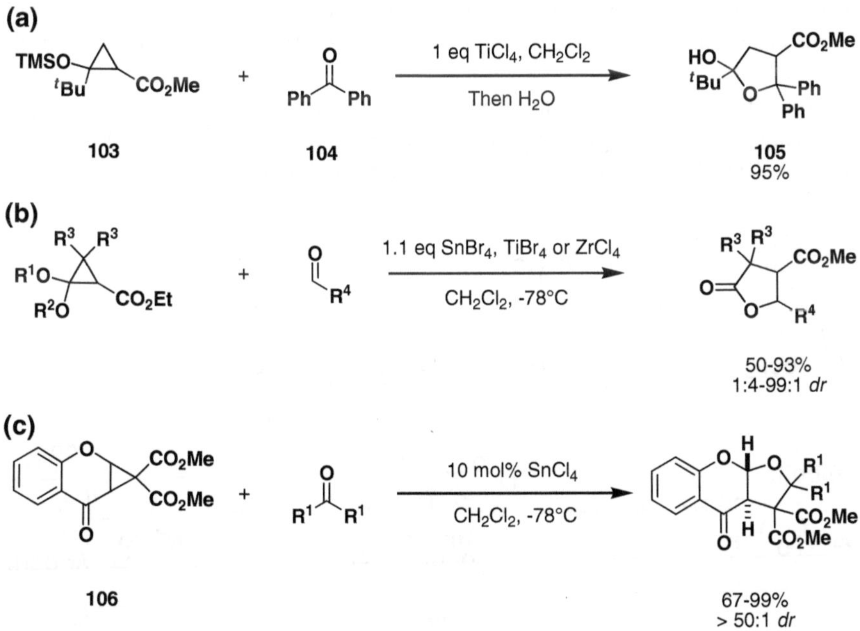

Scheme 1.26 Pioneering [3 + 2] annulations using carbonyl compounds

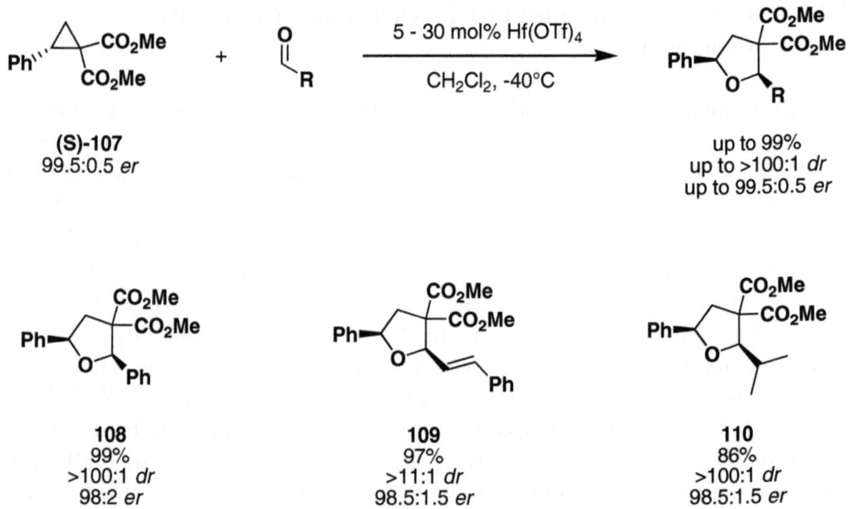

Scheme 1.27 [3 + 2] annulation developed by Johnson and co-workers

for this reaction: An attack by the aldehyde via an S_E^2 process, a S_N^2 mechanism through an intimate ion pair intermediate (II, Scheme 1.28) and a concerted $[{}_\pi 2_s + {}_\sigma 2_a]$ cycloaddition. The deuteration experiment allowed them to rule out the first two options thanks to the comparison of absolute and relative stereochemistry in starting material and product. Competition studies with aldehydes having different electronic properties permitted them to identify the S_N^2 mechanism as the most probable.

Following the work of Johnson and co-workers, numerous methodologies involving [3 + 2] annulations of carbonyls were developed by Yadav (A) [93], Johnson (B) [94, 95], Niggemann (C) [96], Shao (D) [97], Yang [98] and their

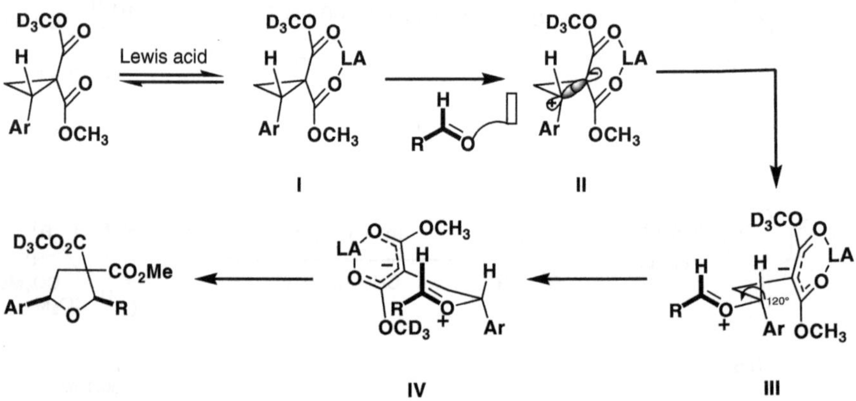

Scheme 1.28 Proposal for the mechanism of the [3 + 2] annulation by Johnson and co-workers

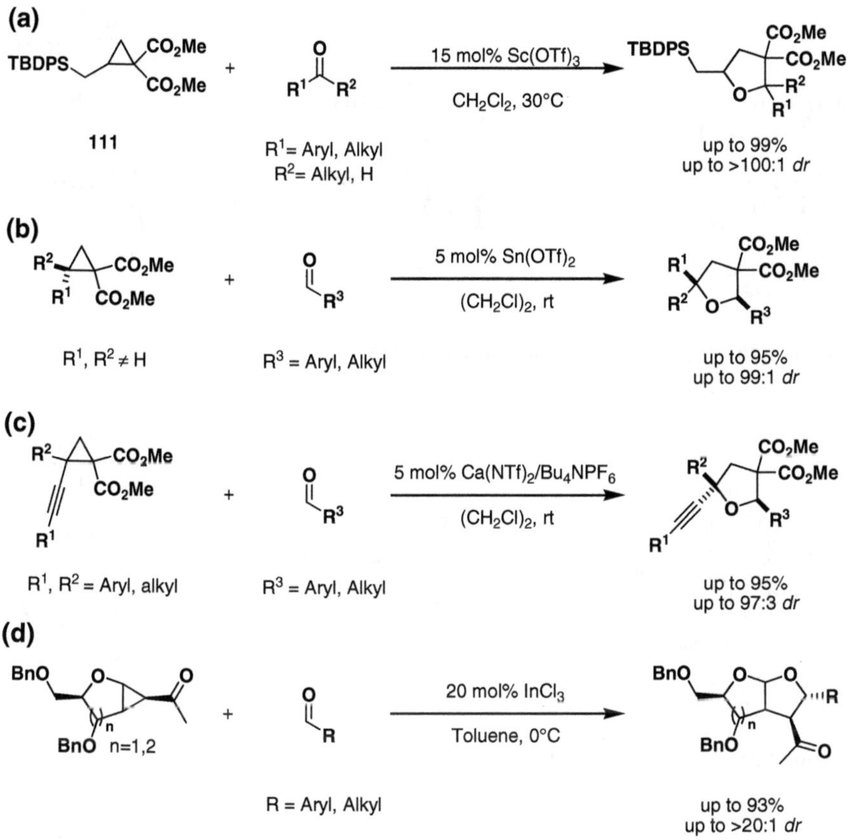

Scheme 1.29 [3 + 2] annulations between carbonyl compounds and DA cyclopropanes

co-workers using different types of DA cyclopropanes (Scheme 1.29). Silicon aryl, alkyl, alkynyl and alkoxy substituted cyclopropanes were employed, giving access to different types of polysubstituted tetrahydrofurans.

In order to highlight the efficiency of [3 + 2] annulations, the total syntheses of various natural products have been reported using such reactions (Scheme 1.30). The enantioselective total synthesis of (+)-polyantheline A (**112**) by Johnson and co-workers involved a [3 + 2] annulation between bicyclic DA cyclopropane **114** and chiral aldehyde **115** using an aluminium based Lewis acid (Scheme 1.31) [99]. The authors managed to optimize the reaction to afford an exceptional level of face selectivity, especially considering the presence of the stereocenter on the linear chain of the aldehyde. The total synthesis was completed in 15 steps in total.

Later, Kerr and co-workers reported a very detailed account of their journey toward the synthesis of (+)-isatisine A (**116**) using a [3 + 2] annulation between enantiopure DA cyclopropane **119** and aldehyde **118** with $Sn(OTf)_2$ as catalyst (Scheme 1.31) [100]. The reaction was selective with various protecting groups on

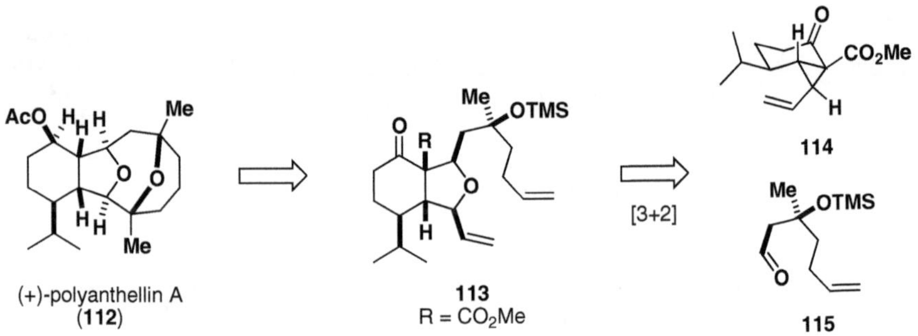

Scheme 1.30 Total synthesis of polyanthellin A (**112**)

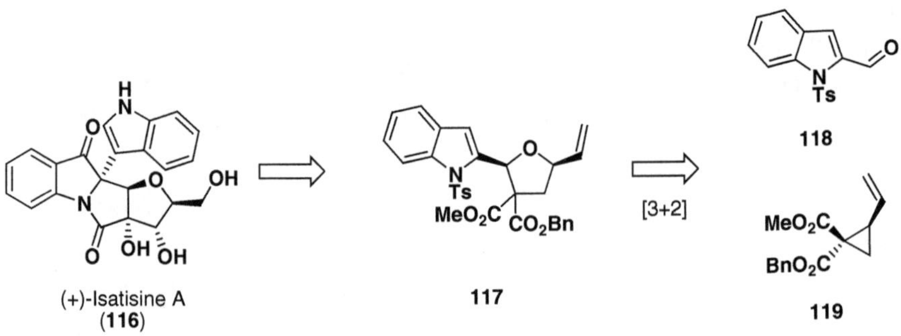

Scheme 1.31 Total synthesis of Isatisine A (**116**)

the indole but the tosyl was crucial for the completion of the synthesis with an overall yield of 5.8 % in 14 steps.

1.4.3.5 Lewis Acid Mediated [3 + 2] Annulations with Other Dipolarophiles

Imines

The pyrrolidine scaffold can be found in a large number of bioactive molecules. It is possible to access pyrrolidines by [3 + 2] annulation between DA cyclopropanes and a nitrogen-containing dipolarophile. Various types of cyclopropanes have been used as formal 1,3 dipoles for the construction of pyrrolidines with imines,[3] but only a few methods that produce racemic products involve DA cyclopropanes. Saigo reported in 1990 the use of excess $TiCl_4$ in combination with DA cyclopropanes and tosyl-aldimines for the synthesis of lactams [112]. In 2005, Kerr and

[3]For acceptor-substituted cyclopropanes: [101–103]. For methylene cyclopropanes: [104–108]. With transition metals: [109, 110]. Intramolecular: [111].

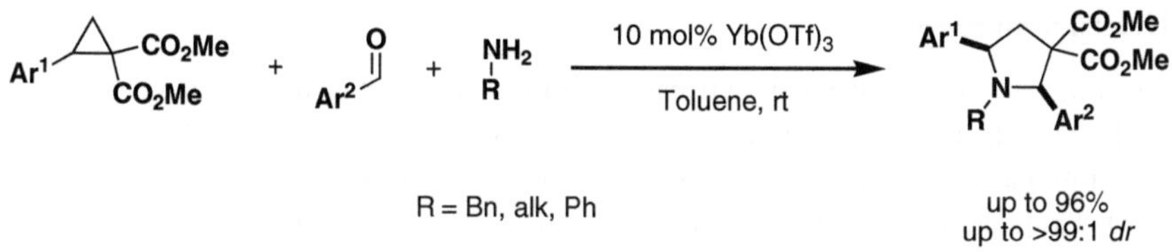

Scheme 1.32 Multi-component reaction reported by Kerr and co-workers

co-workers reported a 3-component reaction between aldehydes, primary amines and aryl substituted DA cyclopropanes for the synthesis of 1,5-*cis* pyrrolidines (Scheme 1.32) [113]. By changing the combination of the three components, the reaction allowed the diversification of pyrrolidine substituents and afforded the products with yields up to 96 % and diastereoselectivities from 1:1 to 99:1. As both the amine and the aldehyde are potentially reactive with the cyclopropane, the aldimines have to be pre-formed in situ.

This work was followed the next year by a method from Tang and co-workers [114]. However, in this case the aldimines were isolated instead of reacted in situ. Christie and co-workers reported the use of the same alkyne DA cyclopropane than described previously (Scheme 1.9) but this time they reacted with aldimines in order to access very useful alkyne substituted pyrrolidines [115].

Nitriles

Nitriles have been successful dipolarophiles in combination with DA cyclopropanes (Scheme 1.33). Their unique reactivity and their combination with various 1,3 formal dipoles gave access to diverse structures such as pyrrolines [116–122] or pyrroles [123–127].

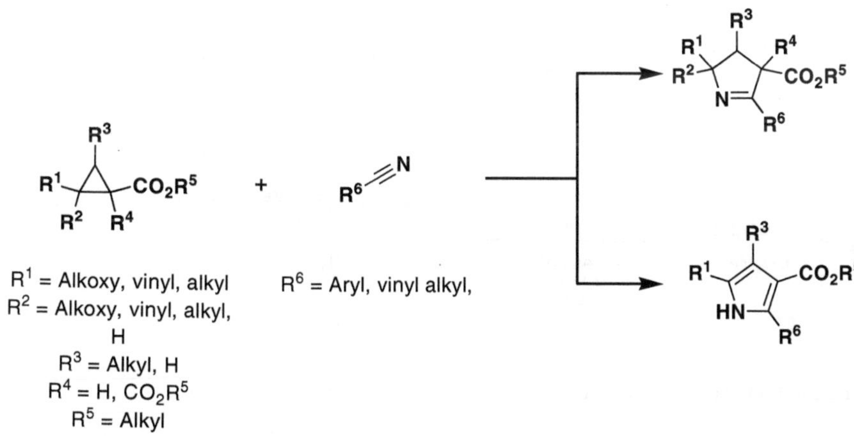

Scheme 1.33 [3 + 2] annulation using nitriles

Heterocumulenes

Isocyanates, isothiocyanates and carbodiimides have been used to access heterocycles such as lactams, thioimidates or amidines [128, 129]. Until recently, their use in combination with DA cyclopropanes has been limited to a couple of examples [130, 131]. Li and co-workers reported in 2012 the synthesis of thiolactams such as **120** from isothiocyanates (Scheme 1.34) [132]. However, Stoltz and co-workers reassigned the obtained product to thioimidate **121** in their study of similar reactions [133].

Other annulation partners such as diazines [134, 135], carbon disulfide [136] or nitrosyl chloride [137], have been less studied and still present some interesting potential for the discovery of new synthetic methods.

1.4.3.6 [3 + 2] Annulations with Other Modes of Activation

As the combination between electron-poor olefins and Lewis acid-activated DA cyclopropanes is limited, alternative methods using Lewis bases or transition metals have been applied to deliver the corresponding annulation products. Marino and co-workers reported the use of silyloxy-cyclopropanes and vinylphosphonium salts in presence of KF and crown ether to access cyclopentenes [138]. Transition metals such as nickel (a, Scheme 1.35) [139] palladium (b, Scheme 1.35) [140–142], iron (c, Scheme 1.35) [143] were used to perform stereoselective annulations with α-β unsaturated carbonyls. More recently, a [3 + 2] annulation using a NHC catalyst was reported by Lupton and co-workers (d, Scheme 1.35) [144] and a method using photoredox activation and involving enol ethers was developed by Yao and co-workers [145].

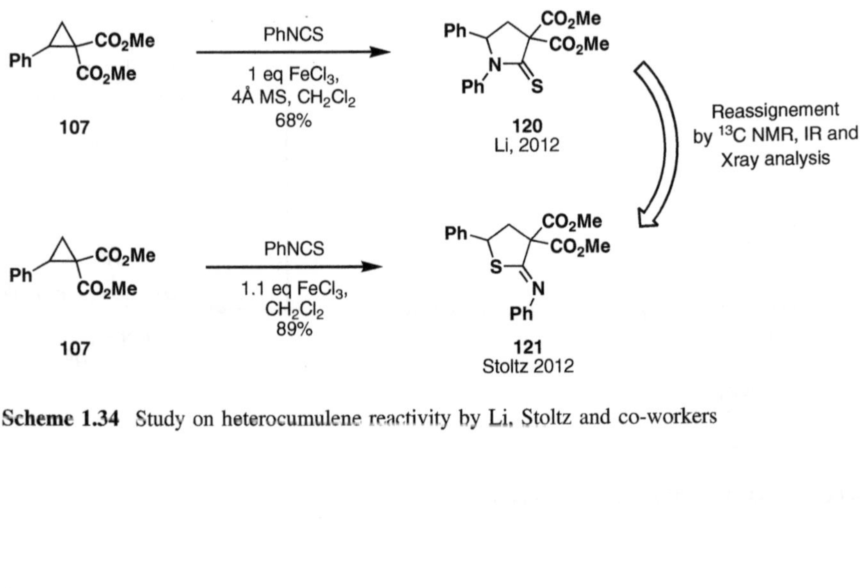

Scheme 1.34 Study on heterocumulene reactivity by Li, Stoltz and co-workers

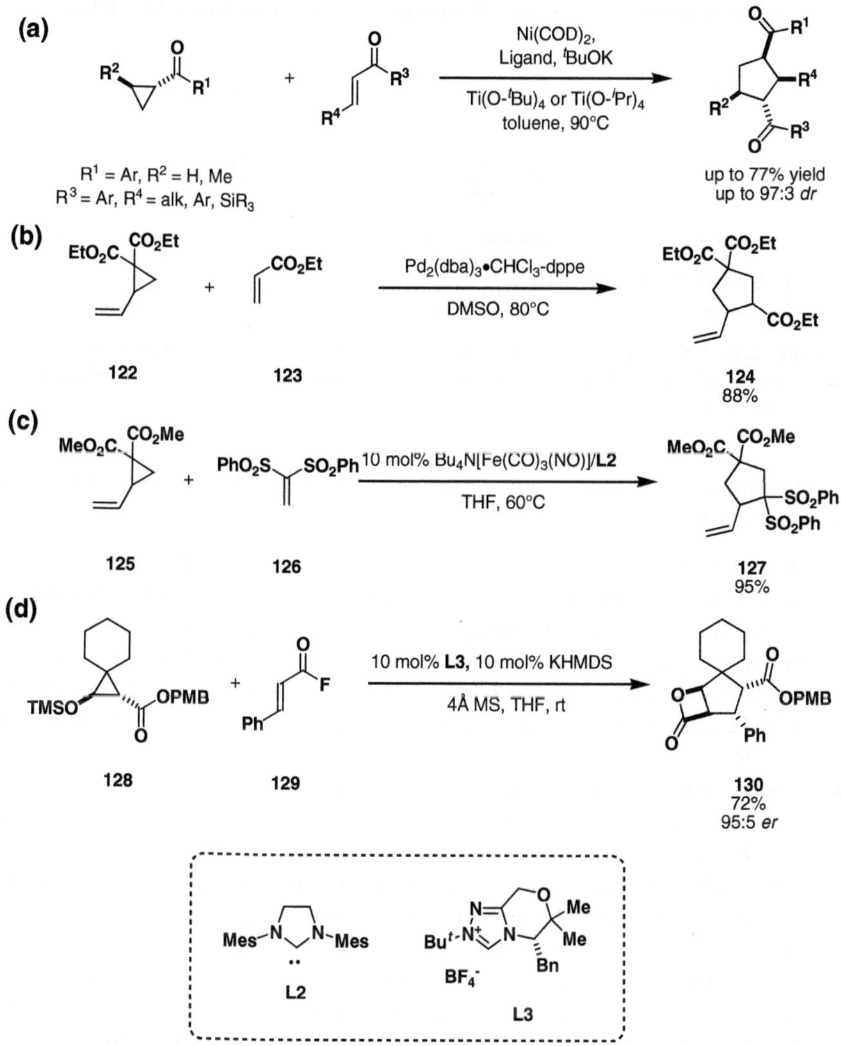

Scheme 1.35 Other modes of DA cyclopropane activation

1.4.3.7 Enantioselective Processes Involving DA Cyclopropanes

The synthesis of enantiopure products is very valuable to the pharmaceutical, fragrance or crop protection industries as enantiomers of bioactive entities do not have the same properties. DA cyclopropanes, in most of the cases, possess at least one stereocenter and are therefore chiral molecules. In order to access a single enantiomer of a product from a chiral cyclopropane, stereospecific methods that require enantiopure starting materials or stereoselective methodologies that can “destroy” the chirality need to be employed. The existing methods developed to access an

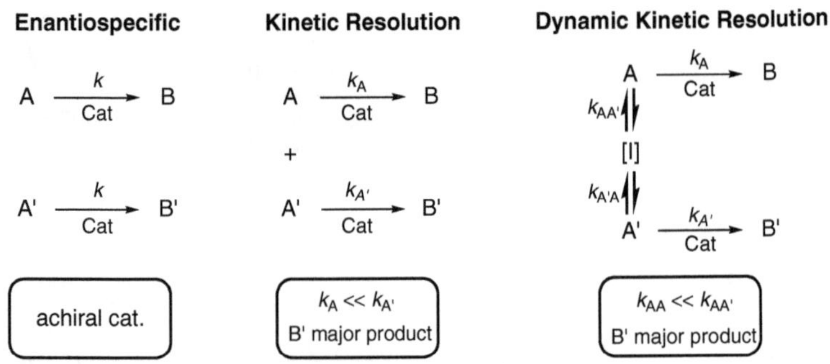

Scheme 1.36 Enantiospecific reaction; Kinetic and Dynamic Kinetic Resolutions (*A, A′* enantiomeric substrates; *B, B′* enantiomeric products; *Cat* catalyst; *[I]* achiral intermediate; *k* rate constants)

enantiopure product from a chiral starting material can be separated into three categories [146].

- Enantiospecific reactions: This is for example the case for the methods developed by Johnson and co-workers [89]. Such a reaction transfers the stereogenic information of a highly enantioenriched starting material to the product using an achiral catalyst. The greatest limitation is the requirement of a highly enantioenriched starting material which can be accessed from enantioselective synthesis in the best case or resolution otherwise. For the latter, the overall yield is limited to 50 %.
- Kinetic resolutions (KR) or dynamic kinetic resolutions (DKR) (Scheme 1.36) [147, 148]: Starting from a racemic mixture, each enantiomer reacts at different rates through diastereomeric intermediates with a chiral catalyst. For a KR, the slow-reacting enantiomer will be recovered enantioenriched and therefore the reaction has a theoretical maximum yield of 50 %. In the DKR, the substrate is able to racemize spontaneously and therefore the racemic mixture is converted to a single enantioenriched product. In order to be selective, the racemization rate has to be equal or higher than the reaction rate of the less reactive enantiomer.
- Dynamic Kinetic Asymmetric Transformations (DyKAT) [147, 148] are also a deracemization process. As for the DKR, the racemate is transformed in one step to a non-racemic product in 100 % theoretical yield (Scheme 1.37). In a DyKAT, the enantiomers coordinated to the chiral catalyst generate a mixture of diastereomeric intermediates (with different energy level and therefore differently populated) that interconvert at different rates. From the experimenter point of view it can be difficult in some case to differentiate DyKAT from DKR.

There are four types of DyKAT; types I and II involve the de-racemization of enantiomers through diastereoisomers and types III and IV involve the de-epimerization of diastereoisomers through diastereoisomers. In this chapter, we will focus only on the type I and II as they are in relation with experiments presented in the result part (Scheme 1.37).

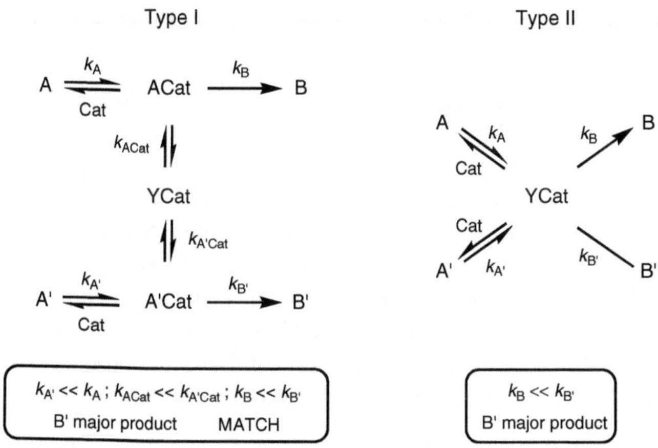

Scheme 1.37 Dynamic Kinetic Asymmetric Transformation (*A, A′* enantiomeric substrates; *B, B′* enantiomeric products; *Cat* chiral catalyst; *ACat, A′Cat* diastereomeric intermediates; *YCat* chiral intermediate; *k* rate constants)

In a type I DyKAT, the substrate enantiomers (A, A′) react with the chiral catalyst at different rates and form diastereomeric mixtures (Acat, A′cat) that are unequally populated. These complex can interconvert trough a chiral intermediate (Ycat) at different rates. They can also irreversibly react with the reagent to afford their corresponding enantiomeric product (B, B′), again with different rates of reaction. This system can be matched or mismatched due to the kinetic difference between the interconversion and the product formation of the two starting enantiomers. If the interconversion to the most reactive diastereomeric intermediate is faster than the formation of the minor product ($k_B \ll k_{B'}$; $k_{ACat} \ll k_{A'Cat}$) then the system is MATCH. The opposite system where the formation of the fast reacting intermediate is slower than the reaction of the minor product formation ($k_B \ll k_{B'}$; $k_{ACat} \gg k_{A'Cat}$) is thus MISMATCH.

In type II DyKAT (Scheme 1.37) the racemate is converted to a chiral intermediate (Ycat), for which the stereogenic information of the starting material has been lost. This intermediate reacts selectively with the reactant through a diastereomeric transition state. The selectivity is imputed to the chiral information from the catalyst and therefore comes only from the difference between $k_{B'}$ and k_B.

In order to have a dynamic process, the difference between k_A and $k_{A'}$ should not be too large, otherwise the process will be associated with a KR. Additionally, in order to have a selective process the rate of product formation should not be faster than the racemization process.

Finally, the advantage of a DyKAT has been very well summed up by Trost and co-workers: "if the act of converting a racemic mixture into a single enantiomeric series is combined with one of the required structural transformations, the dynamic resolution is not an additional step in the synthesis and thereby saves a step [149]."

Kinetic Resolution DKR and Enantioselective Processes Involving DA Cyclopropanes.

Kinetic resolutions have been described using DA cyclopropanes as limiting reagents. Developed by Sibi or Tang and co-workers, it includes [3 + 3] annulations [150–152] as well as homo conjugate addition of nucleophiles [153].

Tang and co-workers reported the diastereo- and enantioselective [3 + 2] annulation between DA cyclopropane **131** and indoles (Scheme 1.38) [154]. The reaction is catalyzed by a C_2-symmetric copper(II)-BOX complex. Various substituted indoles were reacted to access efficiently polycyclic products, and as for the work of Kerr and Pagenkopf [55, 56], a C3 substituent was required to promote annulation. As this method requires the use of DA cyclopropane **131** in excess compared to the indole, it cannot be considered as a DyKAT.

[3 + 2] Annulations Involving Lewis Acid-Catalyzed DyKAT of DA Cyclopropanes.

In 2009 Johnson and co-workers reported the synthesis of enantioenriched 2,5-*cis*-disubstituted tetrahydrofurans starting from racemic carbon-substituted DA cyclopropanes and aldehydes. The dynamic kinetic asymmetric [3 + 2] annulation required the use of a chiral MgI_2-PyBOX complex as catalyst (Scheme 1.39) [155].

In order to achieve high selectivity, strong donating groups that could increase the racemization rate through stabilization of a cationic charge were needed on the cyclopropane. The optimization of the reaction identified the *tert*-butyl-substituted chloro-PyBOX **L5** as ligand of choice for this transformation, affording the products in high enantiomeric ratios. The author tested aryl-, vinyl- as well as alkyl-aldehydes which reacted with yields ranging from 48 to 92 %, diastereoselectivities superior than 9:1 and enantiomeric ratios up to 97:3.

The model for stereoselectivity proposed by the authors implies that the MgI_2-PyBOX catalyst possess an octahedral geometry giving two possible reactive complexes (Scheme 1.40).

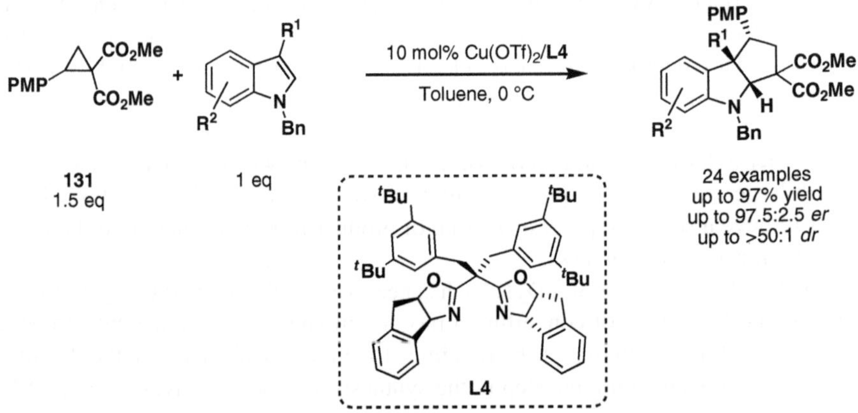

Scheme 1.38 Enantioselective [3 + 2] annulation of DA cyclopropanes and indoles

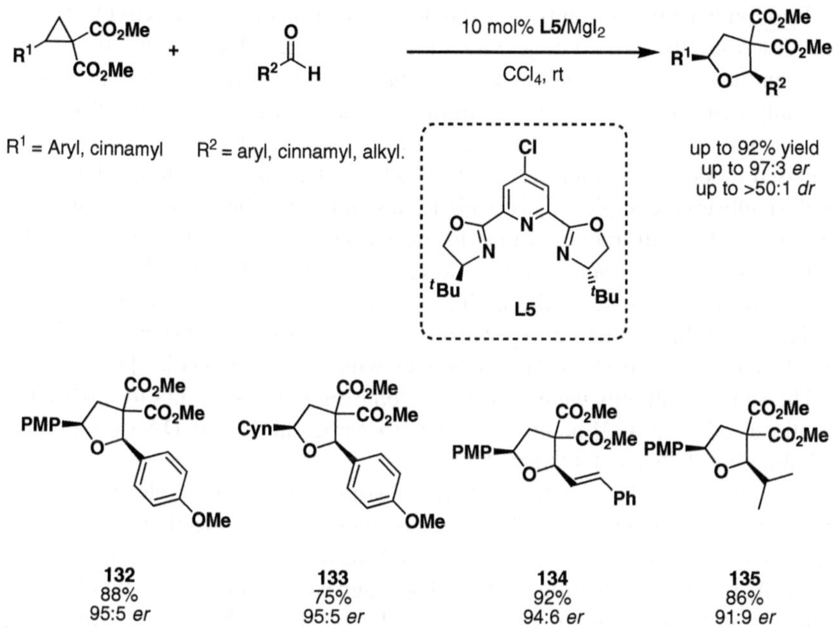

Scheme 1.39 Enantioselective synthesis of tetrahydrofurans. Cyn = Cynnamyl

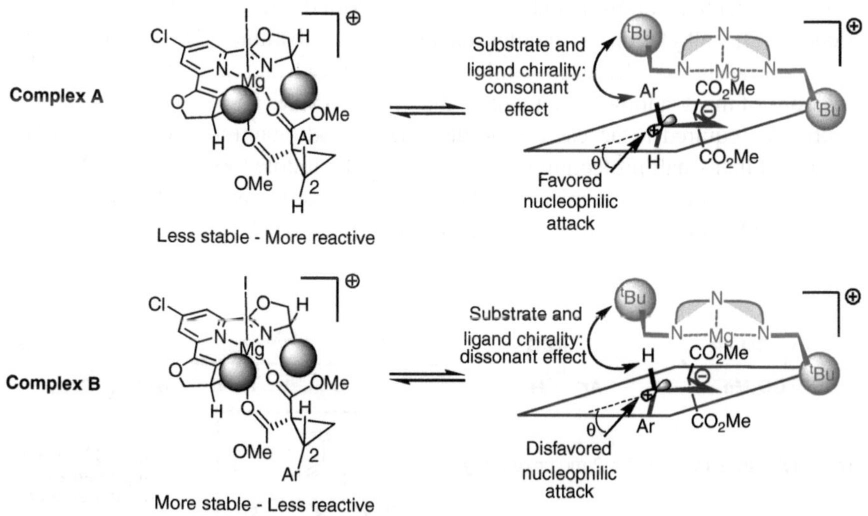

Scheme 1.40 Stereochemical model for the enantioselective [3 + 2] annulation of cyclopropanes

Two fundamental concepts were employed to rationalize the selectivity. The first one describes the drifted trajectory of the aldehyde which allows the overlap between the oxygen lone pair and the p orbital of the carbenium ion in presence of sterically-demanding *tert*-butyl group of the ligand. The second one links the stereoselection to the presence of a less stable, more reactive intermediate. In the proposed model, the complex A confines all the steric bulk in one quadrant of the catalyst-substrate complex. This leads to destabilization but to an easier access to one face of the substrate, and thus to greater stereodifferentiation. Complex B is probably more stable but exhibits a lower reactivity with aldehydes as the steric effects of the substrate and the ligand follow opposite trends. This type of double-stereo-differentiating experiments were performed by Evans in bis(oxazoline)-Cu(II)-catalyzed Diels-Alder reactions with chiral dienophiles [156].

The same conditions were applied by Johnson and co-workers in 2010 for the enantioselective [3 + 2] annulation between carbo-substituted DA cyclopropanes and aldimine (Scheme 1.41) [157].

In a similar way than for the aldehydes, fast-racemizing strong-electron-donating groups were required on the DA cyclopropanes. The protecting group on the aldimine was essential and the electron rich 2-methoxybenzyl group was selected as it is easily removed by hydrogenolysis. The magnesium catalyst was still effective to promote the transformation. However, the ligand had to be modified in order to increase the enantiomeric excess of the products. With the bromo-PyBOX ligand **L6**, the scope was broad and substituted pyrrolidines could be obtained in up to 96 % yield, enantioselectivities superior than 95:5 and up to 92:8 diastereoselectivity.

More recently, Johnson and co-workers reported the first enantioconvergent Friedel-Crafts alkylation of indoles using carbon-substituted DA cyclopropanes (Scheme 1.42) [158]. The reaction tolerated electron-rich and -poor indoles as well as variations of the donating substituent on the cyclopropane. The use of a bulky silicon group was required to access the product with good enantiomeric excess. It was speculated that small protecting groups rendered the indole too reactive and did not leave enough time for racemization. The catalyst complex is the same than for the reaction with aldimines and the authors described this reaction as a type I DyKAT.

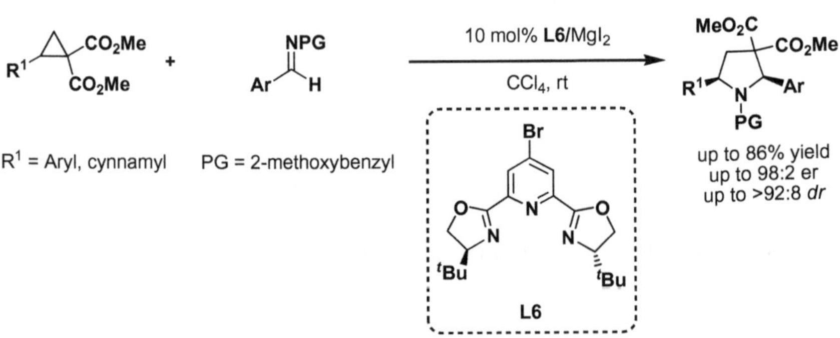

Scheme 1.41 Enantioselective synthesis of 2,5-*cis*-disubstituted pyrrolidines

Scheme 1.42 Enantioselective Friedel-Crafts alkylation of indoles with DA cyclopropanes

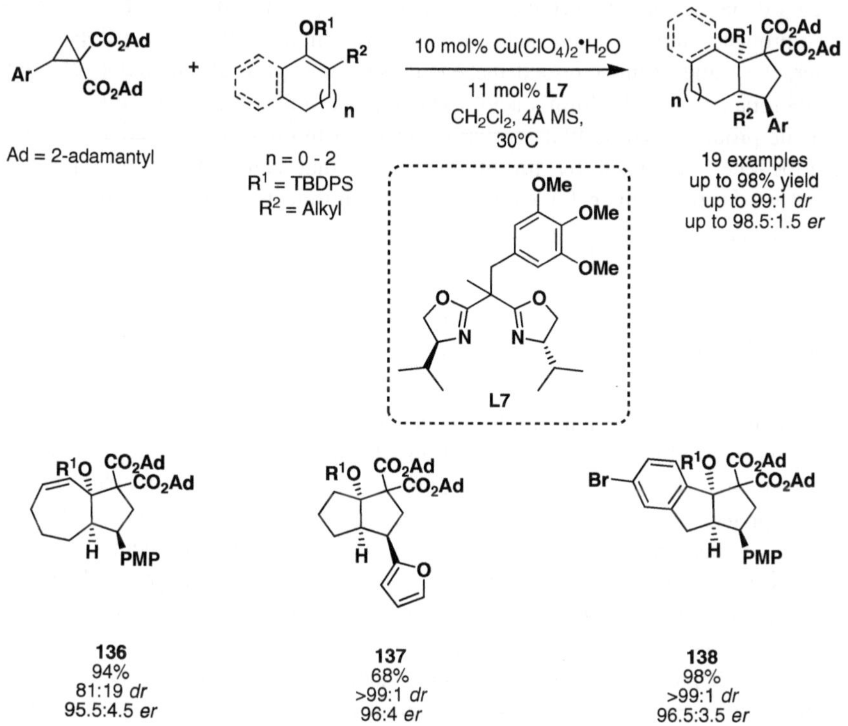

Scheme 1.43 Enantioselective [3 + 2] annulation of DA cyclopropanes and cyclic silyl enol ethers

In 2013, Tang and co-workers reported an enantioselective annulation between carbon-substituted DA cyclopropanes and cyclic silyl enol ethers in presence of a copper(II)-^{i}PrBOX complex (Scheme 1.43) [159]. The reaction is catalyzed by a C_1-symmetric BOX ligand. The use of such ligand can bring improvement in enantioselectivity. However the major drawback lies in the complexity of their synthesis that produces diastereoisomers which need to be separated before use. In order to achieve high enantioselectivity, it was necessary to engineer the side-arms of the BOX ligand. Furthermore, extremely bulky adamantyl esters were required on the cyclopropane as otherwise both enantio- and diastereoselectivity dropped. In order to suppress by-product formation, TBDPS enol ethers were used. The scope is focusing on five- to seven-membered cyclic and benzo-cyclic enol ethers which provided fused bicyclic products in yields ranging from 80 to 99 %, enantiomeric ratios superior than 95:5 and high diastereoselectivities. The author did not describe the reaction as a DyKAT, but it is strongly implied by the stereochemical model that they proposed for the reaction mechanism.

Other modes of activation for enantioselective annulations.
Enantioselective methods involving DA cyclopropanes have also been reported using activation by transition metals.[4] For example, Trost and co-workers developed a dynamic kinetic asymmetric formal [3 + 2] cycloaddition of racemic vinyl cyclopropanes and alkylidene azlactones to afford highly substituted cyclopentanes. Later, the same group reported the enantioselective [3 + 2] annulation of vinylcyclopropanes and Meldrum's acid alkylidenes to afford highly substituted cyclopentane products. These transformations proceed through the formation of π-allyl intermediates catalyzed by transitions metals that can activate both partners, therefore the mechanism for induction of enantioselectivity is different than with Lewis acids.

1.4.4 Conclusion

In conclusion, [3 + 2] annulation between DA cyclopropanes and dipolarophiles has been realized for a large number of possible combinations, but in most cases, it involves the use of carbonyls and olefins as dipolarophiles. In addition, DA cyclopropanes have been used as 1,3 dipoles also in other type of annulations. The variety of dipolarophiles easily accessible allowed the development of [3 + *n*] processes, giving access to a large array of carbo- and heterocycles of different sizes.(See footnote 1) For examples, nitrones have been widely employed to afford tetrahydro-1,2-oxazines derivatives [164–167].

In this chapter, the work on intermolecular annulations involving DA cyclopropanes was presented. There is also a great number of studies that focus on the development of intramolecular annulations [168]. It has allowed the discovery of new reactivities between 1,3 dipoles and dipolarophiles such as cross cycloaddition [169–171]. However, as our work is focusing on intermolecular annulations, we did not enter in the details of these methods.[5]

It is important to remark that [3 + 2] annulations involve mainly carbon- and alkoxy-substituted DA cyclopropanes. Introducing new functionalities to modulate the reactivity and to access new structures represent a challenge as very often the cyclopropanes are difficult to access and do not have the required reactivity. This is

[4]For examples of Pd catalyzed annulations, see [160–163].

[5]For a review on intramolecular annulations, see [172].

probably one of the reasons why annulations using Lewis acid activated DA cyclopropanes having a nitrogen as donating group have never been reported.

1.4.5 *Donor-Acceptor Substituted Cyclobutanes*

As presented in the introduction, cyclobutane has a strain energy almost similar to the one of cyclopropane. However, the use of DA cyclobutanes as formal 1,4 dipoles has been less investigated compared to cyclopropanes. One of the main reasons is probably the lack of efficient access to these structures from simple starting materials. There have been specific methods developed for the synthesis of DA cyclobutanes [173–178] and in the context of this thesis, we will now review the ones that specifically provide access to DA aminocyclobutanes.

1.4.5.1 Synthesis

There are limited options to choose from when it comes to the selective construction of substituted DA aminocyclobutanes. Photocatalysis uses the energy from light to bring molecules into an excited state which triggers reactions that would be impossible under thermal conditions. The formation of cyclobutanes via [2 + 2] photocatalyzed cycloadditions is one of the most encountered example of such reactivity [179, 180]. However, it requires specific and onerous equipment and conditions that are often not suitable for the synthesis of reactive DA cyclobutanes. Furthermore, in order to circumvent the poor regioselectivity associated with photochemical cycloadditions, most of the reactions are performed in an intramolecular fashion. A very efficient method was reported by Aitken and co-workers in 2002 describing the cycloaddition between ethylene and uracil (**139**) (a, Scheme 1.44) [181, 182]. Upon hydrolysis, the protected *cis*-cyclobutane **141** could be isolated in more than 60 % yield over two steps. A variation using a chiral auxiliary was implemented to access enantiomerically pure cyclobutanes (b, Scheme 1.44). The major drawbacks of this method are the use of ethylene, a potentially explosive gas and the lack of substituents that can be introduced. Bach and co-workers reported in 2002 an intramolecular [2 + 2] cycloaddition to access tricyclic constrained scaffolds such as **146** in very good yields (c, Scheme 1.44) [183].

In 2010, White and co-workers irradiated tryptamine derivative **148** in order to access spiro-indoline **149** (Scheme 1.45) [184]. The reaction involved the in situ formation of DA cyclobutane **I** which instantaneously opened via a retro-Mannich reaction to give the desired product **149**. This transformation highlighted the high reactivity of DA cyclobutanes toward ring opening.

In the case of a thermal process, the concerted cycloaddition path is forbidden and therefore harsh conditions are required to cross the activation barrier [185–189]. Uncatalyzed methods employing thermal activation have been hampered by the requirement of strongly reactive partners and therefore were limited to enamines

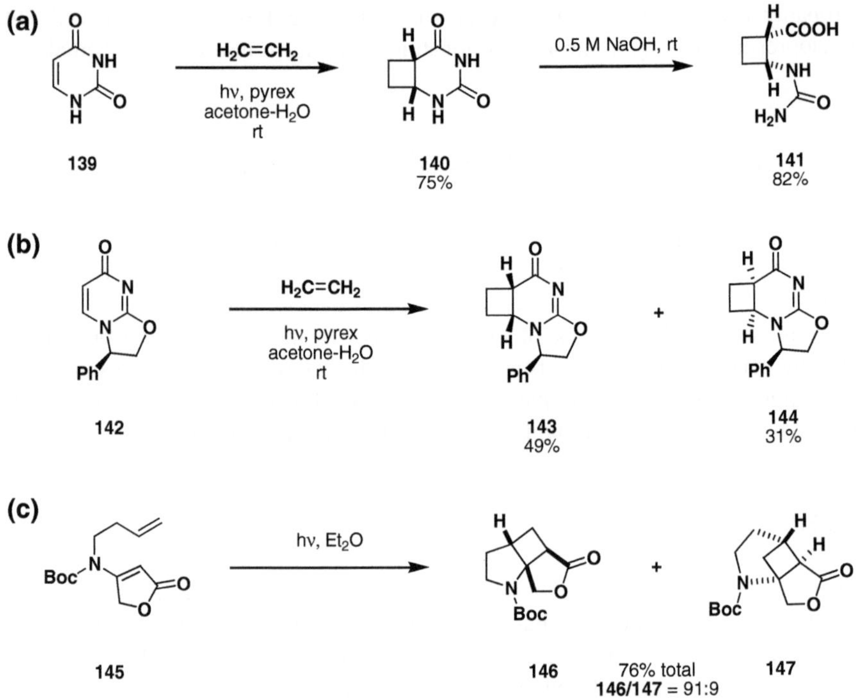

Scheme 1.44 Access to DA aminocyclobutanes via photocatalyzed [2 + 2] cycloadditions

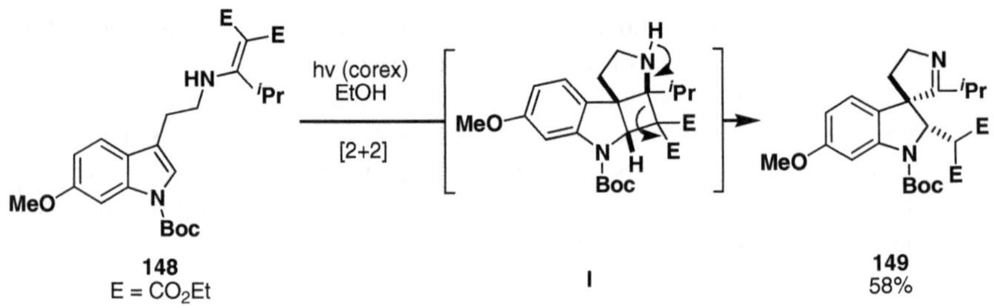

Scheme 1.45 [2 + 2]-Retro Mannich sequence

and electron poor olefins such as tricyanoethylene. As opposed to photocatalysis, the bond-forming events occur in a stepwise fashion and such processes are often called formal [2 + 2] cycloadditions.

More recently, organocatalysis has been used for the first time to access nitro substituted four-membered rings [190]. In this case, dienamine activation provides a nucleophilic π-system that can react with the Michael acceptor and cyclize to generate the four-membered ring. Jorgensen developed a new squaric acid catalyst **150** to promote cyclobutane synthesis starting from aldehydes and nitro olefins (Scheme 1.46) [191].

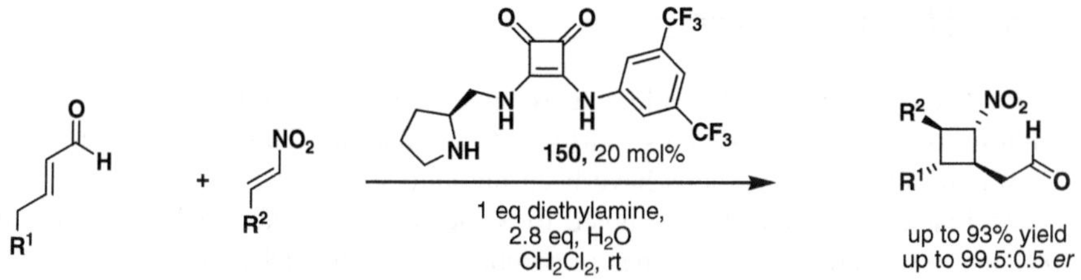

Scheme 1.46 Organocatalyzed formal [2 + 2] cycloaddition

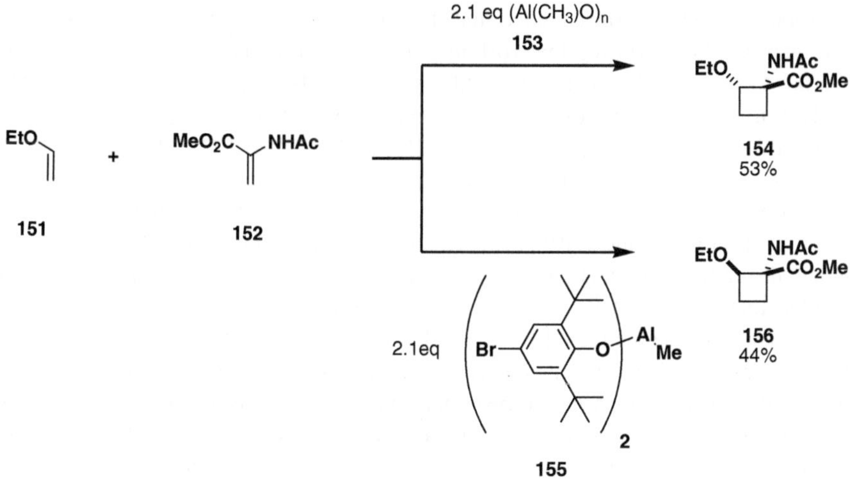

Scheme 1.47 Lewis acid-catalyzed synthesis of aminocyclobutanes

Finally, Lewis acid activation has been used to access DA aminocyclobutanes. Avenoza and co-workers reported the synthesis of nitrogen-substituted DA cyclobutanes such as **154** and **156** in presence of aluminium based Lewis acids **153** or **155** (Scheme 1.47) [192]. It is important to point out that the nitrogen has not a 1,2 relationship with the acceptor. The Lewis acids had to be finely tuned in order to prevent degradation of the product, but it was still necessary to use them in excess.

Finally transformations such as reduction of aminocyclobutenes [193] or Curtius rearrangement of cyclobutyl half-esters [194] were employed to access DA aminocyclobutanes. However as they do not imply the construction of the four-membered ring, we will not describe them further in this thesis.

As we can see the construction of DA aminocyclobutanes has been limited to a few methods using photocatalysis, thermal or Lewis acid activation. It often required harsh conditions, elaborated catalysts or hazardous substances. Additionally, these methods have limitations concerning the selective introduction of substituents on the ring. Considering the versatility of aminocyclobutanes both as structural elements and as reactive intermediates, there is a great need for the development of simple catalytic methods to access them.

1.4.5.2 [4 + 2] Annulations

As for cyclopropanes, DA cyclobutanes have been used as 1,4 dipoles for the synthesis of six-membered rings via [4 + 2] annulations. This chemistry is more recent and only a few methods have been reported so far.

The first example can be attributed to Saigo and co-workers, who reported in 1991 the use of aminocyclobutanes in combination with aldehydes and ketones to access lactols (Scheme 1.48) [195]. The cyclobutanes were accessed via thermal formal cycloadditions and therefore no modification on the core was possible. Furthermore, the requirement of an electron-rich enamine for the synthesis of the cyclobutane had for consequence that the aminal product was very labile and the nitrogen was "lost" during the work up. The reaction required excess of $TiCl_4$ and the products were obtained with yields up to 72 % but poor diastereoselectivities.

During 18 years, DA cyclobutanes have been completely forgotten until 2009 when two reports were published almost simultaneously. The first one was by Johnson and co-workers who developed a very selective and efficient method using carbon-based DA cyclobutanes to access *cis*-2,6-disubstituted tetrahydropyrans heterocycles (a, Scheme 1.49) [196]. The annulation required $Sc(OTf)_3$ for aromatic aldehydes and $MADNTf_2$ (**157**) for aliphatic aldehydes. Building upon these results, they improved the protocol and reported the one pot synthesis of tetrahydropyrans via sequential [[2 + 2] + 2] annulation (b, Scheme 1.49). The cyclobutane was first generated in presence of the Lewis acid, styrene derivative and methylidene methyl malonate (**158**), then the aldehyde was added to complete the annulation. The yields were slightly lower but there were now three bonds formed in one pot and the diastereoselectivity stayed very high. The authors noticed that under these conditions, enantiopure cyclobutanes did not react with transfer of stereogenic information.

The second report is to be credited to Christie and co-workers who adapted their protected alkyne donating group to DA cyclobutane **159** (Scheme 1.50) [197]. The cyclobutane **159** is able to react diastereoselectively with various aldehydes, giving the *cis* tetrahydropyrans in yields ranging from 34 to 95 % and diastereoselectivity superior to 20:1 in almost all of the tested substrates.

Pagenkopf and co-workers studied in depth the reactivity of carbo and alkoxy-substituted DA cyclobutanes (Scheme 1.51). They began their research by improving the synthesis of alkoxy-substituted cyclobutanes. Using ytterbium triflate

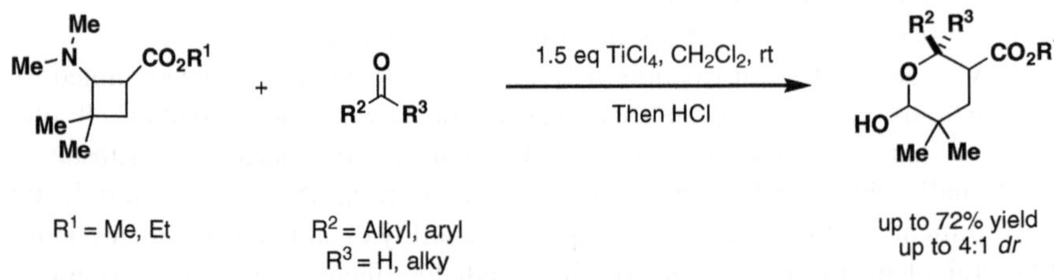

Scheme 1.48 [4 + 2] annulation using aminocyclobutanes reported by Saigo and co-workers

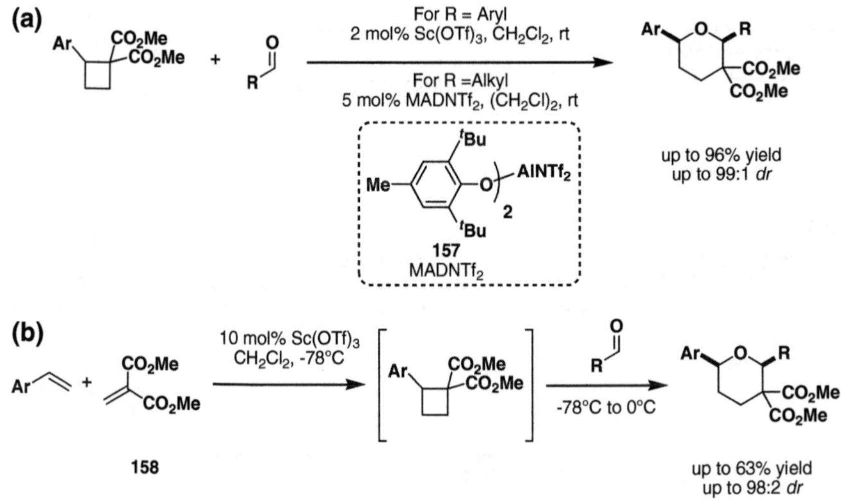

Scheme 1.49 [4 + 2] and [[2 + 2] + 2] annulations reported by Johnson and co-workers

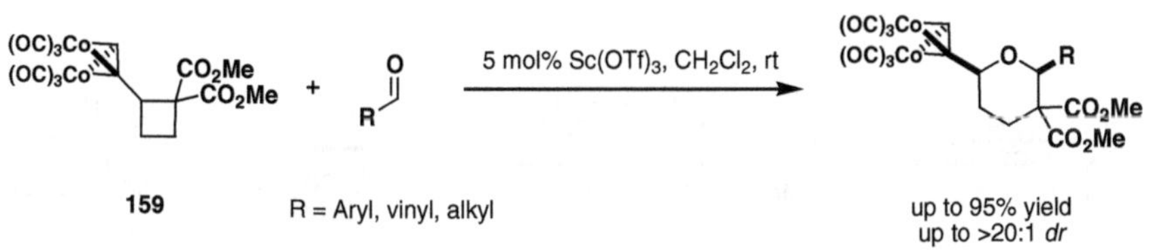

Scheme 1.50 [4 + 2] annulation reported by Christie and co-workers

as catalyst at −78 °C, they were able to react alkyl-enol ethers and methylidene malonates to afford the corresponding alkoxy-cyclobutanes. They then reacted them with various dipolarophiles such as aldimines [198], aldehydes [199] or nitrosoarenes [200] in [4 + 2] annulation to access piperidines or dehydropiperidines, tetrahydropyrans and tetrahydro-1,2-oxazines respectively. Later, they expanded the scope of reaction of DA cyclobutanes using nitrones in [4 + 3] annulations [201]. It can be noted that the tested substrates **161** and **162** failed to react under their conditions.

In 2012 Matsuo and co-workers reported a [4 + 2] annulation between alkoxy substituted DA cyclobutanes and aldehydes or ketones (Scheme 1.52) [202]. The reaction required catalytic amount of $SnCl_4$ and afforded tetrahydropyrans in yields up to 70 % and diastereoselectivity ranging from 1:1 to 10:1 in the case of aldehydes. Only symmetric ketones were tested. However, the formation of a side product could be detected during the course of the reaction. It was identified as the lactonization product **166**. Product **166** could be selectively accessed when changing the catalyst to TMSOTf.

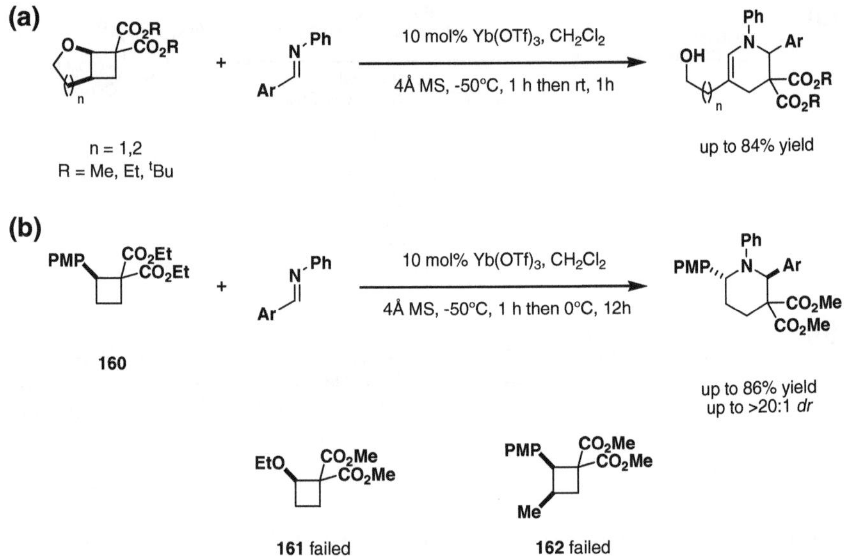

Scheme 1.51 [4 + 2] annulations with aldimines reported by Pagenkopf and co-workers

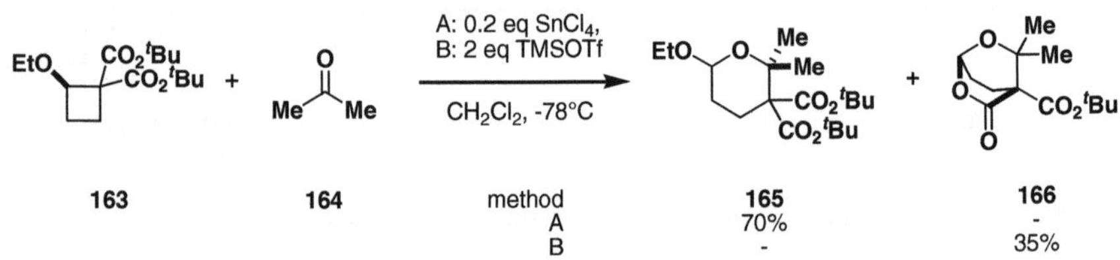

Scheme 1.52 [4 + 2] annulation reported by Matsuo and co-workers

1.4.6 Conclusion

Compared to [3 + 2], the [4 + 2] annulations are at their infancy and the reactivity of cyclobutanes in this area has to be investigated more in depth. Cyclobutanones, for example, can also be used as activated cyclobutanes [203, 204]. The presence of the acceptor part on the four-membered ring allows the ring fragmentation by retro-aldol and subsequent annulation with various dipolarophiles [205]. The annulations that have been developed so far represent good alternative methods to the Diels-Alder cycloaddition for the construction of more saturated six-membered rings. However, there are still important breakthroughs to be achieved in the field of aminocyclobutane annulation as there are no existing methods that can deliver nitrogen-substituted products.

1.5 Reactivity of Nitrogen-Substituted Small Cycloalkanes

In their study concerning the donating ability of various functional groups in DA cyclopropanes, Werz and co-workers found that nitrogen was superior to any other tested heteroatom or carbon group (N > O > S > Se > Ph > Me > P > Cl) [206]. It comes out from the strong donating capacity of nitrogen that the amino-substituted small rings are highly activated and therefore more prone to react by CC bond fragmentation. Deactivation by choosing judicious protecting group is thus necessary to avoid degradation for DA aminocyclopropanes and aminocyclobutanes. I will now shortly describe the different processes involving cyclization of nitrogen-substituted small rings.

1.5.1 Cyclopropanes

Wenkert and co-workers pioneered the synthesis and reactivity of DA aminocyclopropanes in the 1980s [207]. They developed a methodology to access various alkaloids such as **170** or **171** starting from a common aminocyclopropane intermediate **167** (Scheme 1.53).

Another use of cyclopropylamines as reactive intermediates has emerged recently. Ring-opening upon single electron transfer oxidation of tertiary cyclopropylamines initiated radical cascade sequences resulting in the formation of diverse polycyclic structures (Scheme 1.54) [208–210]. The formation of a cationic radical generated by oxidation of the nitrogen atom triggers the selective homolytic cleavage of the C–C bond due to the ring strain of cyclopropane. The fast 5-exo trig cyclization cascade that followed was very rapid and allowed the formation of the cyclic products. Two studies were reported almost at the same time, one using CAN as the oxidant and the second one photoactivation with dicyanobenzene as

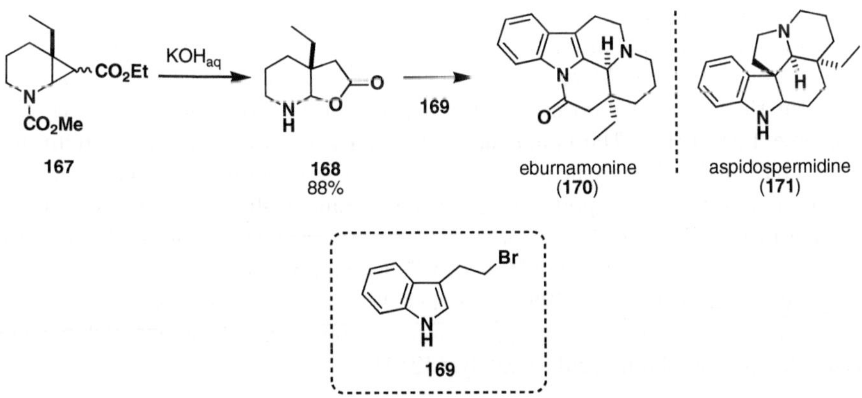

Scheme 1.53 Total syntheses of alkaloids reported by Wenkert and co-workers

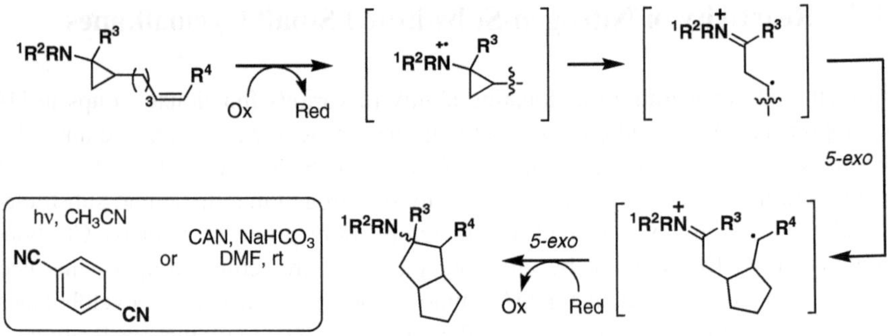

Scheme 1.54 Radical promoted [3 + 2] annulation of cyclopropylamines

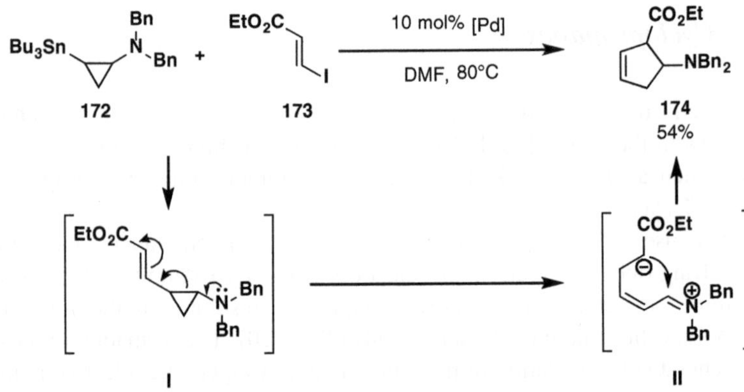

Scheme 1.55 Spontaneous rearrangement of DA aminocyclopropanes

sensitizer [208–210]. More recently, visible-light catalysis was used by Zheng and co-workers to perform intermolecular [3 + 2] annulation between cyclopropylamines and olefins [211].

An isolated example of the high reactivity of DA aminocyclopropanes has been reported by de Meijere, Marek and co-workers as they performed a Suzuki coupling between electron poor iodoacrylate **173** and tin-substituted cyclopropylamine **172** (Scheme 1.55) [212]. The conjugated ester in **I** acts as an activating group and allows the vinylcyclopropane rearrangement to occur under the reaction conditions.

Six and co-workers reported in 2005 the intramolecular cyclization of aminocyclopropanes with electron-rich aromatics under thermal conditions (Scheme 1.56) [213]. By heating substrates **175** and **177** in chlorobenzene, they obtained the tricyclic products **176** and **178** in 67 and 49 % yield respectively. Later, they reported a rearrangement of bicyclic aminocyclopropanes to aminocyclobutanes using camphorsulphonic acid as catalyst [214].

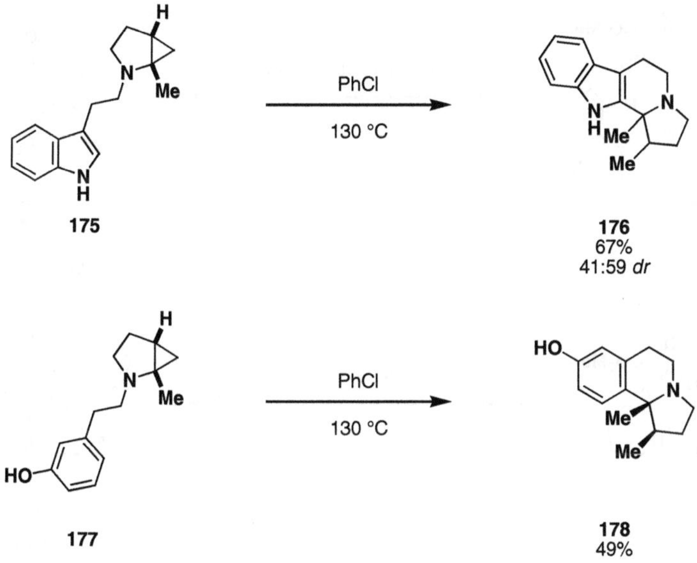

Scheme 1.56 Intramolecular cyclization developed by Six and co-workers

In 2011, France and co-workers developed an indium-catalyzed intramolecular Friedel-Crafts cyclization on DA cyclopropanes [215, 216]. The scope of the reaction involved various donating functionalities and two examples, **179** and **181**, containing nitrogen as a donating group were included (Scheme 1.57). In order to be stable, the aminocyclopropanes had to be protected as an imide or a carbamate.

The first example of the use of transition metal to cyclize cyclopropylamines was reported by Fagnou and co-workers in 2012 (Scheme 1.58) [217–219]. Cyclopropyl-bromo-anilines were reacted in presence of a palladium (II) catalyst and pivalic acid. The concerted-metalation-deprotonation (CMD) mechanism [220] (I–II in Scheme 1.58) was invoked to be at the origin of the palladium C–H insertion. This led to the formation of dihydroquinolines that were too unstable to be isolated. The products were therefore directly oxidized via a one-pot procedure to afford stable quinolines in yields ranging from 52 to 91 % or reduced to tetrahydroquinolines in 45–99 % yields.

A very original work from Bower and co-workers was reported in 2013 using the cyclopropylamine ring strain to promote the insertion of a transition metal (Scheme 1.59). By using rhodium and a urea directing group, they were able to insert a molecule of CO and an alkyne tethered to the nitrogen to produce heterobicyclic enones [221]. The reaction tolerated substitutions on the alkyne tether. In this case, it was moderately diastereoselective and products were obtained in yields up to 71 %. The method also tolerated substitution on the cyclopropane ring. The regioselectivity of the metal insertion was 5:1 when a butyl substituent was present and the diastereoselectivity was superior to 20:1.

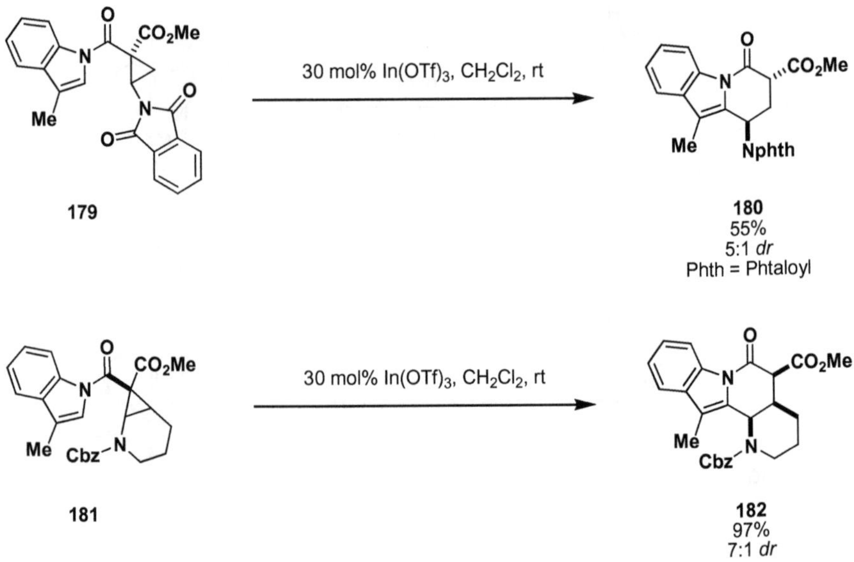

Scheme 1.57 Cyclization on aminocyclopropanes reported by France and co-workers

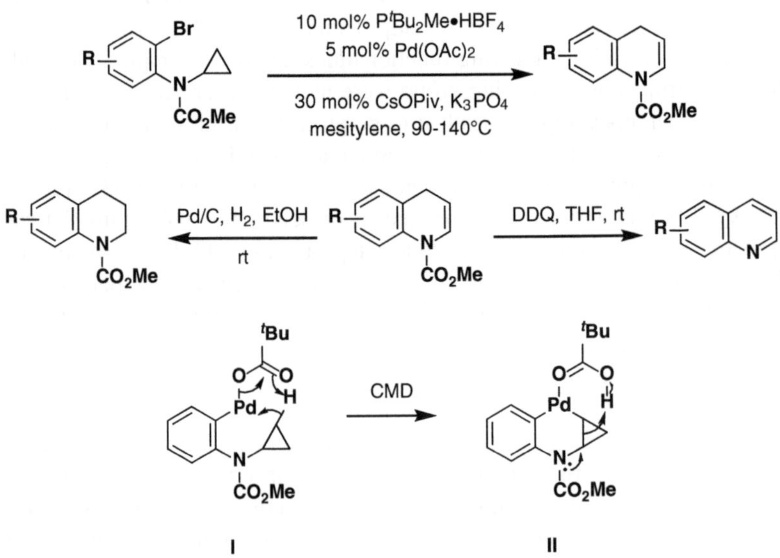

Scheme 1.58 Palladium-catalyzed cyclization of cyclopropyl-anilines

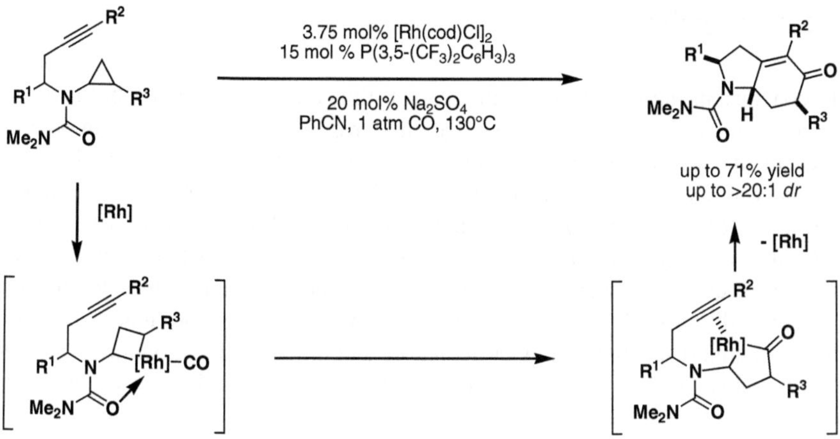

Scheme 1.59 Ring expansion using rhodium catalysis

183 NR_2 = Phthaloyl
185 NR_2 = pyrrolidinone

184 NR_2= Phthaloyl, 51%
186 NR_2 = pyrrolidinone, 87%

Scheme 1.60 Ring expansion to nitronates reported by Werz and co-workers

Werz and co-workers showed that the reaction between electron-rich olefins and nitro diazo esters afforded the cyclized nitronates without isolation of the intermediate DA cyclopropanes (Scheme 1.60) [222]. They reported two examples using nitrogen-substituted olefins bearing a phthalimide or an imidazolidinone functional group. In order to understand the competition between the ester and the nitro group for the ring expansion, the group performed computational studies which showed that the transition state energy for the latter is slightly lower.

1.5.2 *Cyclobutanes*

Release of the ring strain in aminocyclobutanes has also been exploited [29, 184]. However, except from the work of Saigo described previously [195], and isolated examples of ring expansions [223], it has never been thoroughly exploited in ring formation reactions.

1.5.3 Conclusion

In conclusion, the reactivity of nitrogen-substituted cycloalkanes has been studied for more than 30 years. It is known that the nitrogen has to be protected in order to prevent the spontaneous ring opening in the case of DA cycloalkanes. However, considering all the possible configurations of donor and acceptor functionalities, it becomes very difficult to find the balance between stability and reactivity. This is why for so many years the reactivity of such rings has been limited to intramolecular reactions. However, this strategy is limited when it comes to increase structural diversity. In this context, the development of intermolecular annulations between amino-substituted DA cycloalkanes and different dipolarophiles would be highly valuable as it would give access to diverse analogues in a very convergent manner.

1.6 Work in the Group

Our group began to study the reactivity of cyclopropanes in catalytic formal homo-Nazarov reactions in 2009 (Scheme 1.61) [224]. This reaction gave efficient access to polysubstituted six-membered rings using inexpensive and easy-to-handle tosic acid.

This methodology was extended to the total synthesis of indole alkaloids using aminocyclopropanes (Scheme 1.62) [225, 226]. By carefully selecting the reaction conditions, it was possible to control the regioselectivity of the cyclization reaction. Starting from intermediate **187** the formal synthesis of aspidospermidine (**171**) and starting from **189**, the total synthesis of goniomitine (**191**) were completed. The cytotoxicity of goniomitine (**191**) was tested against tumor cell lines but its activity was superior to 10 μM. Later, another natural product jerantinine E (**192**) could be synthesized for the first time using the same strategy. This natural product displayed significant cytotoxicity (1.0 μM) against cancer cells. Its mode of action could be established as an inhibitor of microtubule polymerization with a profile similar to colchicine.

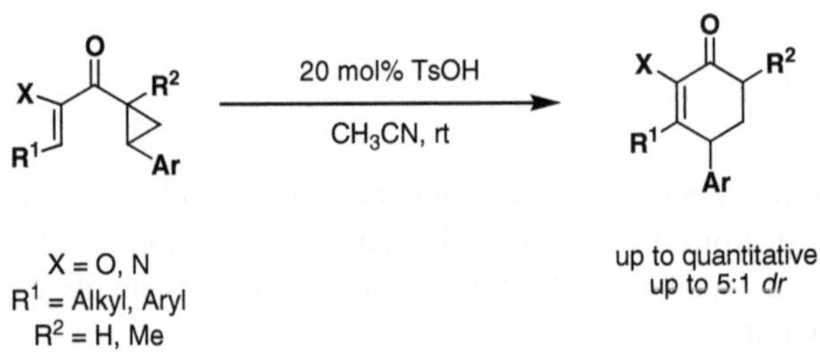

Scheme 1.61 Formal homo-Nazarov reaction reported by Waser and co-workers

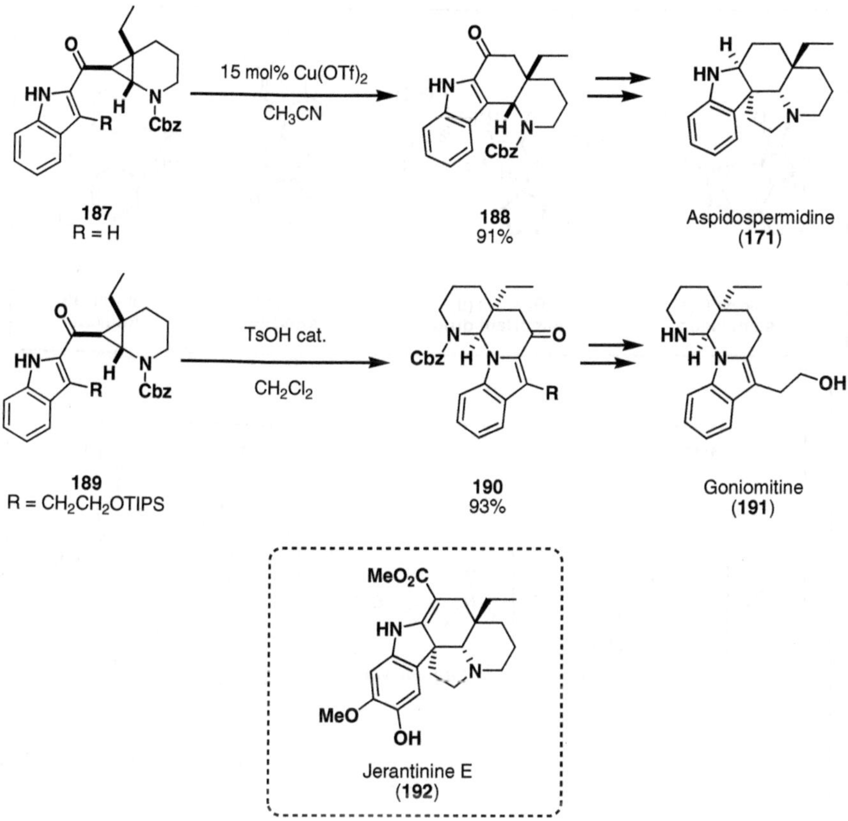

Scheme 1.62 Application of formal homo-Nazarov cyclization to total synthesis

1.7 Goal of the Project

As we have seen in the introduction, nitrogen-substituted rings are omnipresent in natural as well as in synthetic bioactive molecules. In order to discover new potent entities, it is therefore required to develop efficient and selective methods toward these privileged scaffolds.

Three- or four-membered rings have been known to react, when activated by a donor and an acceptor group, to generate 1,3 or 1,4 formal dipoles. These reactive intermediates are able to cyclize efficiently with various types of dipolarophiles to afford the corresponding carbo- or heterocycles. It has been possible to further develop these methodologies in order to access enantioenriched products.

Using the strain of amino-substituted small rings has been a good strategy to access relevant cyclic structures present in natural products or synthetic drugs. However previous works have been limited to intramolecular cyclizations. Such processes are often limited due to the difficult synthesis of a precursor containing all

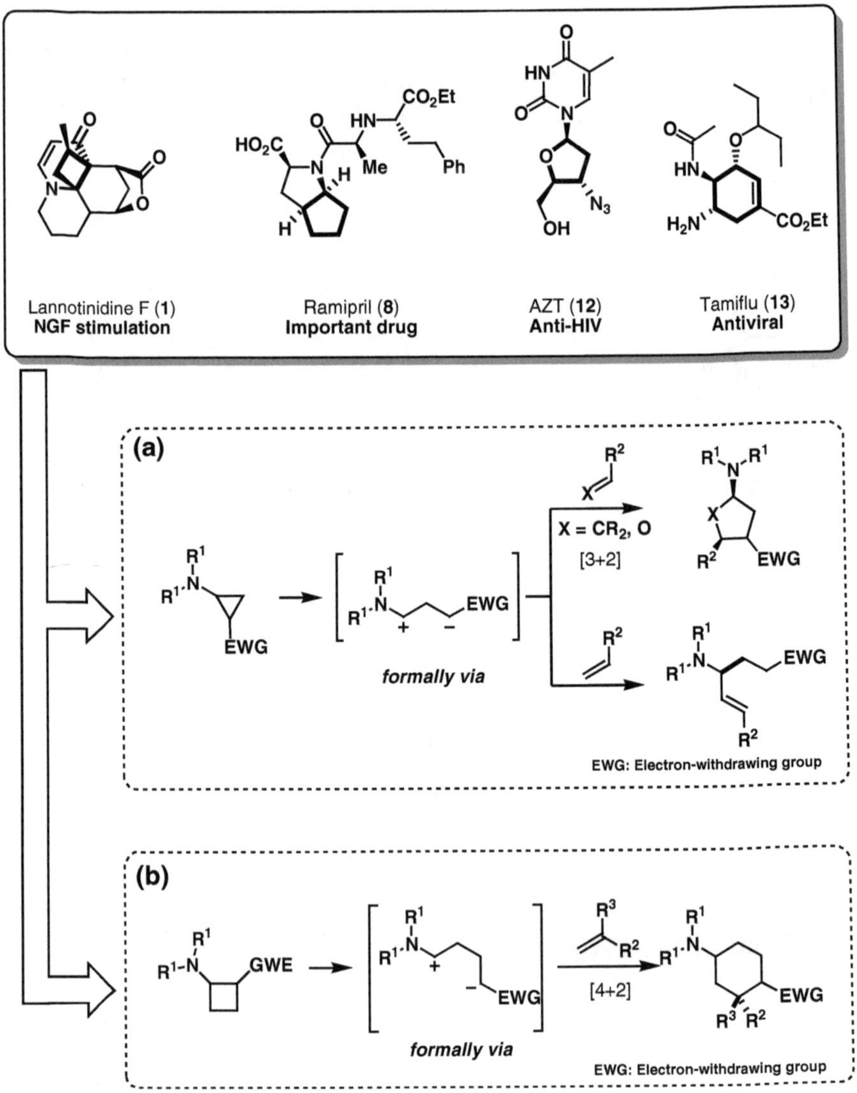

Fig. 1.8 Development of intermolecular annulation and addition reactions of aminocyclopropanes and aminocyclobutanes

the required electrophilic and nucleophilic functionalities. Additionally, there has been no report of the use of small nitrogen rings to access cyclized products in an enantioselective fashion.

Based on our group expertise in the synthesis and use of aminocyclopropanes, we therefore aimed to exploit the potential of amino-substituted small rings in intermolecular annulation reactions (a, Fig. 1.8). First, it was required to find a stable yet reactive aminocyclopropane that could be used with various dipolarophiles to generate substituted five-membered carbo- and heterocycles. The use of Lewis acids to trigger the reaction would allow to combine the use of electrophilic cyclopropanes with electron-rich systems such as olefins or carbonyls. The development of an enantioselective method for the [3 + 2] annulations would strengthen the use of aminocyclopropanes in the construction of amino-substituted five-membered rings.

The use of aminocyclopropanes would also be extended to nucleophilic ring-opening reactions. The Friedel-Crafts alkylation of (hetero)-aromatic would provide an efficient access to linear derivatives of γ-amino-acids.

Our second goal was to develop the use of aminocyclobutanes as precursors for the selective synthesis of amino-substituted six-membered rings (b, Fig. 1.8). As there is no method to access efficiently and selectively substituted aminocyclobutanes, it was required to first develop a protocol toward their synthesis. With the four-membered ring in hand we would finally use them as formal dipoles in [4 + 2] annulations.

Ideally, the development of new annulations method would be followed by the creation of small libraries of compounds which bioactivity could be evaluated against relevant targets.

References

1. Boutureira O, Matheu MI, Diaz Y, Castillon S (2013) Chem Soc Rev 42:5056
2. Risseeuw M, Overhand M, Fleet GJ, Simone M (2013) Amino Acids 45:613
3. Dembitsky V (2008) J Nat Med 62:1
4. Darses B, Greene AE, Poisson J-F (2011) J Org Chem 77:1710
5. Gorrea E, Pohl G, Nolis P, Celis S, Burusco KK, Branchadell V, Perczel A, Ortuño RM (2012) J Org Chem 77:9795
6. Celis S, Nolis P, Illa O, Branchadell V, Ortuno RM (2013) Org Biomol Chem 11:2839
7. Fulop F, Martinek TA, Toth GK (2006) Chem Soc Rev 35:323
8. Kinnel RB, Gehrken HP, Scheuer PJ (1993) J Am Chem Soc 115:3376
9. Brodersen DE, Clemons WM, Carter AP, Morgan-Warren RJ, Wimberly BT, Ramakrishnan V (2000) Cell 103:1143
10. Hanessian S, Vakiti RR, Dorich S, Banerjee S, Lecomte F, DelValle JR, Zhang J, Deschênes-Simard B (2011) Angew Chem Int Ed 50:3497
11. Malinowski JT, Sharpe RJ, Johnson JS (2013) Science 340:180
12. Tan X, Chu CK, Boudinot FD (1999) Adv Drug Delivery Rev 39:117
13. Saxton JE (1997) Nat Prod Rep 14:559
14. O'Connor SE, Maresh JJ (2006) Nat Prod Rep 23:532
15. Hass HB, McBee ET, Hinds GE (1936) Ind Eng Chem 28:1178
16. Wong HNC, Hon MY, Tse CW, Yip YC, Tanko J, Hudlicky T (1989) Chem Rev 89:165
17. Liebman JF, Greenberg A (1976) Chem Rev 76:311
18. de Meijere A (2005) Angew Chem Int Ed 44:7836
19. Coulson CA, Moffitt WE (1947) J Chem Phys 15:151
20. Boese R, Miebach T, De Meijere A (1991) J Am Chem Soc 113:1743
21. Chakrabarti P, Seiler P, Dunitz JD, Schluter AD, Szeimies G (1981) J Am Chem Soc 103:7378
22. Walsh AD (1947) Nature 159:712
23. Walsh AD (1949) Trans Far Soc 45:179

24. Dewar MJS (1984) J Am Chem Soc 106:669
25. Bernett WA (1967) J Chem Educ 44:17
26. The Chemistry of Cyclobutanes (2005) In: Rappoport ZZ, Liebman JF (eds) Wiley, Chichester
27. Wiberg KB (1986) Angew Chem Int Ed 25:312
28. Bartell LS, Andersen B (1973) J Chem Soc Chem Commun 786
29. Namyslo JC, Kaufmann DE (2003) Chem Rev 103:1485
30. Doyle MP, Van Leusen D (1982) J Org Chem 47:5326
31. Lifchits O, Charette AB (2008) Org Lett 10:2809
32. Kunkel E, Reichelt I, Reißig H-U (1984) Liebigs Ann Chem 1984:802
33. Xing S, Li Y, Li Z, Liu C, Ren J, Wang Z (2011) Angew Chem Int Ed 50:12605
34. Seebach D (1979) Angew Chem Int Ed 18:239
35. UPAC (1997) Compendium of chemical terminology, 2nd edn. (the "Gold Book"). In: McNaught AD, Wilkinson A (eds) Blackwell Scientific Publications, Oxford
36. Danishefsky S (1979) Acc Chem Res 12:66
37. Wenkert E (1980) Acc Chem Res 13:27
38. Reissig H-U, Zimmer R (2003) Chem Rev 103:1151
39. Yu M, Pagenkopf BL (2005) Tetrahedron 61:321
40. Agrawal D, Yadav VK (2008) Chem Comm 6471
41. De Simone F, Waser J (2009) Synthesis 2009:3353
42. Kerr MA, Lebold TP (1797) Pure Appl Chem 82
43. Mel'nikov MY, Budynina EM, Ivanova OA, Trushkov IV (2011) Mendeleev Commun 21:293
44. Cavitt MA, Phun LH, France S (2014) Chem Soc Rev 43:804
45. Schneider TF, Kaschel J, Werz DB (2014) Angew Chem Int Ed. doi:10.1002/anie.201309886
46. Carson CA, Kerr MA (2009) Chem Soc Rev 38:3051
47. Tang P, Qin Y (2012) Synthesis 44:2969
48. Wang Z (2012) Synlett 23:2311
49. Lebel H, Marcoux J-F, Molinaro C, Charette AB (2003) Chem Rev 103:977
50. Simmons HE, Smith RD (1958) J Am Chem Soc 80:5323
51. Furukawa J, Kawabata N, Nishimura J (1966) Tetrahedron Lett 7:3353
52. Harrington P, Kerr MA (1997) Tetrahedron Lett 38:5949
53. Grover HK, Lebold TP, Kerr MA (2010) Org Lett 13:220
54. Emmett MR, Kerr MA (2011) Org Lett 13:4180
55. England DB, Woo TK, Kerr MA (2002) Can J Chem 80:992
56. Bajtos B, Yu M, Zhao H, Pagenkopf BL (2007) J Am Chem Soc 129:9631
57. Ivanova OA, Budynina EM, Grishin YK, Trushkov IV, Verteletskii PV (2008) Eur J Org Chem 2008:5329
58. Chagarovskiy AO, Budynina EM, Ivanova OA, Grishin YK, Trushkov IV, Verteletskii PV (2009) Tetrahedron 65:5385
59. Gharpure SJ, Vijayasree U, Reddy SRB (2012) Org Biomol Chem 10:1735
60. Stork G, Grieco PA (1969) J Am Chem Soc 91:2407
61. Corey EJ, Balanson RD (1973) Tetrahedron Lett 14:3153
62. Christie SDR, Davoile RJ, Jones RCF (2006) Org Biomol Chem 4:2683
63. Nicholas KM, Petit R (1971) Tet Lett 12:3474
64. Beal RB, Dombroski MA, Snider BB (1986) J Org Chem 51:4391
65. Saigo K, Shimada S, Shibasaki T, Hasegawa M (1093) Chem Lett 1990:19
66. Adams J, Belley M (1986) Tetrahedron Lett. 27:2075
67. Komatsu M, Suehiro I, Horiguchi Y, Kuwajima I (1991) Synlett 1991:771
68. Horiguchi Y, Suehiro I, Sasaki A, Kuwajima I (1993) Tetrahedron Lett. 34:6077
69. Mathieu B, Ghosez L (2002) Tetrahedron 58:8219
70. Takasu K, Nagao S, Ihara M (2006) Adv Synth Catal 348:2376
71. Mukaiyama T, Narasaka K, Banno K (1011) Chem Lett 1973:2

72. Fang J, Ren J, Wang Z (2008) Tetrahedron Lett 49:6659
73. Qu J-P, Deng C, Zhou J, Sun X-L, Tang Y (2009) J Org Chem 74:7684
74. Qu J-P, Liang Y, Xu H, Sun X-L, Yu Z-X, Tang Y (2012) Chem Eur J 18:2196
75. Kerr MA, Keddy RG (1999) Tetrahedron Lett 40:5671
76. Venkatesh C, Singh PP, Ila H, Junjappa H (2006) Eur J Org Chem 2006:5378
77. Lambert JB, Zhao Y, Emblidge RW, Salvador LA, Liu X, So J-H, Chelius EC (1998) Acc Chem Res 32:183
78. Sugita Y, Yamadoi S, Hosoya H, Yokoe I (2001) Chem Pharm Bull 49:657
79. Yadav VK, Sriramurthy V (2004) Org Lett 6:4495
80. Yadav VK, Sriramurthy V (2004) Angew Chem Int Ed 43:2669
81. Qi X, Ready JM (2008) Angew Chem Int Ed 47:7068
82. Mackay WD, Fistikci M, Carris RM, Johnson JS (2014) Org Lett 16:1626
83. Wolfe JP, Hay MB (2007) Tetrahedron 63:261
84. Brueckner C, Reissig HU (1988) J Org Chem 53:2440
85. Shimada S, Hashimoto Y, Sudo A, Hasegawa M, Saigo K (1992) J Org Chem 57:7126
86. Shimada S, Hashimoto Y, Saigo K (1993) J Org Chem 58:5226
87. Sugita Y, Yokoe I (2000) Heterocycles 53:657
88. Sugita Y, Yokoe I (2001) Heterocycles 55:135
89. Pohlhaus PD, Johnson JS (2005) J Am Chem Soc 127:1601
90. Pohlhaus PD, Johnson JS (1057) J Org Chem 2005:70
91. Pohlhaus PD, Sanders SD, Parsons AT, Li W, Johnson JS (2008) J Am Chem Soc 130:8642
92. Campbell MJ, Johnson JS, Parsons AT, Pohlhaus PD, Sanders SD (2010) J Org Chem 75:6317
93. Gupta A, Yadav VK (2006) Tetrahedron Lett 47:8043
94. Parsons AT, Campbell MJ, Johnson JS (2008) Org Lett 10:2541
95. Smith AG, Slade MC, Johnson JS (1996) Org Lett 2011:13
96. Haubenreisser S, Hensenne P, Schröder S, Niggemann M (2013) Org Lett 15:2262
97. Ma X, Tang Q, Ke J, Yang X, Zhang J, Shao H (2013) Org Lett 15:5170
98. Yang G, Sun Y, Shen Y, Chai Z, Zhou S, Chu J, Chai J (2013) J Org Chem 78:5393
99. Campbell MJ, Johnson JS (2009) J Am Chem Soc 131:10370
100. Karadeolian A, Kerr MA (2010) J Org Chem 75:6830
101. Alper PB, Meyers C, Lerchner A, Siegel DR, Carreira EM (1999) Angew Chem Int Ed 38:3186
102. Huang W, O'Donnell M-M, Bi G, Liu J, Yu L, Baldino CM, Bell AS, Underwood TJ (2004) Tetrahedron Lett 45:8511
103. Qiu L, Guo X, Zhou J, Liu S, Yang L, Wu X, Hu W (2013) RSC Adv 3:20065
104. Lautens M, Han WS (2002) J Am Chem Soc 124:6312
105. Scott ME, Han W, Lautens M (2004) Org Lett 6:3309
106. Taillier C, Bethuel Y, Lautens M (2007) Tetrahedron 63:8469
107. Taillier C, Lautens M (2007) Org Lett 9:591
108. Scott ME, Lautens M (2008) J Org Chem 73:8154
109. Dieskau AP, Holzwarth MS, Plietker B (2012) J Am Chem Soc 134:5048
110. Tombe R, Kurahashi T, Matsubara S (2013) Org Lett 15:1791
111. Jackson SK, Karadeolian A, Driega AB, Kerr MA (2008) J Am Chem Soc 130:4196
112. Saigo K, Shimada S, Hasegawa M (1990) Chem Lett 19:905
113. Carson CA, Kerr MA (2005) J Org Chem 70:8242
114. Kang Y-B, Tang Y, Sun X-L (2006) Org Biomol Chem 4:299
115. Christie SDR, Davoile RJ, Jones RCF (2006) Org Biomol Chem 4:2683
116. Yu M, Pagenkopf BL (2003) J Am Chem Soc 125:8122
117. Chagarovskiy AO, Budynina EM, Ivanova OA, Trushkov IV (2010) Chem Het Comp 46:120
118. Goodell JR, Poole JL, Beeler AB, Aubé J, Porco JA (2011) J Org Chem 76:9792
119. Sathishkannan G, Srinivasan K (2011) Org Lett 13:6002

120. Chagarovskiy AO, Ivanov KL, Budynina EM, Ivanova OA, Trushkov IV (2012) Chem Het Comp 48:825
121. Cui B, Ren J, Wang Z (2013) J Org Chem 79:790
122. Cui B, Ren J, Wang Z (2013) J Org Chem 79:790
123. Yu M, Pagenkopf BL (2003) Org Lett 5:5099
124. Yu M, Pantos GD, Sessler JL, Pagenkopf BL (1057) Org Lett 2004:6
125. Morales CL, Pagenkopf BL (2007) Org Lett 10:157
126. Bajtos B, Pagenkopf BL (1072) Eur J. Org Chem 2009:2009
127. Moustafa MMAR, Pagenkopf BL (2010) Org Lett 12:3168
128. Wang H, Yang W, Liu H, Wang W, Li H (2012) Org Biomol Chem 10:5032
129. Sun Y, Yang G, Chai Z, Mu X, Chai J (2013) Org Biomol Chem 11:7859
130. Graziano ML, Iesce MR (1987) J Chem Res Synop 362
131. Graziano ML, Cimminiello G (1989) J Chem Res Synop 42
132. Li H, Wang W, Liu H, Wang H, Yang W (2012) Org Biomol Chem 10:5032
133. Goldberg AFG, O'Connor NR, Craig RA, Stoltz BM (2012) Org Lett 14:5314
134. Graziano ML, Iesce MR, Cermola F (1996) J Chem Res Synop 82
135. Korotkov VS, Larionov OV, Hofmeister A, Magull J, de Meijere A (2007) J Org Chem 72:7504
136. Bruckner C, Reissig HU (1985) Angew Chem Int Ed Engl 24:588
137. Cermola F, DiGioia L, Graziano ML, Iesce MR (2005) J Chem Res Synop 677
138. Marino JP, Laborde E (1985) J Am Chem Soc 107:734
139. Liu L, Montgomery J (2006) J Am Chem Soc 128:5348
140. Tsuji J, Takahashi H, Morikawa M (1965) Tetrahedron Lett 6:4387
141. Hiroi K, Yamada A (1835) Tetrahedron: Asymmetry 2000:11
142. Goldberg AFG, Stoltz BM (2011) Org Lett 13:4474
143. Dieskau AP, Holzwarth MS, Plietker B (2012) J Am Chem Soc 134:5048
144. Candish L, Forsyth CM, Lupton DW (2013) Angew Chem Int Ed 52:9149
145. Gu X, Li X, Qu Y, Yang Q, Li P, Yao Y (2013) Chem Eur J 19:11878
146. Wolf C (2008) Dynamic stereochemistry of chiral compounds: principles and applications. The Royal Society of Chemistry, Cambridge, pp 331, 359–369, 372
147. Faber K (2001) Chem Eur J 7:5004
148. Steinreiber J, Faber K, Griengl H (2008) Chem Eur J 14:8060
149. Trost BM, Bunt RC, Lemoine RC, Calkins TL (2000) J Am Chem Soc 122:5968
150. Sibi MP, Ma Z, Jasperse CP (2005) J Am Chem Soc 127:5764
151. Kang Y-B, Sun X-L, Tang Y (2007) Angew Chem Int Ed 46:3918
152. Zhou Y-Y, Li J, Ling L, Liao S-H, Sun X-L, Li Y-X, Wang L-J, Tang Y (2013) Angew Chem Int Ed 52:1452
153. Zhou Y-Y, Wang L-J, Li J, Sun X-L, Tang Y (2012) J Am Chem Soc 134:9066
154. Xiong H, Xu H, Liao S, Xie Z, Tang Y (2013) J Am Chem Soc
155. Parsons AT, Johnson JS (2009) J Am Chem Soc 131:3122
156. Evans DA, Miller SJ, Lectka T, von Matt P (1999) J Am Chem Soc 121:7559
157. Parsons AT, Smith AG, Neel AJ, Johnson JS (2010) J Am Chem Soc 132:9688
158. Wales SM, Walker MM, Johnson JS (2013) Org Lett
159. Xu H, Qu J-P, Liao S, Xiong H, Tang Y (2013) Angew Chem Int Ed 52:4004
160. Trost BM, Morris PJ (2011) Angew Chem Int Ed 50:6167
161. Trost BM, Morris PJ, Sprague SJ (2012) J Am Chem Soc
162. Mei LY, Wei Y, Xu Q, Shi M (2012) Organometallics 31:7591
163. Mei LY, Wei Y, Xu Q, Shi M (2013) Organometallics 32:3544
164. Young IS, Kerr MA (2003) Angew Chem Int Ed 42:3023
165. Young IS, Kerr MA (2004) Org Lett 6:139
166. Sibi MP, Ma Z, Jasperse CP (2005) J Am Chem Soc 127:5764
167. Kang YB, Sun XL, Tang Y (2007) Angew Chem Int Ed 46:3918
168. Kerr MA, Lebold TP (2010) Pure Appl Chem 82:1797
169. Bai Y, Tao W, Ren J, Wang Z (2012) Angew Chem Int Ed 51:4112

170. Tang P, Qin Y (2012) Synthesis 44:2969
171. Wang Z, Ren J, Wang Z (2013) Org Lett 15:5682
172. Cavitt MA, Phun LH, France S (2014) Chem Soc Rev 43:804
173. Baar MR, Ballesteros P, Roberts BW (1986) Tetrahedron Lett 27:2083
174. Moustafa MMAR, Pagenkopf BL (2010) Org Lett 12:4732
175. Parsons AT, Johnson JS (2009) J Am Chem Soc 131:14202
176. Takasu K, Ueno M, Inanaga K, Ihara M (2003) J Org Chem 69:517
177. Inanaga K, Takasu K, Ihara M (2005) J Am Chem Soc 127:3668
178. Canales E, Corey EJ (2007) J Am Chem Soc 129:12686
179. Schuster DI, Lem G, Kaprinidis NA (1993) Chem Rev 93:3
180. Jo H, Fitzgerald ME, Winkler JD (2009) Org Lett 11:1685
181. Gauzy C, Pereira E, Faure S, Aitken DJ (2004) Tetrahedron Lett 45:7095
182. Gauzy C, Saby B, Pereira E, Faure S, Aitken DJ (2006) Synlett 2006:1394
183. Basler B, Schuster O, Bach T (2005) J Org Chem 70:9798
184. White JD, Li Y, Ihle DC (2010) J Org Chem 75:3569
185. Brannock KC, Bell A, Burpitt RD, Kelly CA (1964) J Org Chem 29:801
186. Fleming I, Harley-Mason J (1964) J Chem Soc (Resumed) 2165
187. Kuehne ME, Foley L (1965) J Org Chem 30:4280
188. Hall HK, Ykman P (1975) J Am Chem Soc 97:800
189. Evans SB, Abdelkader M, Padias AB, Hall HK (1989) J Org Chem 54:2848
190. Ishihara K, Nakano K (2007) J Am Chem Soc 129:8930
191. Albrecht Ł, Dickmeiss G, Acosta FC, Rodríguez-Escrich C, Davis RL, Jørgensen KA (2012) J Am Chem Soc 134:2543
192. Avenoza A, Busto JH, Canal N, Peregrina JM, Pérez-Fernández M (2005) Org Lett 7:3597
193. de Meijere A, Limbach M, Janssen A, Lygin A, Korotkov VS (2010) Eur J Org Chem 2010:3665
194. Sans M, Illa O, Ortuño RM (2012) Org Lett 14:2431
195. Shimada S, Saigo K, Nakamura H, Hasegawa M (1991) Chem Lett 20:1149
196. Parsons AT, Johnson JS (2009) J Am Chem Soc 131:14202
197. Allart EA, Christie SDR, Pritchard GJ, Elsegood MRJ (2009) Chem Comm 7339
198. Moustafa MMAR, Pagenkopf BL (2010) Org Lett 12:4732
199. Moustafa MMAR, Stevens AC, Machin BP, Pagenkopf BL (2010) Org Lett 12:4736
200. Vemula N, Stevens AC, Schon TB, Pagenkopf BL (2014) Chem Comm 50:1668
201. Stevens AC, Palmer C, Pagenkopf BL (2011) Org Lett 13:1528
202. Okado R, Nowaki A, Matsuo JI, Ishibashi H (2012) Chem Pharm Bull 60:21
203. Seiser T, Saget T, Tran DN, Cramer N (2011) Angew Chem Int Ed 50:7740
204. Belluš D, Ernst B (1988) Angew Chem Int Ed 27:797
205. Kawano M, Kiuchi T, Negishi S, Tanaka H, Hoshikawa T, Matsuo J-I, Ishibashi H (2013) Angew Chem Int Ed 52:906
206. Schneider TF, Werz DB (1848) Org Lett 2011:13
207. Wenkert E, Hudlicky T (1953) J Org Chem 1988:53
208. Ha JD, Lee J, Blackstock SC, Cha JK (1998) J Org Chem 63:8510
209. Takemoto Y, Yamagata S, Furuse S-I, Hayase H, Echigo T, Iwata C (1998) Chem Commun 651
210. Lee HB, Sung MJ, Blackstock SC, Cha JK (2001) J Am Chem Soc 123:11322
211. Maity S, Zhu M, Shinabery RS, Zheng N (2012) Angew Chem Int Ed 51:222
212. Wiedemann S, Rauch K, Savchenko A, Marek I, de Meijere A (2004) Eur J Org Chem 631
213. Larquetoux L, Ouhamou N, Chiaroni A, Six Y (2005) Eur J Org Chem 4654
214. Wasilewska A, Woźniak BA, Doridot G, Piotrowska K, Witkowska N, Retailleau P, Six Y (2013) Chem Eur J 19:11759
215. Patil DV, Cavitt MA, Grzybowski P, France S (2011) Chem Comm 47:10278
216. Patil DV, Cavitt MA, France S (2012) Heterocycles 84:1363
217. Rousseaux S, Liégault B, Fagnou K (2012) Chem Sci 3:244
218. DosSantos A, El Kaïm L, Grimaud L, Ramozzi R (2012) Synlett 438

219. Ladd CL, Roman DS, Charrette AB (2013) Tetrahedron 69:4479
220. Rousseaux S, Gorelsky SI, Chung BKW, Fagnou K (1069) J Am Chem Soc 2010:132
221. Shaw MH, Melikhova EY, Kloer DP, Whittingham WG, Bower JF (2013) J Am Chem Soc 135:4992
222. Schmidt CD, Kaschel J, Schneider TF, Kratzert D, Stalke D, Werz DB (2013) Org Lett 15:6098
223. Hall HK, Ykman P (1975) J Am Chem Soc 97:800
224. De Simone F, Andrès J, Torosantucci R, Waser J (1023) Org Lett 2009:11
225. De Simone F, Gertsch J, Waser J (2010) Angew Chem Int Ed 49:5767
226. Frei R, Staedler D, Raja A, Franke R, Sasse F, Gerber-Lemaire S, Waser J (2013) Angew Chem Int Ed 52:13373

Chapter 2
Ring-Opening Reactions of Aminocyclopropanes

When I began the project, the intermolecular annulations using DA aminocyclopropanes were unknown. I will present in a first time the investigations undertaken on the design of new aminocyclopropanes that can be used as 1,3 dipole precursors. The different projects that were developed involving the annulation and the Friedel-Crafts alkylation of such dipoles will then be presented and discussed.

2.1 [3 + 2] Annulation with Enol Ethers [1]

2.1.1 Discovery of the Reaction and Optimization

It was expected that DA aminocyclopropanes had to be protected in order to decrease the strong donating effect of the nitrogen. We selected several protecting group associating the donating ability of the nitrogen atom to its pKa values (Fig. 2.1).[1] A low pKa value indicates a more deactivated nitrogen due to the poorer electron density. Lactam, benzyl carbamate, oxazolidinone and phthalimide were chosen as they have been used in a large number of synthetic methods and possess different electronic properties.

The acceptor part was divided into two types of functionalities: mono- and *gem*-disubstituted esters. In the same way as for nitrogen, the activating capacity could be evaluated considering the stabilization of the charge resulting from bond fragmentation. Additionally, due to the possibility of chelation with some Lewis acids, 1,3 dicarbonyl compounds can be more efficiently activated.

In order to access the desired cyclopropanes, the combination of an electron-rich olefin and an electron-poor carbenoid seemed the most reliable (Scheme 2.1). Electron-poor diazo derivatives are easy to access, stable and generate electrophilic carbenoid precursors in presence of transition metals. On the other side, electron-rich enamides or enimide are commercially available or easily accessed via reported methods.

---

[1] http://www.chem.wisc.edu/areas/reich/pkatable/index.htm.

F. de Nanteuil, *Synthesis and Reactivity of Donor-Acceptor Substituted Aminocyclopropanes and Aminocyclobutanes*, Springer Theses,
DOI 10.1007/978-3-319-23006-1_2

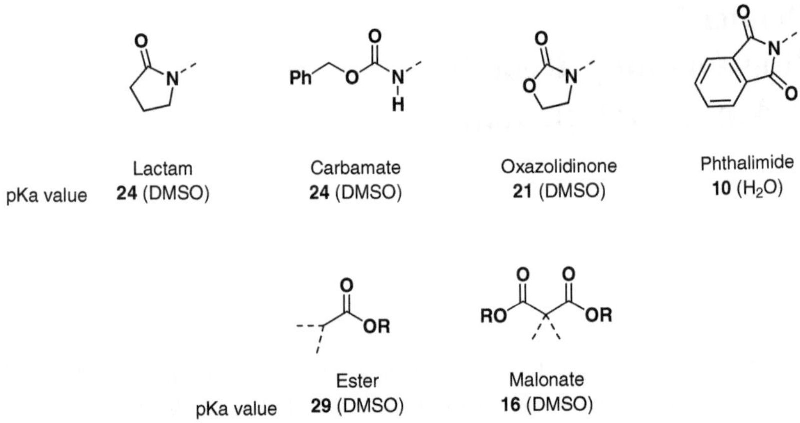

Fig. 2.1 pKa values of donating and accepting groups

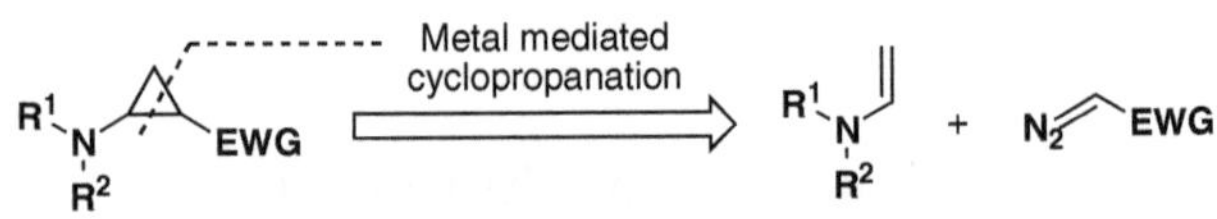

Scheme 2.1 Retrosynthesis of DA aminocyclopropanes

Carbamate protected enamine **195** (a, Scheme 2.2) was synthesized by a Curtius rearrangement on acryloyl chloride (**193**) [2]. Vinyl oxazolidinone **185** (b, Scheme 2.2) was synthesized in one step using a palladium-catalyzed vinyl transfer [3]. The malonic ester diazo derivatives **199** and **201** were prepared following a very simple protocol involving diazo transfer from tosyl azide under basic conditions, (c, Scheme 2.2) [4]. Starting materials such as ethyl diazoacetate (**202**), vinyl phthalimide (**183**) or vinyl pyrrolidinone were commercially available.

The non-optimized reaction of the diazo compounds in presence of catalysts such as Rh(II) (a, Scheme 2.3) or Cu(I) (b, Scheme 2.3) furnished the desired mono-ester DA cyclopropanes **203–206**. Except for the oxazolidinone-substituted cyclopropane **204**, which was obtained as a mixture, only the *trans* isomer of the product was isolated.

It has been shown by our group that simple Brønsted acids such as *p*-toluenesulfonic acid are strong enough to activate DA cyclopropanes [37, 38]. Therefore, the newly synthesized aminocyclopropanes were reacted under various acidic conditions in order to study their reactivity.

Direct subjection of the monoester-substituted cyclopropanes **203–205** to catalytic amounts of acids in presence of diverse olefins and carbonyls resulted in recovery of the starting cyclopropanes (Scheme 2.4). However, under specific conditions, the isomerization from *cis* to *trans* of the donor and the acceptor substituents could be observed in **204**. Depending on the solvent, and on the strength of the acid, the isomerization process gave information on the reactivity. For example,

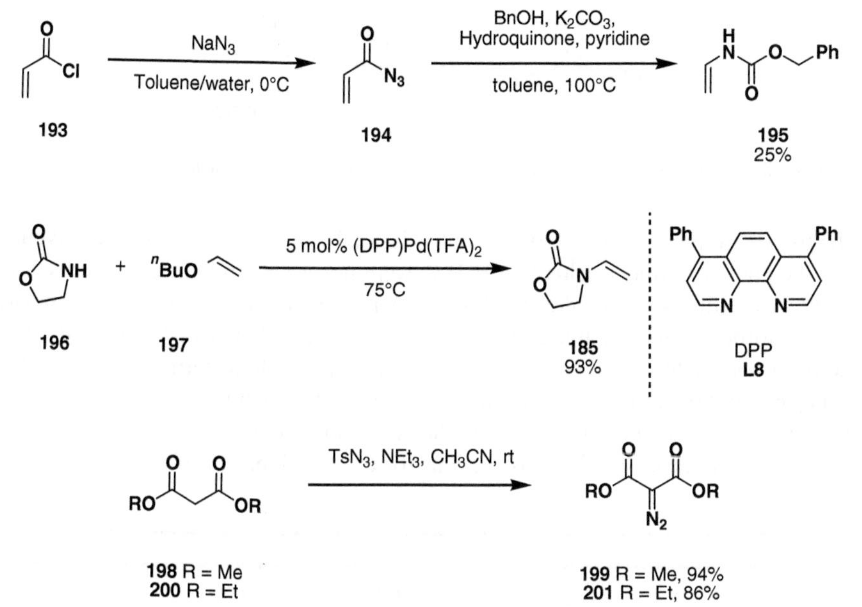

Scheme 2.2 Synthesis of aminocyclopropane precursors

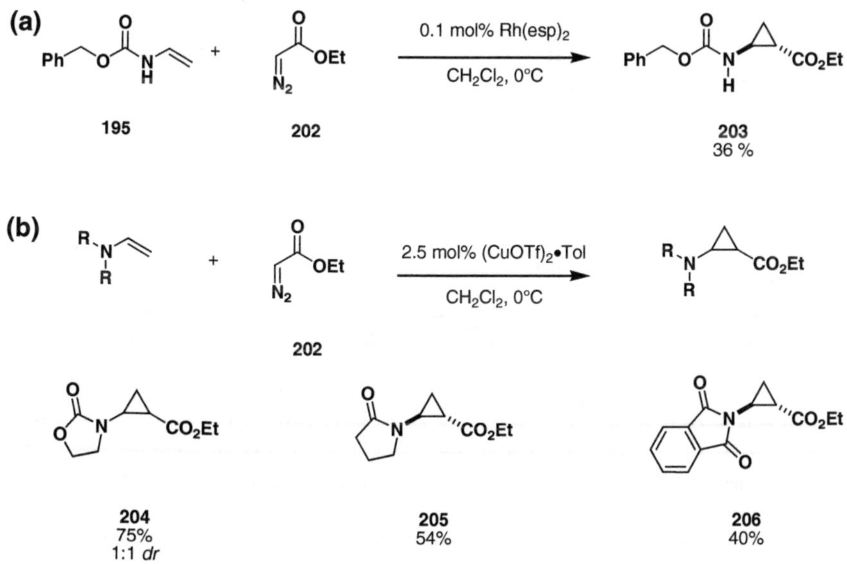

Scheme 2.3 Synthesis of mono-ester DA aminocyclopropanes

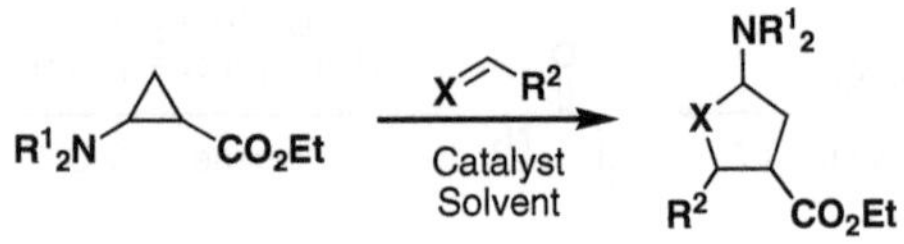

Scheme 2.4 Acids, solvents and dipolarophiles tested under Brønsted acid catalysis

the use of strong acids like HBF_4 resulted in total isomerization in all the solvents. On the contrary, milder acids like H_3PO_4 were never able to induce isomerization. More interestingly, acids like PTSA or CSA isomerized the product in acetonitrile or dichloromethane but not in toluene. It can be concluded that Brønsted acids activation is unable to trigger the reaction with a nucleophilic partner. However, the catalyst is not the only factor to take into account. The stabilization of the resulting reactive polarized intermediate by choosing an adequate solvent will be an important parameter to take into account for the development of the reaction.

Partners

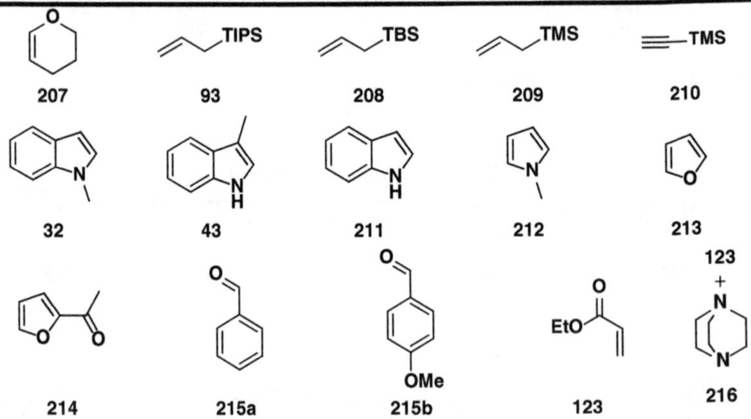

Brønsted acids				
H_3PO_4	Camphorsulfonic acid[b]	TFA	PTSA[b]	HBF_4[a]

Solvents			
Toluene, rt, 100°C	CH_2Cl_2, rt	CH_3CN, rt, 80°C	MTBE, rt

(a) Isomerization occurred with all solvents. (b) Isomerization occurred in CH_3CN and CH_2Cl_2

Electron-rich olefins or carbonyls were often unstable under acidic conditions, resulting in degradation of the dipolarophile. Taking these results into account, we envisioned that diester-substituted aminocyclopropanes would be more suitable as DA aminocyclopropanes. The presence of a bidentate functionality allows chelation

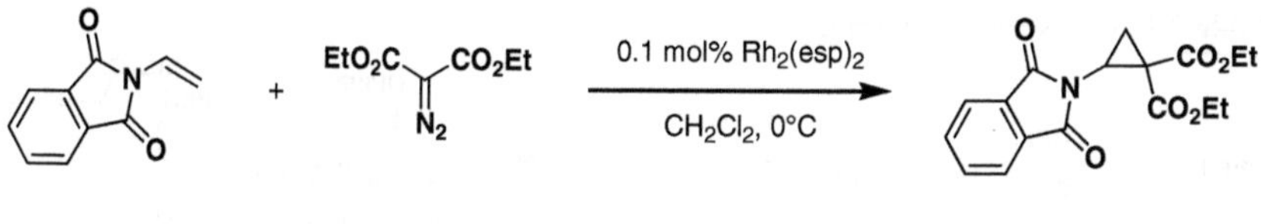

183 201 217 90%

Scheme 2.5 Synthesis of diester DA aminocyclopropanes

to the catalyst and an increased reactivity due to the second electron-withdrawing group. As a consequence, milder reaction conditions should be possible with these substrates. The use of the malonate derivatives also simplifies the system due to the elimination of diastereomeric mixtures in the starting material as well as in the product.

The synthesis of diester derivatives was attempted. However, the products where too reactive and degradation occurred. In the case of vinyl carbamate **195** neither product nor starting material could be detected in the crude mixture of the reaction. Vinyl oxazolidinone analogue **185** and lactam **204**, could be cyclized but degradation occurred during purification, indicating the low stability of these aminocyclopropanes. Eventually, the phthalimide derivative **217** was the only one that could be isolated in very good yield (Scheme 2.5).

As we were interested in developing an efficient access to cyclopentylamines from aminocyclopropanes, we decided to investigate the use of olefins as annulation partners. Consequently, silyl enol ether **219a** was chosen as dipolarophile as we expected that an electron-rich olefin would react faster in presence of DA aminocyclopropane **217**. Additionally, they can be very easily accessed on large scales starting from ketone **218a** using an electrophilic silylating reagent (Scheme 2.6).

In order to avoid dipolarophile degradation, we oriented our choice of catalyst toward Lewis acids. We investigated their influence on the [3 + 2] annulation between aminocyclopropane **217** and silyl enol ether **219a**. Brønsted acids (entries 1–3) and some of the tested Lewis acids (entries 4–9) did not promote conversion of the starting material under these conditions. In the case of $Ni(ClO_4)_2$ and $Zn(OTf)_2$, partial conversion of the cyclopropane was observed (entries 10–11). However, when using triflimide (entries 12–13), triflate (entries 14–20) or chloride (entries 21–23) based Lewis acids full conversion was observed. Whenever conversion to the product **220a** was obtained, the side product **221** resulting from a retro-aldol

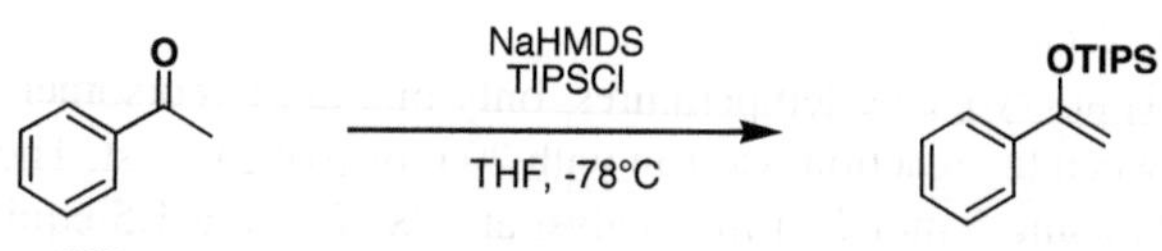

218a 219a 77%

Scheme 2.6 Synthesis of enol ether **219a**

Table 2.1 Lewis acid optimization

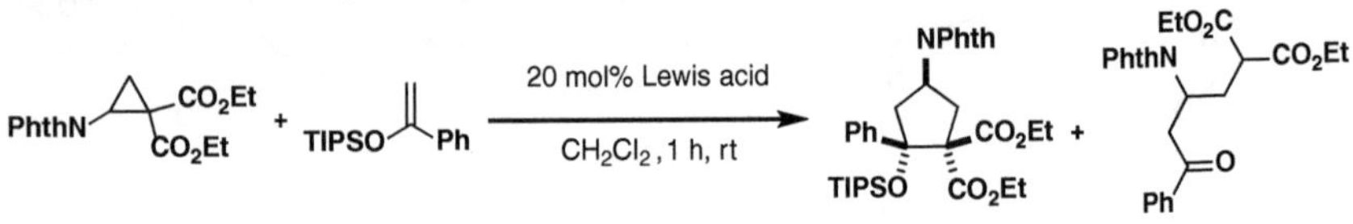

217 219a 220a 221

Entry[a]	Lewis acid	Conversion[b]	**220a/221**[b]
1	Trifluoroacetic acid	0	nd
2	Camphorsulfonic acid	0	nd
3	H_3PO_4	0	nd
4	$^{n}Bu_2BOTf$	0	nd
5	$BF_3 \cdot OEt_2$	0	nd
6	MgI_2	0	nd
7	Me_2AlCl	0	nd
8	AgOTf	0	nd
9	$CuCl_2$	0	nd
10	$Ni(ClO_4)_2$	30	10:1
11	$Zn(OTf)_2$	40	1:2
12	$Zn(NTf)_2$	100	1:2.5
13	$HNTf_2$	100	>1:20
14	$Hf(OTf)_4$, H_2O	100	1:2
15	$Sc(OTf)_3$	100	1:3
16	$In(OTf)_3$	100	>1:20
17	$(CuOTf)_2 \cdot Tol$	100	1:1
18	$Cu(OTf)_2$	100	5:1
19	$Yb(OTf)_3$	100	5:1
20	$Sn(OTf)_2$	100	2:1
21	$AlCl_3$	100	1:2.5
22	$SnCl_4$	100	1:1
23	**$SnCl_4$ (–78 °C)**	**100**	**>20:1**

[a]*Reaction conditions* 1 eq cyclopropane, 1.5 eq enol ether, 0.2 mol% Lewis acid, CH_2Cl_2 (0.08 M), 60 min, room temperature
[b]Determined by ^1H NMR of the crude reaction

fragmentation was also detected (entries 11–22). Only when $SnCl_4$ was used at –78 °C, the desired product **220a** was obtained selectively (entry 23). In some cases **221** was found as the major (entries 11, 12, 14, 15, 21) or the sole (entries 13, 16) product (Table 2.1).

Using $SnCl_4$ at cryogenic temperatures, only one diastereoisomer was obtained in 95 % yield when the reaction was run with 20 mol% of catalyst. These conditions were optimized using 5 mol% of tin catalyst at –78 °C using 1.5 equivalents of the enol ether in dry dichloromethane (Scheme 2.7). Quenching the reaction with triethylamine has to be done at –78 °C otherwise open product **221** is observed.

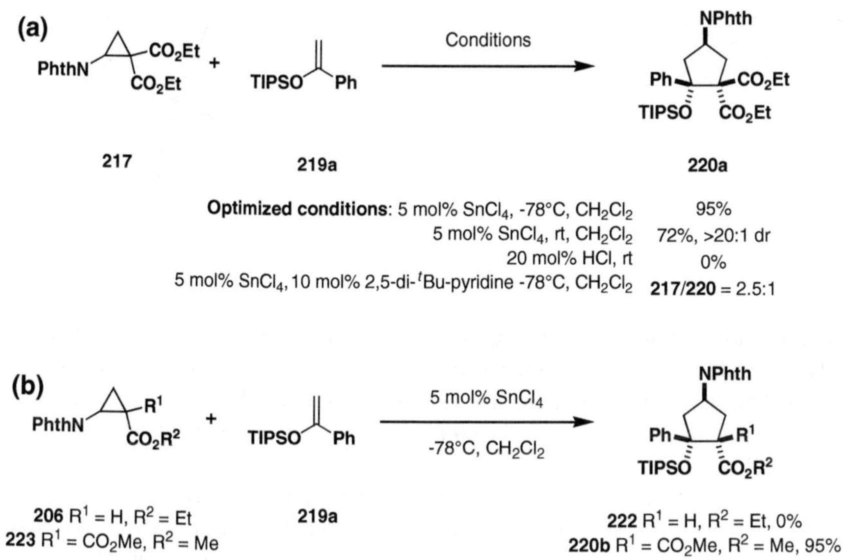

Scheme 2.7 Control experiments

Several control experiments where performed in order to gather information on the catalyst as well as on the cyclopropane reactivity. The reaction performed at room temperature gave a lower isolated yield of 72 % of the product **220a**.[2] In order to exclude the possibility that some traces of HCl resulting from tin tetrachloride decomposition could have catalysed the reaction, we performed the reaction using 20 mol% HCl in dioxane. No conversion could be observed. Adding 10 mol% di-^{t}Bu-pyridine as a proton scavenger to the standard reaction conditions did not inhibit the reaction even if full conversion was not obtained after 1 h. This could be explained by pyridine poisoning of the catalyst or by neutralization of a synergistic effect between the tin catalyst and traces of acid. The influence of the structure of the aminocyclopropane was investigated next. Switching to the dimethyl malonate derivative **223** did not alter the reactivity and using the monoester analogue **206** synthesized earlier left the starting material unreacted.[3] The relative configuration of the product was confirmed by X-ray diffraction analysis (Fig. 2.2).

2.1.2 Scope and Limitations

The promising results obtained with silyl enol ether **219a** motivated us to investigate the tolerance of the reaction to differents silyl enol ethers. They were synthetized using two different methods (Scheme 2.8): the first one generating the

[2]Overall yield of the reaction is 96 % with 24 % yield of open product.

[3]**204** and **205** were also tested unsuccessfully.

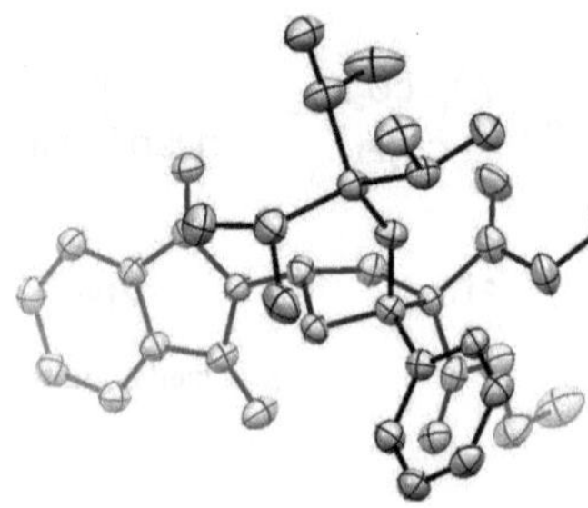

Fig. 2.2 Structure of **220a** determined by X-ray diffraction analysis

sodium enolate with NaHMDS and subsequently quenching with the corresponding silyl chloride at –78 °C; the second one using a combination of triethylamine and silyl triflate at room temperature. These two methods afforded silyl enol ethers from electron rich and poor aromatic, heteroaromatic as well as aliphatic ketones in very good yields.

For this purpose, enol ethers derived from acetophenone (**218a**) were prepared with different silicon substitution (entries 1, 2, Table 2.2). Electron-withdrawing as well as donating substituents were introduced at diverse positions of the aromatic ring (entries 3–8). Cyclic (entries 9, 10) and silyl enol ethers substituted in β position (entries 11, 12) were also prepared. Silyl enol ether **219n** with a heterocyclic substituent was synthesized in order to test the influence of a basic functionality on the reaction (entry 13). Finally, silyl enol ethers derived from alkyl ketones were prepared (entries 14, 15).

In the case of **218n** the regioselectivity and the diastereoselectivity were not total and the product **219p** was obtained as a 10:1.3:0.3 mixture of the di- and Z/E trisubstituted isomers respectively. **219o** was isolated as a 8:1 mixtures of Z/E isomers.

We then turned our attention to the scope and limitations of the method. The silyl enol ethers previously synthesized were evaluated (Table 2.3). The use of different silicon derivatives did not decrease the yield and even the more sensitive TMS enol ethers **219q** could be employed (entries 1–4). Changing the ester group did not change the yield on preparative scale as the ethyl ester derivative was obtained in 98 % yield (entry 5). Electron-poor as well as electron-rich aromatics gave the product in yields superior than 92 % and only one diastereoisomer could be detected (entries 6–11). Only in the case where a cyano substituent was present (entry 8), a lower yield of 70 % was obtained at full conversion. The reaction also tolerated more challenging substrates such as hindered *o*-bromo- or napthyl-substituted silyl enol ethers **219d** and **219e** (entries 11, 12).

In the previous reports of [3 + 2] annulations between DA cyclopropanes and silyl enol ethers, most of the methods were limited by the substitutions of the olefin [39].

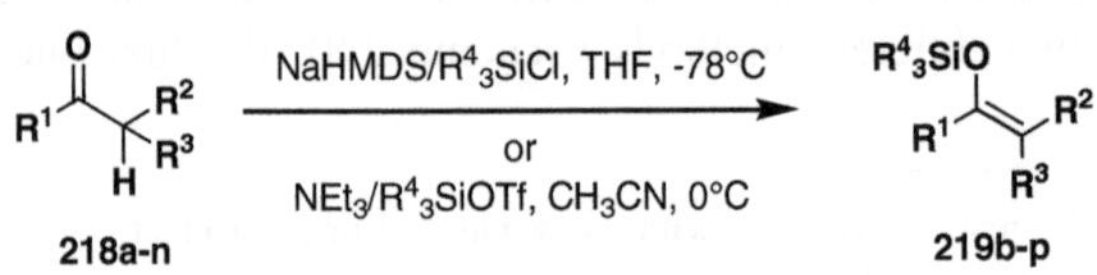

Scheme 2.8 Synthesis of silyl enol ethers

Table 2.2 Synthesis of silyl enol ethers

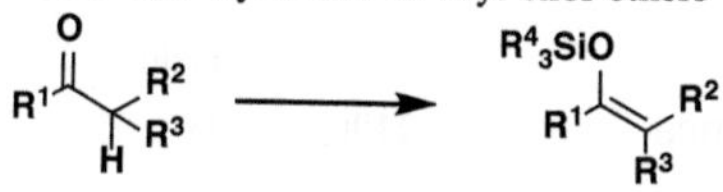

218a-n **219b-p**

Entry	Substrate		Product		Yield (%)[a]
1	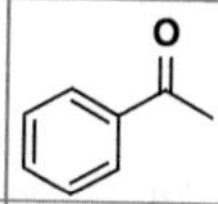	**218a**	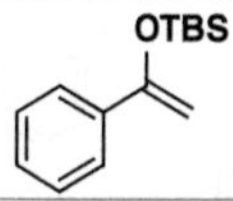	**219b**	85
2	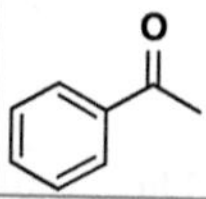	**218a**	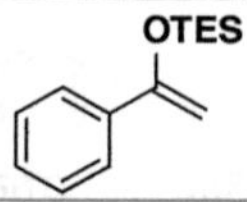	**219c**	77
3	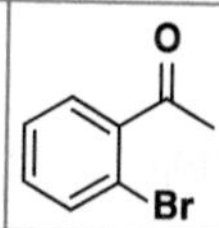	**218b**	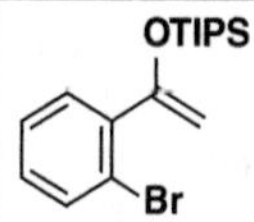	**219d**	88
4	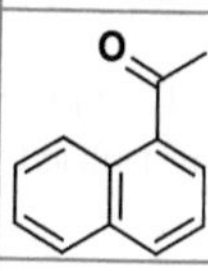	**218c**	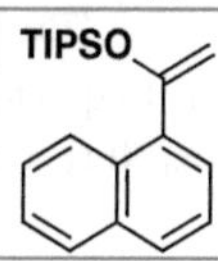	**219e**	91
5	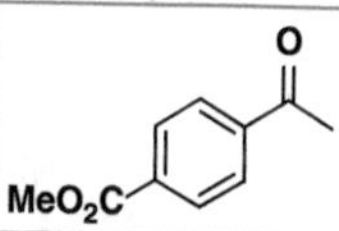	**218d**	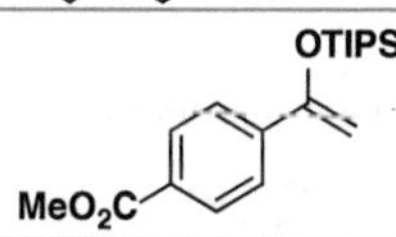	**219f**	91
6	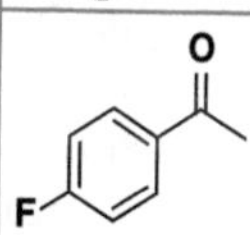	**218e**	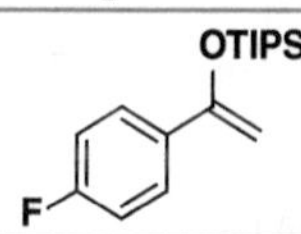	**219g**	73
7	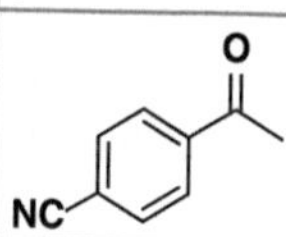	**218f**	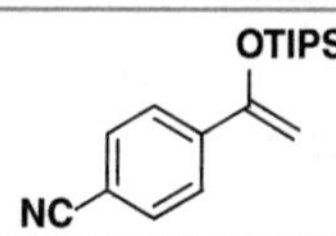	**219h**	97
8	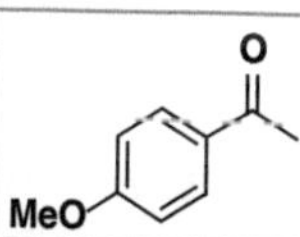	**218g**	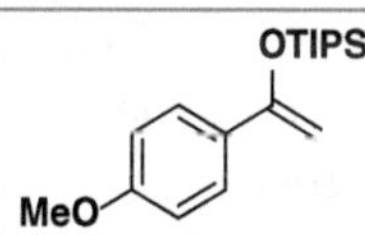	**219i**	87
9	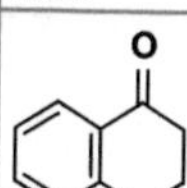	**218h**	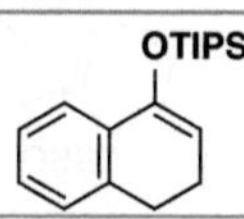	**219j**	87
10	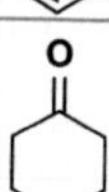	**218i**	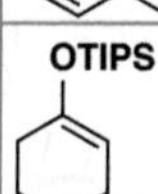	**219k**	99

(continued)

Table 2.2 (continued)

Entry	Substrate		Product		Yield (%)[a]
11	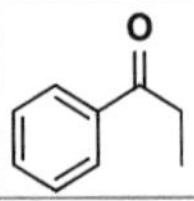	**218j**	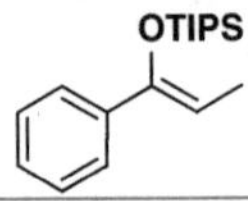	**219l**	98
12	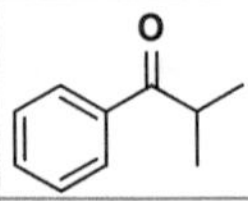	**218k**	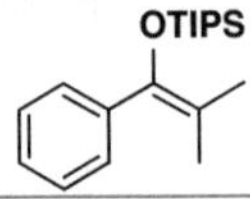	**219m**	98
13	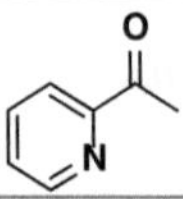	**218l**	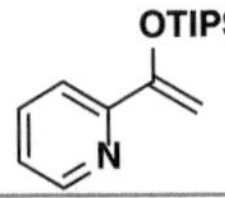	**219n**	58
14	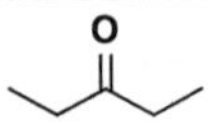	**218m**	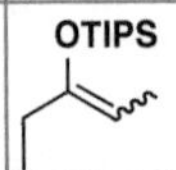	**219o**	61
15	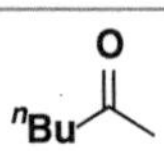	**218n**	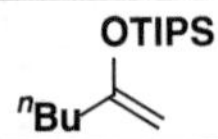	**219p**	92

[a]Reaction conditions entries (1–3, 8–15): 1 eq ketone, 1.22 eq NaHMDS, 1.2 eq $ClSiR^4_3$, THF, –78 °C or entries (4–7) 1 eq ketone, 1.8 eq NEt_3, 1.2 eq $TfOSiR^4_3$, CH_2Cl_2, rt. Isolated yields after column chromatography

Table 2.3 Scope with aromatic di-substituted silyl enol ethers

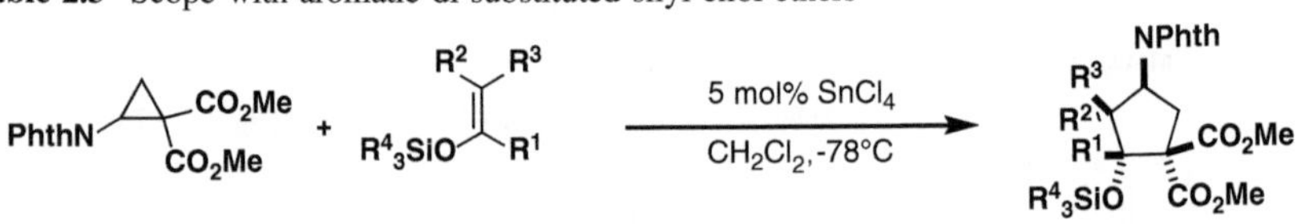

223 **219a-i,q** **220a-k**

Entry	Substrate		Product		Yield (%)[a]
1	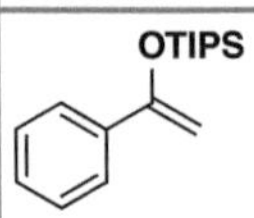	**219a**	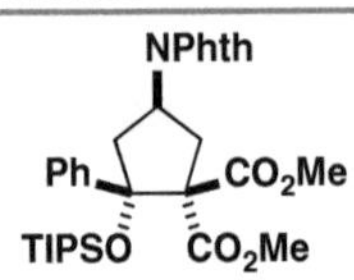	**220b**	95 (>20:1 *dr*)
2	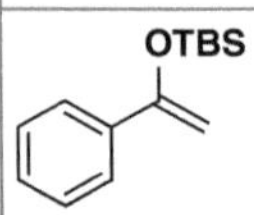	**219b**	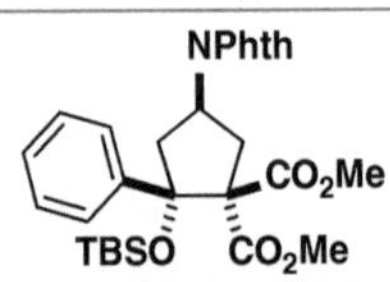	**220c**	98 (>20:1 *dr*)
3	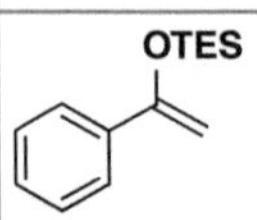	**219c**	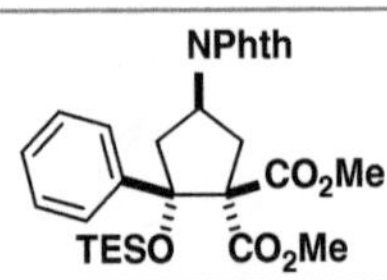	**220d**	99 (>20:1 *dr*)
4	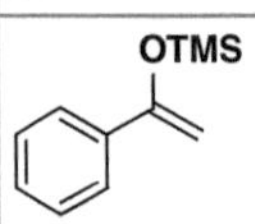	**219q**	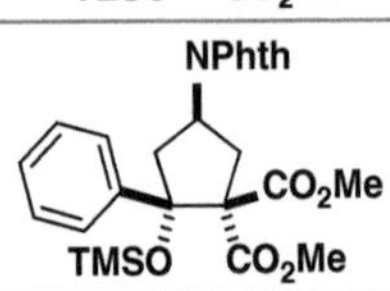	**220e**	95 (>20:1 *dr*)

(continued)

Table 2.3 (continued)

Entry	Substrate		Product		Yield (%)[a]
5	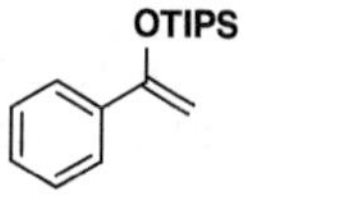	**219a**	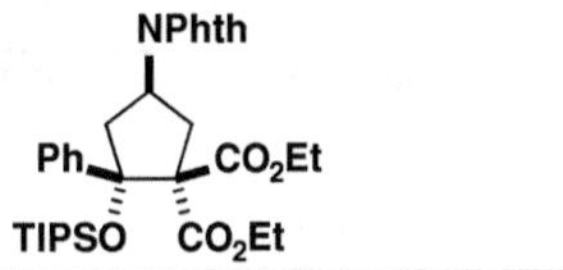	**220a**	98[b] (>20:1 *dr*)
6	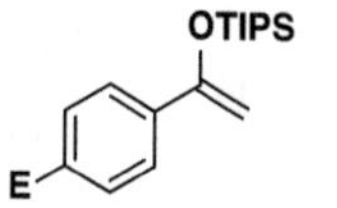 E = CO_2Me	**219f**	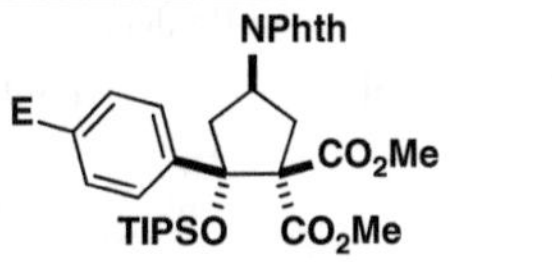 E = CO_2Me	**220f**	95 (>20:1 *dr*)
7	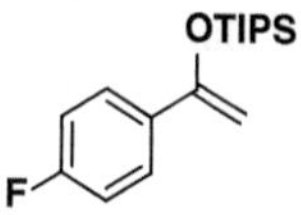	**219g**	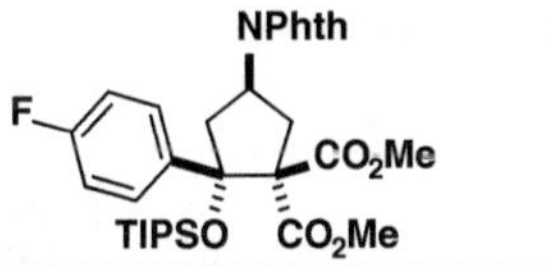	**220g**	91 (>20:1 *dr*)
8	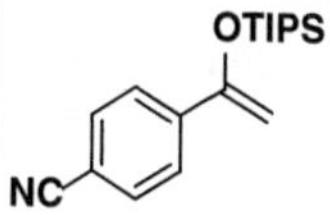	**219h**	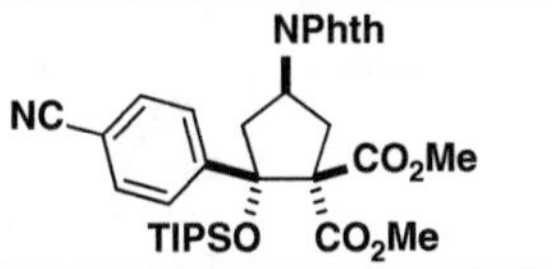	**220h**	70 (>20:1 *dr*)
9	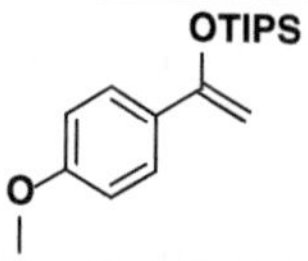	**219i**	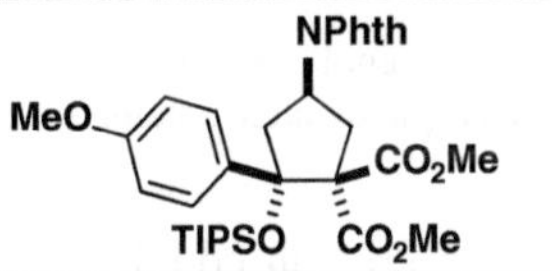	**220i**	99 (>20:1 *dr*)
10	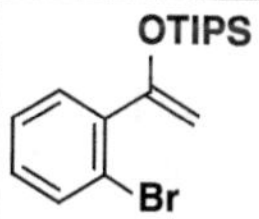	**219d**	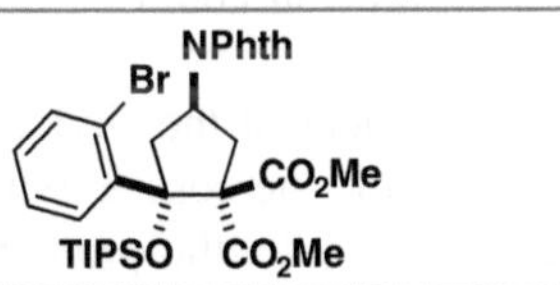	**220j**	92 (>20:1 *dr*)
11	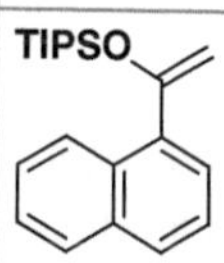	**219e**	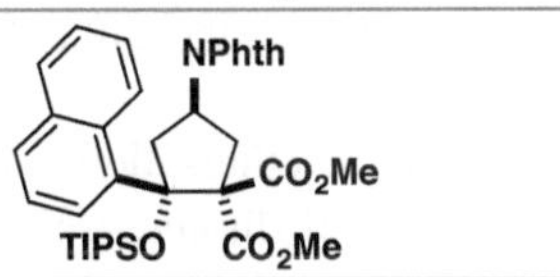	**220k**	91 (>20:1 *dr*)

[a]Reaction conditions: 0.45 mmol enol ether, 0.30 mmol cyclopropane, 0.015 mmol $SnCl_4$, in 2 mL CH_2Cl_2, 1 h, –78 °C. Isolated yields after column chromatography
[b]Cyclopropane **217** was used

However, it was possible with our method to react tri-and even tetrasubstituted olefins in very good yields and with high diastereospecificity (entries 1–3, Table 2.4).

X-ray analysis showed that the relative stereochemistry in **220m** was opposite to the one observed in other products (Fig. 2.3).[4] Again, the diastereoselectivity

[4]The diastereospecificity of the reaction will be discussed in the mechanism part of this chapter.

Table 2.4 Scope with substituted silyl enol ethers

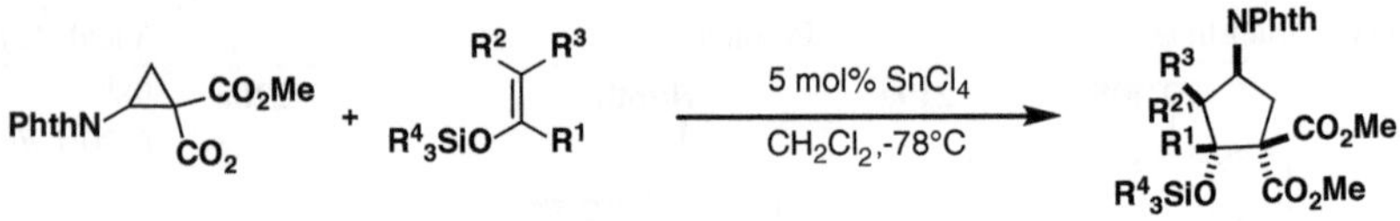

223 219j,l,m 220l-n

Entry	Substrate		Product		Yield (%)[a]
1	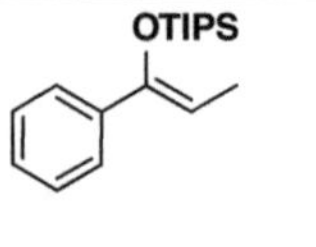	219l	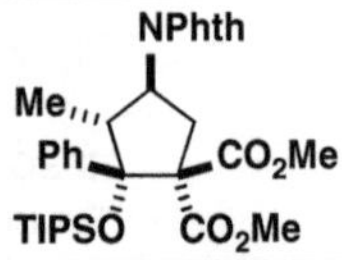	220l	92 (20:1 *dr*)
2	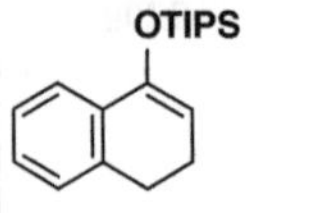	219j	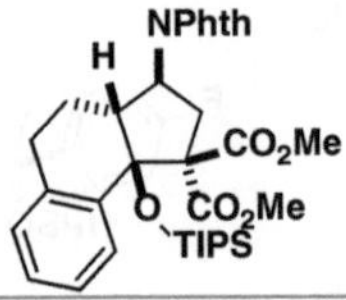	220m	91 (10:1 *dr*)
3	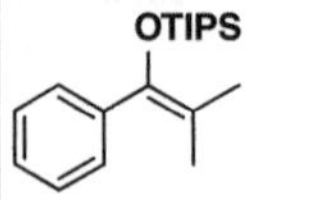	219m	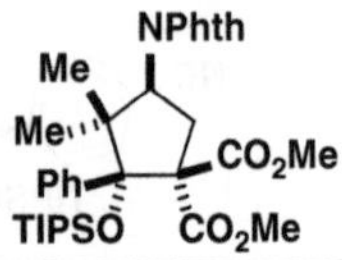	220n	81 (4:1 *dr*)

[a]Reaction conditions: 0.45 mmol enol ether, 0.30 mmol cyclopropane, 0.015 mmol $SnCl_4$, in 2 mL CH_2Cl_2, 1 h, –78 °C. Isolated yields after column chromatography

remained high, even if an erosion could be noticed in the case of hindered tetra-substituted enol ether.

Replacing the aryl substituent of the silyl enol ether by an alkyl group did not decrease the reactivity nor the stereoselectivity (entries 1–3, Table 2.5). As the reaction is highly diastereospecific, the products **220o** and **220p** were isolated as mixture of diastereoisomers due to the difficulty to obtain the pure isomeric enol ethers (entries 1, 2).

Simple silyl and alkyl enol ethers derived from acetaldehyde gave cyclized products in excellent yields and diastereoselectivity (entries 4, 5). It is worth to notice that it is the first time that alkyl enol ethers are used for [3 + 2] annulations with DA cyclopropanes. Finally, the acid-sensitive dihydropyran **207** was also successfully reacted in quantitative yield (entry 6).

The relative stereochemistry of **220q** was assigned by ROE experiment and the one of **220t** by X-ray analysis (Fig. 2.4).

Pyridine derived silyl enol ether **219n** and vinyl acetate **224** did not react under these conditions (Fig. 2.5). The former is probably deactivating the catalyst while the latter is not nucleophilic enough.

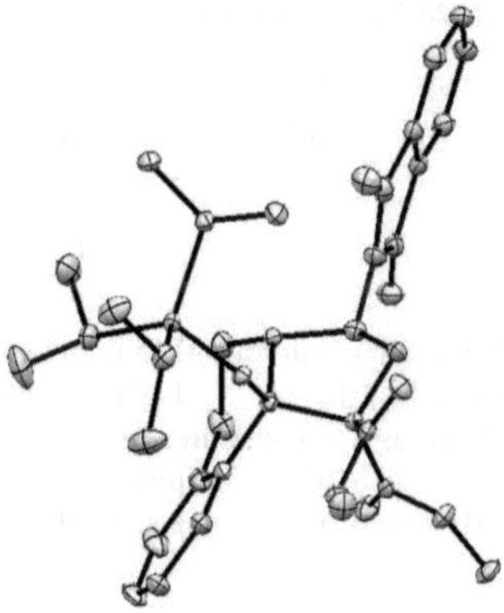

Fig. 2.3 Structure of **220m** determined by X-ray diffraction analysis

Table 2.5 Scope with alkyl-substituted silyl and alkyl enol ethers

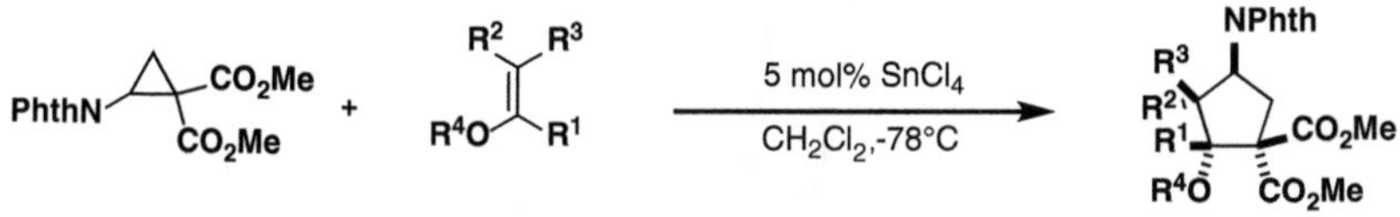

223 **219k,o,p,r 197, 207** **220o-t**

Entry	Substrate		Product		Yield (%)[a]
1	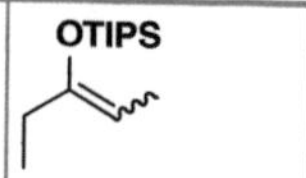	**219o**[b]	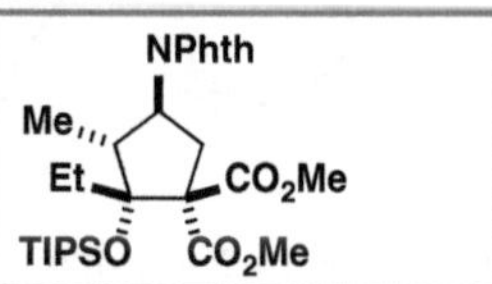	**220o**	96 (11:4:1 *dr*)
2	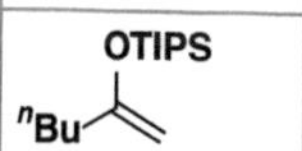	**219p**[c]	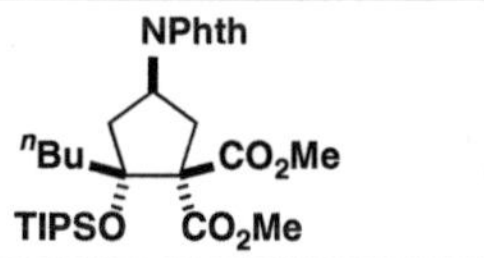	**220p**	98[d] (17:1 *dr*)
3	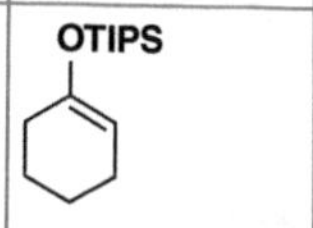	**219k**	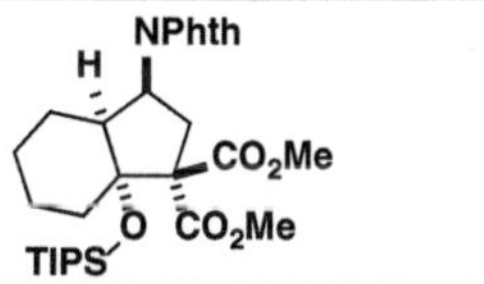	**220q**	90 (4:1 *dr*)
4	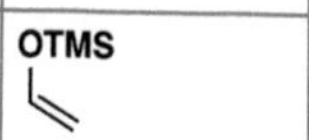	**219r**	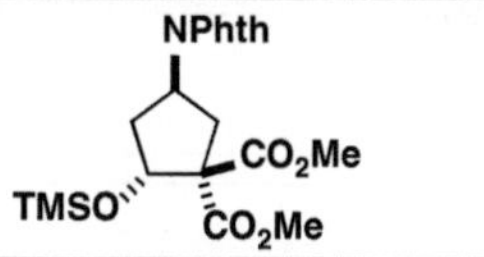	**220r**	77 (20:1 *dr*)
5	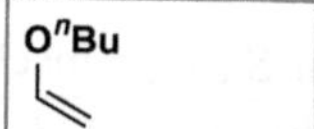	**197**	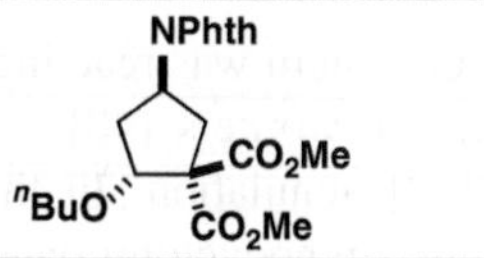	**220s**	99 (20:1 *dr*)

(continued)

Table 2.5 (continued)

Entry	Substrate		Product		Yield (%)[a]
6	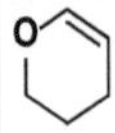	**207**	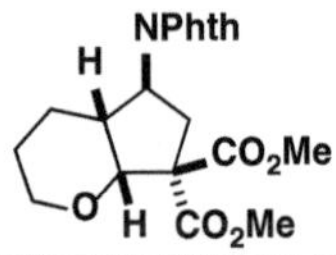	**220t**	99 (2:1 *dr*)

[a]Reaction conditions: 0.45 mmol enol ether, 0.30 mmol cyclopropane, 0.015 mmol $SnCl_4$, in 2 mL CH_2Cl_2, 1 h, −78 °C. Isolated yields after column chromatography
[b]Used as a 8:1 mixtures of Z/E isomers
[c]Used as a 10:1.3:0.3 mixture of the di- and Z/E trisubstituted isomers respectively
[d]Mixture of regioisomers. NMR yield of major product = 60 %

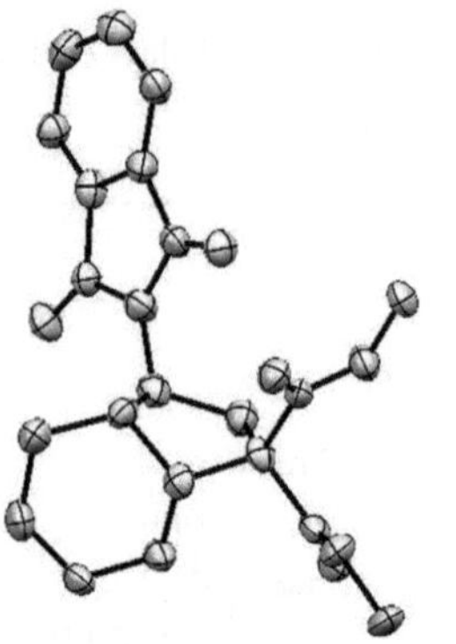

Fig. 2.4 Structure of **220t** determined by X-ray diffraction analysis

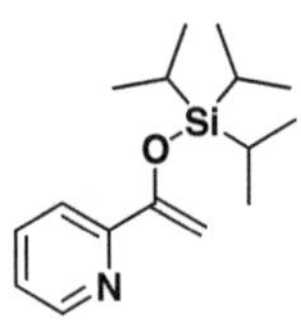

219n

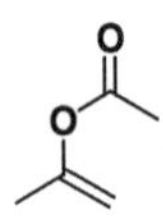

224

Fig. 2.5 Unreactive enol ethers

2.1.3 *Mechanism*

In order to determine if our system was reacting in a similar S_N^2 type mechanism as described by Johnson and co-workers [40], we decided to investigate the enantiospecificity of the [3 + 2] annulation. In this context, we isolated one of the enantiomers of the starting aminocyclopropane **217**. We had to use chiral phase chromatography separation for this purpose as asymmetric synthesis of such

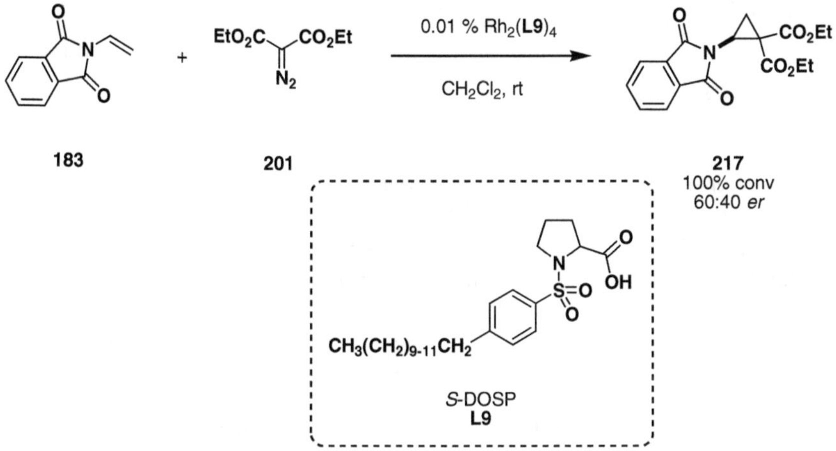

Scheme 2.9 Attempt towards enantioselective synthesis of **217**

aminocyclopropanes was not reported and using chiral rhodium complex did not lead to promising results (Scheme 2.9).

With enough enantiomerically enriched material in hand, we submitted various silyl enol ethers to the reaction conditions. Except for the slow reacting *o*-bromo-substituted enol ether **219d**, all the products were obtained with complete transfer of the stereogenic information (Scheme 2.10). This information allowed us to propose that the reaction mechanism involves a S_N^2 type substitution on the aminocyclopropane with transfer of the stereogenic information [40].

The tin catalyst is very effective to promote stereospecific substitution/ring closure sequence. However, in the case of a more hindered nucleophile such as **219d**, racemization of the starting aminocyclopropane is occurring, probably via an open zwitterionic intermediate. More studies have been undertaken to determine if the reaction occurred with retention or inversion of the absolute stereochemistry. However, it was not possible to obtain so far the crystalline material required for the analysis.

Based on these result, a catalytic cycle of the reaction is presented in Scheme 2.11. Aminocyclopropane **223** is first activated by chelation with the tin catalyst, resulting in the formation of intermediate **I** presenting a stable intimate ion pair between the two substituted carbons. The silyl enol ether attacks at the electrophilic center with inversion of the absolute configuration via a S_N^2 type process. The cyclopropyl strain is released by C–C bond fragmentation which leads to the formation of zwitterionic intermediate **II**. C–C bond rotation allows the good alignment between the malonic negative charge and the silyl-oxonium moiety in **III** and further ring closure to give the desired product.

In the case where the nucleophilic attack is slower (for example with hindered silyl enol ether **219d**), the C–C bond fragmentation occurs without nucleophilic

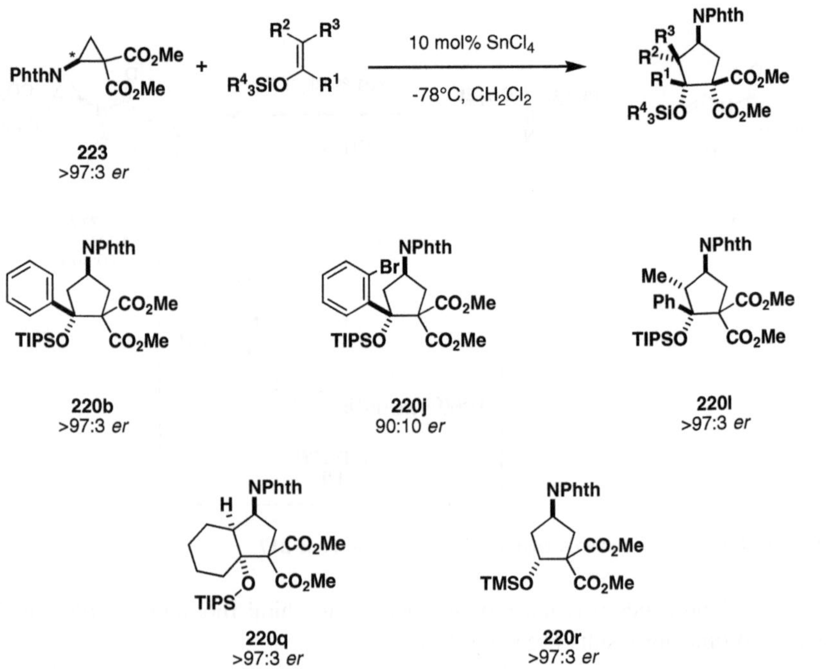

Scheme 2.10 Enantiospecific synthesis of cyclopentylamines

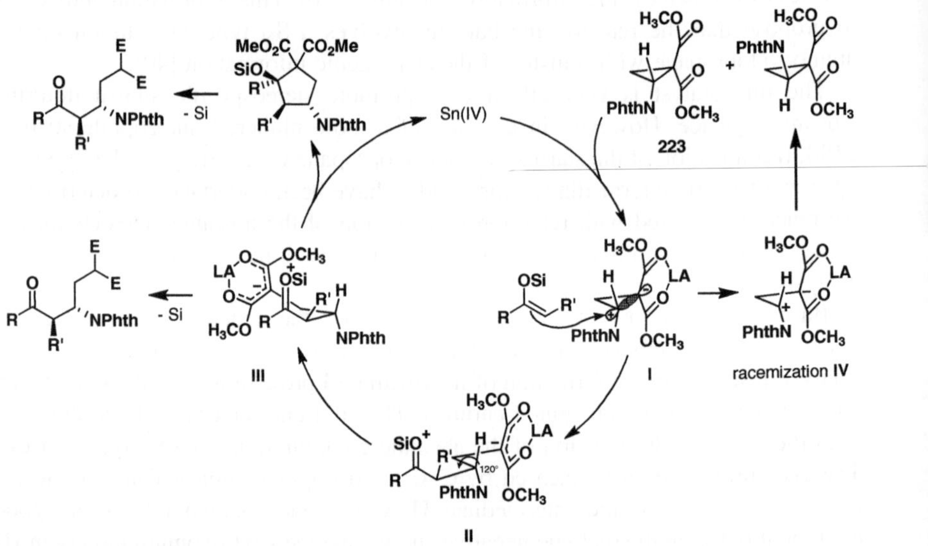

Scheme 2.11 Speculative catalytic cycle for the [3 + 2] annulation

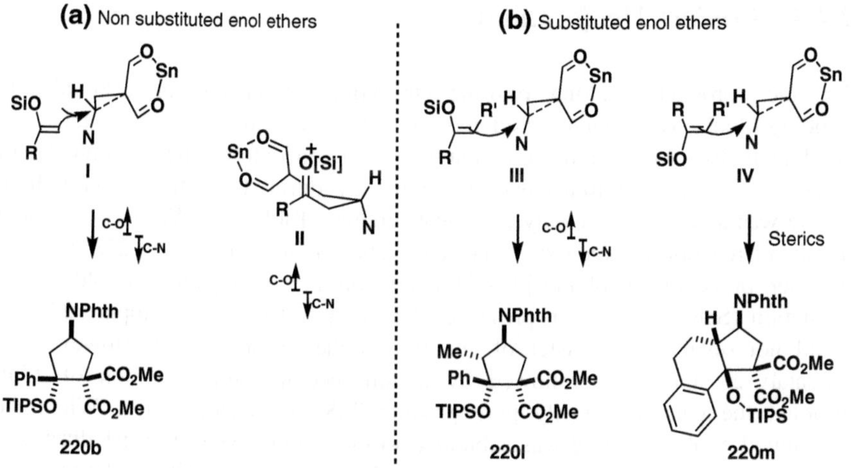

Scheme 2.12 Tentative rationalization of the diastereoselectivity

attack, resulting in the formation of the zwitterionic intermediate **IV**.[5] This leads to the loss of the stereogenic information in the starting material and therefore in the product.

As seen earlier, side product such as **221** are formed due to a retro-aldol fragmentation (Scheme 1.15). From the proposed catalytic cycle, another origin for **221** could be the loss of the silicon group directly from intermediate **III**.

The stereochemistry of the reaction with unsubstituted enol ethers arises at two different stages (a, Scheme 2.12). At first, the facial selectivity is defined during the nucleophilic attack by the opposite orientation of the C–N and C–O dipoles. Following the C–C bond rotation (**II** to **III**, Scheme 2.11), the orientation of the same dipoles in opposite directions in intermediate **II** positions the phthalimide and the silyloxy functionalities in *trans* relationship after ring closure.

When applying the same model to *cis*-trisubstituted silyl enol ether (R' ≠ H, b, Scheme 2.12), the substituent R′ is positioned in front of the small hydrogen. The reduction of the steric interactions and the dipoles orientation works in a synergistic manner. However when the silyl enol ether **219j** derived from tetralone is used, the reduction of the steric effects by orientating the R′ substituent, now in *trans*-position to the enol ether, in front of the hydrogen is predominant. Therefore the selectivity is inverted in the product **220m** and the *cis* product is obtained.

[5]The same phenomenon was observed in the project with ketones.

2.1.4 Product Modifications

The linear products resulting from the retro-aldol fragmentation are also valuable building blocks as nitrogen-substituted homo-Michael adducts. Such products can be obtained by a Mannich reaction, but high diastereocontrol using substituted enol ethers is a difficult task to achieve [5]. As we saw during the optimization, indium triflate was able to deliver only the linear product (Table 2.1). We thus compared the direct reaction between silyl enol ether and aminocyclopropane **223** with the two step procedure involving [3 + 2] annulation and subsequent retro-aldol fragmentation (Scheme 2.13). In presence of indium and using enantiopure **223** we could not obtain full transfer of chirality to the linear product. However the tin-catalyzed annulation followed by indium-mediated fragmentation of **220b** afforded the enantiomerically pure product **225** (a, Scheme 2.13). When the substituted enol ether **219l** was submitted to the indium catalyst, a 4:1 diastereomeric ratio was obtained for linear product **226**. Again, superiority of the two step protocol was highlighted by the isolation of only one diastereoisomer of **226** in 78 % yield from **220l** (b, Scheme 2.13).

The synthetic versatility of the cyclic product was investigated through different transformations (Scheme 2.14). The free amine **227** was accessed in good yield by

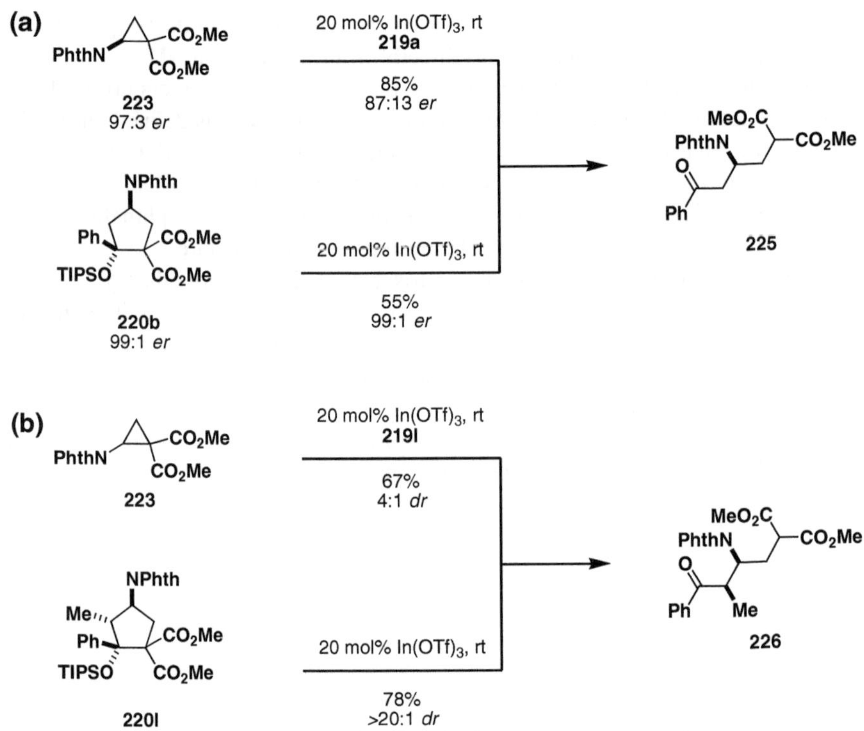

Scheme 2.13 Access to linear building blocks

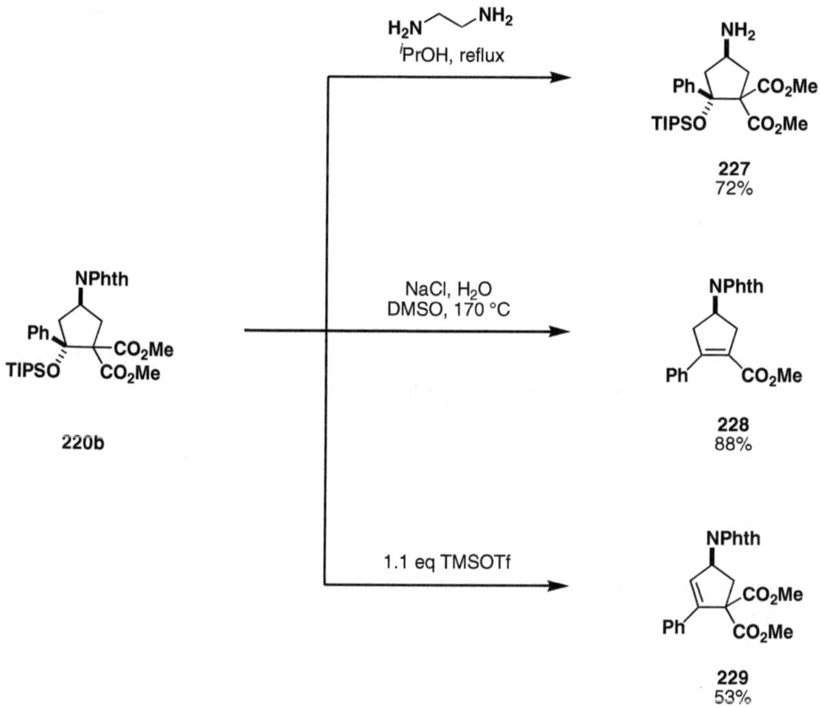

Scheme 2.14 Synthetic modification of cyclic products

removal of the phthalimide using ethylene diamine in isopropanol. Under Krapcho conditions [6], decarboxylation occurred with elimination of the neighboring silanol, affording the corresponding conjugated cyclopentene **228**. Another cyclopentenyl amine isomer **229** was obtained in 53 % yield from β-elimination using TMSOTf. The structural complexity of these building blocks could be potentially increased via further functionalizations such as oxidation, reduction or conjugate addition on the double bond for example.

2.2 [3 + 2] Annulation with Aldehydes[6]

As we saw in the introduction, aldehydes have been very versatile dipolarophiles for [3 + 2] annulation with DA cyclopropanes. In our hands, such process would lead to the formation of substituted amino-tetrahydrofurans. This scaffold is the

[6]The [3 + 2] annulation of aminocyclopropanes with aldehydes has been studied in our group by Dr Fides Benfatti after preliminary results obtained during my project. Text between quotation marks ("") is taken out of our publication: [7].

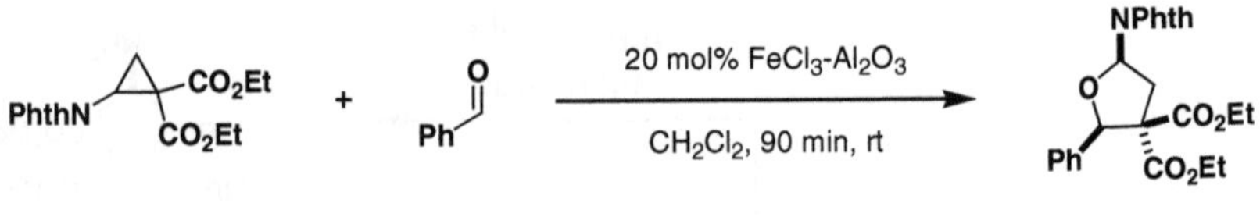

Scheme 2.15 Optimal result for [3 + 2] annulation using iron trichloride

central core of natural and synthetic nucleoside analogues, and therefore represents a very important target for organic synthesis [8].

As the catalytic system developed for the enol ethers was not compatible with aldehydes, we performed a new optimization of Lewis acid catalysts. Amongst all the tested combinations, we selected $FeCl_3–Al_2O_3$,[7] which afforded the product in quantitative NMR yield and more than 20:1 diastereoselectivity (Scheme 2.15). Additionally, working with iron had other advantages as it is an abundant and therefore inexpensive as well as a non-toxic catalyst. Despite its wide use as Lewis acid, it is the first time that iron was reported to catalyze the [3 + 2] annulation with DA cyclopropanes [11, 12].

"The scope of the reaction with DA aminocyclopropanes **217** and **223** was explored, using 5 mol% of iron(III) chloride on alumina, and 1.5 equiv. of aldehydes **215a–m** at room temperature. In general, excellent yields were obtained in short reaction time (2 h), while employing aldehydes with diverse steric and electronic properties."

First, the use of aromatic aldehydes was investigated (Table 2.6) "No difference in reactivity/selectivity was observed between the ethyl diester **217** and the methyl diester **223** in the reaction with benzaldehyde (**215a**) (entries 1, 2). The electron-rich *para*- and *ortho*-anisaldehydes (**215b** and **215c**) and thiophene-2-carboxaldehyde (**215d**) displayed modest stereoselectivities at rt (up to 6:1 dr, entries 3–5), nevertheless increased dr's were obtained at a lower temperature. Interestingly, no detrimental effect on yield and stereoselectivity of *ortho* versus *para* substituent was observed with anisaldehyde **215b** versus **215c** (entries 3, 4).

In all cases, the two isomers could not be separated by flash chromatography. However, the pure *cis* isomer could be obtained by means of a single recrystallization, except for products **230m** and **230n.** The X-ray diffraction analysis performed on aminotetrahydrofuran **230c** allowed the unambiguous attribution of the 2,5-*cis* relative stereochemistry (Fig. 2.6).

The annulation reaction was not limited to aromatic aldehydes, and a good yield was also obtained in the case of cinnamyl aldehyde **215g** (entry 1, Table 2.7), albeit

[7]Iron trichloride gave comparable results, but the alumina-supported reagent was preferred because it is easier to handle and known to be a scavenger of adventitious traces of water and acid. For examples on the use of $FeCl_3–Al_2O_3$, see [9, 10].

Table 2.6 Scope of the [3 + 2] annulation using aromatic aldehydes

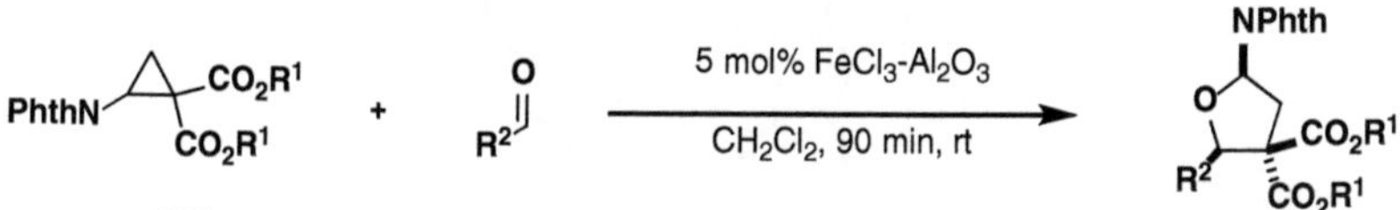

217, 223 **215a-f** **230a-g**

Entry	Substrate		Product		Yield (%)[a]
1	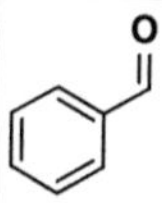	**215a**	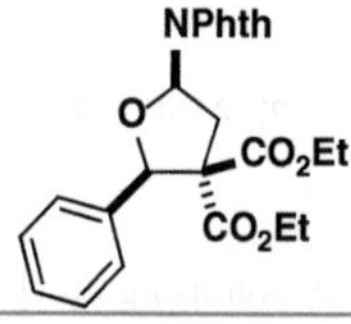	**230a**	94 (>20:1 *dr*)
2	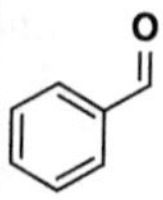	**215a**	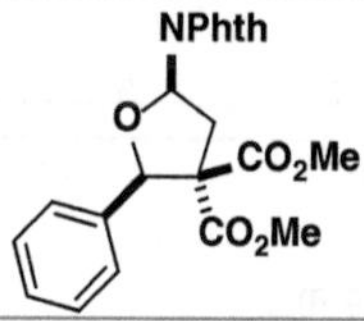	**230b**	95 (>20:1 *dr*)
3	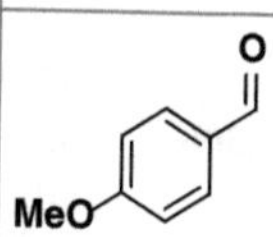	**215b**	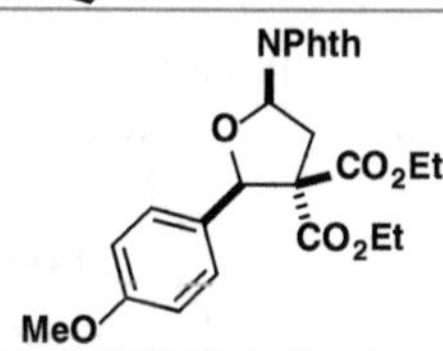	**230c**	98[b] (9:1 *dr*) (20:1 *dr*)[c]
4	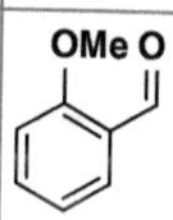	**215c**	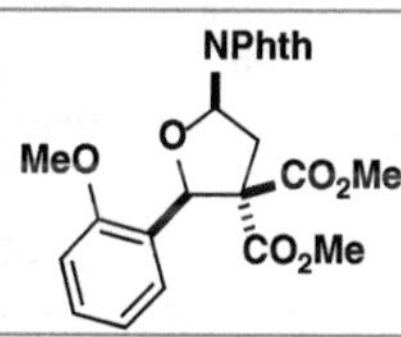	**230d**	92[b] (17:1 *dr*)
5	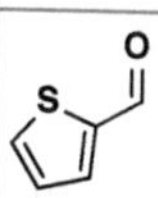	**215d**	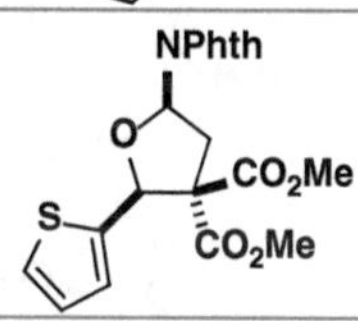	**230e**	89[b] (>20:1 *dr*)
6	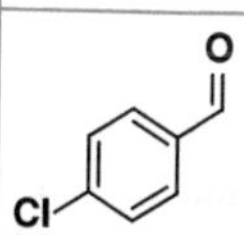	**215e**	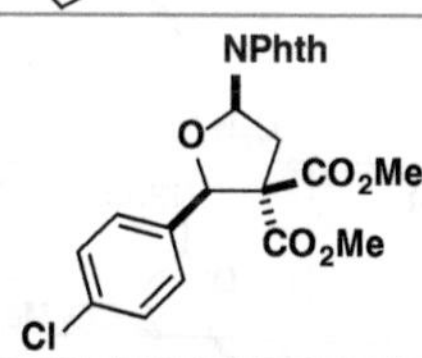	**230f**	91 (>20:1 *dr*)
7	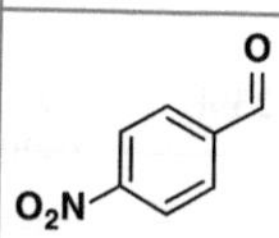	**215f**	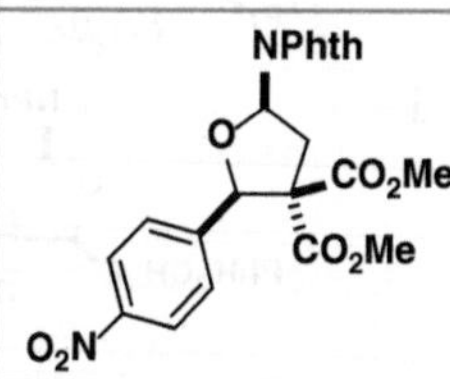	**230g**	71 (>20:1 *dr*)

[a]Reaction conditions: 0.20 mmol cyclopropane, 0.30 mmol aldehyde, 5 mol% $FeCl_3$–Al_2O_3, in CH_2Cl_2, 0.1 M, 2 h, rt. Isolated yields after column chromatography
[b]Yields and diastereoselectivities are reported for the reaction at −10 °C
[c]Diastereomeric ratio after recrystallization

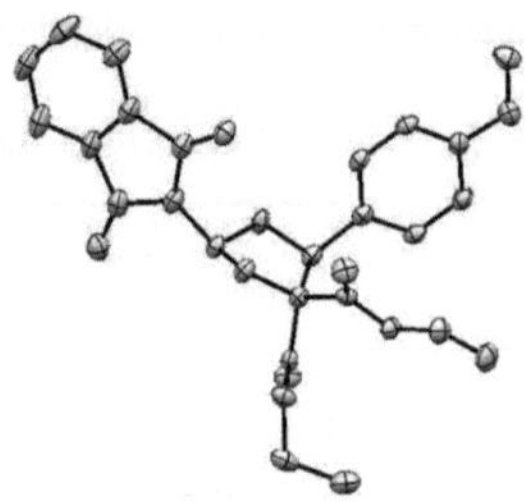

Fig. 2.6 Structure of **230c** determined by X-ray diffraction analysis

Table 2.7 Scope of the [3 + 2] annulation using alkyl and vinyl aldehydes

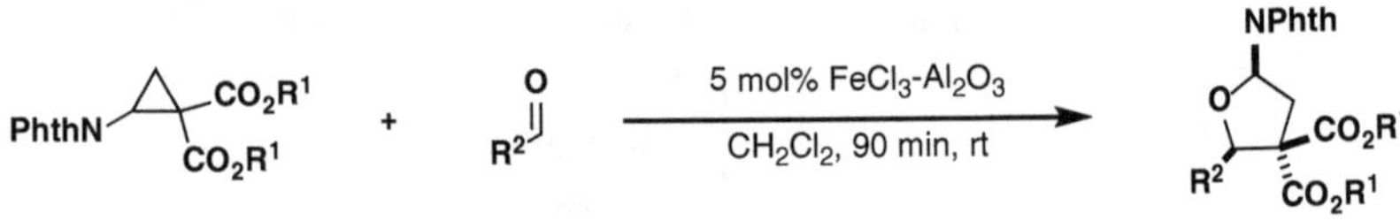

217, 223 **215g-m** **230h-o**

Entry	Substrate		Product		Yield (%)[a]
1	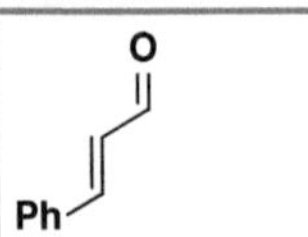	**215g**	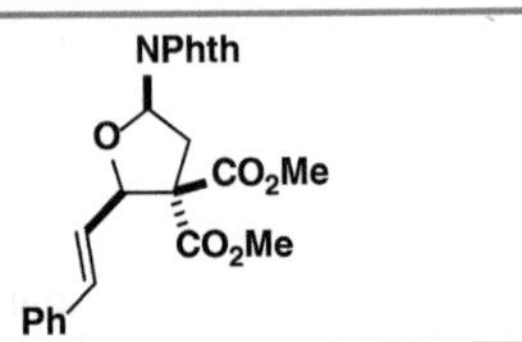	**230h**	95[b] (10:1 *dr*) (>20:1 *dr*)[c]
2	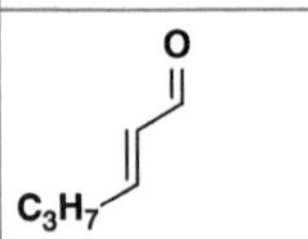	**215h**	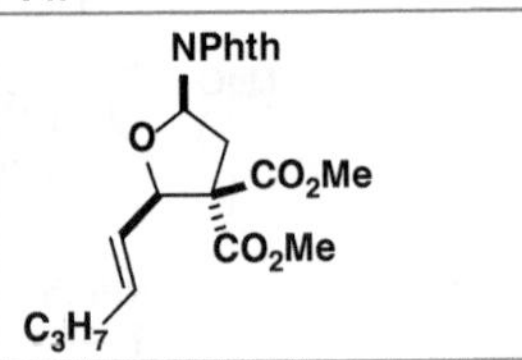	**230i**	95 (7:1 *dr*) (>20:1 *dr*)[c]
3	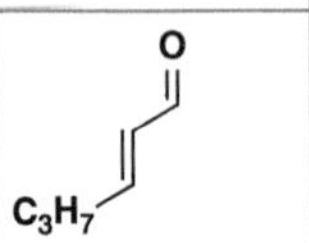	**215h**	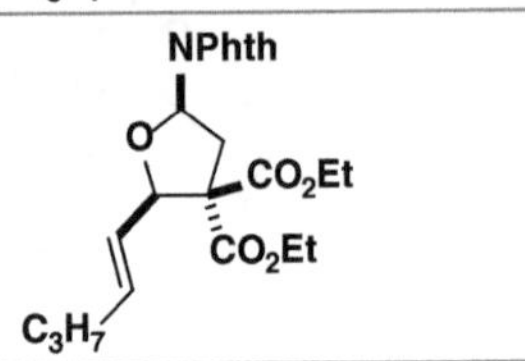	**230j**	99 (10:1 *dr*) (>20:1 *dr*)[c]
4	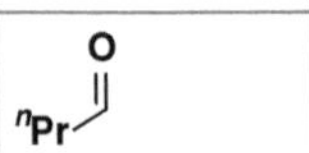	**215i**	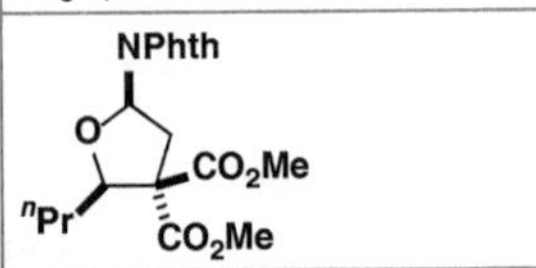	**230k**	94 (9:1 *dr*) (>20:1 *dr*)[c]
5	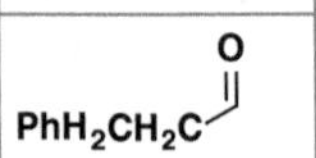	**215j**	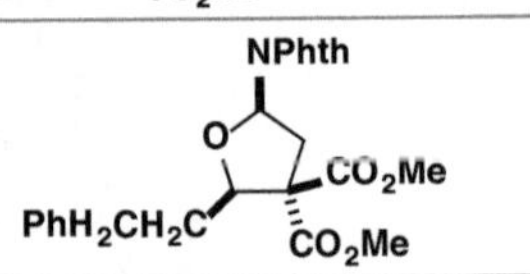	**230l**	89 (>20:1 *dr*)

(continued)

Table 2.7 (continued)

Entry	Substrate		Product		Yield (%)[a]
6	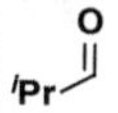	**215k**	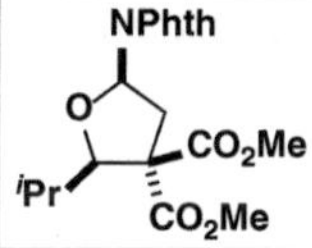	**230m**	99 (7:1 *dr*)
7	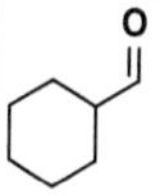	**215l**	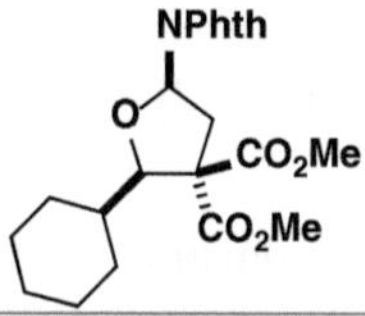	**230n**	90 (7:1 *dr*)
8	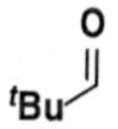	**215m**	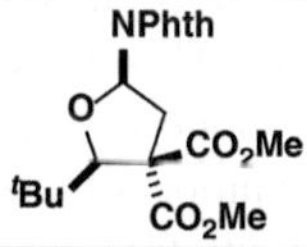	**230o**	92 (>20:1 *dr*)

[a]Reaction conditions: 0.20 mmol cyclopropane, 0.30 mmol aldehyde, 5 mol% $FeCl_3–Al_2O_3$, in CH_2Cl_2, 0.1 M, 2 h, rt. Isolated yields after column chromatography
[b]Yields and diastereoselectivities are reported for the reaction at –10 °C
[c]Diastereomeric ratio after recrystallization

with a low diastereoselectivity. Once again, the *cis*-selectivity could be increased by lowering the temperature to –10 °C. The reaction was also successful for unsaturated aldehyde **215h**, both with aminocyclopropanes **223** and **217** (entries 2, 3).

Aliphatic aldehydes are generally challenging substrates for Lewis acid-catalyzed reactions, as they are prone to undergo aldol side reactions. Gratifyingly, this was not an issue under our mild reaction conditions, and aliphatic aldehydes with linear (entries 4,5) or branched substituents (entries 6–8) afforded the corresponding amino tetrahydrofurans in outstanding yields (89–99 %) and good diastereoselectivities (up to >20:1 *cis*:*trans*).

As a further evidence of the efficiency of our methodology, the reaction of aminocyclopropane **217** with an equimolar amount of benzaldehyde (**215a**) in the presence of 1 mol% loading of iron catalyst, afforded compound **230a** in 88 % yield and excellent dr on a 1 mmol scale (Scheme 2.16).

To gain more knowledge on the mechanism of our [3 + 2] annulation, enantioenriched **217** (er = 99:1) was reacted with **215a** under the standard reaction

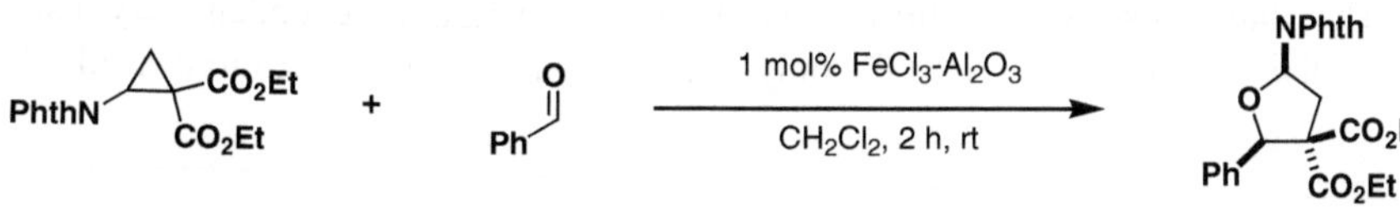

217
1.0 mmol

215a
1.0 mmol

230a
88%
>20:1 *dr*

Scheme 2.16 Annulation on 1 mmol scale using 1 mol% catalyst

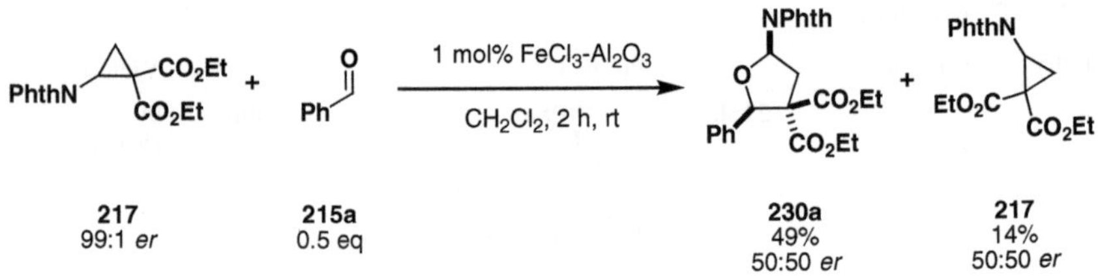

Scheme 2.17 Annulation using enantiopure **217**

conditions. The iron-catalyzed annulation of **217** with benzaldehyde (**215a**) at rt resulted in a complete loss of the stereochemical information, as tetrahydrofuran **230a** was isolated in a racemic form. This evidence suggests that, upon Lewis acid activation, the aminocyclopropane undergoes fast racemization via an open, zwitterionic species. While the result hampers the synthesis of enantioenriched amino THFs, it is a potential starting point for the development of a dynamic kinetic asymmetric transformation (Scheme 2.17)."

2.3 [3 + 2] Annulation with Ketones[8]

If the [3 + 2] annulation of DA cyclopropanes with aldehyde has been widely explored, the use of ketones has been more limited. Only a few works reported the use of ketones and in all the cases the lack of diastereoselectivity highlighted the difficulty of facial discrimination when non-symmetrical ketones were used [14].

The [3 + 2] annulation of DA aminocyclopropanes with ketones would offer a solution to access aminotetrahydrofurans substituted by a quaternary center at C5. These scaffolds have been poorly explored as potential bioactive molecules, probably due to the lack of methods providing an efficient access to them.

The catalytic system implemented for the annulation with aldehydes was not suitable and had to be re-optimized. A set of condition was tested and it turned out that the tin-catalyzed protocol that was used with enol ethers was the most efficient to deliver the products in good yield and selectively. The optimized reaction conditions made use of 5 mol% of $SnCl_4$ at −78 °C in presence of aminocyclopropane **217** and acetophenone **218a** to afford the 2,5-*cis*-aminotetrahydrofuran **231a** in 99 % yield (Scheme 2.18).

The diastereoselectivity turned out to be temperature dependent. By using 20 mol% of $SnCl_4$ at –10 °C, only the *trans*-aminotetrahydrofuran could be

[8]The [3 + 2] annulation of aminocyclopropanes with ketones has been studied in our group by Dr Fides Benfatti after preliminary results obtained during my project. Text between quotation marks ("") is taken out of our publication [13].

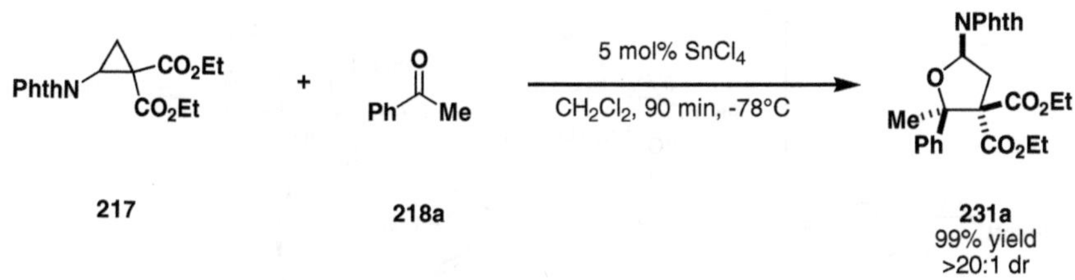

Scheme 2.18 Optimized catalytic system for the [3 + 2] annulation of ketones

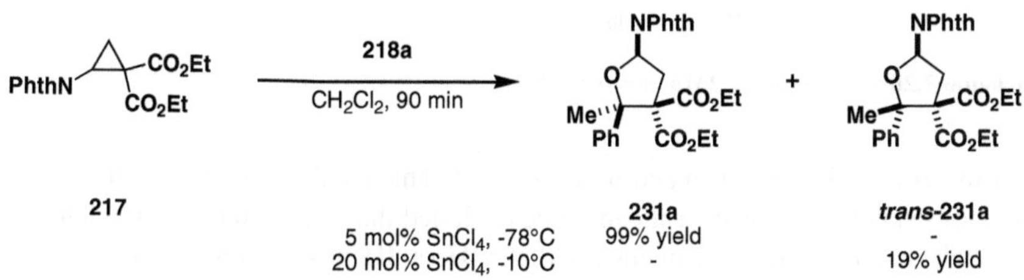

Scheme 2.19 Temperature-dependent diastereoselectivity

detected in the crude reaction (Scheme 2.19). Although the high catalyst loading induced significant degradation, the product could be obtain in 19 % yield.

The stereochemistry of the products was confirmed by X-Ray diffraction analysis (Fig. 2.7).

In order to understand the mechanism behind these results, we treated **231a** with 20 mol% $SnCl_4$ at –10 °C. We could observe partial conversion to the starting cyclopropane **217** via a retro-[3 + 2] annulation. The isomerization might therefore result from a sequence of retro-[3 + 2]/[3 + 2] annulation where intermediate such as an intimate ion pair **Ia** or a zwitterion **Ib** can be considered. When **231a** was submitted to the reaction condition in presence of 1 equivalent of acetophenone

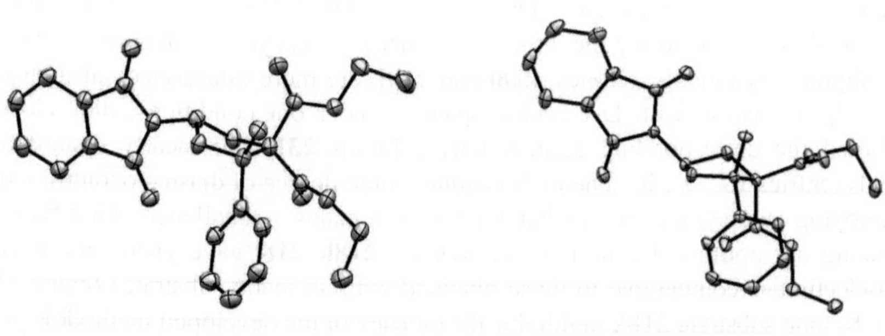

231a *trans*-**231a**

Fig. 2.7 Structure of **231a** and *trans*-**231a** determined by X-ray diffraction analysis

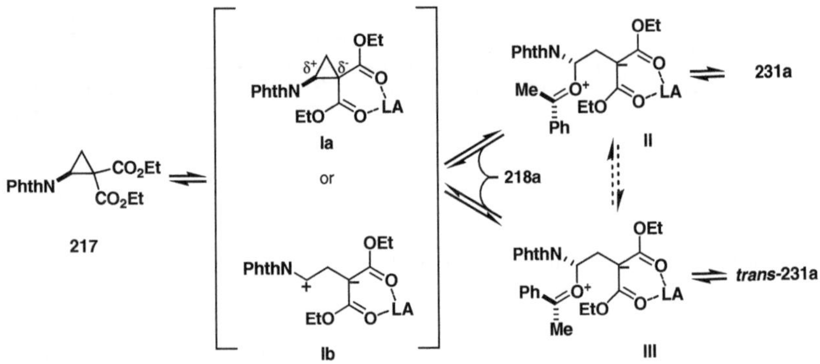

Scheme 2.20 Formation of **231a** and *trans*-**231a**

(**218a**), *trans*-**231a** was isolated in 45 % yield. This result is in agreement with the previous statement, however it cannot be excluded that the interconversion happens through direct substitution of intermediates **II** or **III** at the aminal center by acetophenone (Scheme 2.20).

The scope and limitations of the reaction were evaluated by submitting aromatic, heteroaromatic and aliphatic ketones to the optimal reaction conditions. "DA cyclopropanes **217** and **223** displayed a similar reactivity toward acetophenone (**218a**), affording aminotetrahydrofurans **231a** and **231b** in excellent yields and diastereoselectivity (entries 1, 2, Table 2.8). A lower yield (79 %, entry 3) was obtained in the case of 1'-acetonaphthone (**218b**), most likely due to the unfavorable *ortho* substitution. Electron-rich aromatic and heteroaromatic ketones **218c** and **218d** showed lower diastereoselectivities for the [3 + 2] annulation (entries 4,5). Nevertheless, the diastereomeric ratio could be improved through a single recrystallization. Electron-poor aromatic ketones **218e–218f** were also tested, and they gave the corresponding aminotetrahydrofurans **231f** and **231g** in high yields, as a single diastereoisomer (entries 6, 7). Excellent stereochemical discrimination between the phenyl and the ethyl substituent was observed also with propiophenone (**218g**), demonstrating the versatility of our methodology (entry 8). 1-Tetralone (**218h**) displayed excellent reactivity and selectivity, delivering **231i** in high yield and diastereoselectivity, but favoring the 2,5-*trans* isomer for this cyclic system (entry 9).

Aliphatic symmetric ketones (**218i** and **218j**) are more established substrates in [3 + 2] annulation with DA cyclopropanes. Under our conditions, they cleanly afforded the corresponding aminotetrahydrofurans **231j–l** in nearly quantitative yields (entries 10–12). In general, obtaining a high degree of diastereocontrol when employing non-symmetric aliphatic ketones remains a challenge. Gratifyingly, utilizing our optimized conditions on ketones **218k–218l** gave yields and diastereoselectivities comparable to those obtained with aromatic substrates (entries 13, 14). Ketone substrate **218k** highlights the efficacy of the developed methodology as a good diastereomeric ratio (10:1, entry 13) was obtained with two carbonyl substituents, methyl and ethyl, possessing only a small difference in size.

Table 2.8 Scope of [3 + 2] annulations between DA aminocyclopropanes and ketones

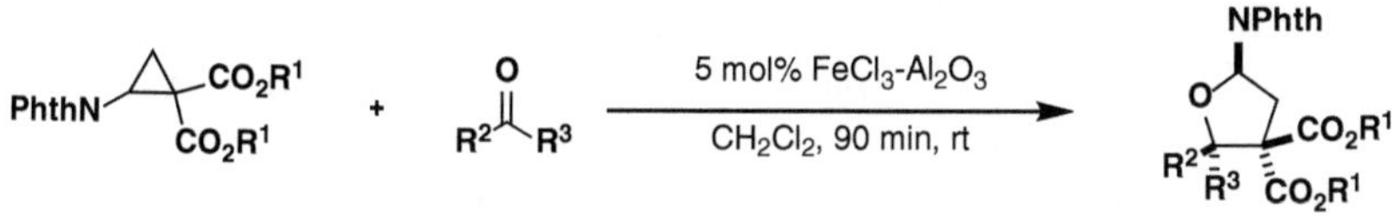

217, 223 218a-l 231a-n

Entry	Substrate		Product		Yield (%)[a]
1	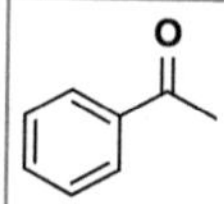	**218a**	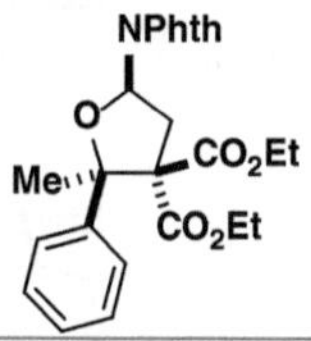	**231a**	99 (>20:1 *dr*)
2	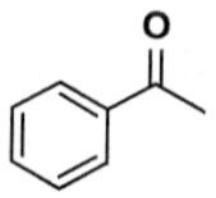	**218a**	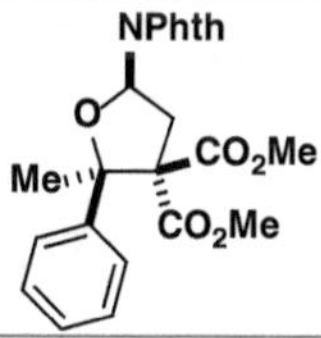	**231b**	96 (>20:1 *dr*)
3	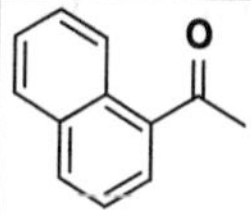	**218b**	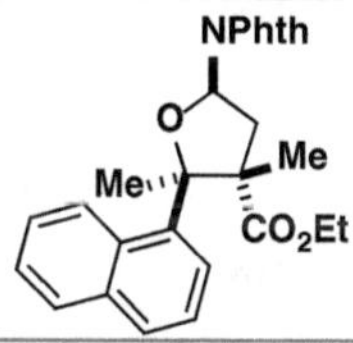	**231c**	79 (>20:1 *dr*)
4	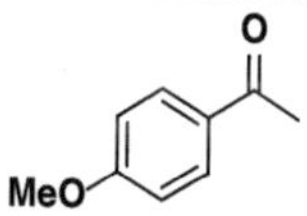	**218c**	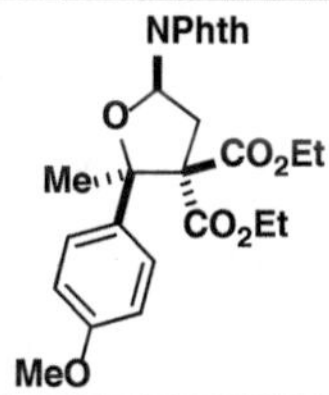	**231d**	90 (16:1 *dr*) (>20:1 *dr*)[b]
5	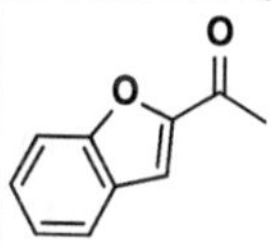	**218d**	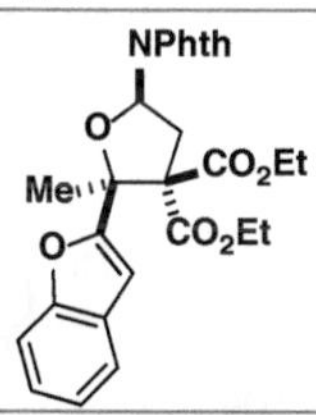	**231e**	95 (4:1 *dr*) (14:1 *dr*)[b]
6	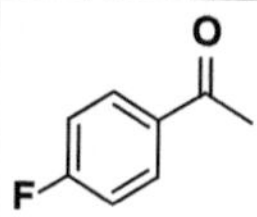	**218e**	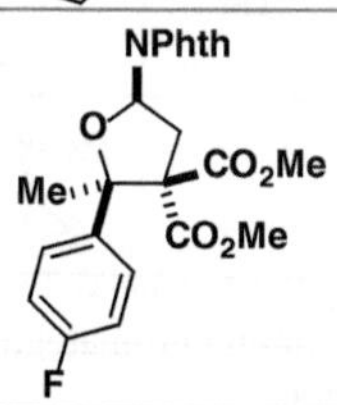	**231f**	93 (>20:1 *dr*)

(continued)

Table 2.8 (continued)

Entry	Substrate		Product		Yield (%)[a]
7	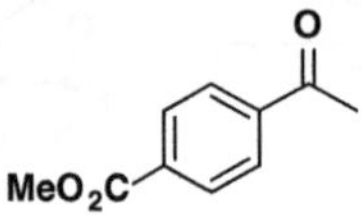	**218f**	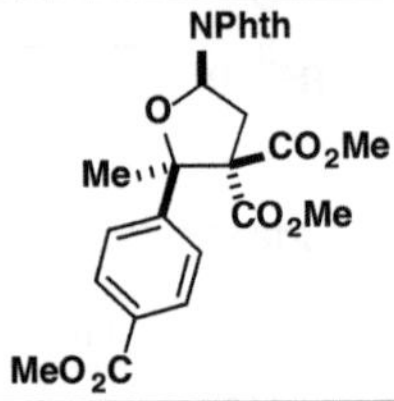	**231g**	99 (>20:1 *dr*)
8	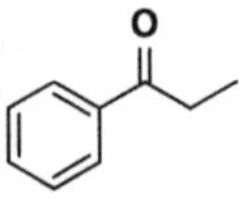	**218g**	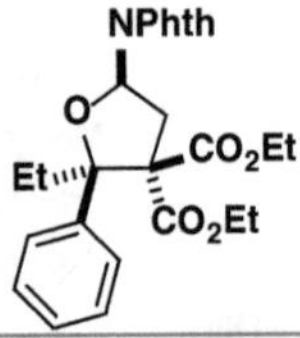	**231h**	95 (>20:1 *dr*)
9	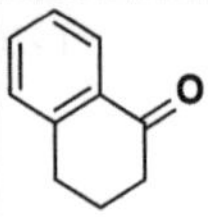	**218h**	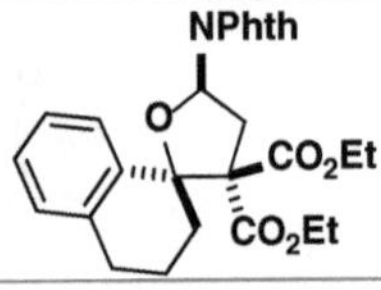	**231i**	94 (>20:1 *dr*)
10	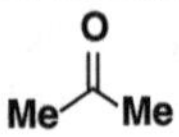	**218i**	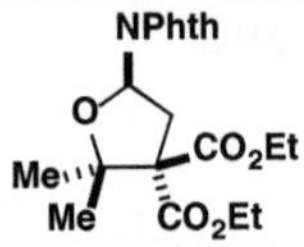	**231j**	94
11	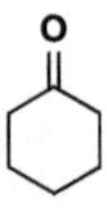	**218j**	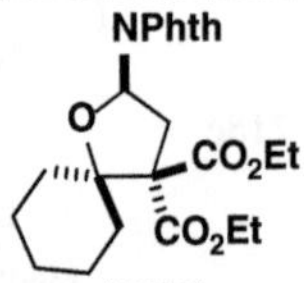	**231k**	99
12		**218j**	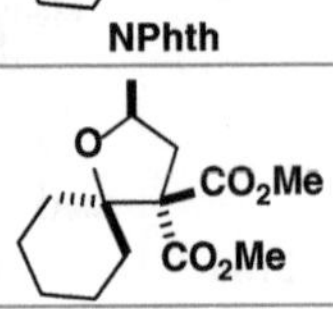	**231l**	99
13	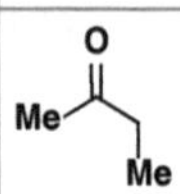	**218k**	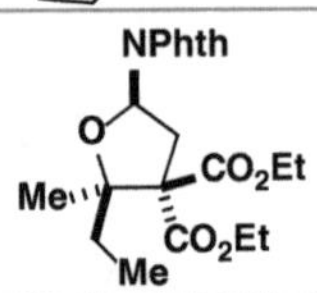	**231m**	89 (10:1 *dr*) (16:1 *dr*)[b]
14	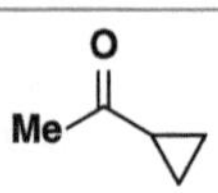	**218l**	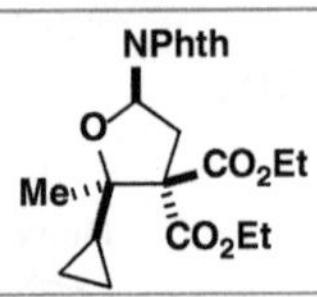	**231n**	96 (>20:1 *dr*)

[a]Reaction conditions: 0.2 mmol cyclopropane, 0.3 mmol ketone, 5 mol% $SnCl_4$, in CH_2Cl_2, 0.1 M at −78 °C. Isolated yields after column chromatography
[b]Obtained after one recrystallization

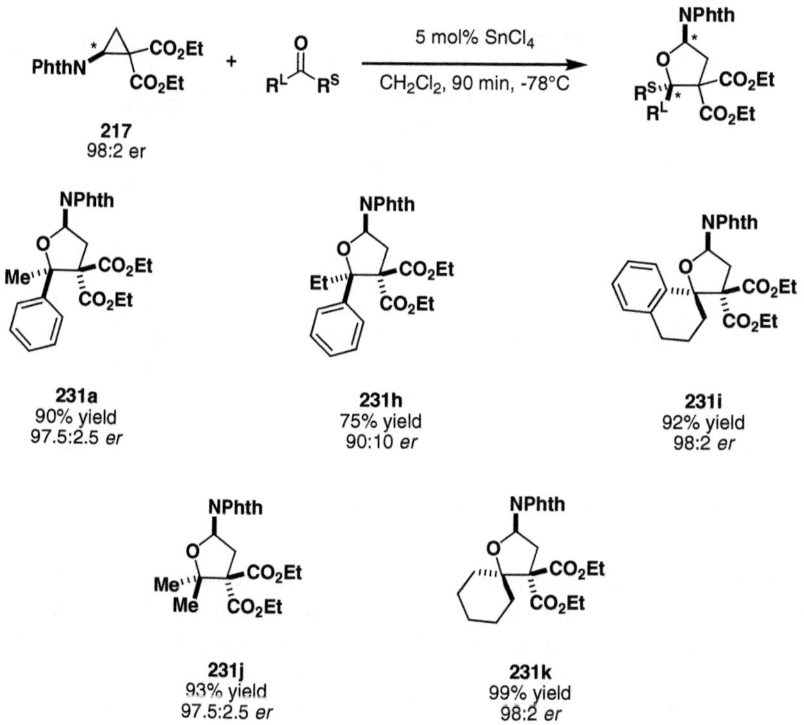

Scheme 2.21 Enantiospecific [3 + 2] annulation. R^L = Larger substituent. R^S = Smaller substituent

To assess the stereospecificity of the tin(IV)-catalysed [3 + 2] annulation, a selection of ketones was reacted with enantioenriched phthaloyl cyclopropane **217** at –78 °C (Scheme 2.21). Under these conditions, no loss of stereochemical purity was observed with acetophenone (**218a**) and ketones **218g–i**, while aminotetrahydrofuran **231h** was isolated with a slightly decreased enantiomeric excess.

Based on the high enantiospecificity and diastereoselectivity of the reaction, we wondered if the reaction of racemic **217** with a chiral ketone would allow for a kinetic resolution to take place. For example, the reaction of cyclopropane **217** with (-)-menthone (**232**) should in principle favor the formation of the two diastereoisomers **233** and **233'**, as both have the phthalimide group in *cis* relationship to the more bulky group (Scheme 2.22). Both products would be obtained as single enantiomers, as enantiopure (-)-menthone (**232**) is used as starting material. The opposite absolute stereochemistry at the nitrogen center would result from the enantiospecific reaction of both enantiomers of **217**. However, a severe steric interaction between the ester and the isopropyl group of (-)-menthone (**232**) is present only in **233'**: the formation of this diastereoisomer is consequently expected to be slower (mismatched case), allowing a kinetic resolution with re-isolation of

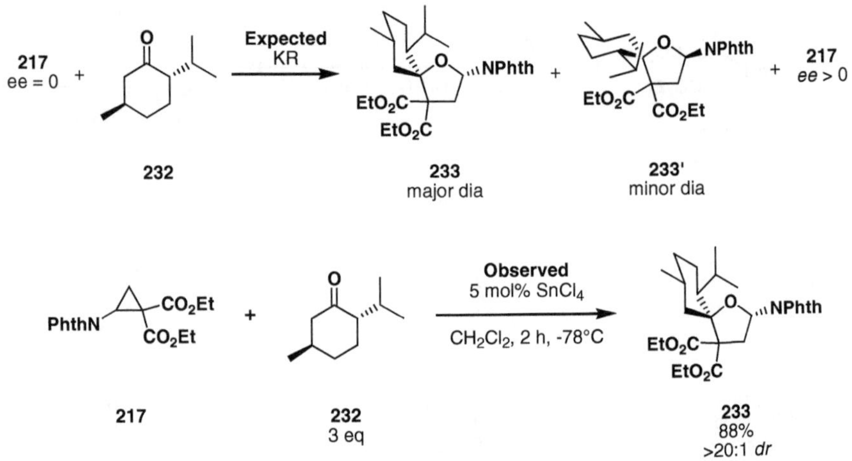

Scheme 2.22 [3 + 2] Annulation with menthone (**232**)

enantioenriched **217**. Unfortunately, the kinetic resolution of **217** using sub-stoichiometric amount of (-)-menthone (**232**) could not be accomplished, mainly due to the sluggish reactivity observed in this case. When increasing to 3 equivalents the amount of **232**, the reaction was accelerated, giving full conversion after 2 h and the annulation product **233** was isolated as a single diastereoisomer in 88 % yield. Consequently, the reaction was not enantiospecific, but stereoconvergent.

Different rationales could account for this result: $SnCl_4$ is either active in the racemization of the DA aminocyclopropane **217** or in the isomerization of the product **233**. To obtain additional clues, the loss of enantiomeric purity of enantioenriched **217** (*ee* = 94 %) in the presence of 5 mol% $SnCl_4$ was monitored at –78 °C. After 1 h, **217** was recovered with an *ee* = 75 %, while after 2.5 h almost all its stereochemical information was lost (*ee* = 20 %). This result supports the hypothesis that the observed dynamic kinetic resolution could take place via racemization of the aminocyclopropane **217**, probably via an open zwitterionic intermediate **Ib** (Scheme 2.20) The results obtained with (-)-menthone (**232**) could be explained by a limited lifetime for a tight ion-pair **Ia**: If the following annulation reaction is fast, an enantiospecific reaction takes place, but if the desired reaction is slow, as for the mismatched case with (-)-menthone (**232**), dissociation would have time to occur, which would lead to racemization even at –78 °C and to the stereoconvergent reaction observed. In contrast, the *cis-trans* isomerization described in Table 1.1 would require higher temperature to proceed. We note that further experiments would be required to confirm this interpretation."

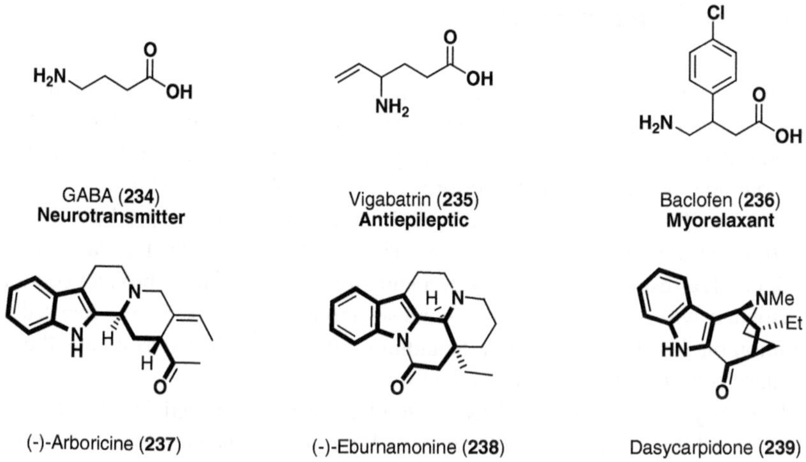

Fig. 2.8 GABA and molecules containing the GABA scaffold

2.4 Friedel-Craft Alkylation of Indoles[9] [15]

After focusing on DA aminocyclopropanes as 1,3 dipoles for [3 + 2] annulations, we wondered if they could be used as efficient *homo*-Michael acceptors with aromatic nucleophiles. The products resulting from Friedel-Crafts alkylation of aminocyclopropanes will possess a structure comparable to γ-aminobutyric acid (GABA). This motif is found in natural as well as synthetic drugs such as neurotransmitters [16, 17], in peptidomimetics [18] or in the core of alkaloid natural products (Fig. 2.8). Vigabatrin (**235**) for example is an antiepileptic drug that inhibits irreversibly GABA transaminase and baclofen (**236**) is an agonist of $GABA_B$ receptor used to treat spasticity. Alkaloids **237–239** exhibit interesting bioactivities. For example Arboricine (**237**) has a moderate ability to reverse MDR in vincristine resistant cells [19].

The combination of aromatic and heteroaromatic nucleophiles with DA aminocyclopropanes represent a straightforward access to GABA analogues via Umpolung of the reactivity. Compared to other methods that used the nucleophilic opening of carbo-substituted DA cyclopropanes with amines [20–25], this method present a strong advantage as it allows the creation of a large diversity of analogues starting from a single aminocyclopropane.

[9]This project has been carried out in collaboration with master student Joachim Loup from EPFL for his master thesis under my supervision.

2.4.1 Preliminary Results

The reaction was first tested between *N*-methyl-indole (**32**) and aminocyclopropane **223**. A range of Lewis acids were screened at 10 mol% loading in dichloromethane at room temperature (Table 2.9). N-methyl indole (**32**) was chosen first in order to rule out the possibility of N-substitution. Under all the tested conditions, full conversion of starting **223** was observed. In almost all cases, two products could be identified in the crude reaction mixture: The FC alkylation adduct **240a** as well as product **241a** resulting from a [3 + 2] annulation. As the cyclized product could not be isolated with both chemo- and stereoselectivity (entries 1–5), we focused our attention on the FC product **240a**. Scandium triflate was selected as it gave the best chemoselectivity for the linear product (entry 6).

Two major issues appeared as we tried to use unprotected indole (**211**) and 5-chloro indole (**242a**) for the FC reaction. First, in the crude mixture of the former (entry 7), the product **243** arising from the bis addition of the indole could be observed (Fig. 2.9). Then, the use of electron poor 5-chloro indole (**242a**) afforded a complex mixture of FC adduct **240c**, cyclized product **241c** and unidentified side products (entry 8).

Table 2.9 Optimization of the FC alkylation

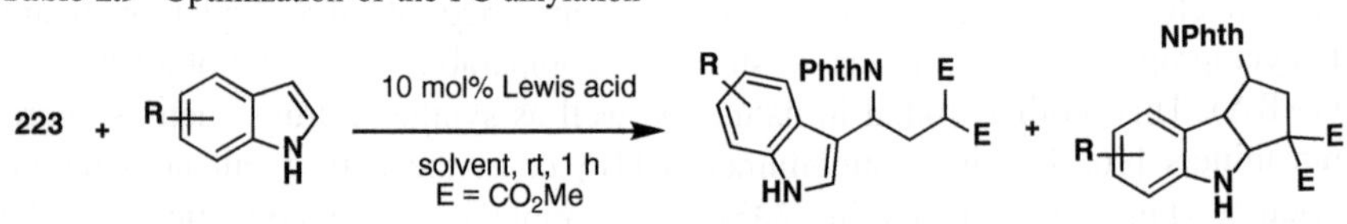

Entry[a]	Indole	Lewis acid	Solvent	240/241	*dr*[b]
1	**32**	$Cu(OTf)_2$	DCM	1:19	1.2:1
2	**32**	$FeCl_3$-Al_2O_3	DCM	2:1	3.5:1
3	**32**	$Sn(OTf)_2$	DCM	1:1.4	6:1
4	**32**	$SnCl_4$-SiO_2	DCM	2:1	>20:1
5	**32**	$In(OTf)_3$	DCM	1:1.3	3.5:1
6	**32**	$Sc(OTf)_3$	DCM	>20:1	nd
7	**211**	$Sc(OTf)_3$	DCM	>20:1	nd
8	**242a**	$Sc(OTf)_3$	DCM	complex	nd
9	**211**	$Sc(OTf)_3$	$MeNO_2$	2:1[c]	nd

[a]Reaction conditions: 0.034 mmol cyclopropane, 1.5 eq indole, 10 mol% cat., 0.2 mL solvent, rt, 1 h

[b]Determined by ^{1}H NMR of the crude reaction mixture

[c]The cyclized product **241c** was not isolated. The NMR ratio was determined by analogy with the cyclized product **241a**

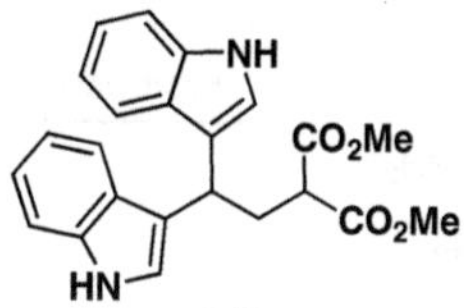

Fig. 2.9 Side product observed with indole (**211**)

Solvent screening allowed identifying nitromethane as best solvent to suppress the formation of the side product **243**. However, a 2:1 mixture of cyclized **240b** and FC products **241b** was observed (entry 9).

The over-reactivity of the free indole (**211**) and the poor chemoselectivity of indole **242a** required a new round of optimization. For this purpose, we decided to investigate the influence of the cyclopropane structure on the outcome of the reaction.

2.4.2 *Synthesis of New Cyclopropane Analogues*

We synthesized a range of new aminocyclopropanes having different electronic as well as steric properties (Fig. 2.10). Both the donor and the acceptor parts would be modified, requiring the synthesis of the analogues from the starting imides.

We began our investigation by synthetizing new vinyl phthalimide derivatives that could then be submitted to a rhodium-catalyzed cyclopropanation. Imides derivatives **245** and **246** were not all commercially available and had to be synthesized from the corresponding phthalic anhydrides in presence of formamide (**244**) under microwave heating (Scheme 2.23) [26]. All anhydrides could be purchased.

We attempted to synthesize the vinyl derivatives using a dibromoethane addition/elimination sequence [27]. However the reaction was giving poor yields and was not reproducible. We turned our attention to trans-vinylation using vinyl acetate (**224**) catalyzed by sodium-tetrachloropalladate, but poor conversion was observed when **245** was submitted to the reaction conditions (a, Scheme 2.24). We attempted to replace the sodium-tetrachloropalladate complex by palladium acetate and for this purpose, we tested the reaction with phthalimide. The reaction failed and formation of inactive palladium black was observed. Eventually, we used a combination of inexpensive $PdCl_2$ with LiCl at 80 °C using vinyl acetate as solvent

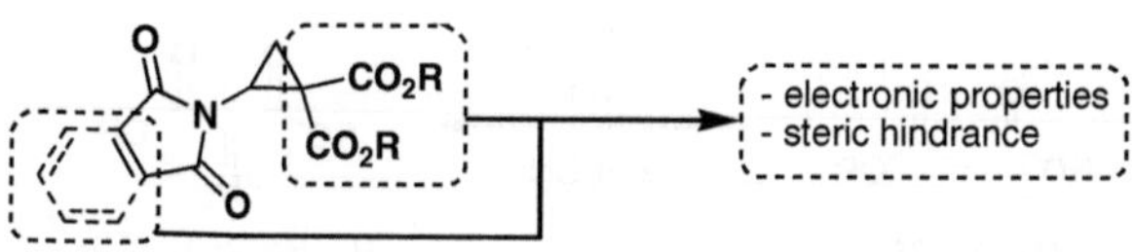

Fig. 2.10 Structural modifications of aminocyclopropanes

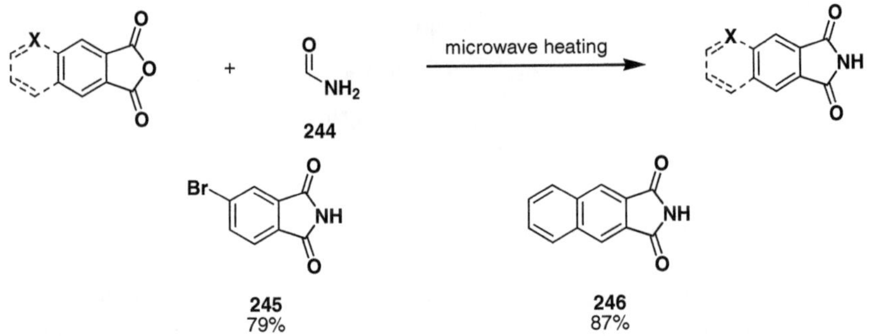

Scheme 2.23 Synthesis of imides

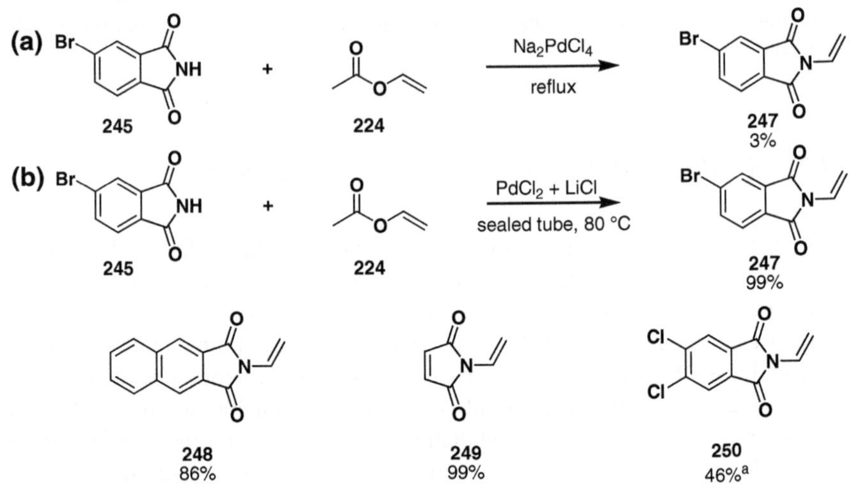

Scheme 2.24 Synthesis of vinyl imides. **a** Na_2PdCl_4 was used

to access the product **247** in excellent yields (b, Scheme 2.24). These inexpensive conditions have never been reported before and provided desired product **248–250** in high yields.

New diazomalonate derivatives were prepared using 4-acetamidobenzenesulfonyl azide (*p*-ABSA) as the diazo transfer reagent (Scheme 2.25). Benzyl malonate was selected in order to study the influence of the

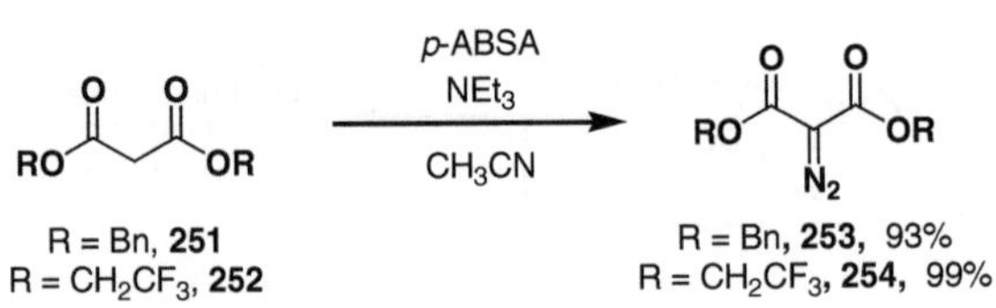

Scheme 2.25 Synthesis of diazomalonates

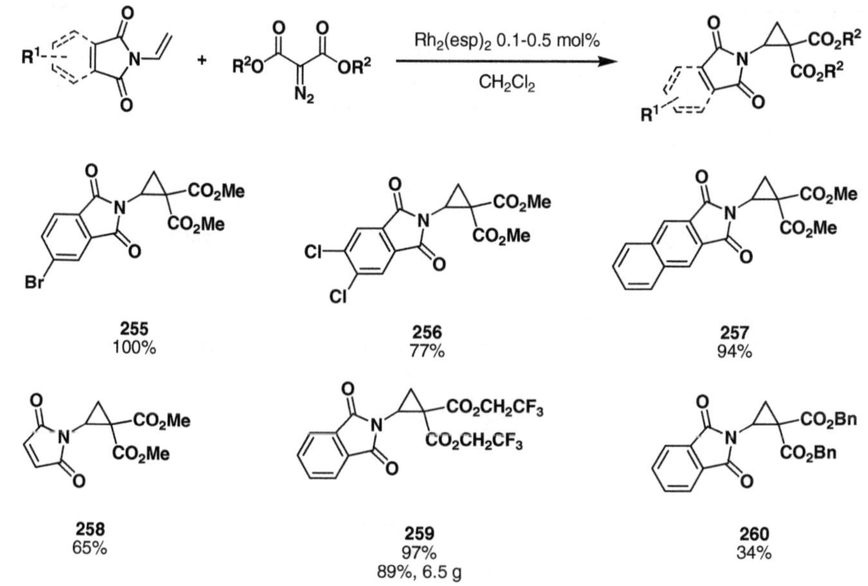

Scheme 2.26 Synthesis of aminocyclopropanes

steric interactions induced by the aromatic groups. Trifluoroethyl malonate was chosen for its lower electron density and therefore more stabilizing capacity of the negative charge resulting from cyclopropane fragmentation.

The unoptimized cyclopropanation of new vinyl-imides and diazomalonates using catalytic amount of rhodium(II) afforded aminocyclopropanes **255–259** in very good yields (65–100 %) and **260** was obtained in 34 % yield (Scheme 2.26). The synthesis of **259** was performed on a multigram scale affording the cyclopropane in 89 % yield.

2.4.3 *Optimization of the Reaction*

With the aminocyclopropanes in hand, we began to screen various aminocyclopropanes in combination with indole (**211**) in presence of scandium triflate in nitromethane (Table 2.10). The modification on the nitrogen had no influence on the selectivity of the reaction (entries 1–4) with a ratio of desired product to bis-addition of two to one. On the other hand, modification of the acceptor part had more impact on the chemoselectivity. When benzyl-substituted cyclopropanes **260** was submitted to the reaction conditions, a slight improvement could be observed (entry 5) and when bis-trifluoroethyl ester-substituted cyclopropane **259** was used, a complete conversion to the desired product was obtained (entry 7). We thus selected **259** to continue the study and with this new system, a last round of

Table 2.10 Screening of aminocyclopropanes in the Friedel-Crafts reaction with indole (**211**)

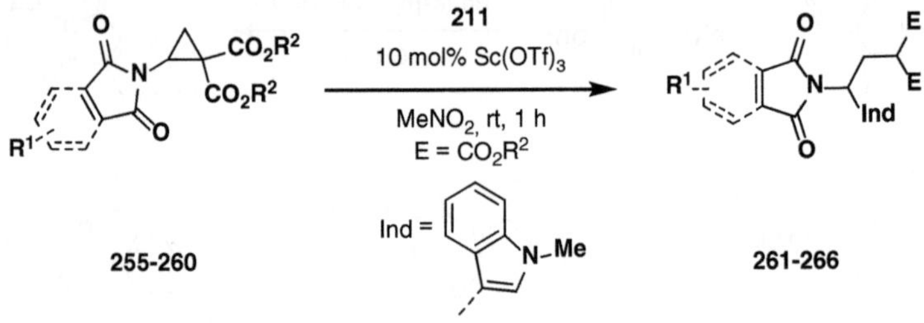

Entry[a]	Cyclopropane	Product		Mono/Bis[b,c]
1	**255**	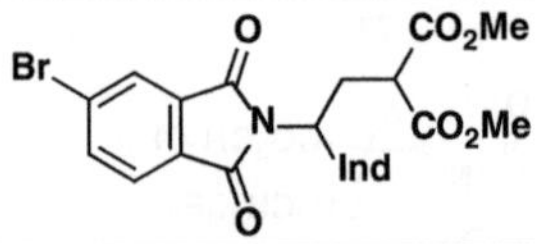	**261**	2.4:1
2	**256**	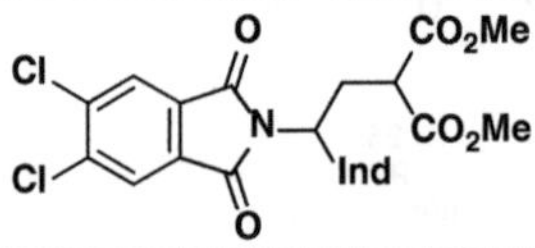	**262**	2:1
3	**257**	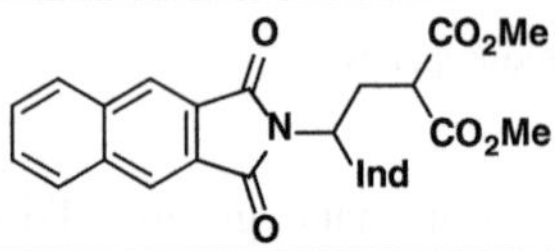	**263**	2:1
4	**258**	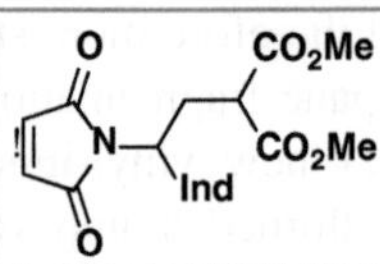	**264**	2:1
5	**260**	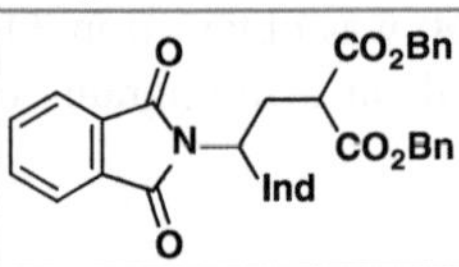	**265**	4:1
6	**259**	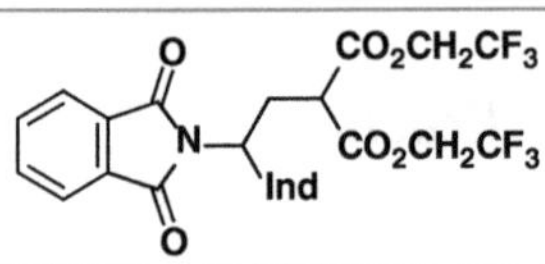	**266a**	>20:1

[a]Reaction conditions: 1 eq cyclopropane, 1.5 eq **211**, 10 mol% $Sc(OTf)_3$, 0.2 mL nitromethane, rt, 1 h
[b]Determined by ^1H NMR of the crude reaction mixture. The products **261–265** were not isolated and solely identified in the ^1H crude NMR
[c]The bis-indole adducts were not isolated. The NMR ratios were determined by analogy with **243**

optimization[10] was performed. Scandium turned out to be the best catalyst and we managed to replace toxic nitromethane by diethyl ether as solvent. During the optimization, the formation of compound **267** due to the presence of water

[10]See supporting information.

Fig. 2.11 Side product **267**

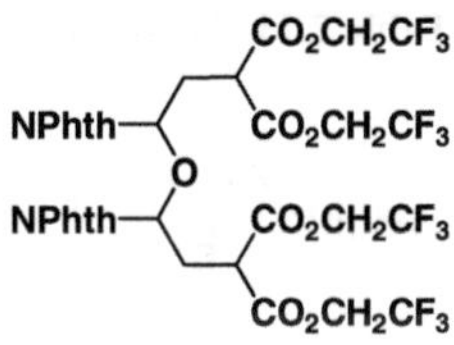

267

(Fig. 2.11) could be observed. It is therefore important to run the reaction under anhydrous conditions (dry solvent, N_2 atmosphere).

2.4.4 *Scope of the Reaction*

With these conditions in hand, we evaluated the scope and the limitations of the reaction. For this purpose, indole **242g** was prepared by N-alkylation and indoles **242j** [28] and **242 l** [29] were prepared via direct and selective C–H alkynylation developed in our lab. We first tested various unprotected indoles (Table 2.11). The reaction tolerated electron-poor as well as electron-rich substitution on the arene part, affording the products in very good yields (70–85 %, entries 1–6). Important functionalities such as halide (entries 2, 5) or a sensitive boronic ester (entry 4) that would allow further elaboration of the products were tolerated under the reaction conditions. The reaction could be scaled up to gram scale with indole (**32**) and **43** and no erosion of the yield was observed (entries 1, 7). The product resulting from the reaction with the deactivated nitro indole **242e** had to be trans-esterified in situ to the dimethyl ester as otherwise degradation occurred during purification. Therefore the product **266f** is obtained with a slightly lower yield of 58 % over the two steps (entry 6).

N-Methyl substitution of the indole was tolerated and afforded **266g** in almost quantitative yield (entry 7). Adding an ester on the C6 position on **242f** only lowered the yield to 86 % (entry 8) despite its deactivating effect and using a silyloxyethyl substituent on the nitrogen afforded the product **266i** in a good 80 % yield (entry 9).

Substitution on the C2 position with methyl and phenyl substituents provided the FC products **266j** and **266k** in very good yields (entries 10, 11). The product **266l** containing an alkyne functionality could be obtained in 52 % yield (entry 12). This sensitive product is similar to the one used previously by Kerr and co-workers for Conia-ene cyclization [41]. As for Kerr, Pagenkopf and co-workers, substituents in C3 position promoted the selective C2 alkylation of indole derivatives. It can be assumed that the reaction occurs via a C3 to C2 1,2 shift but the products resulting from the migration of the initial C3 substituents were never observed. This could be explained by the stabilizing effect of a positive charge by the nitrogen during the alkyl group migration. If the product **266m** substituted with a methyl was obtained in good yield (entry 13), the alkynyl product **266n** was obtained in a lower 49 % yield (entry 14). It has to be mentioned that for these substrates, the temperature of

Table 2.11 Scope of the FC with unprotected indoles

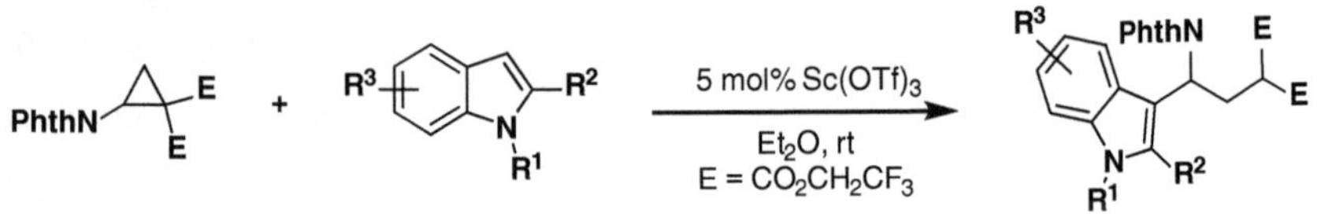

259 **32, 43, 211, 242a-m** **266 a-o**

Entry	Substrate		Product		Yield (%)[a]
1	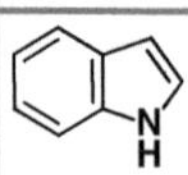	**32**	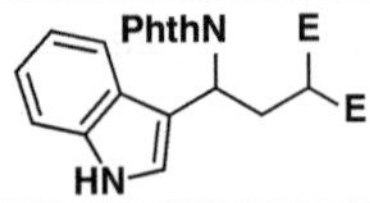	**266a**	85 87[b]
2	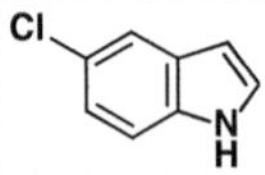	**242a**	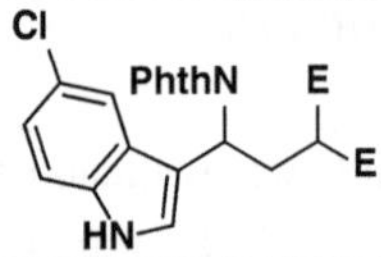	**266b**	82
3	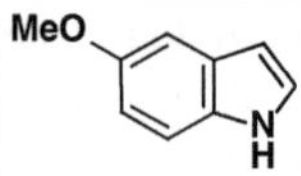	**242b**	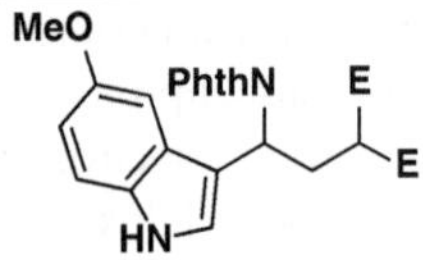	**266c**	68
4	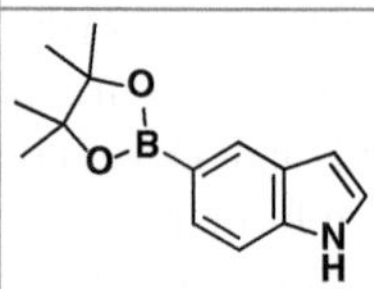	**242c**	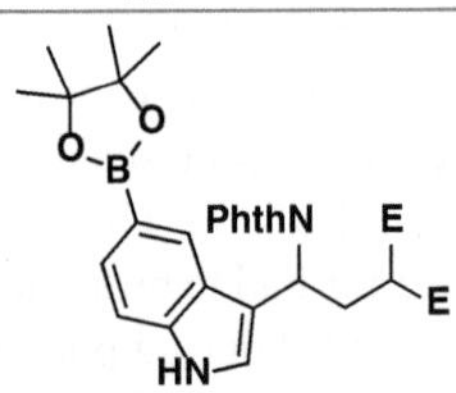	**266d**	63
5	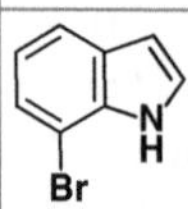	**242d**	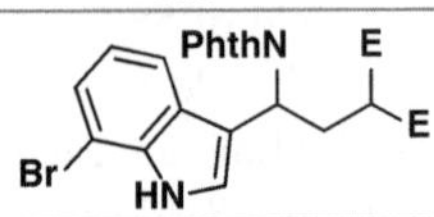	**266e**	70
6	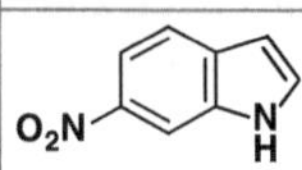	**242e**	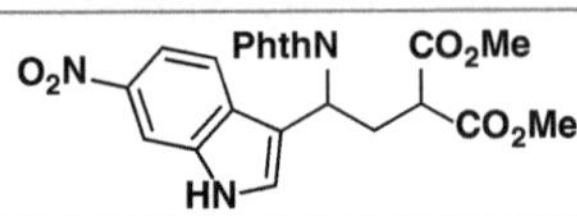	**266f**	58[c]
7	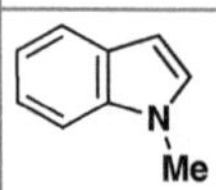	**43**	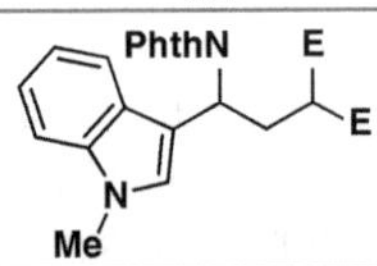	**266g**	94 83[d]
8	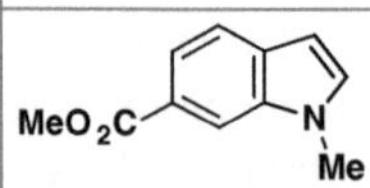	**242f**	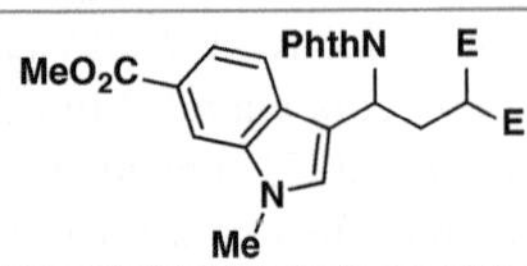	**266h**	86

(continued)

Table 2.11 (continued)

Entry	Substrate		Product		Yield (%)[a]
9	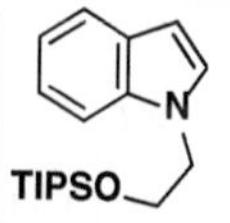	**242g**	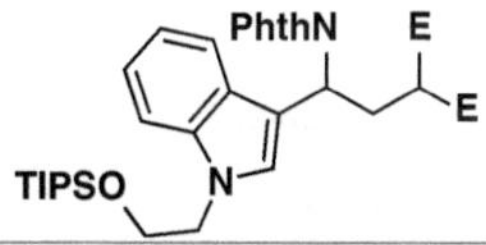	**266i**	80
10	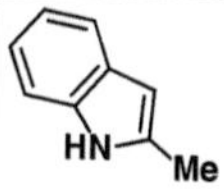	**242h**	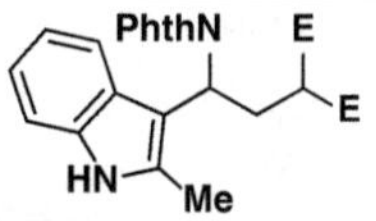	**266j**	97
11	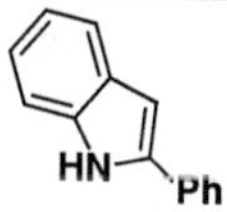	**242i**	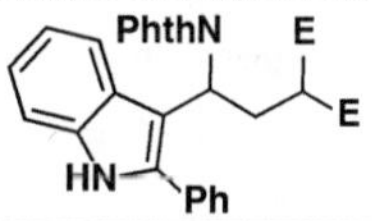	**266k**	93
12	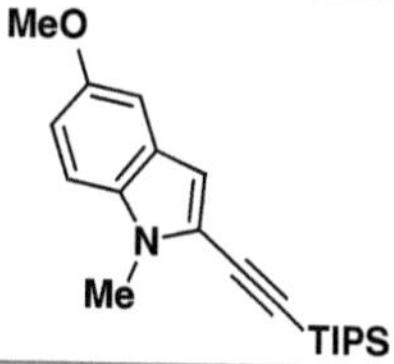	**242j**	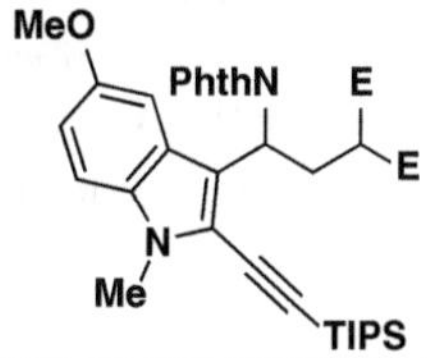	**266l**	52
13	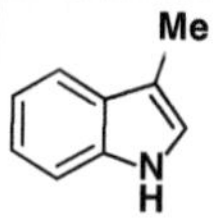	**242k**	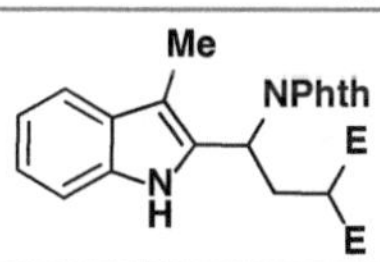	**266m**	60 82[c] 67[f]
14	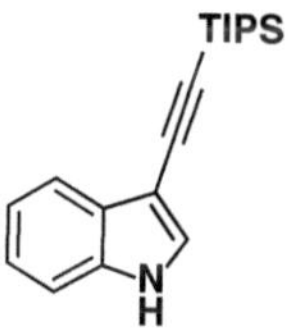	**242l**	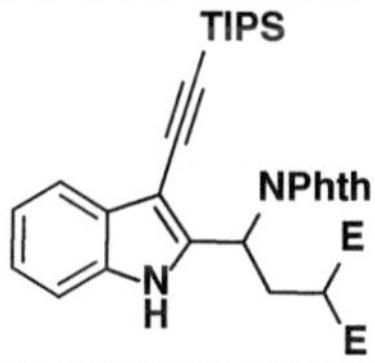	**266n**	49[f]
15	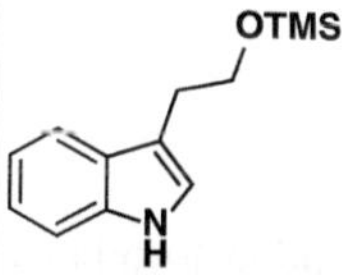	**242m**	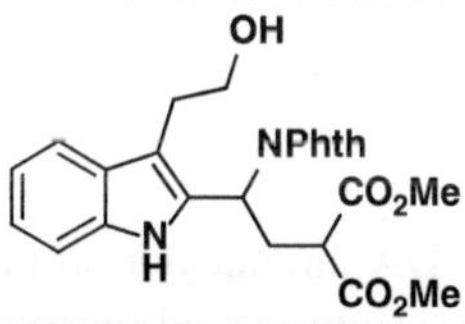	**266o**	72[c]

[a]Reaction conditions: 0.20 mmol cyclopropane, 0.22 mmol indole, 5 mol% $Sc(OTf)_3$, 1.2 mL Et_2O, rt. Isolated yields after column chromatography
[b]Yield over two steps after transesterification (K_2CO_3, MeOH, 0 °C)
[c]Reaction in Et_2O at 35 °C
[d]Reaction in toluene at 60 °C

the reaction had to be increased (35 °C for **266m** and 60 °C in toluene for **266n**) and therefore substantial degradation occurred. The product **266o** was obtained from the TMS protected tryptophol **242m** and had to be isolated after trans-esterification

Table 2.12 Scope of other aromatics and heteroaromatics

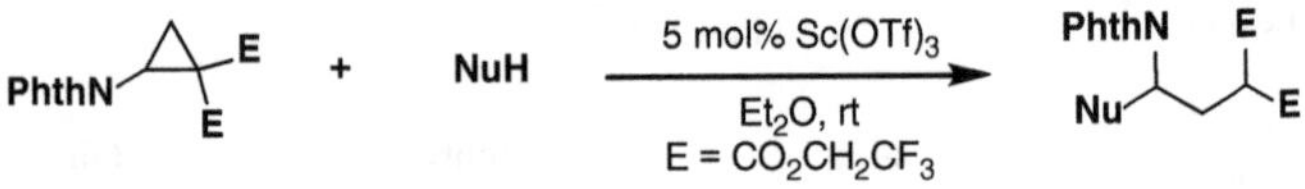

259 268-271 272-275

Entry	Substrate		Product	Yield (%)[a]
1	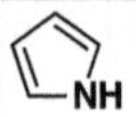	**268**	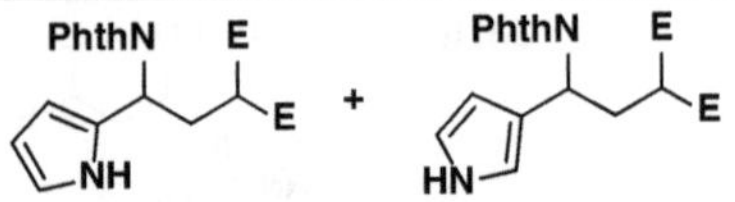 **272a/272b**	61/30
2	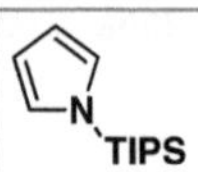	**269**	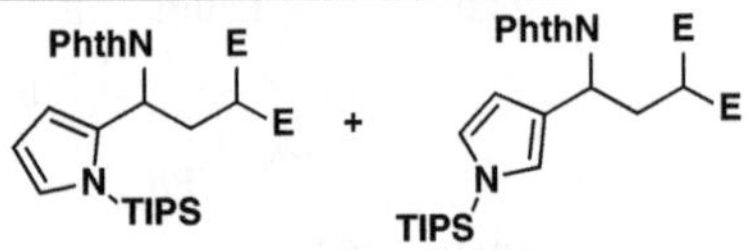 **273a/273b**	88[b]
3	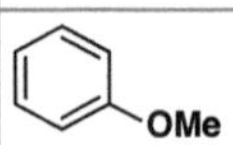	**270**	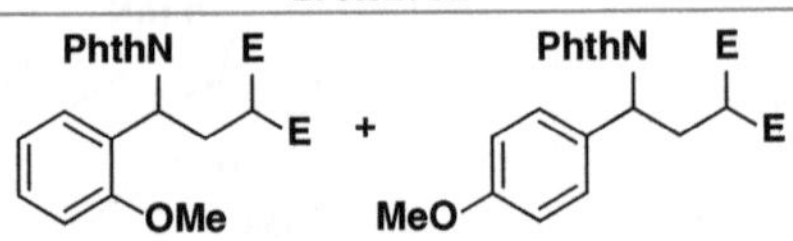 **274a/274b**	60[c]
4	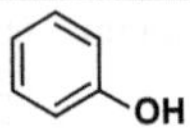	**271**	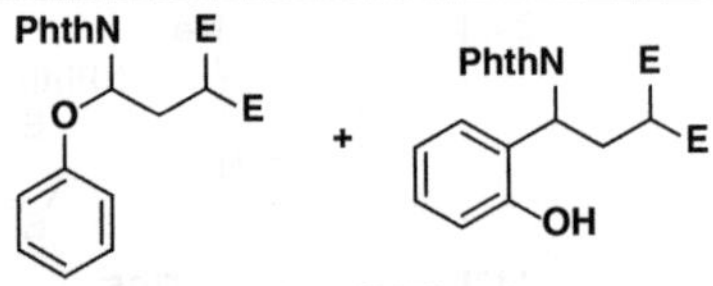 **275a/275b**	74/12

[a]Reaction conditions: 0.20 mmol cyclopropane, 0.22 mmol NuH, 5 mol% $Sc(OTf)_3$, 1.2 mL Et_2O, rt. Isolated yields after column chromatography
[b]7:1 mixture of inseparable isomers
[c]1:1 mixture of inseparable isomers

(entry 15). The reactivity of the C2 and the C3 substituted indoles allows a complementary solution to access regioisomers of structurally related products such as **266l** and **266n**.

The FC alkylation of DA aminocyclopropanes was extended to other aromatic nucleophiles (Table 2.12). Unprotected pyrrole **268** could be reacted in good yield, affording the products **272a** and **272b** with a poor C2/C3 selectivity in favor of C2 (entry 1). The use of the bulky N-protecting group TIPS allowed inverting the regioselectivity with a ratio of 7:1 for the C3 position without erosion of the yield (entry 2). When anisole was tested, a 1:1 mixture of C2 and C4 alkylation could be obtained in 60 % yield (entry 3). The use of phenol afforded a 6:1 mixture of *O*-alkylation product **274a** and *C*-alkylation **274b** (entry 4). The fact that a poor regioselectivity is observed for pyrrole and the preference for hard *O*-nucleophiles indicates the presence of a hard reactive electrophilic intermediate.

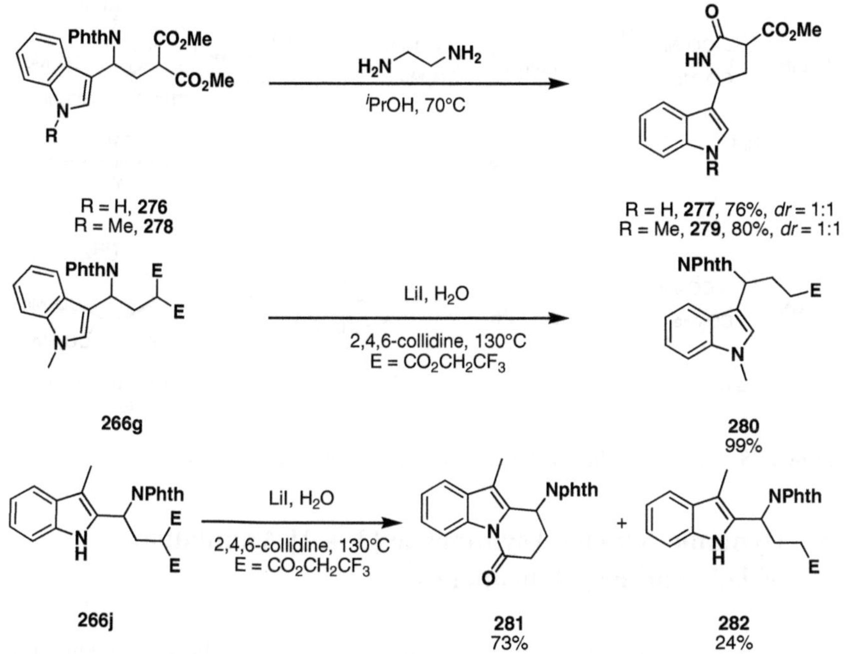

Scheme 2.27 Product transformations

2.4.5 *Product Transformations*

In order to highlight the synthetic versatility of the FC adducts, chemical transformations were performed with various substrates (Scheme 2.27). Phthalimide could be deprotected using the dimethyl ester **276** or **278** in presence of ethylenediamine in isopropanol. The deprotected nitrogen spontaneously cyclized to form the lactam, affording a mixture of diastereoisomers in 76 % yield for unprotected indole and 80 % yield for the *N*-methyl derivative. Krapcho decarboxylation of the FC adducts had to be optimized as standard conditions (NaCl, DMSO, 150 °C) were not effective[11]. Using lithium iodide in collidine at 130 °C afforded the product **280** in quantitative yield. To the best of our knowledge, this constitutes the first report of a Krapcho decarboxylation on a trifluoroethyl ester. When using the C2 substituted analogue **266j**, cyclized product **281** was obtained in 73 % yield together with 24 % of linear product **282**. Extended reaction time did not increase the ratio of the two products toward the cyclized product, probably because cyclization had occurred before decarboxylation.

[11]See experimental part for more details.

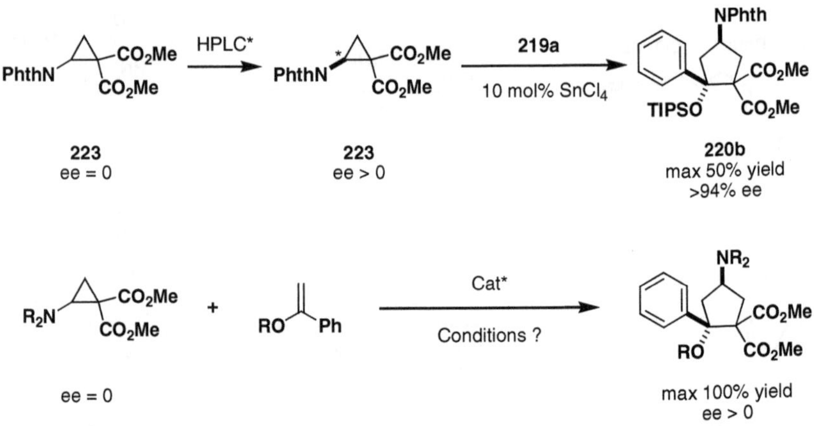

Scheme 2.28 DyKAT for the enantioselective synthesis of cyclopentylamines

2.5 Dynamic Kinetic Asymmetric [3 + 2] Annulation of DA Aminocyclopropanes[12]

We have shown in Sect. 2.1 that the reaction between enol ethers and enantiopure aminocyclopropane **223** afforded the cyclopentylamines with transfer of stereogenic information. From the work of Johnson and co-workers, this result can be attributed to the presence of a tight ion pair intermediate reacting in a S_N^2-type fashion. However, in our case the difficult access to enantiopure starting material limits considerably the applications of the method. By developing an enantioselective method that would allow the complete conversion of a racemic mixture of aminocyclopropanes to an enantiopure cyclopentylamine in a single step would be therefore highly valuable. This would be possible in the case of a dynamic kinetic asymmetric transformation (DyKAT) involving aminocyclopropanes (Scheme 2.28).

In this hypothesis, the use of a chiral Lewis acid complex would play two roles: First, to promote the interconversion of the two enantiomers of the aminocyclopropanes. Second, to differentiate the reactivity between the two substrate enantiomers with an appropriate dipolarophile. In the case of a reaction occurring via a tight ion pair intermediate such as **I** or **III**, the DyKAT involved would therefore be of type one (Scheme 1.37 and 2.29).

[12]Text in quotation marks '' has been taken from our publication [30].

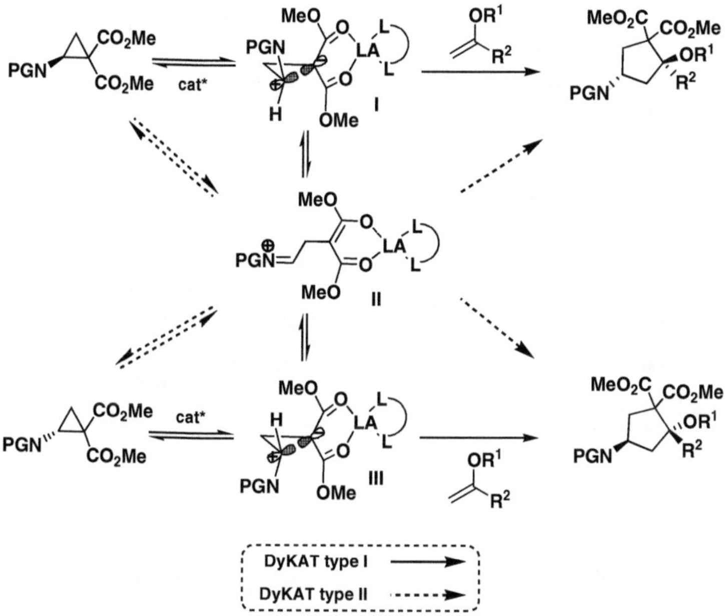

Scheme 2.29 Type 1 and 2 DyKat using DA aminocyclopropanes

2.5.1 DyKAT: 1st Generation

Initial optimization was performed on the method described for the racemic [3 + 2] annulation using aminocyclopropane **223** and silyl enol ether **219a**. The crude reaction was analyzed by ^{1}H-NMR to calculate the ratio of generated product over the remaining starting material and the open by-product **225** resulting from retro-aldol. In a first time various reported combinations of chiral Lewis acid complexes were tested (Fig. 2.12).[13]

From this first broad screening, copper(II)/bisoxazoline complexes were the most promising. Therefore, a more in depth screening of ligands and conditions was investigated using various copper sources and ligands (Table 2.13, Fig. 2.13).

Copper triflate/BOX complexes using ligands **L10, L16, L17** were evaluated first and the best result was obtained with 60 % ee of the desired product using the ^{i}Pr-BOX ligand **L10** (entries 1–3) but the crude reaction contained a mixture of the product and the undesired side product **225**. Other ligands **L12, L18–L20** were tested, however either poor conversion or racemic products were obtained (entries 4–7). We supposed that the open product could arise from acidic or water traces present during the reaction. Thus, drying agents such as molecular sieves or salts were tested, unfortunately with no improvement of the reaction outcome (entries

[13]See supporting information for more details.

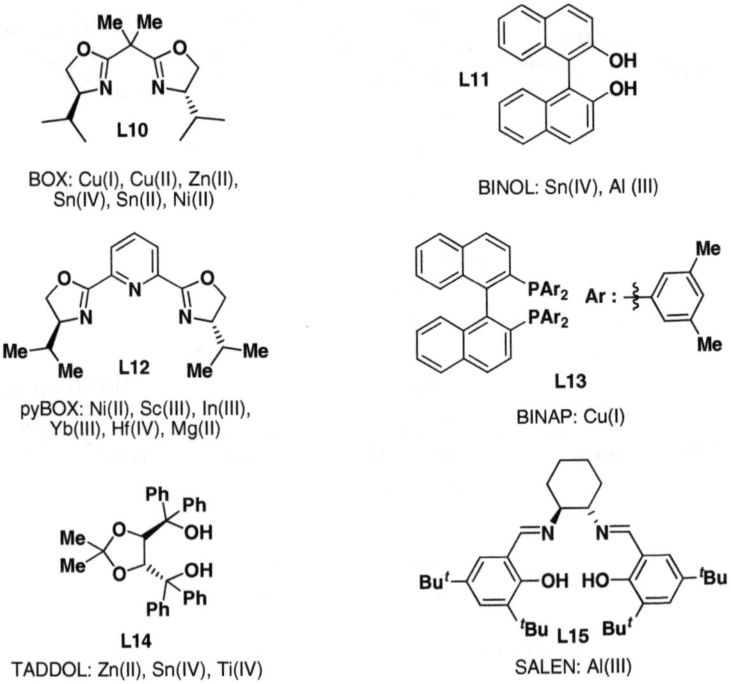

Fig. 2.12 Metal-ligand complexes tested for the DyKAT

8–11). Using the proton scavenger di-*tert*-butyl pyridine (entry 12) did not change the selectivity. In contrary, adding Lewis acid (entry 13) or water (entry 14) completely shut down the reaction. Lowering the temperature or the catalyst loading resulted in uncompleted conversion (entries 15, 16). Screening counter ions allowed to identifying SbF_6^- as a way to suppress formation of the side product, however, without improvement of the enantiomeric excess (entries 17–21). Finally, the other BOX ligands were tested with this new Lewis acid and except the CH_2-indaBOX ligand **L21** that matched the selectivity of the ^{i}Pr-BOX **L10**, no improvement was obtained.

As we went forward in the optimization, we noticed that some results turned out to be irreproducible. The color of the copper complex was an indication of the presence of water in the media. A deep green color was observed for the dry complex, whereas blue was observed when two molecules of water were present in the metal coordination sphere [31]. Formation of the bis(aquo) complex was obtained despite the use of anhydrous conditions and hampered the reactivity as well as the enantioselectivity.

After some investigations, we found that running the complex formation and the reaction in presence of pre-activated 3 Å molecular sieves produced reproducible results.

Table 2.13 Copper catalyst optimization

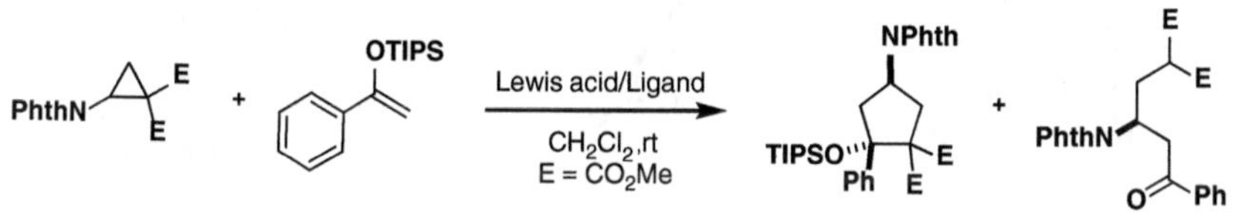

Entry[a]	Lewis acid	Ligand	Conv.[b]	er (%)[c]	**220b/225**[d]
1	$Cu(OTf)_2$	^{i}Pr-BOX **L10**	100	80:20	Mixture
2	$Cu(OTf)_2$	Ph-BOX **L16**	100	75:25	Mixture
3	$Cu(OTf)_2$	^{t}BuBOX **L17**	100	–	**225**
4	$Cu(OTf)_2$	SPRIX **L18**	<100	rac	Mixture
5	$Cu(OTf)_2$	^{t}Bu-QUINOX **L19**	<100	rac	Mixture
6	$Cu(OTf)_2$	^{i}Pr-PyBOX **L12**	0	–	–
7	$Cu(OTf)_2$	PHOX **L20**	0	–	–
8	$Cu(OTf)_2$ + 4 Å MS	^{i}Pr-BOX **L10**	100	80:20	Mixture
9	$Cu(OTf)_2$ + 3 Å MS	^{i}Pr-BOX **L10**	100	80:20	Mixture
10	$Cu(OTf)_2$ + K_2CO_3	^{i}Pr-BOX **L10**	0	–	–
11	$Cu(OTf)_2$ + $MgSO_4$	^{i}Pr-BOX **L10**	100	80:20	Mixture
12	$Cu(OTf)_2$ + ^{t}Bupyridine	^{i}Pr-BOX **L10**	100	80:20	Mixture
13	$Cu(OTf)_2$ + TIPSOTf	^{i}Pr-BOX **L10**	0	–	–
14	$Cu(OTf)_2$ + H_2O	^{i}Pr-BOX **L10**	0	–	–
15	$Cu(OTf)_2^{e}$	^{i}Pr-BOX **L10**	0	–	–
16	$Cu(OTf)_2^{f}$	^{i}Pr-BOX **L10**	<100	80:20	Mixture
17	$Cu(PF_6)_2$	^{i}Pr-BOX **L10**	0	–	–
18	$Cu(BF_4)_2$	^{i}Pr-BOX **L10**	0	–	–
19	$Cu(BARF)_2$	^{i}Pr-BOX **L10**	0	–	–
20	$Cu(NTf_2)_2$	^{i}Pr-BOX **L10**	100	80:20	Mixture
21	$Cu(SbF_6)_2$	^{i}Pr-BOX **L10**	100	80:20	**220b**
22	$Cu(SbF_6)_2$	CH_2-indaBOX **L21**	100	80:20	**220b**
23	$Cu(SbF_6)_2$	Ph-BOX **L16**	100	60:40	**220b**
24	$Cu(SbF_6)_2$	^{t}Bu-BOX **L17**	100	78:22	**220b**

[a]Reaction conditions: 1 eq **223**, 1.2 eq **219a**, 20 mol% catalyst, 22 mol% ligand, 1.0 mL CH_2Cl_2, 1 h, rt
[b]Conversion of **220b** determined by ^{1}H-NMR as the ratio in % of product over the remaining starting material
[c]Determined by Chiral HPLC
[d]Determined by ^{1}H-NMR in the crude mixture
[e]–78 to –10 °C
[f]5 mol% cat. are used

With this problem solved, we could begin to study the effect of the solvent and the temperature on the reaction using Cu(^{i}PrBOX)$(SbF_6)_2$ as chiral Lewis acid (Table 2.14).

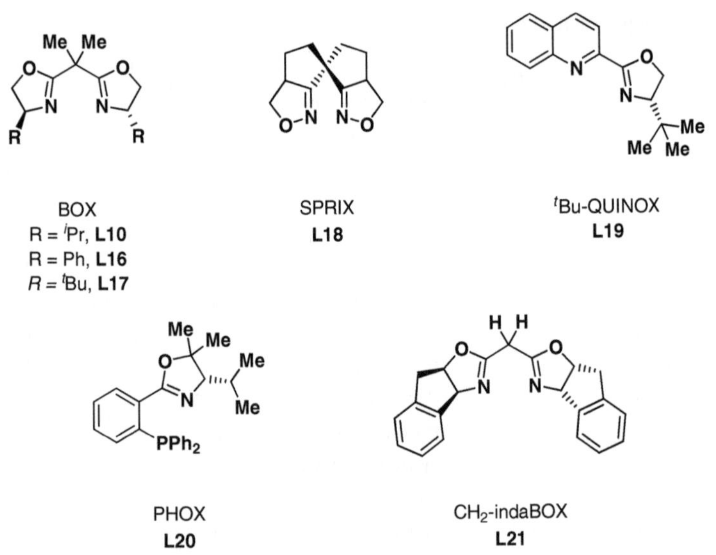

Fig. 2.13 Screened ligands

Table 2.14 Solvent optimization

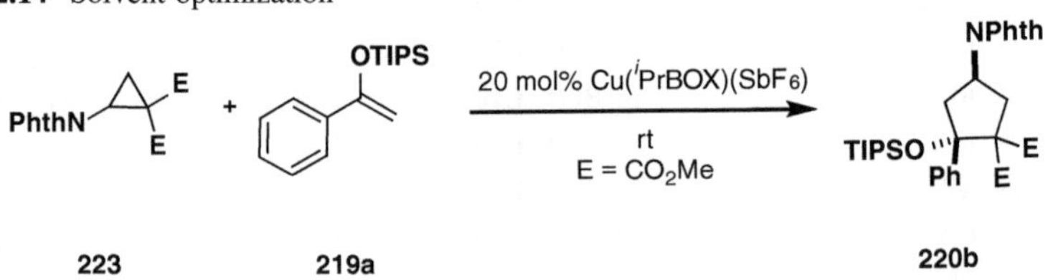

Entry	Solvent[a]	220b[b] (%)	er 220b	ee 223[c]
1	Toluene	0	–	rac
2	Et_2O	0	–	58:42
3	THF	0	–	54:46
4	CH_3CN	0	–	Rac
5	CH_3NO_2	100	67:33	–
6	CCl_4	20	92:8	59:41
7	$CHCl_3$	77	75:25	92:8
8	DCE	86	75:25	rac
9	Trifluorotoluene	100	79:21	–
10	CH_2Cl_2	100	80:20	–

[a]Reaction conditions: 1 eq **223**, 1.2 eq **219a**, 20 mol% Cu(^{i}PrBOX)(SbF_6)$_2$, 1.0 mL solvent, 1 h, rt, MS 3Å beads 10 % w/v

[b]Ratio of product/(starting material left + open product **225**), determined by ^{1}H-NMR in the crude mixture

[c]Determined by Chiral HPLC

Table 2.15 Temperature optimization

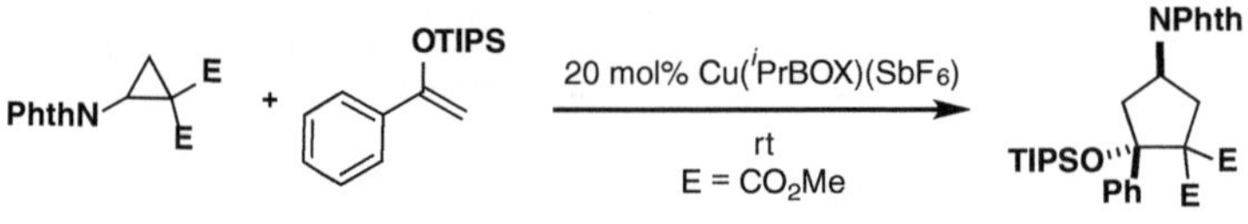

223 219a 220b

Entry	T[a]	**220b**[b] (%)	er **220b**[c]	er **223**[c]
1	–20 °C	65	75:25	79:21
2	0 °C	74	77:23	52:48
3	rt	100	80:20	rac
4	40 °C	79	71:29	87:13

[a]Reaction conditions: 1 eq **223**, 1.2 eq **219a**, 20 mol% Cu(^{i}PrBOX)(SbF_6)$_2$, 1.0 mL solvent, 1 h, rt, MS 3 Å beads 10 % w/v

[b]Ratio of product/(starting material left + open product **225**), determined by ^{1}H-NMR in the crude mixture

[c]Determined by Chiral HPLC

Switching the solvent from dichloromethane to toluene shut down the reaction (entry 1). Polar solvents such as diethyl ether, tetrahydrofuran or acetonitrile were not compatible with the reaction (entries 2–4). Nitromethane and halogenated solvents were compatible with the reaction. However, the outcome of the reaction was not improved compared with dichloromethane (entries 5–10). When carbon tetrachloride was employed (entry 6), the enantiomeric excess increased to 84 %. However, the reaction was very slow and the product was obtained with a low yield.

The reaction was then run at different temperature, from –20 to 40 °C (Table 2.15). At 0 and –20 °C, the reaction was slower and starting material could be detected. In light of the enantiomeric excess and conversion of the starting material and the product, it seems that the reaction exhibited a kinetic resolution profile at –20 °C. When heating at 40 °C, the degradation of the catalyst might explain the lower conversion as we observed a change of the solution color to dark yellow.

2.5.2 *DyKAT: 2nd Generation*

So far, the different parameters of the reaction that were evaluated did not lead to sufficient stereoselectivity. In this context, we decided to investigate the use of aminocyclopropanes that were synthesized during the project on FC alkylation and to synthesize new ones.

For the diester part, methyl (**223**), benzyl (**260**) and trifluoroethyl (**259**) analogues were chosen to study the influence of their steric and electronic effects. We also extended the range of structures for the donating part by preparing new

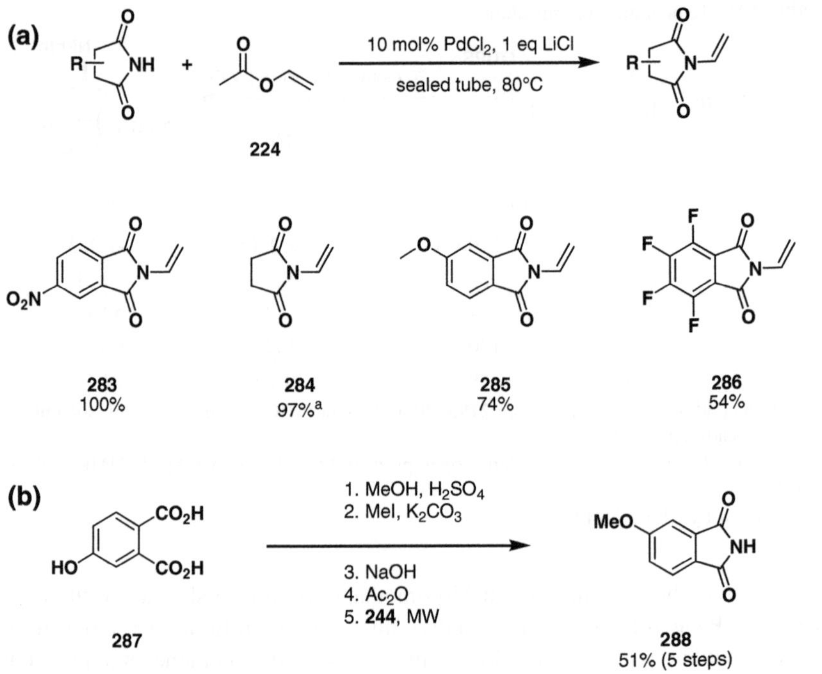

Scheme 2.30 Synthesis of enimides. a Na_2PdCl_4 was used

imide-substituted cyclopropyl derivatives (a, Scheme 2.30). Electron-poor nitro and tetrafluoro phthalimide as well as electron-rich methoxy-substituted phthalimide derivatives were chosen to study the influence of the electronic density on the phthalimide moiety. The saturated succinimide was also considered in order to study its influence in comparison to the “flat” phthalimide.

As for the project concerning the FC alkylation, the synthesis of new amino-cyclopropanes requires vinyl enimides and diazomalonates derivatives that were synthesized using the same methods than previously. The electron-rich phthalimide derivative 5-methoxyisoindoline-1,3-dione (**288**), was prepared in five steps from the corresponding diacid **287** (b, Scheme 2.30).

New cyclopropanes were also prepared using the same rhodium catalyst, affording the products in very good yields. Electron poor nitro-**289** and tetrafluoro **290** phthalimide analogues were prepared in good yields as well as the succinimide and methoxy derivative **291** and **292** (Scheme 2.31).

With the new analogues in hand, we evaluated the influence of the ester substituents on the outcome of the reaction. Compared to the previous screening, two new ligands deriving from the inda BOX skeleton were integrated to the study (**L22** and **L23**, Fig. 2.14) as well as the Bn-BOX **L24**. **L22** and **L23** were prepared from alkylation of the methylene carbon of **L21.**

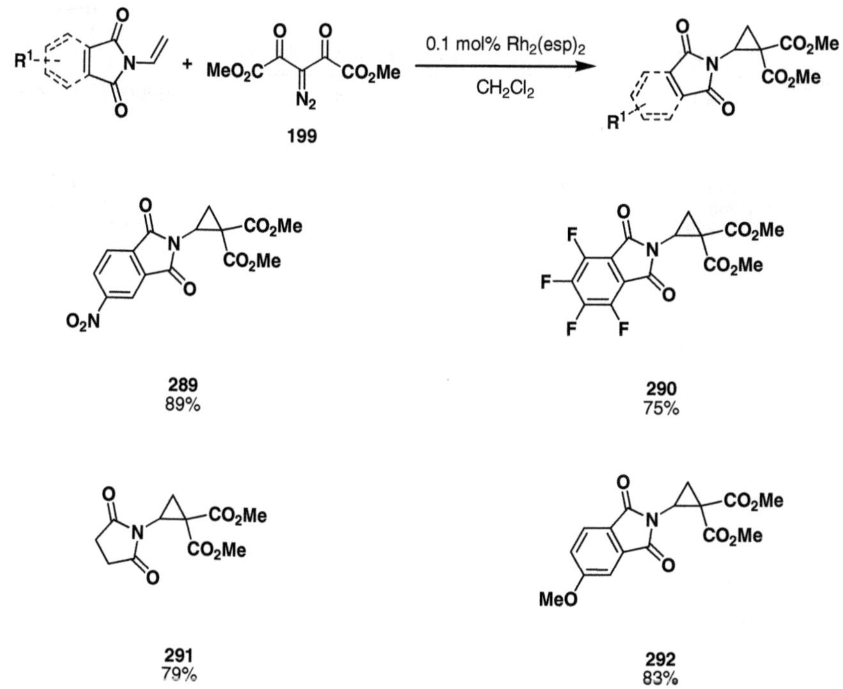

Scheme 2.31 Synthesis of aminocyclopropanes

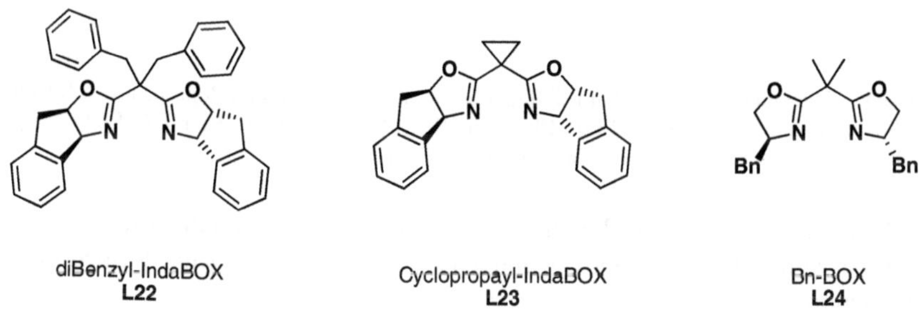

Fig. 2.14 BOX Ligands

The results are summarized in Fig. 2.15.[14] As we can see, aminocyclopropane **223** afforded excellent conversion with all the ligands and best enantioselectivity was obtained with ^{t}BuBOX **L17**. The benzyl substituted aminocyclopropanes **260** turned out to be more sensitive to steric hindrance of the ligand as lower conversion was observed using ^{t}BuBOX. Enantiomeric excess were generally lower compared

[14]See experimental part for the full table of results

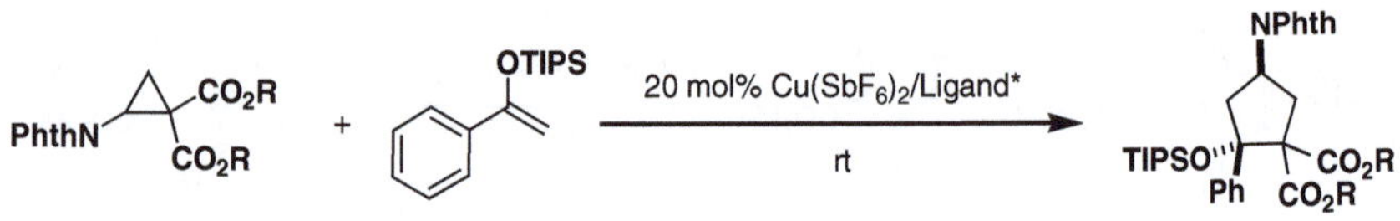

R = Me, **223**
R = Bn, **260**
R = CH_2CF_3, **259**

219a

R = Me, **220b**
R = Bn, **293**
R = CH_2CF_3, **294**

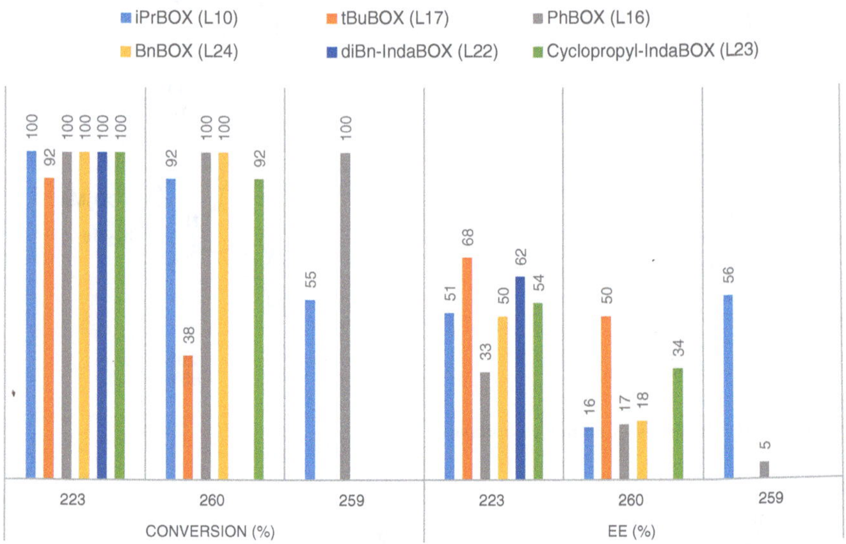

Fig. 2.15 Screening of ligands with different ester-substituted aminocyclopropanes. **a** Reaction conditions: 1 eq cyclopropane, 1.2 Eq **219a**, 20 mol% Cu(ligand)$(SbF_6)_2$, 1.0 mL CH_2Cl_2, MS 3 Å beads 10 % w/v, 1 h, rt. ee: Determined by chiral HPLC

to **223**. Aminocyclopropane **259** did not bring improvement and the results could be analyzed with only two ligands as unidentified side products were formed during the reaction.

Next, the modification of the imide part was investigated. In the first series of analogue tested, different electron-withdrawing functionalities were present on the phthalimide moiety (Fig. 2.16). The use of the ^{t}BuBOX ligand gave better enantioselectivities than with aminocyclopropane **223** for all the tested derivatives. The best selectivities were obtained in the case of the nitro substituted phthalimide **289** with a peak at 83 % *ee*. However, incomplete conversion was observed.

Finally, imides substituents having different steric properties were tested (Fig. 2.17). In all the cases, conversion was equal or superior to 80 %. Naphthyl-substituted aminocyclopropane **257** afforded the product with lower enantiomeric excess than the two others. The structural difference between maleimide and succinimide lies in one unsaturation, however, this small difference has a great

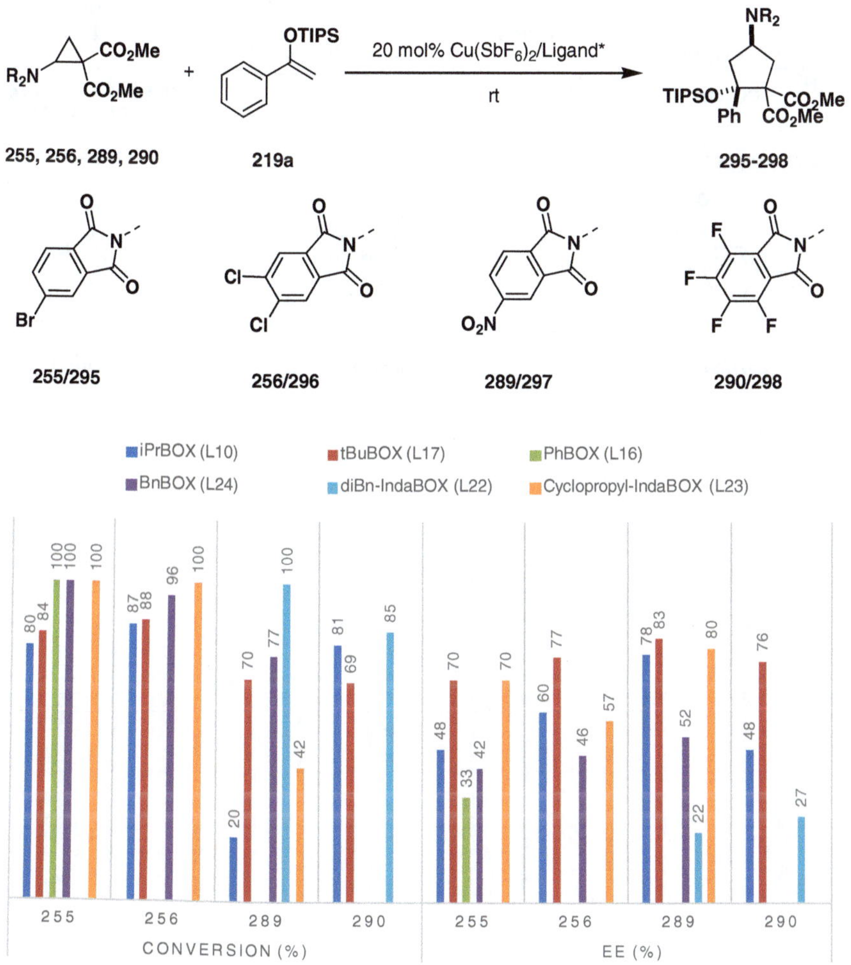

Fig. 2.16 Screening of ligands with different imide-substituted aminocyclopropanes. **a** Reaction conditions: 1 eq cyclopropane, 1.2 eq **219a**, 20 mol% Cu(ligand)$(SbF_6)_2$, 1.0 mL CH_2Cl_2, MS 3 Å beads 10 % w/v, 1 h, rt. ee: Determined by chiral HPLC

influence on the selectivity. Indeed, succinimide afforded the best enantioselectivity so far with 88 % (vs. 76 % with maleimide) obtained in presence of the ^{t}BuBOX ligand.

From this study on the influence of the structure of aminocyclopropanes, it resulted that the nitro-phthalimide and succinimide derivatives **289** and **291** were the two most promising precursors. Compared to the initial test conducted with phthalimido aminocyclopropane **223**, it is interesting to notice that the nitrophthalimide derivative **302** possess a lower pKa of 8.5 (Fig. 2.18). This difference can have influence on the rate of the interconversion between enantiomers due to a poorer stabilization of the zwitterionic intermediate II (Scheme 2.29) as well as on

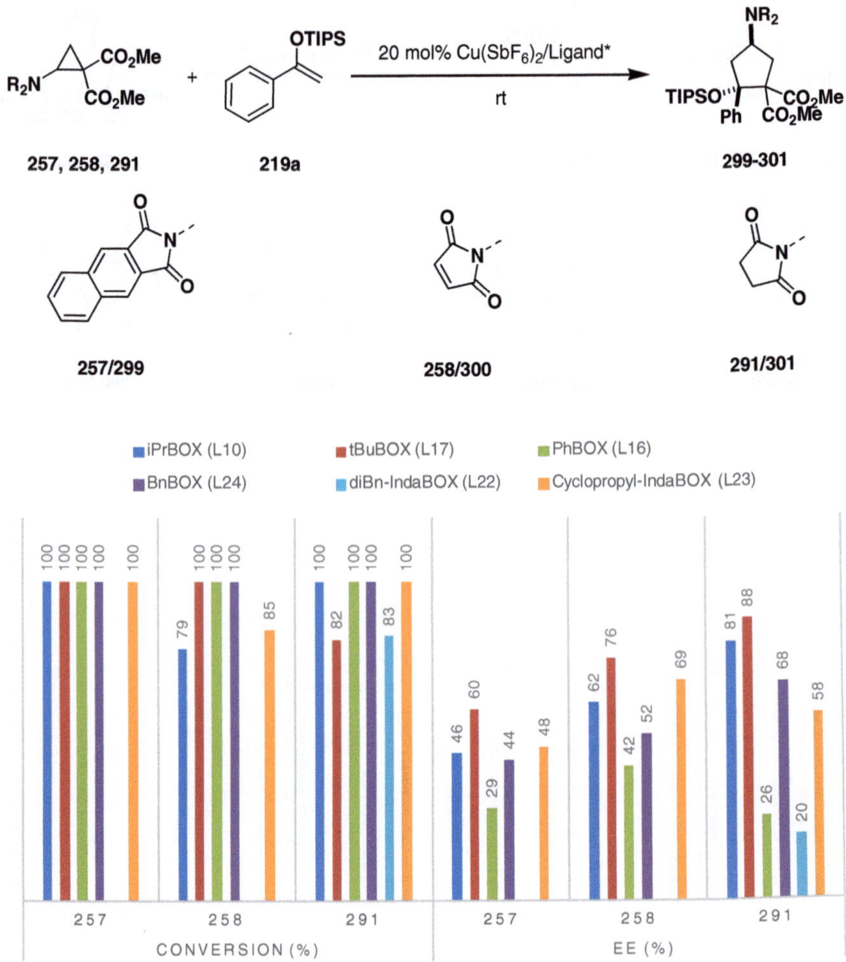

Fig. 2.17 Screening of ligands with different imide-substituted aminocyclopropanes. **a** Reaction conditions: 1 eq cyclopropane, 1.2 eq **219a**, 20 mol% Cu(ligand)$(SbF_6)_2$, 1.0 mL CH_2Cl_2, MS 3 Å beads 10 % w/v, 1 h, rt. ee: Determined by chiral HPLC

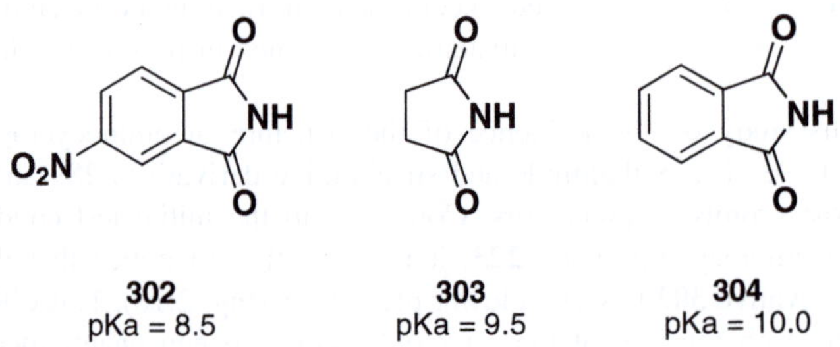

Fig. 2.18 pKa of different imides

their reactivity with the nucleophilic partner. Succinimide (**303**) has a smaller pKa difference with phthalimide than nitrophthalimide. Even if it cannot be excluded that it has some impact on the reactivity, the difference of enantioselectivity can most probably be imputed to the fact that succinimide possesses a different tridimensional structure due to its sp^3 carbons.

When performed on a 0.2 mmol scale, the reaction with **289** and **291** afforded the products in 40 and 34 % yield respectively. The reaction did not go to full conversion for both and more problematic was the large amount of open side product. It was hypothesized that degradation of silyl enol ethers or silicon impurities resulting from a retro-aldol reaction inhibited the catalyst. Unfortunately, further optimization of counter anions, temperature and additives with **289** and **291** did not led to any improvement.

To summarize this section, the [3 + 2] annulation between aminocyclopropanes and silyl enol ether afforded promising enantioselectivities up to 88 % using succinimide derivative **291**. However, irreproducibility of the reaction on preparative scale hampered the development of a useful method. Consequently, the use of silyl enol ethers was abandoned for this reaction.

2.5.3 DyKAT: 3rd Generation

As the main issue identified for the [3 + 2] annulation of aminocyclopropanes was the lability of the silicon group, we investigated the use of alkyl enol ethers as dipolarophiles. Alkyl enol ethers were successfully used for the racemic reaction and possess a more robust C–O bond which should prevent retro aldol side-reaction during the annulation. If the products are expected to be more stable, the access to alkyl enol ethers is less straightforward than the silylated equivalents as they are more sensitive to hydrolysis.

The reaction was optimized using enol ether **305a**. The benzylated analogue of **219a** was prepared using the Petasis reagent [32] and evaluated in combination with various aminocyclopropanes (Table 2.16). We selected four different precursors in order to cover the reactivity range in a minimum of experiment. Phthalimido substituted cyclopropane **223** was used as reference. The two previously successful succinimide **291** and nitro-phthalimide **289** as well as the electron rich methoxy-phthalimide **292** analogues were screened. The work of Tang and co-workers on annulation of silyl enol ethers [39] was reported during this time using a copper(II) perchlorate complex, we began our investigation using this catalyst in combination with the ^{t}BuBOX ligand that gave the best result with silyl enol ethers.

The reaction afforded the desired cyclopentylamines with full conversion, however, an undesired side product (**310–313**) resulting from the attack of the benzyl alcohol was observed in different ratios. Except for the nitrophthalimide derivative, **289** (entry 3), good diastereoselectivity could be achieved (entries 1–4). The stereoselectivity was in accordance with the previous results we had obtained

Table 2.16 Screening of aminocyclopropane

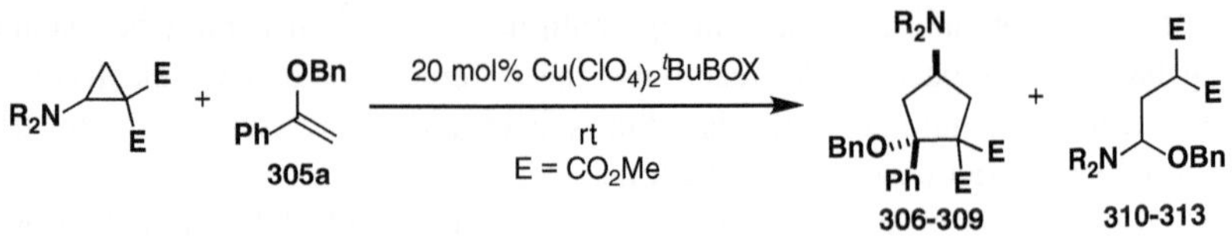

Entry[a]	Imide	(SM/Product/O-alk)	C: O[b]	d.r.[b]	er[c]
1	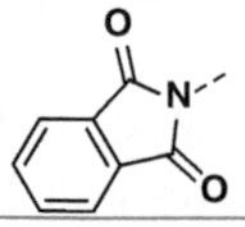	**223/306/310**	4.6:1.0	>10:1	77:23
2	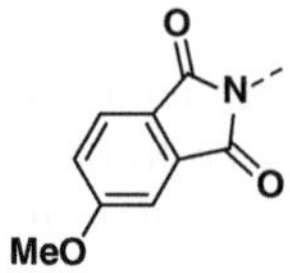	**292/307/311**	5.4:1.0	10:1	74:26
3	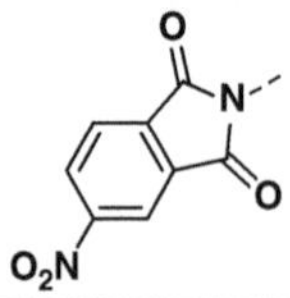	**289/308/312**	3.4:1.0	4:1	92:8
4	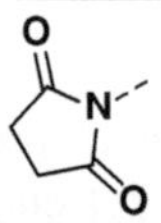	**291/309a/313**	7.3:1.0	>10:1	95:5

[a]Reaction conditions: 1 eq aminocyclopropane, 2.5 eq **305a**, 20 mol% Cu(tBuBOX)$(ClO_4)_2$, 1.0 mL DCM, MS 3 Å beads 10 % w/v, rt, two steps formation of the catalyst
[b]Ratio determined by ^{1}H-NMR
[c]Determined by Chiral HPLC, in %

with silyl enol ethers as nitrophthalimide and succinimide derivatives afforded the best enantioselection (entries 3, 4).

Following these results, succinimide derivative **291** and enol ether **305a** were selected to continue the optimization. Counter-ions, ligands, solvents and temperature were evaluated (Fig. 2.19).[15]

Amongst the tested counterions and BOX ligands, the combination of ^{t}Bu-BOX with $[ClO_4^-]$ stayed the most promising. Modification of the solvent did not improve the outcome of the reaction. Using carbon tetrachloride had the same effect than described previously and shut down the reaction. The influence of the temperature was also examined. By performing the reaction at 0 °C, similar enantiomeric and diastereomeric ratios were observed. However, the yield dropped to 80 %. When the reaction was run at 40 °C, the yield and enantiomeric excess stayed

[15]See experimental part for full tables of optimization. Concentration and stoichiometry were tested without significant improvements.

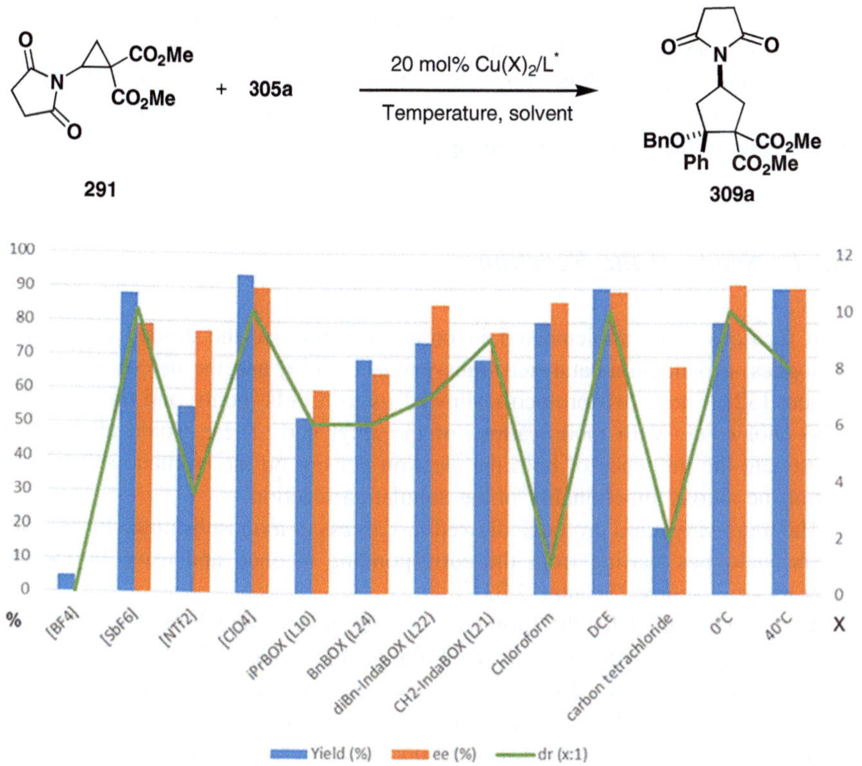

Fig. 2.19 Optimization of the DyKAT parameters. ^{t}BuBox was used for counterions screening and $Cu(ClO_4)_2$ was used for ligand screening

similar, but this time the diastereomeric ratio decreased to 8:1. Finally, the best system was found using copper(II) perchlorate complex, with ^{t}Bu-BOX as ligand, in dichloromethane at room temperature. The aminocyclopropane was reacted with two equivalents of enol ether with 10 mol% of the catalyst and 3 Å MS under argon atmosphere to afford the product in a 95:5 enantiomeric ratio, 10:1 diastereomeric ratio and 94 % NMR yield (Scheme 2.32). The relative anti stereochemistry was confirmed by 2D NMR experiments. It was found that strict exclusion of water was

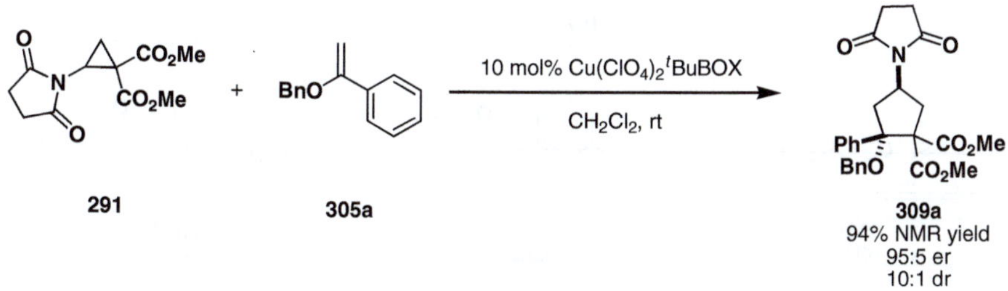

Scheme 2.32 Optimized conditions for the [3 + 2] annulation

necessary to avoid a diminution of the enantioselectivity due to the formation of bis-aquo copper complex. For this reason, the starting materials were dried by dissolution in benzene followed by evaporation under high vacuum just before their use and the atmosphere was replaced by argon from gas tank instead of N_2 from general lines which was not dry enough.

2.5.4 *Scope of the Reaction*

Two methods were implemented in order to synthesize the desired enol ethers analogues: a palladium-catalyzed trans-vinylation [33] and the use of the Petasis reagent [32]. These two protocols allowed us to access mono- and di-substituted alkyl enol ethers with modifications of the alkyl part. These would allow us to investigate the influence of both the electron density on the aromatic substituents and of the alkoxy functionality in the annulation reaction.

The trans-vinylation protocol allowed fast access to monosubstituted benzyl enol ethers analogues starting from the corresponding alcohol, and using butyl-vinyl ether as vinyl source (Scheme 2.33).

The ester required for the formation of enol ethers using Petasis reagent were synthesized starting from the corresponding acyl chloride (Scheme 2.34).

The Petasis reagent **318** was prepared via treatment of titanium dichloride complex with methyl Grignard and used as a solution with different esters (Scheme 2.35) [34]. The products were obtained after a first purification on basic alumina in good yields. Nevertheless, a second purification was always performed prior to the [3 + 2] annulation in order to eliminate traces of acetophenone.

"On preparative scale, cyclopentylamine **309a** could be obtained in quantitative yield with a 96:4 er and 7:1 dr (entry 1, Table 2.17). Variation of the oxygen substituent was examined first: A methyl enol ether (entry 2) and a more electron-withdrawing trifluoroethyl group (entry 3) both worked in the annulation reaction, but for the latter the diastereoselectivity of the reaction was lost. Variation

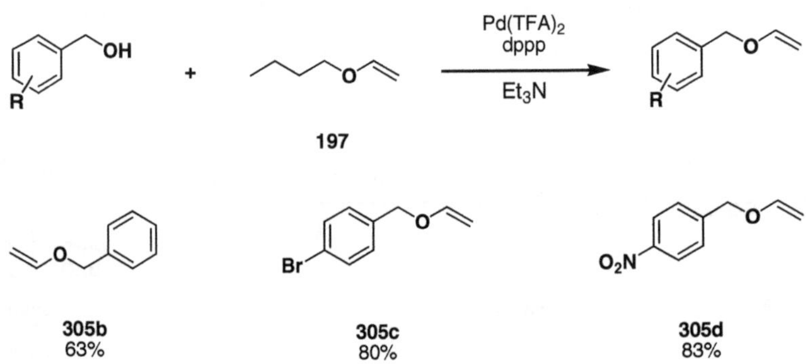

Scheme 2.33 Synthesis of benzyl mono-substituted enol ethers

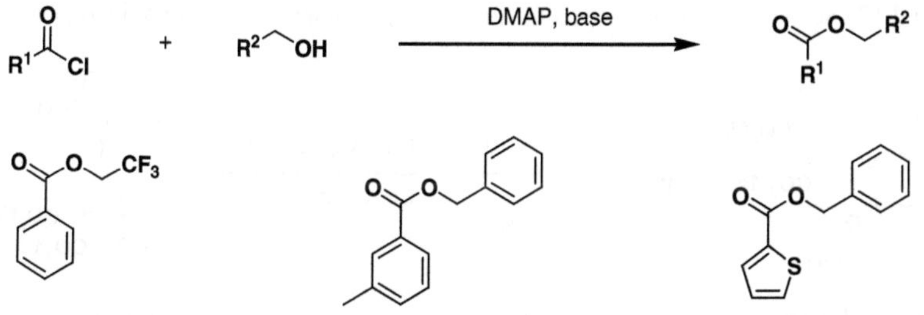

314 53% **315** 97% **316** 96%

Scheme 2.34 Synthesis of ester precursors

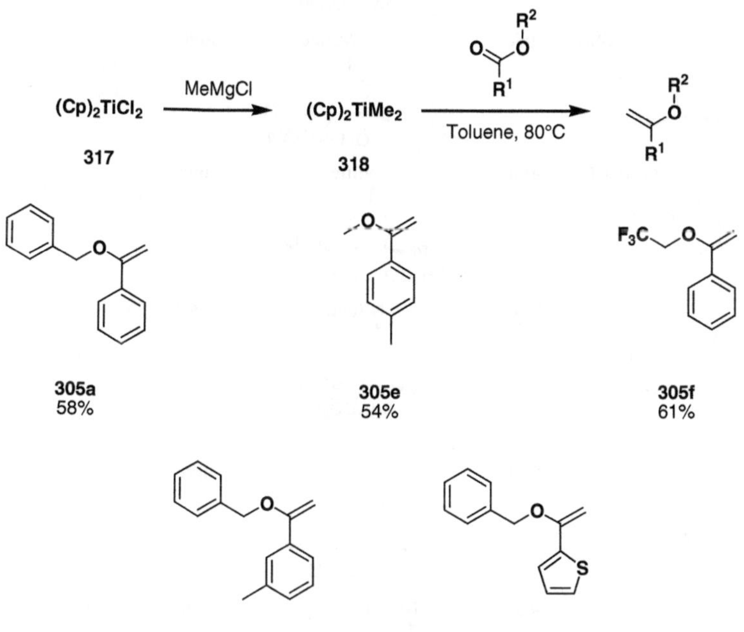

317 **318**

305a 58% **305e** 54% **305f** 61%

305g 45% **305h** 58%

Scheme 2.35 Synthesis of disubstituted enol ethers using Petasis reagent

of the aromatic substituent on the olefin gave comparable enantioinduction for both a *meta* methyl-substituted phenyl ring (entry 4) and a thiophene heterocycle (entry 5). The annulation reaction was not limited to the synthesis of tertiary ethers: unsubstituted benzyl ethers **305b–d** also gave the desired products **309f–h** with useful selectivity (entries 6–8). On a 1 mmol scale, product **309a** was obtained in 80 % yield and a 95.5:4.5 er (entry 6).

Table 2.17 Scope of alkyl enol ethers for the [3 + 2] annulation of aminocyclopropanes

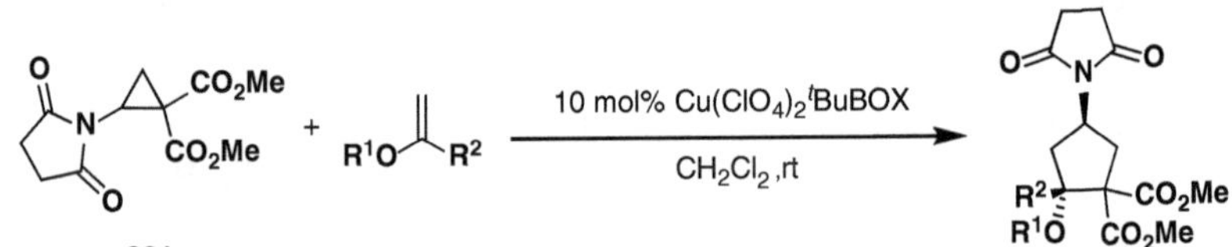

291

Entry	Substrate		Product		Yield (%)[a] *er*[b] *dr*[c]
1	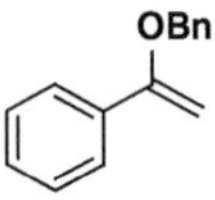	205a	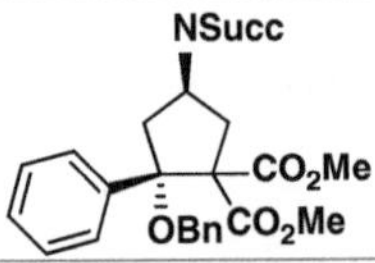	309a	97 96:4 *er* 7:1 *dr*
2	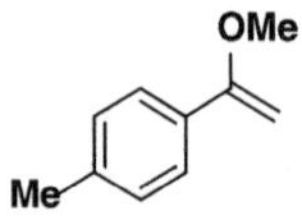	305e	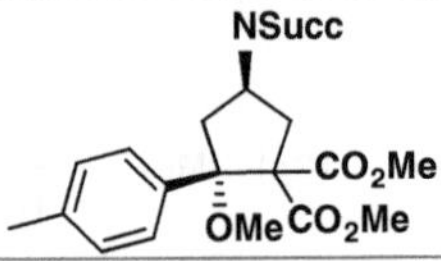	309b	95 94.5:5.5 *er* 20:1 *dr*
3	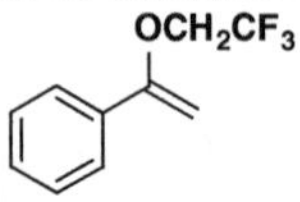	305f	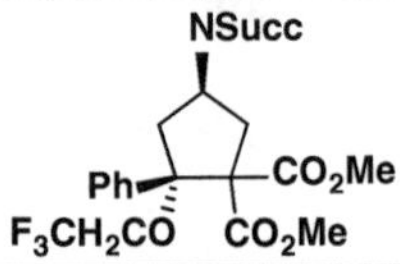	309c	88 95.5:4.5 *er*[d] 1.5:1 *dr*
4	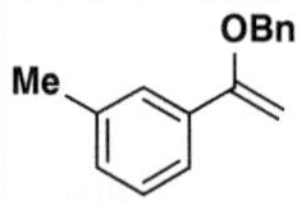	305g	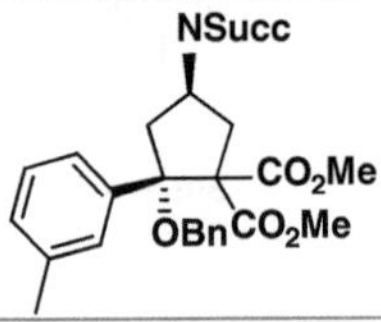	309d	99 95:5 *er* >20:1 *dr*
5	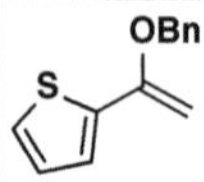	305h	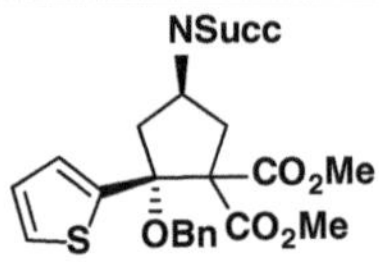	309e	94 94:6 *er* 8:1 *dr*
6	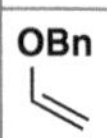	305b	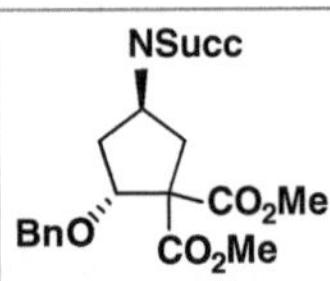	309f	96 (80)[e] 96.5:3.5 (95.5:4.5) *er* 4:1 (4:1) *dr*
7	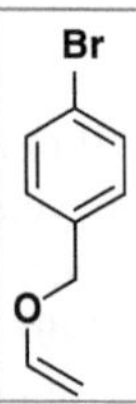	305c	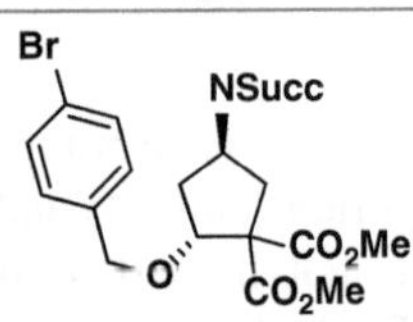	309g	73 94.5:5.5 *er* 5:1 *dr*

(continued)

Table 2.17 (continued)

Entry	Substrate		Product		Yield (%)[a] *er*[b] *dr*[c]
8		**305d**	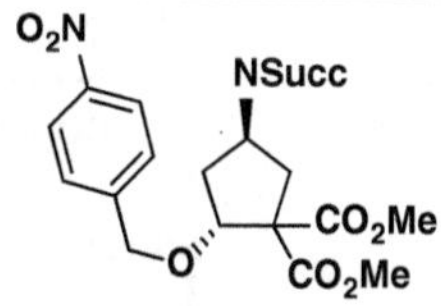	**309h**	82 98:2 *er* 5:1 *dr*

[a]Reaction conditions: 0.20 mmol cyclopropane, 0.40 mmol enol ether, 0.02 mmol catalyst, 3Å MS in dichloromethane, rt, under argon. Isolated yields after column chromatography
[b]Determined by chiral phase HPLC
[c]Determined by analysis of crude ^{1}H NMR
[d]Value for major anti diastereoisomer, syn diastereoisomer: er = 96.5:3.5
[e]Values in brackets correspond to the result on 1 mmol scale

Achieving high selectivity in DYKAT processes is challenging and the catalytic system often has to be optimized for each class of substrates. Nevertheless, when benzaldehyde (**215a**) was used in the [3 + 2] annulation process with aminocyclopropane **291**, the DYKAT process was successful and gave the desired tetrahydrofurylamine **319a** with a 92:8 er and a 13:1 dr (entry 1, Table 2.18). Even

Table 2.18 Scope of aldehydes for the [3 + 2] annulation of aminocyclopropanes

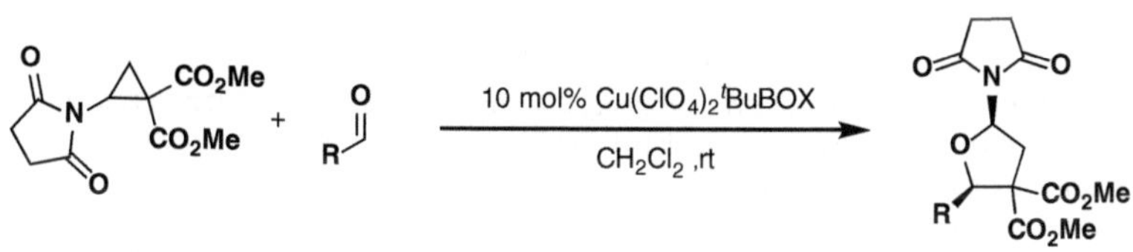

291

Entry	Substrate		Product		Yield (%)[a]/er[b]/dr[c]
1	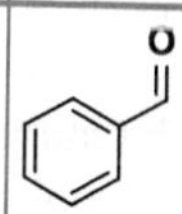	**215a**	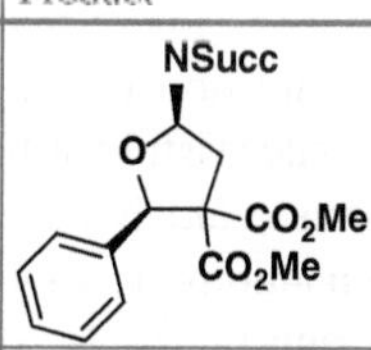	**319a**	82 92:8 er 13:1 dr
2	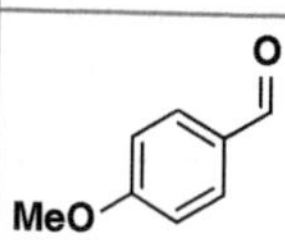	**215b**	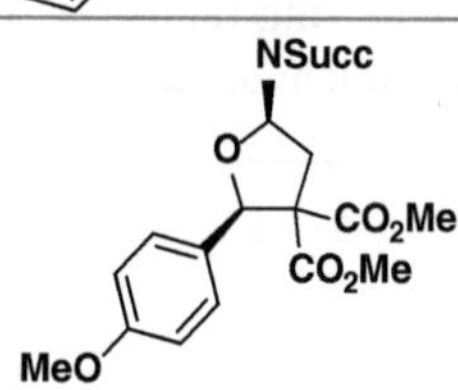	**319b**	69 96:4 er >20:1 dr

(continued)

Table 2.18 (continued)

Entry	Substrate		Product		Yield (%)[a]/er[b]/dr[c]
3	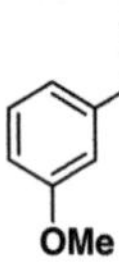	**215c**	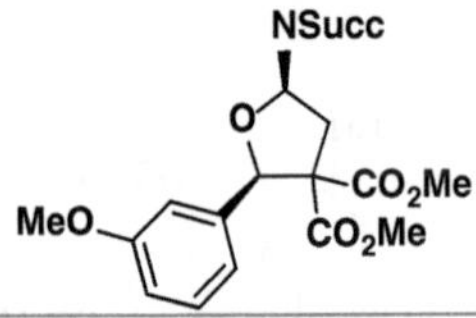	**319c**	84 93:7 er 10:1 dr
4	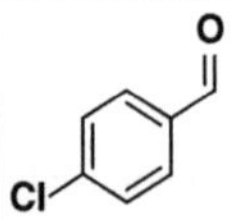	**215e**	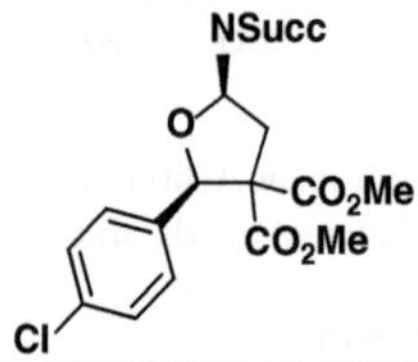	**319d**	90 91:9 er 14:1 dr
5	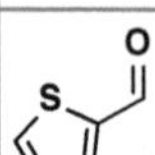	**215d**	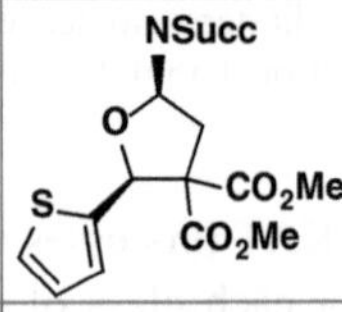	**319e**	97 95:5 er >20:1 dr
6	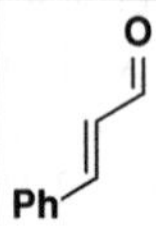	**215g**	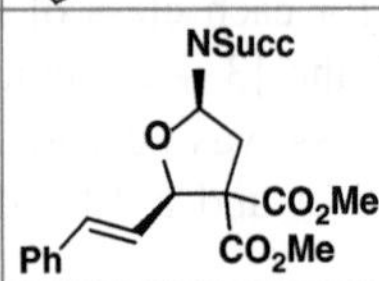	**319f**	96 94:6 er 14:1 dr
7	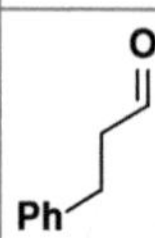	**215j**	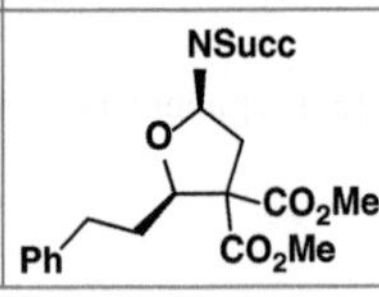	**319g**	85 91.5:8.5 er 13:1 dr

[a]Reaction conditions: 0.20 mmol cyclopropane, 0.40 mmol aldehyde, 0.02 mmol catalyst, 3Å MS in dichloromethane, rt, under argon. Isolated yields after column chromatography
[b]Determined by chiral phase HPLC
[c]Determined by analysis of crude ^{1}H NMR

though the observed enantioselectivity is lower than for enol ethers, it is already impressive considering that no optimization of the reaction conditions was conducted. The annulation reaction was successful for both electron-rich (entries 2 and 3) and electron-poor (entry 4) aromatic aldehydes, as well as for thiophene carboxaldehyde (entry 5). The best enantiomeric ratio (96:4) was observed for the *para*-methoxy substituted benzene ring (entry 2). The reaction was not limited to aromatic aldehydes: both cinnamaldehyde (**215g**) and aliphatic aldehyde **215j** could be used (entries 6, 7)."

2.5.5 *Origin of Asymmetric Induction*

X-Ray diffraction analysis allowed us to confirm the absolute *S* configuration of the center bearing the nitrogen (Fig. 2.20). It also confirmed the *trans* relationship of the imide and the benzyloxy substituent in **309f**.

In order to get information on the selectivity, a crystal of the copper/ligand complex **320** has been analyzed. In this system, two molecules of water are present at the coordination site (Fig. 2.21). It has been shown that 1,3-dicarbonyl compounds position their oxygen at a similar place when chelated to the complex [35]. It can therefore be proposed that the active catalyst presents a distorted square planar geometry with an angle value of *ca* 30° which influences the position of the cyclopropane relatively to the *tert*-butyl substituents.

The trajectory of the nucleophilic attack is deviated from the C–C bond trajectory in order to get a better overlap with the σ* orbital of **291** [40]. Due to the

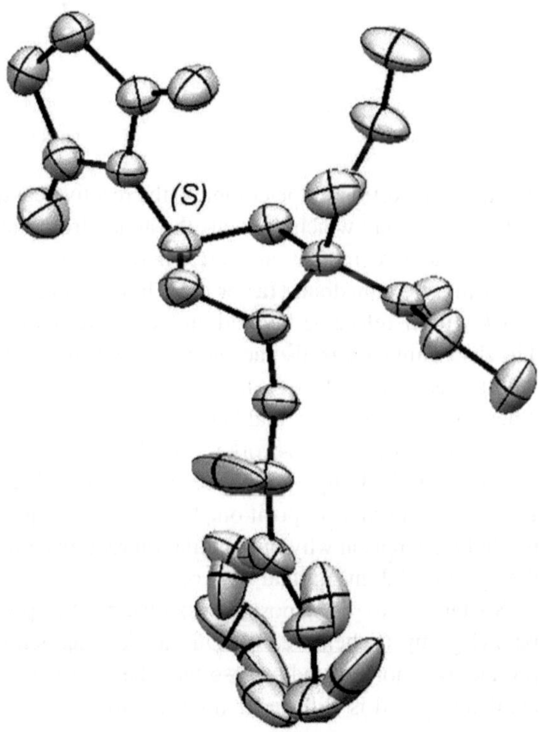

Fig. 2.20 Structure of **309f** determined by X-ray diffraction analysis, with determination of absolute configuration

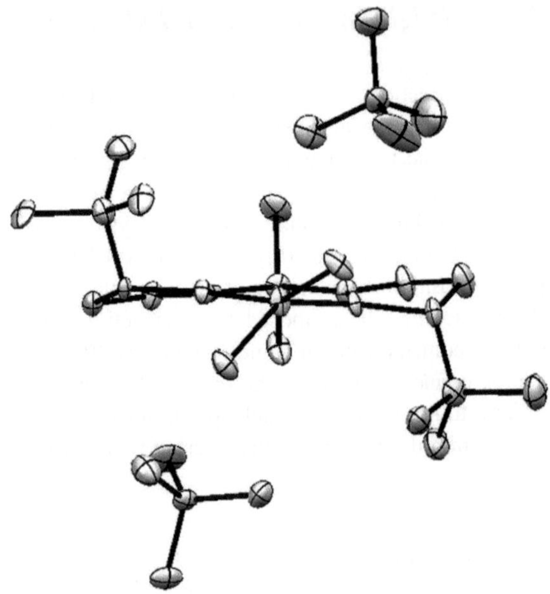

Fig. 2.21 Structure of complex **320** determined by X-ray diffraction analysis

geometry distortion of the complex, the reactive center on the aminocyclopropane is oriented in a way which prevents the attack from the lower quadrants (a, Fig. 2.22). Two productive models are left where the position of the succinimide group is either in a free quadrant (**Ia**) or in an hindered one (**Ib**) (b, Fig. 2.22). Nucleophilic attack will therefore be favored on the less destabilized diastereomeric complex **Ia**. The discrimination is also accentuated from the fact that the complex **Ia** leaves the "north east quadrant" free for nucleophilic attack as opposed to the complex **Ib** where both "north-east" and "north-west" are blocked. This model differs from the one of Johnson and co-workers as the productive complex is not the one that requires that the bulky group faces the ligand substituent (such as in **Ib**). Instead in **Ia**, the phthalimide is positioned in an empty quadrant. This "matching" system might be the reason why our annulation requires a few hours compared to days with the system of Johnson and co-workers.

So far, it has been possible to observe the product **321** via reduction of the benzyl group (Scheme 2.36). Due to decomposition during isolation process, the product was not isolated pure so far. The removal of the succinimide has been more challenging and is still under investigation.

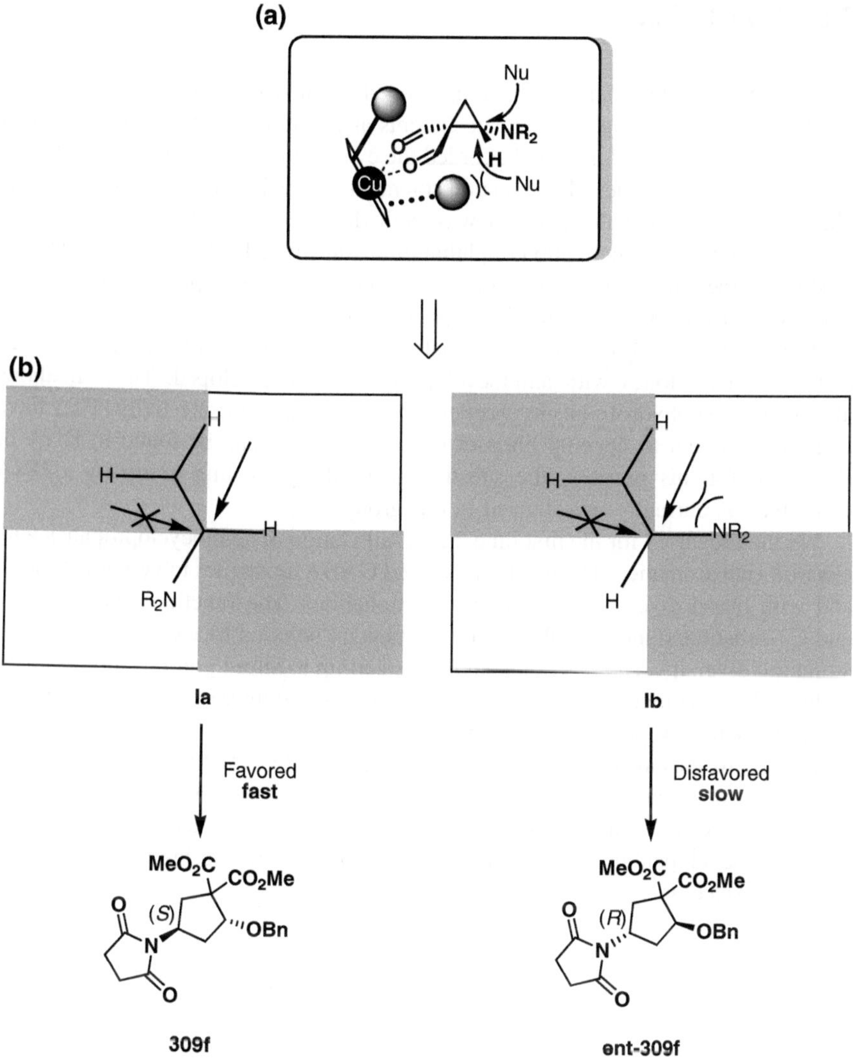

Fig. 2.22 Rational for the enantioselectivity

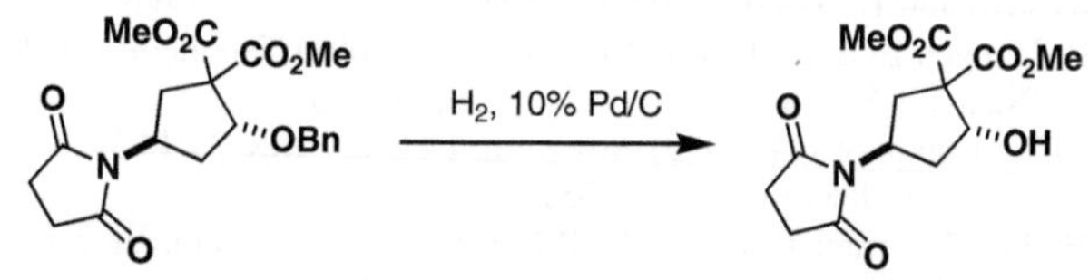

309f 321

Scheme 2.36 Benzyl removal

2.6 Conclusion

To conclude this first part, we have developed the first [3 + 2] annulation of DA cyclopropanes involving nitrogen substituents as donating group. The reaction was catalyzed by tin tetrachloride and afforded the cyclopentylamines in excellent yields and diastereoselectivities. The reaction tolerated a broad scope of silyl as well as alkyl enol ethers and the reaction was both diastereo- and enantiospecific. The success of the [3 + 2] annulation of phthalimido-substituted DA cyclopropanes with silyl enol ethers to deliver selectively cyclopentylamines prompted us to study the behaviour of this system with other dipolarophiles.

The first iron-catalyzed [3 + 2] annulation of aldehydes and tin-catalyzed [3 + 2] annulation of ketones with aminocyclopropanes was developed. These methods afford aminotetrahydrofurans in excellent yields and diastereoselectivity. They have a strong potential to develop libraries of structures that can be found in DNA or RNA and for this purpose, the substitution of the protecting group by a DNA carbo-base is currently investigated in our group [36].

We then reported for the first time the FC alkylation of aminocyclopropanes with electron-rich aromatics. The reaction afforded GABA analogues in very good yields and with broad diversity of tolerated functionalities. The functionalization of C2 and C3-substituted indoles offers a complementary solution to access products with structural diversity. The reaction could be scaled-up to afford gram quantities of FC adducts. Various modifications of the products revealed their synthetic potential for the construction of polycyclic structures.

Finally, we managed to develop a new enantioselective method involving [3 + 2] annulation of DA aminocyclopropanes. The reaction is based on the use of a chiral copper-bisoxazoline catalyst and can be applied for the first time to the formation of both aminocyclopentanes and aminotetrahydrofurans, keeping high levels of reactivity and selectivity.

References

1. De Nanteuil F, Waser J (2011) Angew Chem, Int Ed 50:12075
2. am Ende DJ, DeVries KM, Clifford PJ, Brenek SJ (1998) Org Process Res Dev 2:382
3. Brice JL, Meerdink JE, Stahl SS (1845) Org Lett 2004:6
4. Wyatt P, Hudson A, Charmant J, Orpen AG, Phetmung H (2006) Org Biomol Chem 4:2218
5. Arend M, Westermann B, Risch N (1044) Angew Chem, Int Ed 1998:37
6. Krapcho AP, Jahngen EGE Jr, Lovey AJ, Short FW (1091) Tetrahedron Lett 1974:15
7. Benfatti F, de Nanteuil F, Waser J (2011) Org Lett 14:386
8. Tan X, Chu CK, Boudinot FD (1999) Adv Drug Delivery Rev 39:117
9. Tietze LF, Beifuss U (1988) Synthesis 5:359
10. Tietze LF, Beifuss U, Antel J, Sheldrick GM (1988) Angew Chem, Int Ed 27:703
11. Hilt G, Janikowski J (2008) Iron catalysis in organic chemistry: Reactions and applications. Wiley, Weinheim (Chapter 9)
12. Bolm C, Legros J, Le Paih J, Zani L (2004) Chem Rev 104:6217

13. Benfatti F, de Nanteuil F, Waser J (2012) Chem Eur J 18:4844
14. Shi M, Xu B (2003) Tetrahedron Lett 44:3839
15. de Nanteuil F, Loup J, Waser J (2013) Org Lett 15:3738
16. Johnston GAR (1996) Pharmacol Ther 69:173
17. Chebib M, Johnston GAR (2000) J Med Chem 43:1427
18. Vasudev PG, Chatterjee S, Shamala N, Balaram P (2010) Chem Rev 111:657
19. Lim KH, Komiyama K, Kam TS (2007) Tetrahedron Lett 48:1143
20. Zhou Y-Y, Wang L-J, Li J, Sun X-L, Tang Y (2012) J Am Chem Soc 134:9066
21. Emmett MR, Grover HK, Kerr MA (2012) J Org Chem 77:6634
22. So SS, Auvil TJ, Garza VJ, Mattson AE (2012) Org Lett 14:444
23. Lifchits O, Charette AB (2008) Org Lett 10:2809
24. Lindsay VNG, Nicolas C, Charette AB (2011) J Am Chem Soc 133:8972
25. Blanchard LA, Schneider JA (1986) J Org Chem 51:1372
26. Kacprzak K (2003) Synth Commun 33:1499–1507
27. Muller PM, Albrecht, (Ed: B Aktiengesellschaft) (2001)
28. Brand JP, Charpentier J, Waser J (2009) Angew Chem, Int Ed 48:9346
29. Tolnai GL, Ganss S, Brand JP, Waser J (2012) Org Lett 15:112
30. de Nanteuil F, Serrano E, Perrotta D, Waser J (2014) J Am Chem Soc 136:6239
31. Evans DA, Peterson GS, Johnson JS, Barnes DM, Campos KR, Woerpel KA (1998) J Org Chem 63:4541
32. Petasis NA, Bzowej EI (1990) J Am Chem Soc 112:6392
33. Bosch M, Schlaf M (2003) J Org Chem 68:5225
34. Payack JF, Hughes DL, Cai D, Cottrell IF, Verhoeven TR (2002) Org Synth 79:19
35. Desimoni G, Faita G, Jørgensen KA (2006) Chem Rev 106:3561
36. Racine S, de Nanteuil F, Serrano E, Waser J, Angew Chem, Int Ed (in press)
37. De Simone F, Gertsch J, Waser J (2010) Angew Chem, Int Ed 49:5767
38. Frei R, Staedler D, Raja A, Franke R, Sasse F, Gerber-Lemaire S, Waser J (2013) Angew Chem, Int Ed 52:13373
39. Qu J-P, Liang Y, Xu H, Sun X-L, Yu Z-X, Tang Y (2012) Chem Eur J 18:2196
40. Pohlhaus PD, Johnson JS (2005) J Am Chem Soc 127:16014
41. Emmett MR, Kerr MA (2011) Org Lett 13:4180

Chapter 3
Synthesis and [4 + 2] Annulation of Aminocyclobutanes

In comparison to the cyclopropanes, the reactivity of DA cyclobutanes has been less explored. It is probably because the methods to access DA cyclobutanes are scarce, as explained in the introduction chapter 1.4.5.1 [1].

After discovering the reactivity of DA aminocyclopropanes as 1,3 formal dipoles, we were wondering whether aminocyclobutanes would provide in a similar fashion the corresponding 1,4 formal dipoles and react with a dipolarophile to afford the corresponding six-membered rings. In order to test this hypothesis, it was necessary first to find a synthetic access to such DA cyclobutanes as there was no precedent in the literature.

3.1 Discovery of the [2 + 2] Cycloaddition

A few methods have been reported to access DA cyclobutanes, making use of an electron-rich olefin and an electrophilic alkylidene malonates in presence of a Lewis acid (Sect. 1.4.5.1). This method seemed appropriate for our system as it allows modulating the structure of each partner (Scheme 3.1).

3.1.1 Synthesis of Methylidene Malonates

Methylidenes malonates (R^4 = H, Scheme 3.1) are very unstable building blocks. They are highly electrophilic and traces of nucleophiles such as water or alcohols promote their polymerization. De Keyser and co-workers reported a method for their synthesis involving a multi-step protocol (Scheme 3.2) [2]. The reaction involves a copper-catalyzed Knoevenagel condensation, followed by in situ trapping of the product by anthracene via a Diels-Alder reaction. The stable adduct is separated from copper salts and unreacted anthracene by crystallization and cracked at 225 °C in paraffin oil with maleic anhydride to trap anthracene. A distillation is then performed in acid-washed glassware to afford the methylidene malonate **323a** in 30 % yield. This very tedious/low-yielding protocol was performed once. It was

F. de Nanteuil, *Synthesis and Reactivity of Donor-Acceptor Substituted Aminocyclopropanes and Aminocyclobutanes*, Springer Theses,
DOI 10.1007/978-3-319-23006-1_3

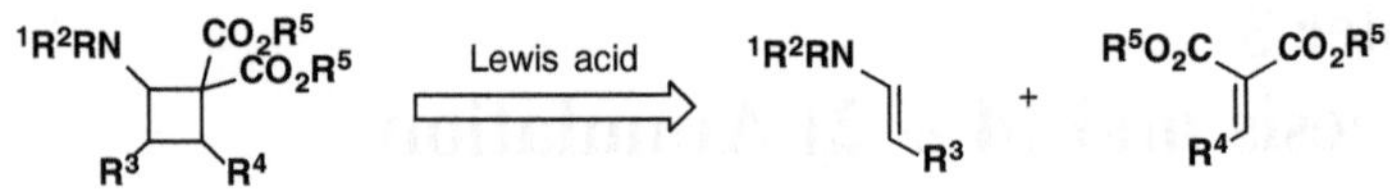

Scheme 3.1 Retrosynthesis of DA aminocyclobutanes

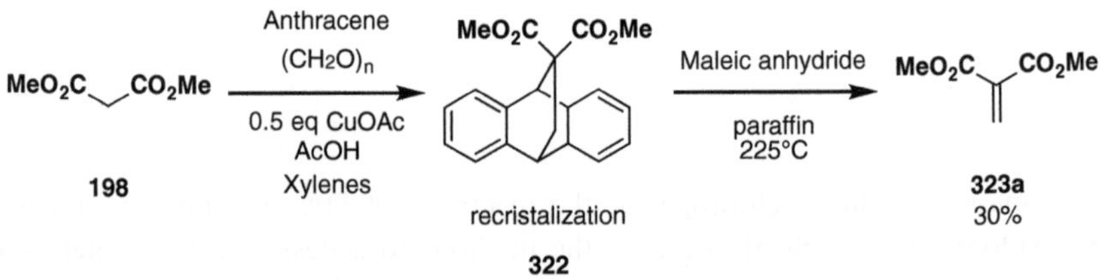

Scheme 3.2 Synthesis of **323a** reported by De Keyser and co-workers

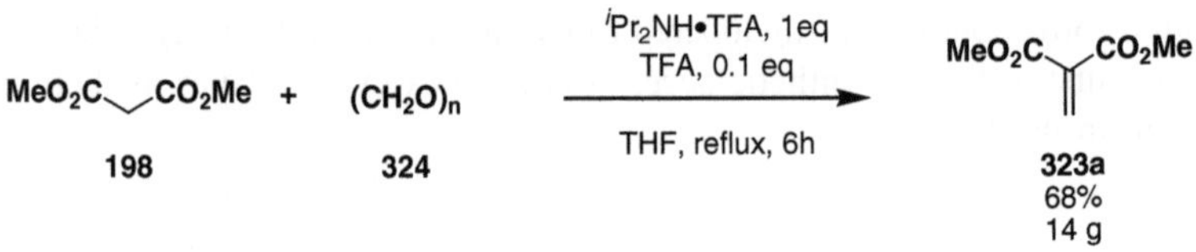

Scheme 3.3 Single-step synthesis of **323a**

found that the product is not stable for more than 2 weeks when stored at −20 °C and therefore had to be distilled every time before use. Considering the key role of this building block in our methodology, and the large quantities required, we decided to improve its synthesis.

After some unfruitful experimentations, we found that adapting the simple method developed by Connel and co-workers [3], it was possible to obtain the product **323a** in 68 % yield on multigram scale in a single step after distillation (Scheme 3.3).

3.1.2 Discovery and Optimization of the [2 + 2] Cycloaddition

Based on our results with cyclopropanes, we expected that deactivation of the donating nitrogen would be necessary. We then began our investigations by submitting vinyl phthalimide (**183**) and vinylidene malonate (**323a**) to known conditions for [2 + 2] cycloaddition (entries 1–5, Table 3.1) [2, 4]. Using zinc, ytterbium or scandium Lewis acids at low temperatures did not afford any desired product

Table 3.1 Optimization of the Lewis acid

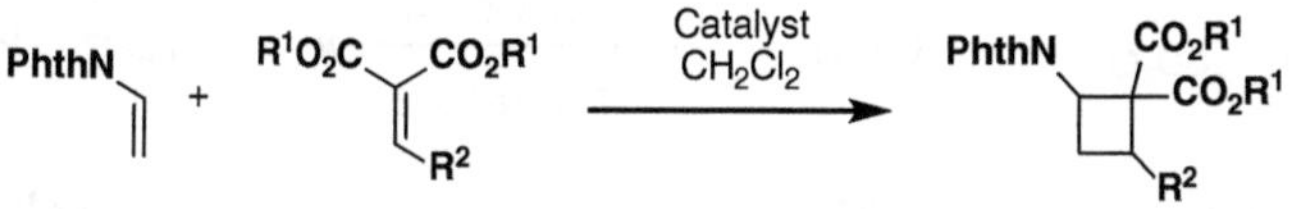

Entry	Substrate	Catalyst (mol%)	t (h)	T (°C)	Ratio **325/183**[a]
1	**323a**	$ZnBr_2$ (100)[b]	1	−78	0:1
2	**323a**	$Yb(OTf)_3$ (10)[b]	0.75	−78	0:1
3	**323a**	$Sc(OTf)_3$ (10)[b]	0.75	−78	0:1
4	**323a**	$Sc(OTf)_3$ (10)[b]	0.75	−30	tr. of **325a**
5	**323a**	$Sc(OTf)_3$ (10)[b]	0.75	0	1:0
6	**323b**	$Sc(OTf)_3$ (20)[c]	0.5	rt	0:1
7	**323b**	$In(OTf)_3$ (20)[c]	12	rt	1.5:1
8	**323b**	$FeCl_3$–Al_2O_3 (20)[c]	12	rt	1:0

[a]Monitored by ^1H NMR spectroscopy
[b]Known procedures were followed. See Experimental part
[c]0.2 eq Lewis acid, 1 eq alkylidene malonate and 1.2 eq **183** added dropwise, CH_2Cl_2, 0.1 mM.
tr. = traces

(entries 1–4). It was possible to detect traces of product when scandium triflate was used at −30 °C (entry 4) and full conversion to the cyclobutane was obtained at 0 ° C (entry 5). We tested less reactive alkylidene malonate **323b** with this protocol and we found that reactivity was shut down in this case (entry 6). We tested another series of Lewis acids and found that indium was able to catalyze the reaction but starting material was left (entry 7). We eventually found that the same iron salt $FeCl_3$–Al_2O_3 that catalyzed the annulation between aminocyclopropanes and aldehydes (Sect. 2.2) was very efficient to deliver the substituted cyclobutane **325b** (entry 8). A control experiment confirmed that iron was the active catalyst, as only alumina did not afford any traces of product.

3.2 Scope and Limitations

3.2.1 Synthesis of Alkylidene Malonates

Using the method developed for vinylidene malonate **323a**, it was possible to easily access vinylidene derivatives such as **323c** in good yields (Scheme 3.4).

Substituted alkylidene malonates are more stable and therefore their synthesis is more straightforward. In most of the cases, heating the aldehydes and the dicarbonyl compounds in acetic anhydride or acetic acid afforded the products in useful yields (a, Scheme 3.5). It is worth to notice that the synthesis of trifluoro-substituted analogue **323f** was reported before and required three steps and multiple

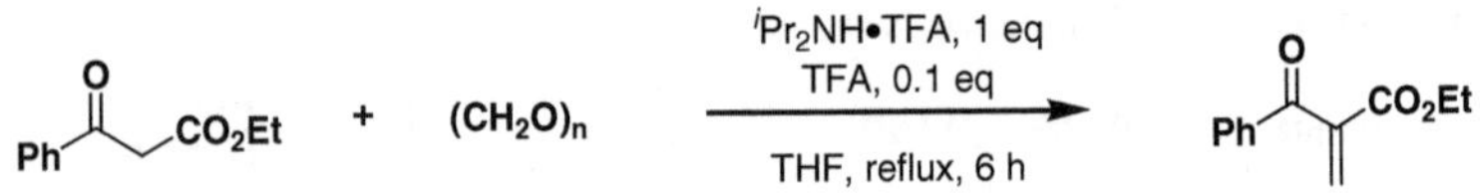

326 324 323c 66% 1.75 g

Scheme 3.4 Synthesis of **323c**

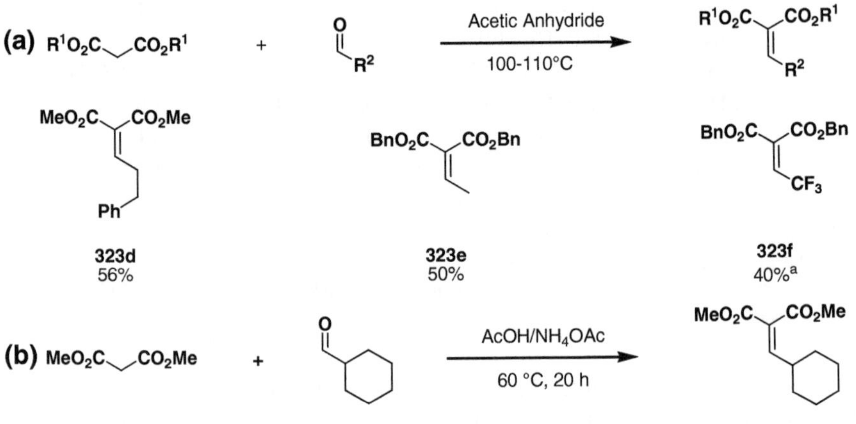

198 215l 323g 76%

Scheme 3.5 Synthesis of alkylidene malonates. **a** The ethyl hemiacetal of the aldehyde was used

purifications [5]. This protocol allows a simple access to **323f** in 40 % yield. For the synthesis of cyclohexyl derivative **323g**, another method using an ammonium salt as catalyst was used (b, Scheme 3.5).

3.2.2 *Synthesis of Enimides*

In order to investigate their influence in the reaction, various vinyl imides were prepared. Modification of the imide part by different deactivating groups would give information on the influence of electronic density on the nitrogen. The substitution on the olefin part with alkyl and aryl groups would give information on the stereoselectivity as well as on the tolerance of the reaction. For this purpose, different synthetic methods were required for each type of substrate. Vinyl imides **183**, **185**, **249, 284** were already synthesized while studying the reactivity of aminocyclopropanes. Protected vinyl thymine **327** was synthesized in two steps, affording the product in 33 % overall yield (Scheme 3.6) [6].

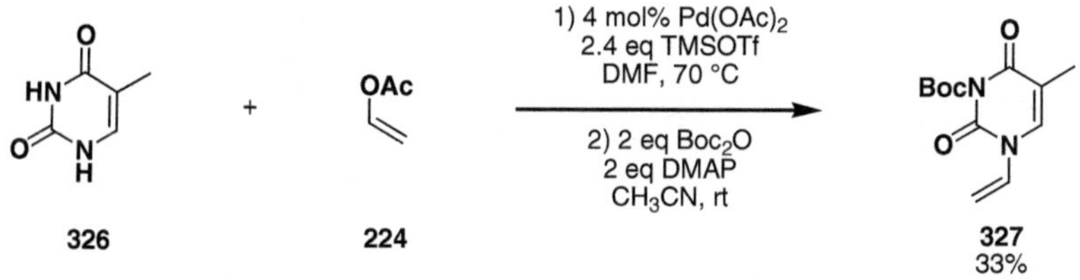

Scheme 3.6 Synthesis of **327**

Substituted vinyl phthalimides **331** and **334** were prepared according to reported procedures (Scheme 3.7) [7, 8]. Allylation followed by isomerization using a ruthenium catalyst afforded the methyl substituted analogue **331**. Chan-Lam coupling between phthalimide (**332**) and boronic acid **333** afforded the substituted product **334** in 33 % yield.

Diverse (*E*)-alkyl and (*Z*)-aryl and alkyl enimides were synthesized selectively using a method developed by Gooßen and co-workers [9]. Changing the substituents on the phosphine from butyl to isopropyl allowed complete inversion of stereoselectivity (a and b, Scheme 3.8). This method proved to be very useful for this project as it provides a stereoselective access to enimides and requires only simple starting materials and catalyst. The products **335–340** bearing aryl or alkyl substituents could be obtained with yields ranging from 13 to 87 %.

Aryl-substituted enimides were synthesized using a palladium-catalyzed Heck reaction between vinyl phthalimide (**183**) and aryl iodides (Scheme 3.9) [10]. The method afforded the product in useful yields and tolerated a large array of functionalities such as halides that would allow efficient post-modification of the products as well as electron—withdrawing and donating groups. Bromo derivatives were obtained with a lower yield, probably due to side-reactivity with palladium.

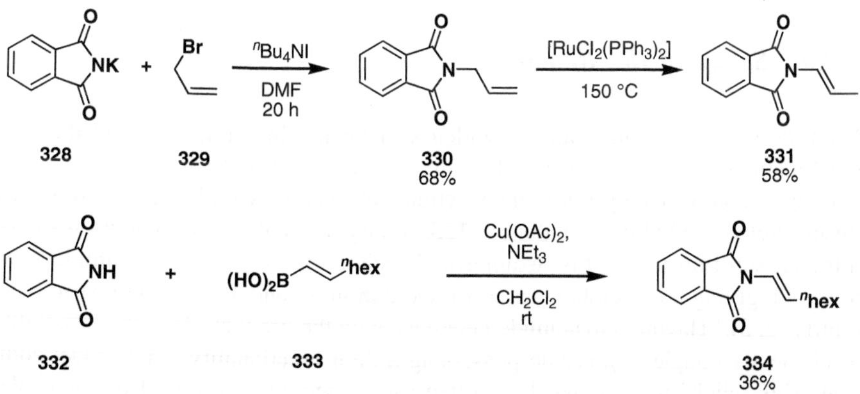

Scheme 3.7 Synthesis of substituted vinyl phthalimides

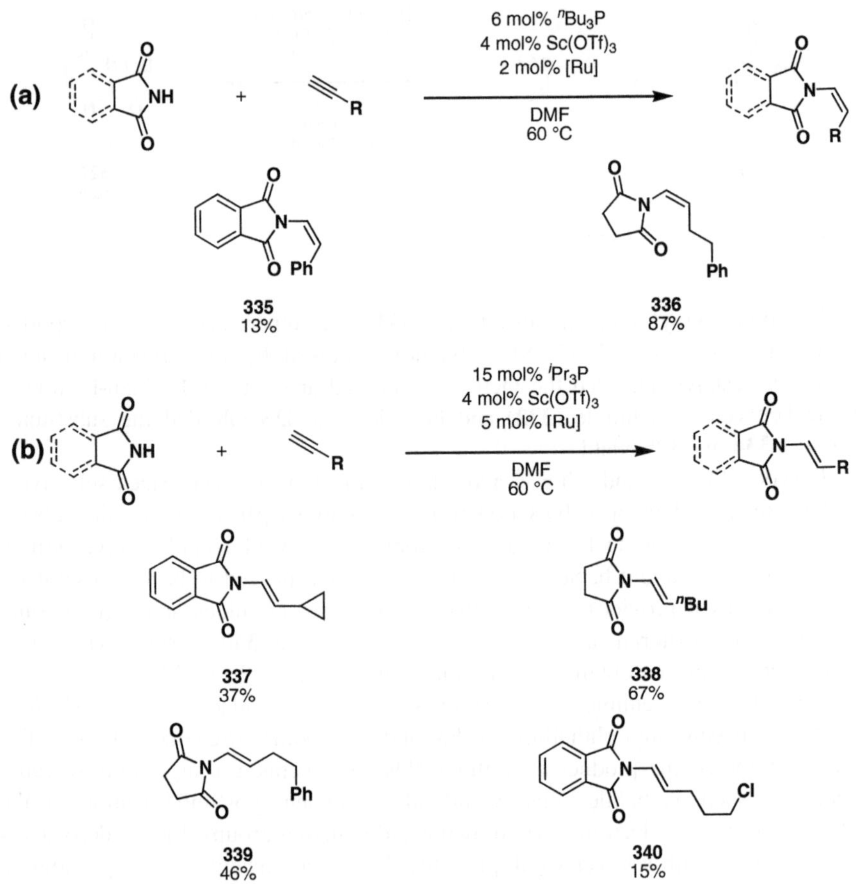

Scheme 3.8 Stereoselective synthesis of (*Z*) and (*E*)-enimides

3.2.3 Scope of the Reaction

With the various enimides and alkylidenes malonates in hand, we tested different combinations. 'On preparative scale using 10 mol% of iron trichloride on alumina was also a good catalyst for the reaction between vinylphthalimide (**183**) and unsubstituted methylidene malonate **325a** (entry 1, Table 3.2).[1] Variation of the nitrogen substituent was first examined. Succinimide as well as maleimide were tolerated, giving the cyclobutanes **347a** and **348** in 91 and 48 % yield respectively (entries 2, 3).' Having a maleimide heterocycle on the product offers the possibility to efficiently couple a molecule possessing a thiol functionality via Michael addition. This "click" reaction has been often use for the engineering of biomolecules

[1]Text in quotation marks '' is taken form our publication [1].

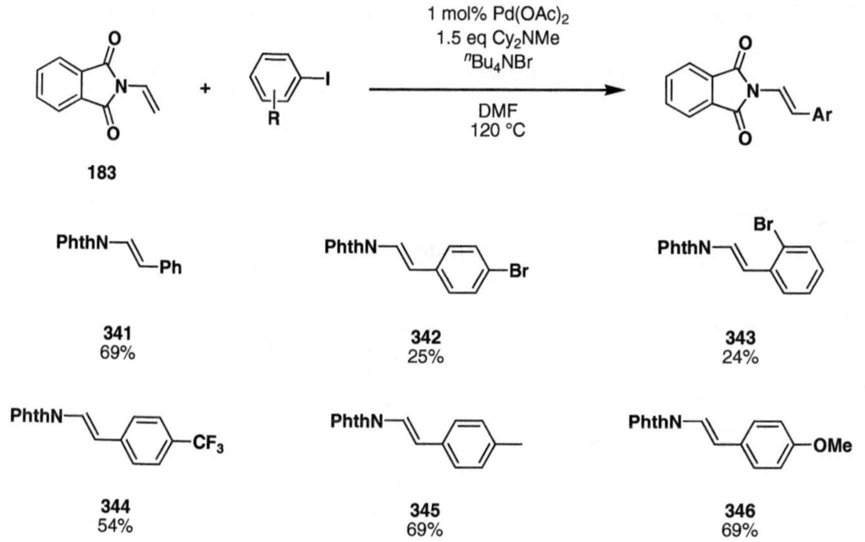

Scheme 3.9 Synthesis of aryl-substituted vinyl phthalimides

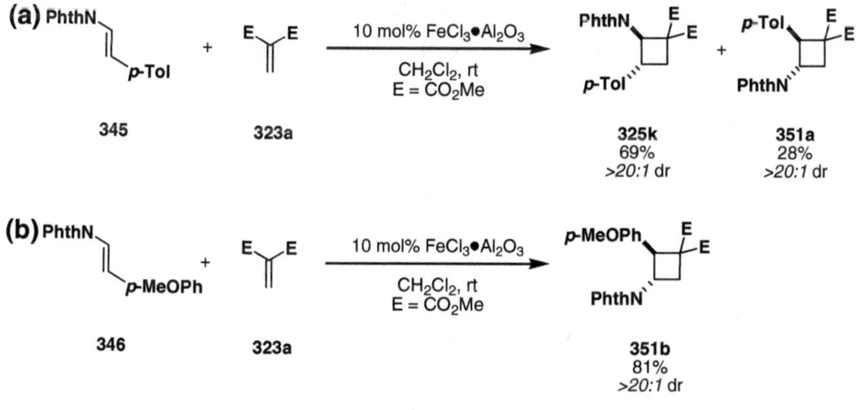

Scheme 3.10 Inverted regioselectivity with electron-rich substituents

[11]. The maleimide functionality has also been widely used as dipolarophile in cycloadditions. 'The reaction also allowed the formation of Boc protected thymine cyclobutane **349** in 76 % yield (entry 4).' Accessing such structure was important to us as it belongs to the family of carbocyclic analogues [12]. 'The use of a N-vinyl-oxazolidinone failed to deliver product **350** due to decomposition of the starting material (entry 5).'

'At this point, we turned to the synthesis of multi-substituted aminocyclobutanes. The use of (*E*)-enimides, was examined first. Enimides substituted with a

Table 3.2 Scope and limitations of the [2 + 2] cycloaddition: variation of the enimide

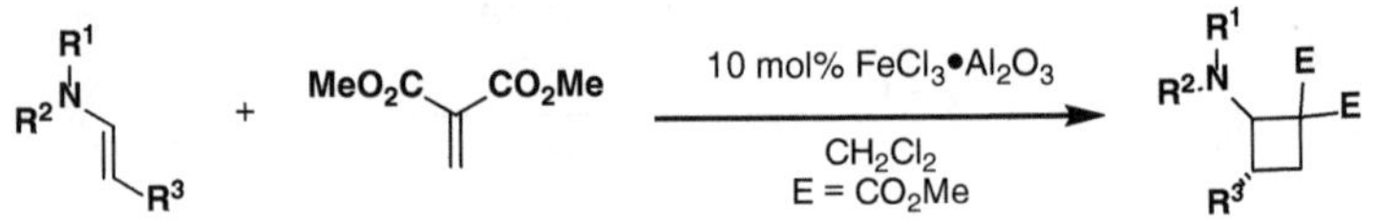

323a

Entry	Substrate	Product	Yield (%)[a]
1	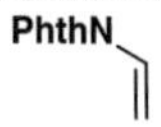 **183**	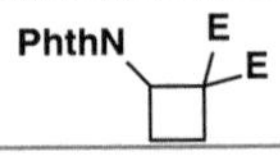 **325a**	96
2	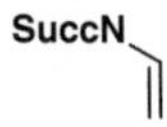 **284**	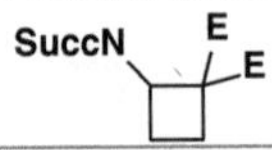 **347a**	91
3	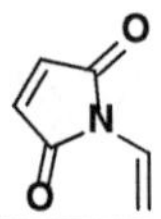 **249**	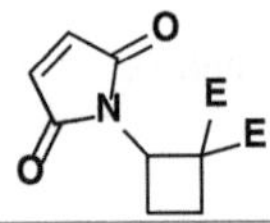 **348**	48
4	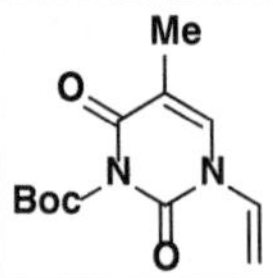 **327**	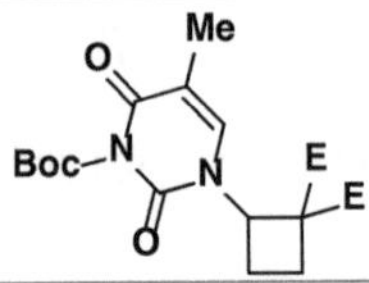 **349**	76
5	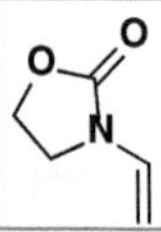 **185**	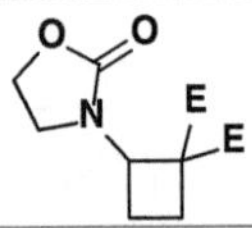 **350**	0
6	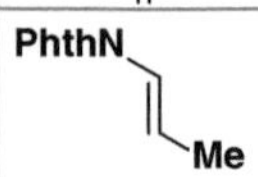 **331**	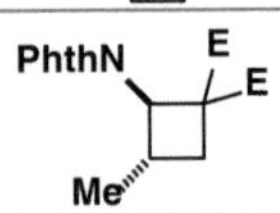 **325c**	83 (>20:1 *dr*)
7	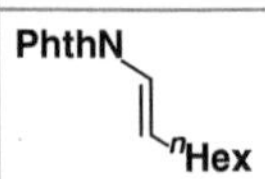 **334**	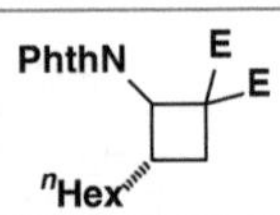 **325d**	85 (>20:1 *dr*)
8	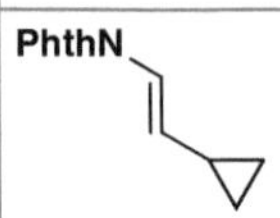 **337**	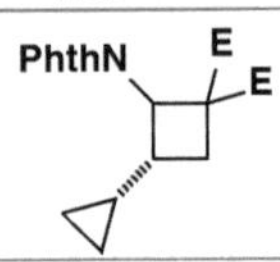 **325e**	74 (>20:1 *dr*)
9	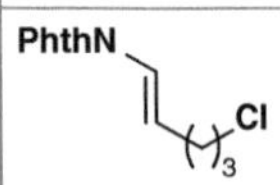 **340**	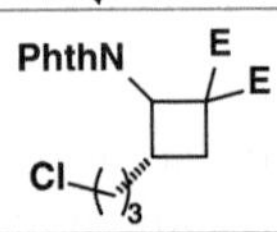 **325f**	95 (>20:1 *dr*)
10	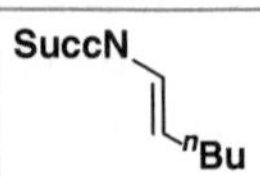 **338**	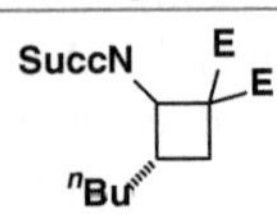 **347b**	81 (>20:1 *dr*)

(continued)

Table 3.2 (continued)

Entry	Substrate	Product	Yield (%)[a]
11	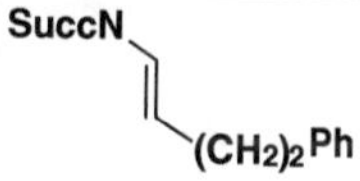 339	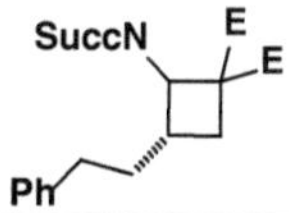 347c	83 (8:1 *dr*)
12	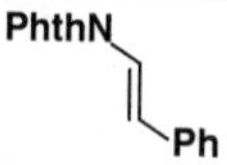 341	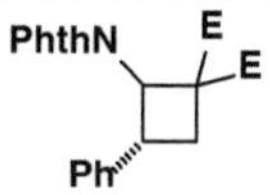 325g	90 (>20:1 *dr*)
13	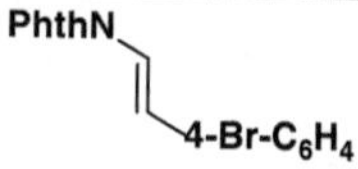 342	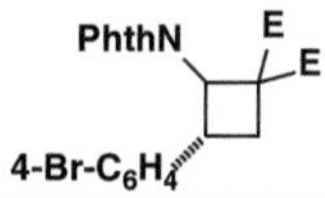 325h	65[b, c] (>20:1 *dr*)
14	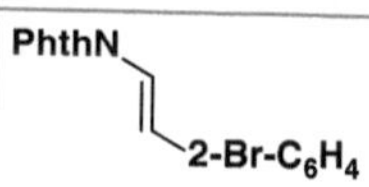 343	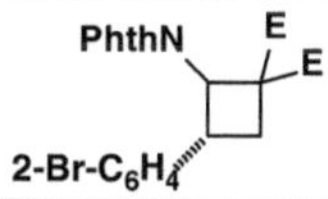 325i	93 (>20:1 *dr*)
15	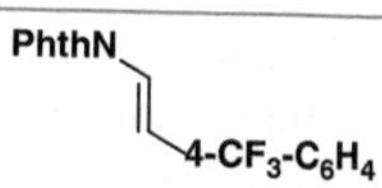 344	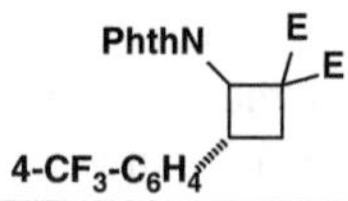 325j	45[b, d] (>20:1 *dr*)

[a]Reaction conditions: 1 eq Enimide, 2 eq alkylidene malonate, 0.1 eq Fe catalyst (1 mmol/g) in CH_2Cl_2 (1 mL) for 0.2–5 h. Yields after purification on column chromatography
[b]4 equivalents of methylidene malonate were used
[c]Based on recovered starting material, 50 % isolated yield
[d]Based on recovered starting material, 38 % isolated yield

methyl, a hexyl or a cyclopropyl group afforded the corresponding cyclobutanes **325c**, **325d** and **325e** in 74–85 % yield (entries 6–8). An aliphatic chloro substituent was also compatible with the reaction conditions (entry 9). Succinimide substituted cyclobutanes **347b**, **347c** could also be obtained in 81–83 % yield (entries 10, 11). Importantly, in all the experiments involving (*E*)-enimides except for the formation of cyclobutane **347c**, only one cyclobutane diastereoisomer could be detected in the crude mixture of the reaction. Aromatic substitution of the enimide was next investigated. The reaction delivered a single diastereoisomer of cyclobutane **325g** bearing a phenyl substituent in 90 % yield (entry 12). Adding a *para*-bromo substituent on the benzene ring slowed down the reaction and no full conversion to cyclobutane **325h** was observed (entry 13). However, decreasing the conjugation of the benzene ring with the enimide by moving the bromine atom to the *ortho* position [13] restored the reactivity and product **325i** could be obtained in 93 % yield (entry 14). Finally, in the presence of a trifluoromethyl group, the reaction was slower, but cyclobutane **325j** could still be obtained in 45 % yield based on recovered starting material (entry 15). X-Ray analysis confirmed the *trans*-relationship between the phthalimide group and the methyl substituent in product **325c** (Fig. 3.1).

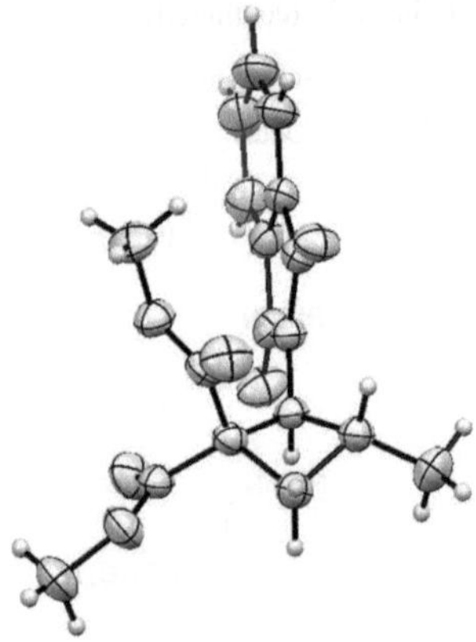

Fig. 3.1 Structure of **325c** determined by X-ray diffraction analysis

When we switched to electron-rich aromatic rings as substituents, we observed a different regioselectivity (Scheme 1.3). When tolyl-substituted enimide **345** was used, a 2.5:1 mixture of "normal" and "inverted" products **325k** and **351a** was obtained (a, Scheme 3.10). *p*-Methoxy benzene-substituted substrate **346** gave only the "inverted" product **351b** (b). This switch in regioselectivity is interesting as it gives access to equally important γ-amino acid cyclobutane derivatives [14]. Additionally, it allowed assigning the relative donating ability of the phthaloyl compared to electron-rich aromatic groups in the [2 + 2] cycloaddition, which may be also useful in the future for the design of other transformations.

Next, we investigated the modifications of the electron-poor partner (Table 3.3). Modification of the ester substituents afforded isopropyl analogue **325l** and allyl analogue **325m** in 66 and 49 % yield respectively (entries 1, 2). Cycloaddition of keto ester substrates was possible, affording cyclobutane **325n** in 62 % yield (entry 3, Table 3.3). 'Less reactive substituted alkylidene malonates also afforded cyclobutanes **325b, o, p** in 59–71 % yields, but usually without diastereoselectivity when using alkyl ester substrates (entries 4–6). The use of benzyl substituted alkylidene malonates **323e** allowed the formation of the product **325q** in a better 3:1 diastereomeric ratio and 62 % yield (entry 7). In the case of trifluoromethyl-substituted aminocyclobutane **325r** only the *trans*-diastereoisomer was obtained in 76 % yield (entry 8). This is an important result, as methods for the synthesis of trifluoromethyl substituted aminocyclobutanes are rare, require numerous steps and usually lack diastereoselectivity [15–17]. No reaction was observed when both the enimide and the alkylidene malonate were β-substituted.'

As explained previously, methylidene malonate **323a** is difficult to access very unstable and had to be re-purified before each use. We therefore implemented a sequential protocol that would avoid the sensitive distillation and afford the product on larger scale. By simply performing a mild acidic workup after the condensation reaction, the crude solution could be used directly to generate the desired aminocyclobutanes (Scheme 3.11). The reaction works with simple methyl malonate

Table 3.3 Scope and limitations: variation of the alkylidene malonate

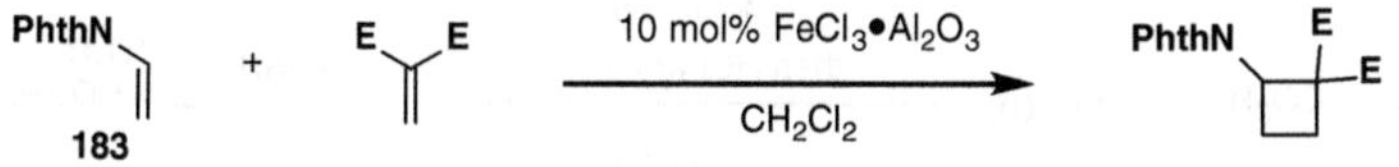

Entry	Substrate	Product	Yield (%)[a]
1	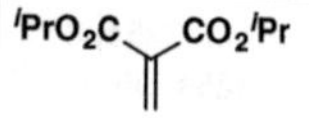 323h	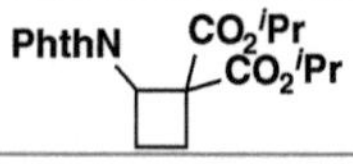 325l	66[b]
2	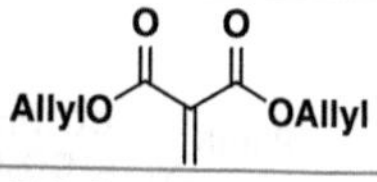 323i	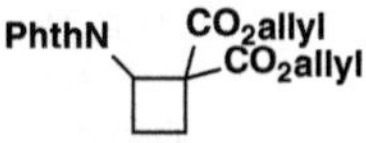 325m	49[b]
3	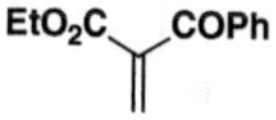 323c	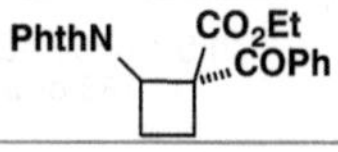 325n	62 (2.5:1 *dr*)
4	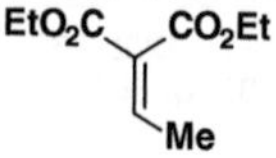 323b	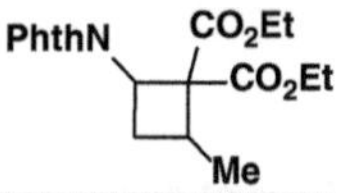 325b	63 (1.2:1 *dr*)
5	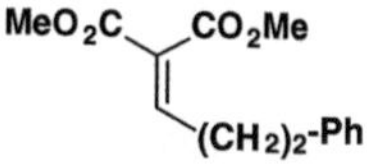 323d	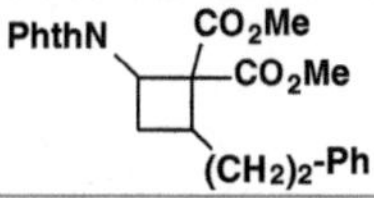 325o	71 (1.3:1 *dr*)
6	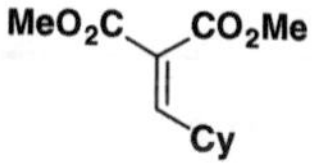 323g	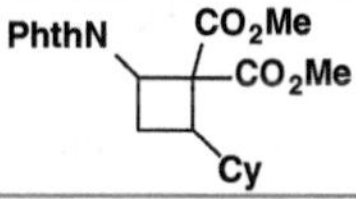 325p	59 (1.2:1 *dr*)
7	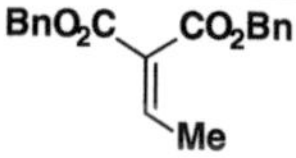 323e	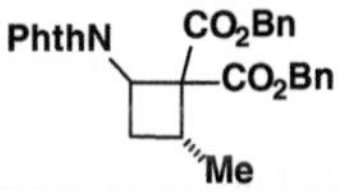 325q	62 (3:1 *dr*)
8	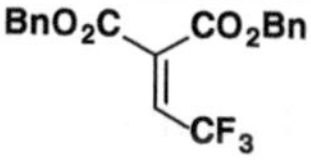 323f	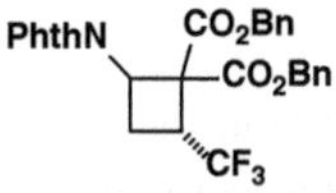 325r	76 (*dr* > 20:1)

[a]Reaction conditions: 1 eq 183, 2 eq alkylidene malonate, 0.1 eq Fe catalyst (1 mmol/g) in 1 mL CH_2Cl_2. Yields after purification on column chromatography
[b]10 mol%, $Sc(OTf)_3$ were used

derivatives to afford products **325a** or **325s** in yields superior than 85 % on multigram scales. Another advantage is the possibility to use benzyl malonates derivatives, for which the corresponding vinylidene malonates decomposed either during column chromatography or distillation. Vinyl phthalimide (**183**) and substituted analogue **331** reacted with the crude mixture of benzyl-substituted vinylidene malonate to afford aminocyclobutanes in good yields and selectivity. It is important to access substituted substrates on a larger scale as they can be interesting precursors for the [4 + 2] annulation, giving information on the stereochemical outcome of such reaction. The benzyl substrate is also an important precursor of β–amino acids as we will see in the next part of this chapter.

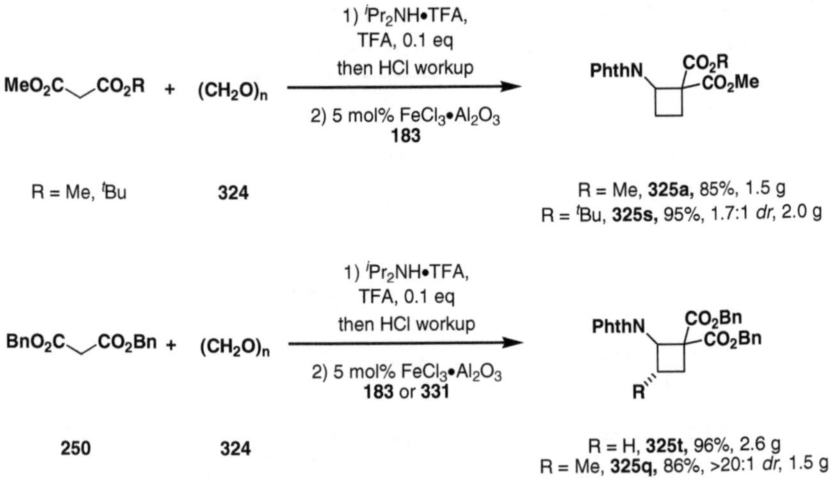

Scheme 3.11 Gram-scale synthesis of aminocyclobutanes

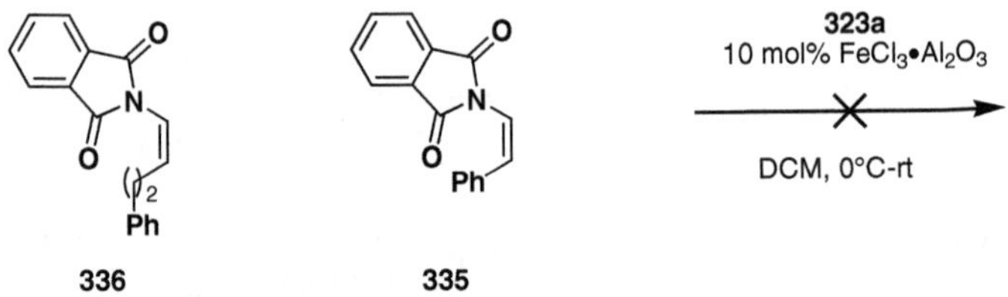

Scheme 3.12 Unreactive (*Z*)-enimides

'In order to better understand the reaction mechanism, it would be important to know if the reaction is stereospecific in relation to the geometry of the enamide or if the observed high *trans*-diastereoselectivity is due only to thermodynamic control. When subjecting (*Z*)-substituted enimides **335** and **336** to the reaction conditions, no conversion was detected, even after prolonged reaction time (Scheme 3.12). The lack of reactivity in the case of the (*Z*) isomer could be tentatively attributed to the loss of hyperconjugation between the nitrogen p orbital electrons and the anti-bonding π* of the olefin.' The lower nucleophilicity of the system is a direct consequence of the allylic strain contained in these molecules.

'To answer the question of stereospecificity, deuterated enimide **355** was consequently synthesized through a three step protocol (Scheme 3.13).' First, synthesis of the phthalimide ynimide **353** was performed via a reported oxidative coupling affording the product in 68 % (a) [18]. The deuterium atom was inserted by transfer from deuterium oxide during alkyne desilylation (b). In order to obtain a good ratio of deuterated alkyne, it was necessary to premix the TBAF with D_2O and to add it

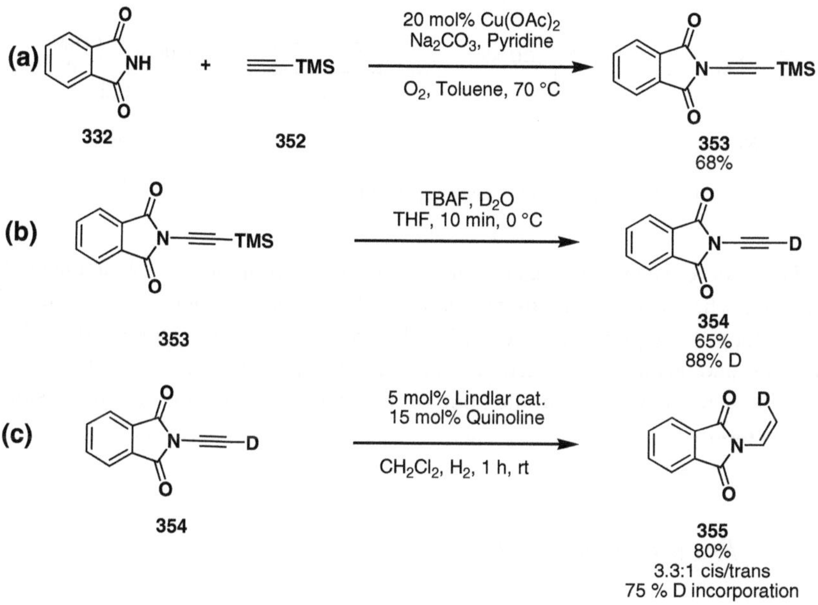

Scheme 3.13 Synthesis of deuterated analogue **355**

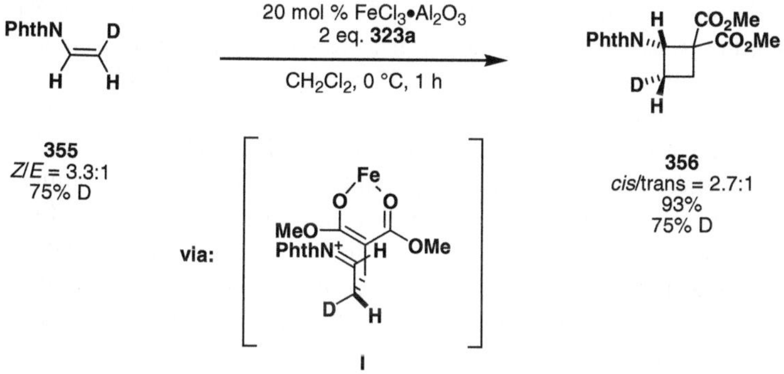

Scheme 3.14 Reaction with deuterated enimide **355**

dropwise to a cold solution of the ynamide in THF/D_2O. Reduction with Lindlar catalyst afforded the product with a 3.3:1 selectivity for the *cis*-product (c).

The deuterated enimide **355** was submitted to the reaction conditions (Scheme 3.14). 'Only a slight loss of stereoinformation was observed during the reaction. This result supported a stepwise mechanism via a zwitterionic intermediate **I**, but also indicated a fast ring-closure, which could compete with single bond rotation.'

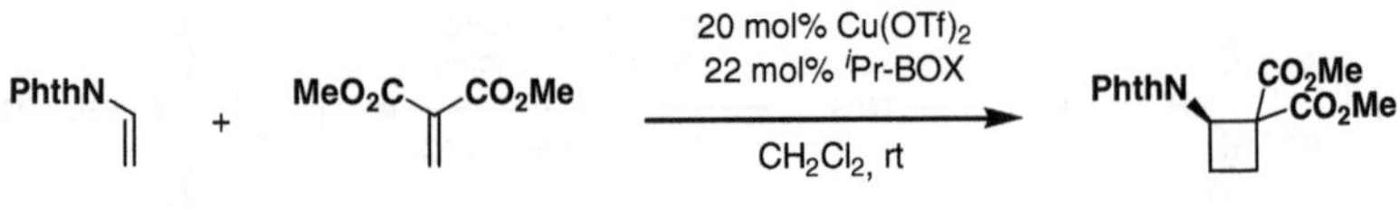

183 323a 325a

Scheme 3.15 Proof of concept for enantioselective [2 + 2]

Proof of concept for enantioselective cycloaddition of the cyclobutane was obtained when vinyl phthalimide (**183**) and **225a** were combined in presence of a chiral copper complex (Scheme 3.15). The crude mixture was submitted to HPLC analysis and enantioenrichment of the product was observed. This preliminary result is very encouraging in the optic of developing an enantioselective version of this method.

3.2.4 Product Modification

In order to show that the synthetized aminocyclobutanes can be used as β-amino acid surrogates, the synthesis of dipeptide **357** was realized (Scheme 3.16). Debenzylation of **325t** occurred very smoothly under hydrogenolysis conditions. Further decarboxylation of the formed diacid was performed using a copper catalyst affording the *trans*-product with a 5:1 diastereomeric ratio. Amide coupling afforded the cyclobutane-glycine dipeptide **357** in very good yield and same *dr* than for the starting material. This protocol offers a very efficient and straightforward method to access peptidic analogues that can be further exploited for applications in peptidomimetics. So far, it was not possible to remove the phthalimide on the dipeptide **357**. However preliminary studies showed that it is possible to remove it on other intermediates. This is currently studied in our group and will be reported in due time. It is also interesting to notice that amide substituted cyclobutanes have found more recently applications in total synthesis of natural products as directing groups for CH activation [19, 20].

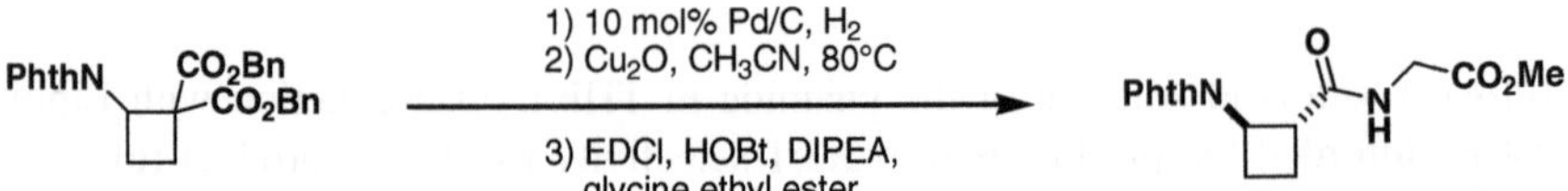

325t

357
78% (3 steps)
5:1 *dr*

Scheme 3.16 Transformation of cyclobutane **325t**

3.3 [4 + 2] Annulation of Aminocyclobutanes

In a similar manner than for aminocyclopropanes, we wanted to test the potential of aminocyclobutanes to react as formal 1,4 dipoles in annulation reactions. On the basis of the excellent results obtained with aminocyclopropanes, and from the observation that no [4 + 2] annulation using anolefin as dipolarophile were reported, we decided to use silyl enol ethers as partners.

3.3.1 *Optimization of the Reaction*

Aminocyclobutane **325a** and TIPS silyl enol ether **219a** were combined in presence of a catalytic amount of Lewis acids (Table 3.4). Metal chloride salts were unable to deliver the product and only starting material was observed in the crude reaction (entries 1–3). Copper (I) and (II) complexes as well as ytterbium triflate were also

Table 3.4 Catalyst optimization for the [4 + 2] annulation

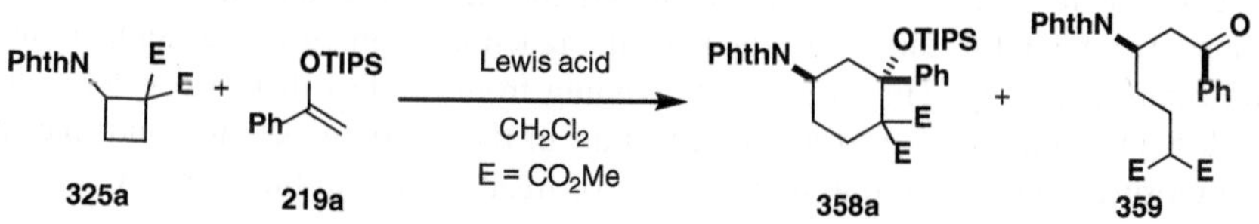

Entry[a]	Lewis acid (20 mol%)	Temperature	Time	**358a/*cis*-358a/359**
1	$InCl_3$	rt	1 h	0/0/1
2	AuCl	rt	1 h	0/0/1
3	$FeCl_3$	rt	1 h	0/0/1
4	CuOTf	rt	1 h	0/0/1
5	$Cu(OTf)_2$	rt	1 h	0/0/1
6	$Yb(OTf)_3$	rt	1 h	0/0/1
7	$Hf(OTf)_4$	rt	1 h	0/1/10
8	$Sn(OTf)_2$	rt	1 h	0/1/10
9	$In(OTf)_3$	rt	1 h	1/3/10
10	$Sc(OTf)_3$	rt	1 h	1/4/10
11	$SnCl_4$	−78 °C	1 h	0/0/1
12	$SnCl_4$	−50 °C	1 h	1/0/0[b]
13	$SnCl_4$	−20 °C	30 min	7/5/1[c]
14	$SnCl_4$	0 °C	30 min	1/3/0[c]
15	$SnCl_4$	−40 °C	1 h	1/0/0[d]

[a]Reaction conditions: 1 eq cyclobutane, 1.5 eq enol ether, 20 mol% catalyst, 1 mL CH_2Cl_2
[b]1:1 mixture **325a/358a**
[c]Vinyl phthalimide **183** resulting from retro [2 + 2] cycloaddition was present as the major product
[d]With 4 Å MS

unable to promote the annulation (entries 4–6). When hafnium or tin (II) catalysts were used, low conversion to the ring-opened side product **359** was observed (entries 7, 8). The product **359** result in a similar manner than in the case of the [3 + 2] annulation from a retro aldol fragmentation or silyl-oxonium hydrolysis after addition of silyl enol ether on the cyclobutane. Indium triflate as well as scandium triflate were more promising as the desired product **358a** could be detected, however as mixtures with **359** (entries 9, 10). The successful conditions for the reaction with aminocyclopropanes ($SnCl_4$, −78 °C) were then tested for this system (entry 11). However, no conversion was observed. It was expected that the reactivity of cyclobutane would be lowered compared to cyclopropane and therefore experiments at higher temperature were done (entries 12–14). At −50 °C, the reaction afforded a clean 1:1 mixture of starting material **325a** and one diastereoisomer of the product **358a** (entry 12). At −20 and 0 °C, the conversion increased but open side product **359** began to appear and diastereoselectivity decreased (entries 13, 14). It is important to notice that in these last two experiments, vinyl phthalimide **183** was detected in amounts superior that 50 %, resulting from the Lewis acid catalyzed retro [2 + 2] reaction. It was later found that in order to prevent ring-opening, 4 Å molecular sieves were required. The reaction was finally performed at −40 °C and full conversion could be observed (entry 15).

Tin tetrachloride was selected as catalyst for the [4 + 2] annulation of cyclobutanes. Compared to the cyclopropanes, the reaction required higher temperatures and catalyst loading. This is probably coming from the fact that the cyclobutanes are less electrophilic due to the lower strain in their carbon skeleton and that the ring closure occurs with one more degree of freedom due to the extra CC bond.

3.3.2 *Synthesis of Substrates*

Silyl enol ethers having different electronic densities were synthesized using common procedure. It involved deprotonation with NaHMDS and quenching with corresponding silyl chloride (Scheme 3.17). Acetophenone derived enol ethers **360a–360d** were obtained in yields ranging from 22 to 71 % after Kugelrohr distillation.

3.3.3 *Scope and Limitations*

Using the optimized conditions, we submitted the aminocyclobutanes and various enol ethers to catalytic amounts of tin tetrachloride (Table 3.5). On a preparative scale, the reaction with TIPS enol ether afforded one diastereoisomer of the product **358a** in 82 % yield (entry 1). Changing the ester groups to benzyl on the cyclobutane afforded **358b** in 64 % yield (entry 2). Switching to TBS enol ether **219b** allowed us to isolate the product **358c** with a better yield (entry 3). In this case, the

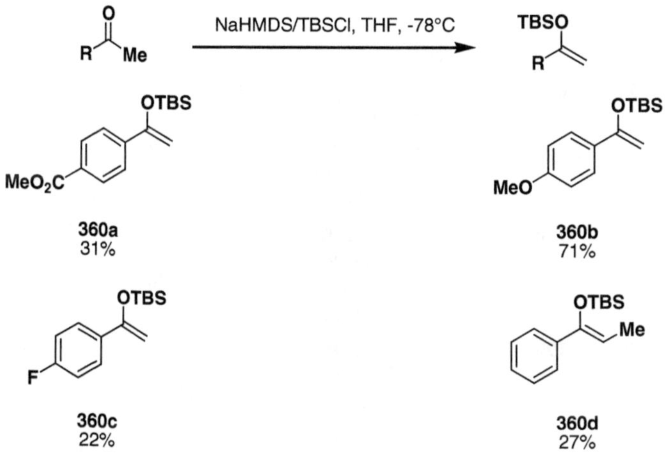

Scheme 3.17 Synthesis of enol ethers

Table 3.5 [4 + 2] annulation of aminocyclobutanes

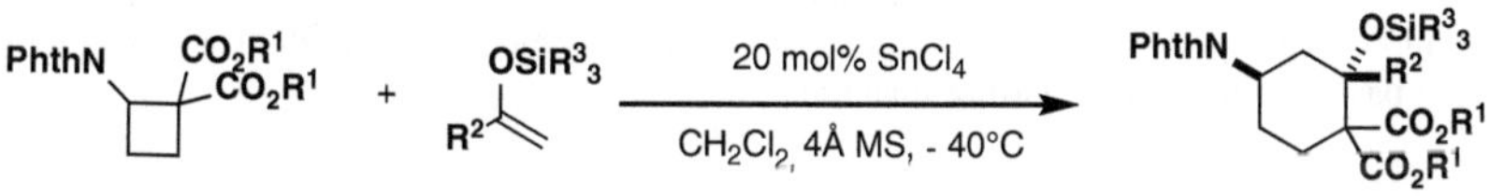

Entry	Substrate		Product	Yield (%)[a]
1	**325a**	**219a**	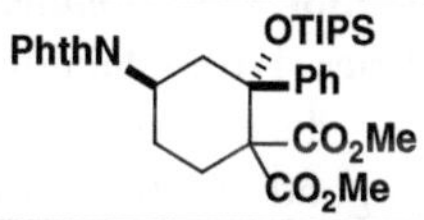 **358a**	82 (*dr* > 20:1)
2	**325u**	**219a**	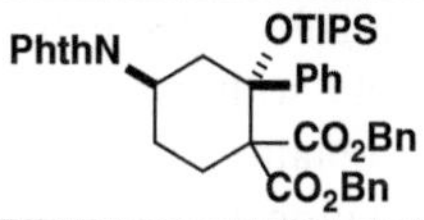 **358b**	64 (*dr* > 20:1)
3	**325a**	**219b**	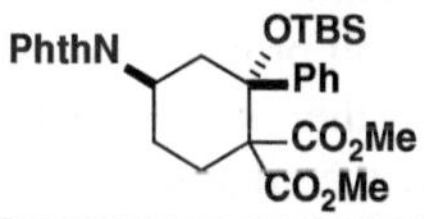 **358c**	95 (*dr* > 20:1)
4	**325a**	**360b**	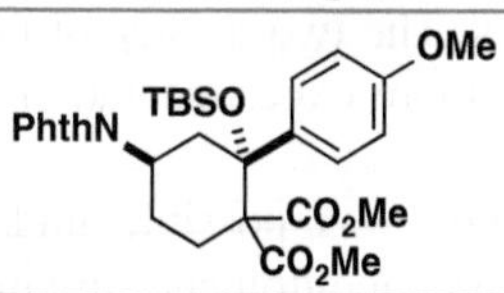 **358d**	90 (*dr* > 20:1)
5	**325a**	**360c**	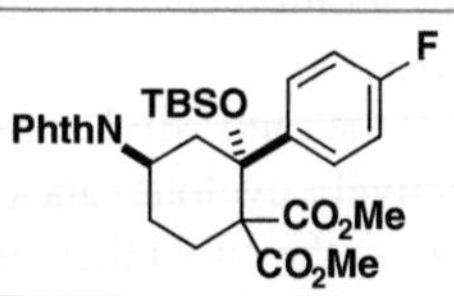 **358e**	99 (*dr* > 20:1)

(continued)

Table 3.5 (continued)

Entry	Substrate		Product		Yield (%)[a]
6	**325a**	**360a**	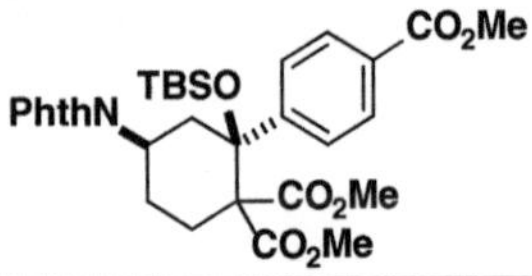	**358f**	82 (*dr* = 2:1)
7	**325u**	**219b**	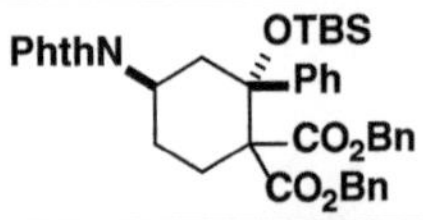	**358g**	72 (*dr* > 20:1)
8	**325l**	**219b**	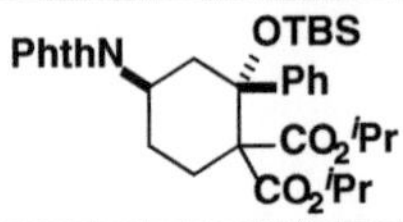	**358h**	55 (68 brsm) (*dr* > 20:1)
9	**325a**	**360d**	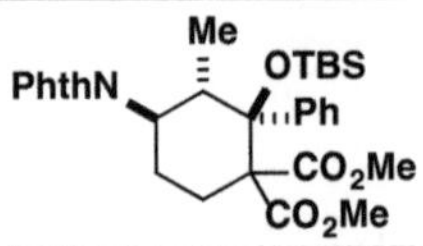	**358i**	90 (*dr* = 2:1)

[a]Reaction conditions: 0.20 mmol cyclobutane, 0.30 mmol enol ether, 0.04 mmol catalyst, 4 Å MS in dichloromethane, −40°C, under nitrogen. Isolated yields after column chromatography. Brsm = based on recovered starting material

reaction tolerated electron-rich and -poor substituents on the aromatic part of the enol ether: Aminocyclohexanes **358d–358f** were obtained in excellent yields (90–99 %) as a single diastereoisomer (entries 4–6). Changing the ester substituents to isopropyl or benzyl afforded one diastereoisomer of the corresponding six-membered rings **358g** and **358h** in good yields (entries 7, 8). In contrast to the [3 + 2] annulation, the reaction with a trisubstituted enol ether proceeded with good stereoselectivity for the β-substituent but with scrambling during the ring closure (entry 9). The product **358i** was thus obtained in good yield as a 2:1 mixture of diastereoisomers at the center in α-position to the malonate. The relative stereochemistry was established as *trans* for all products using 2D-NMR analysis except for **358f** and **358i**. In these two cases the major diastereoisomers possessed a *cis* relationship between phthalimide and the silanol. At this stage of the project it is difficult to rationalize these results. The two diastereoisomers represent the kinetic and thermodynamic products but more experimentation will be required to find a mechanistic explanation.

Deprotection of the phthalimide was possible under aminolysis conditions. Using ethylene diamine in isopropanol/toluene mixture, the free amine was obtained in 87 % yield (Scheme 3.18).

The annulation of aminocyclobutane and silyl enol ether was successfully implemented in order to access selectively cyclohexanes substituted with nitrogen. So far, the reaction tolerated variation on the enol ether substituent and on the esters of the cyclobutanes. Current investigations in our group are done toward the use of

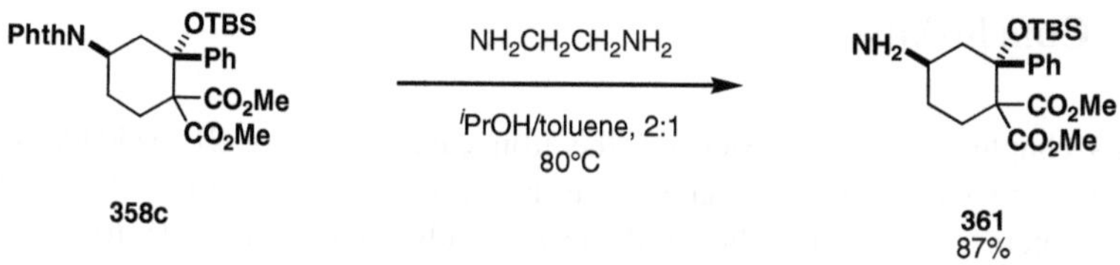

Scheme 3.18 Phthalimide deprotection of **358c**

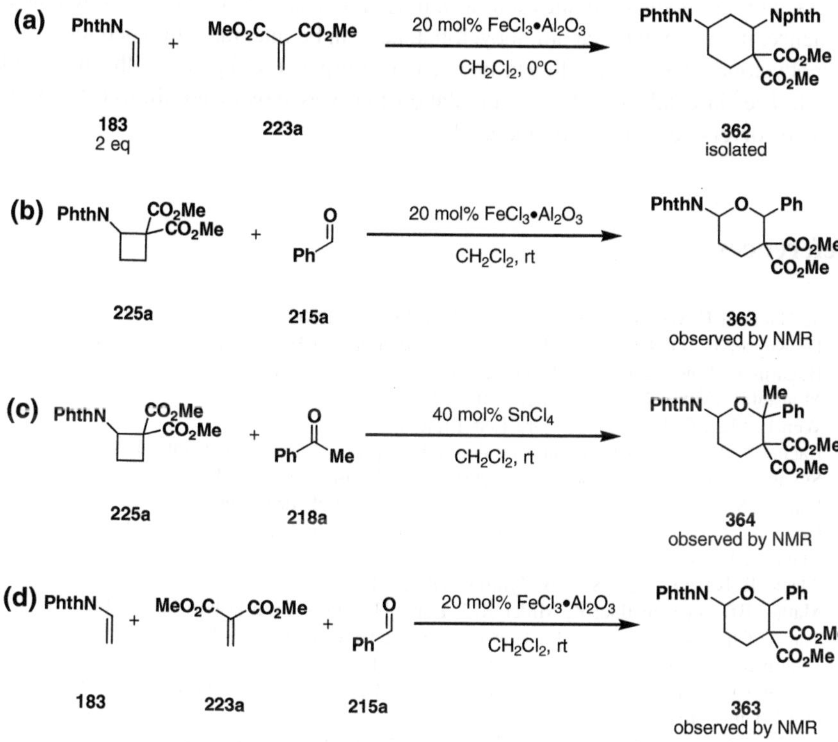

Scheme 3.19 Preliminary results of annulations with enimides and carbonyls

polysubstituted aminocyclobutanes as 1,4 dipoles which should extend the structural diversity accessed by these reactions.

During the optimization of the reaction, side product **362** was observed and identified as the result of [2 + 2 + 2] cycloaddition (a, Scheme 3.19). Such process has been rarely observed but would be very valuable as it gives a direct access to polysubstituted diaminocyclohexanes [21, 22]. Preliminary results indicate that the use of carbonyls as dipolarophiles is able to deliver the corresponding aminotetrahydropyran (b and c) and that [2 + 2 + 2] cycloaddition seems also possible (d). The development of these methods is currently studied in our group.

3.4 Conclusion

In this chapter, we have described the iron catalyzed [2 + 2] cycloaddition of enimides and vinylidene malonates. The reaction tolerated various substituents on both partners and afforded the products in yields ranging from 45 to 96 % and diastereoselectivity up to 20:1. A carbonucleoside analogue could be directly synthesized using a thymine derivative. The synthesis of very reactive methylidene malonates was improved, and simplified. A sequential procedure was developed to access the aminocyclobutanes on a multigram scale. Following a three step sequence, it was shown that aminocyclobutanes can be integrated into peptides. The aminocyclobutanes could efficiently react as formal 1,4 dipoles with silyl enol ethers. The tin-catalyzed [4 + 2] annulation that was developed allowed the synthesis of diverse cyclohexylamine analogues.

References

1. de Nanteuil F, Waser J (2013) Angew Chem Int Ed 52:9009
2. De Keyser JL, De Cock CJC, Poupaert JH, Dumont P (1988) J Org Chem 53:4859
3. Bugarin A, Jones KD, Connell BT (2010) Chem Comm 46:1715
4. Moustafa MMAR, Pagenkopf BL (2010) Org Lett 12:4732
5. Wen L, Shen Q, Lu L (2010) Org Lett 12:4655
6. Racine S, de Nanteuil F, Serrano E, Waser J (in press) Angew Chem Int Ed
7. Stojanovic A, Renaud P, Schenk K (2004) Helv Chim Acta 81:268
8. Lam PYS, Vincent G, Bonne D, Clark CG (2003) Tetrahedron Lett 44:4927
9. Gooßen LJ, Blanchot M, Brinkmann C, Gooßen K, Karch R, Rivas-Nass A (2006) J Org Chem 71:9506
10. Alacid E, Nájera C (2008) Adv Synth Catal 350:1316
11. Mather BD, Viswanathan K, Miller KM, Long TE (2006) Prog Polym Sci 31:487
12. Tan X, Chu CK, Boudinot FD (1999) Adv Drug Delivery Rev 39:117
13. Dhami KS, Stothers JB (1965) Can J Chem 43:510
14. Aguilera J, Cobos JA, Gutiérrez-Abad R, Acosta C, Nolis P, Illa O, Ortuño RM (2013) Eur J Org Chem 2013:3494
15. Proskow S, Simmons HE, Cairns TL (1966) J Am Chem Soc 88:5254
16. Savino TG, Konicki Chenard L, Swenton JS (1983) Tetrahedron Lett 24:4055
17. Gauzy C, Saby B, Pereira E, Faure S, Aitken DJ (2006) Synlett 2006:1394
18. Alford JS, Davies HML (2012) Org Lett 14:6020
19. Parella R, Gopalakrishnan B, Babu SA (2013) J Org Chem 78:11911
20. Gutekunst WR, Baran PS (2014) J Org Chem 79:2430
21. Hall HK, Ykman P (1975) J Am Chem Soc 97:800
22. Avenoza A, Busto JH, Canal N, Garcia JI, Jimenez-Oses G, Peregrina JM, Perez-Fernandez M (2007) New J Chem 31:224

Chapter 4
Conclusions and Outlook

The research work presented in this thesis was aimed at the development of intermolecular ring-opening reaction of donor-acceptor amino-substituted cycloalkanes. In this context, the use of aminocyclopropanes as reactive formal 1,3 dipoles was studied first. It was shown that the structure of both the donor and the acceptor sides had to be finely tuned, in order to generate a stable yet reactive formal 1,3 dipole precursor. A first generation of cyclopropanes bearing a phthalimide protecting group on the nitrogen and a vicinal malonic ester was used for this purpose (Fig. 4.1).

With these substrates, we managed to develop the first diastereo—and enantiospecific [3 + 2] annulation with silyl and alkyl enol ethers as dipolarophiles. The reaction was catalyzed by a tin-based Lewis acid. The cyclopentylamines were obtained in high yields and stereoselectivity. The reaction tolerated a wide range of electron-poor and—rich aromatic and alkyl-substituted silyl enol ethers. Hindered tetrasubstituted olefins could react stereoselectively under these conditions. Alkyl enol ethers were also successfully used for the first time as partner in [3 + 2] annulation with DA cyclopropanes. Deprotection of the phthalimide protecting group and access to cyclopentenylamines were successfully implemented.

In order to expand the diversity of accessible chemical space using [3 + 2] annulations, we then used carbonyls as dipolarophiles. We developed two protocols for the annulation between aminocyclopropanes and aldehydes or ketones. The resulting aminotetrahydrofurans, which are found in numerous natural molecules, were obtained in very good yields and diastereoselectivity. Aldehydes could be reacted stereoselectively using a user-friendly iron catalyst. Ketones required a different set of conditions ($SnCl_4$, −78 °C) in order to achieve a very high degree of facial selectivity. As in the case of aldehydes, racemization of the cyclopropane occurred in presence of the Lewis acid, the products could not be accessed enantiospecifically. In contrast, enantiopure aminotetrahydrofurans bearing quaternary centers could be obtained when ketones were employed.

The reaction between aminocyclopropanes and indoles in presence of a Lewis acid afforded the linear ring-opened products. The reaction required a modified acceptor functionality on the cyclopropane: The alkyl esters had to be changed to trifluoroethyl esters in order to access the GABA analogues in good yields. The reaction tolerated a wide range of indoles derivative as well as carbo—and

F. de Nanteuil, *Synthesis and Reactivity of Donor-Acceptor Substituted Aminocyclopropanes and Aminocyclobutanes*, Springer Theses,
DOI 10.1007/978-3-319-23006-1_4

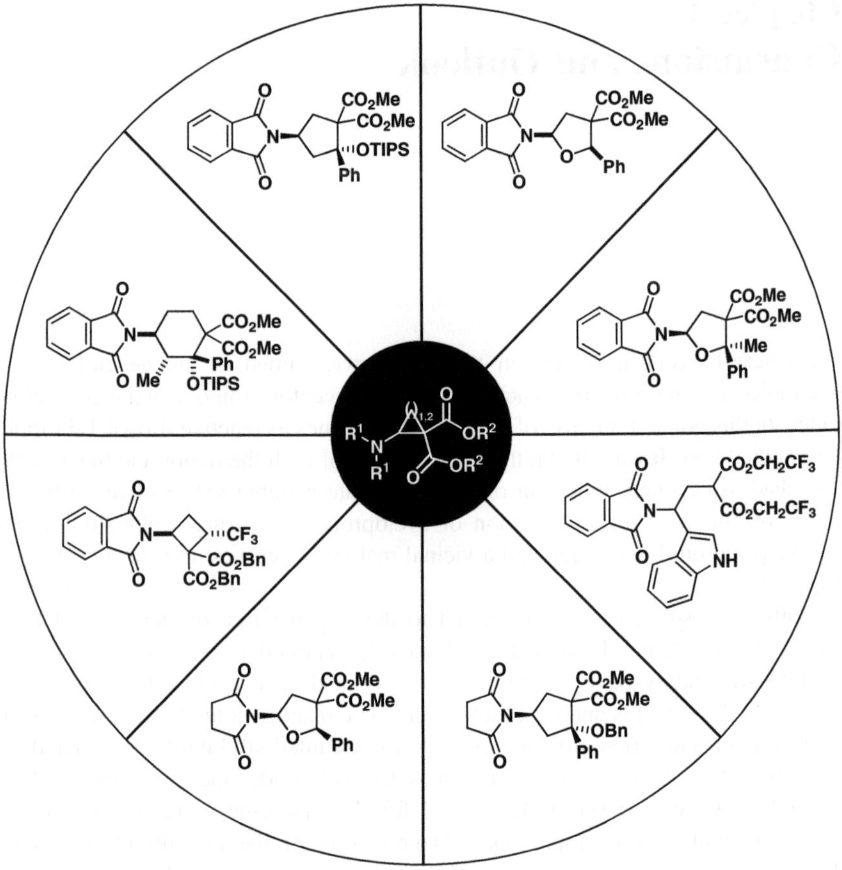

Fig. 4.1 Products obtained by intermolecular ring-opening reaction of donor-acceptor amino-substituted cycloalkanes

heteroaromatics as nucleophiles. The synthetic modifications of the products afforded building blocks that can provide a useful platform for the construction of bio-relevant molecules.

The use of cyclopropanes bearing a succinimide protecting group on the nitrogen allowed us to develop a DyKAT for the enantioselective synthesis of cyclopentylamines and aminotetrahydrofurans. Both products were obtained with an enantiomeric ratio up to 98:2 using a commercially available ^{t}Bu-BOX ligand and a copper-perchlorate salt. As silyl enol ethers were not effective for this process, they were replaced by alkyl enol ethers.

In a second project, the reactivity of strained aminocycloalkanes was extended to aminocyclobutanes. At the beginning of this study, it became obvious that a new synthetic method had to be developed to access DA aminocyclobutanes analogues. We discovered that using again an iron catalyst, the reaction between vinyl imides and alkylidene malonates afforded the aminocyclobutanes in very good yields up to 96 %. The reaction tolerated substitution on each position of the cyclobutane and afforded stereospecifically the four-membered rings. Carbonucleoside analogues could be directly accessed using vinyl thymine derivatives as starting materials. In order to understand more precisely the selectivity, the [2 + 2] cycloaddition outcome was studied using a labelled compound. This experiment highlighted the stereospecificity of the reaction as the product was obtained with retention of the stereogenic information. Important efforts were focused on simplifying the access to the starting materials and to provide multigram quantities of the products. The protocols involved the formation of sensitive alkylidene malonates and its sequential reaction as a crude mixture with the catalyst and the vinyl imide.

Similarly to aminocyclopropanes, the cyclobutane analogues reacted in presence of a Lewis acid and a nucleophilic olefin via [4 + 2] annulation to afford the corresponding six-membered rings. The reaction tolerated silyl enol ethers derived from acetophenone with electron-rich as well as electron-poor functionalities on the aromatic ring.

In conclusion, through these 4 years of Ph.D., I managed to design and optimize a toolbox of reactions making use of strained amino substituted carbocycles. It was the first time that DA aminocyclopropanes and aminocyclobutanes were used as formal dipoles in intermolecular [*n* + 2] annulations to access cyclopentylamines and cyclohexylamines. The reactions developed allowed so far the efficient and stereoselective synthesis of nitrogen substituted four-, five- and six-membered rings. These methods are currently finding applications in diverse projects that are ongoing in our group.

Outlook:

This work has set the base for several projects inside as well as outside our group.

- A chemical diversity project has been funded by the NCCR (National Center of Competence in Research) in order to expand the methodology of [3 + 2] annulations to the synthesis of carba—and heteronucleoside analogues. By using the protocols developed for the annulations with silyl enol ethers, aldehydes and ketones, it was possible to access a range of promising DNA-based scaffolds such as **365–366** (Fig. 4.2) [1].
- A Ph.D. grant has been funded by the SNSF (Swiss National Science Foundation) in order to continue the studies on [4 + 2] as well as [2 + 2 + 2]

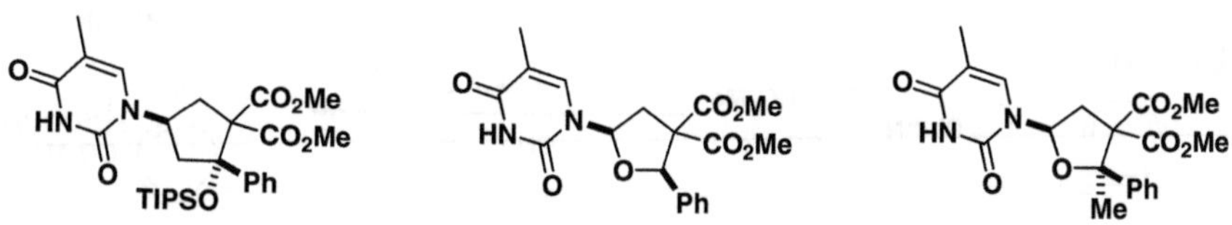

365 **366** **367**

Fig. 4.2 Synthesis of nucleoside analogues

annulations. The scope of the silyl enol ether should be extended to polysubstituted aminocyclobutanes in order to study amongst other the stereoselectivity of the reaction. The reaction with aldehydes, ketones and enimides have to be optimized as they provide efficient access to bio-relevant scaffolds. Finally, the development of [2 + 2 + 2] annulations would represent a very efficient way to synthesize small libraries of amino substituted six-membered rings.

- In 2013, two methods appeared consecutively reporting the [8 + 3] annulations of DA cyclopropanes. The first study conducted by Fernandez, Sierra and co-workers is using the first generation aminocyclopropane **217** in combination with substituted tropones (Scheme 4.1) [2]. The reaction requires catalytic amounts of tin and afforded regio—and diastereoselectively the cyclized product in good yields. The second report by Carretero and co-workers is studying various DA cyclopropanes with unsubstituted tropones in presence of a Nickel Lewis acid [3]. They also successfully included the phthalimide substituted cyclopropane **217** in their study (Scheme 4.1) and obtained a different isomer as they used a non-substituted tropone.
- Johnson and co-workers have shown that cyclopropane **217** did not react under their conditions for the FC alkylation of indoles (Scheme 1.42).

Due to their specific structures, the synthetic potential of DA aminocyclopropanes and aminocyclobutanes is very high and many strategies can be explored based on their reactivity. In the future, I can envisage the use of transition metal catalysts, the development of enantioselective synthesis of amino substituted cyclic structures, the synthesis of peptidomimetics or natural products as examples where aminocyclopropanes and aminocyclobutanes could find applications. The field of annulations with aminocyclopropanes and aminocyclobutanes could be enlarge by the use of new partners such as imines or alkynes, which would give access to important molecular scaffolds.

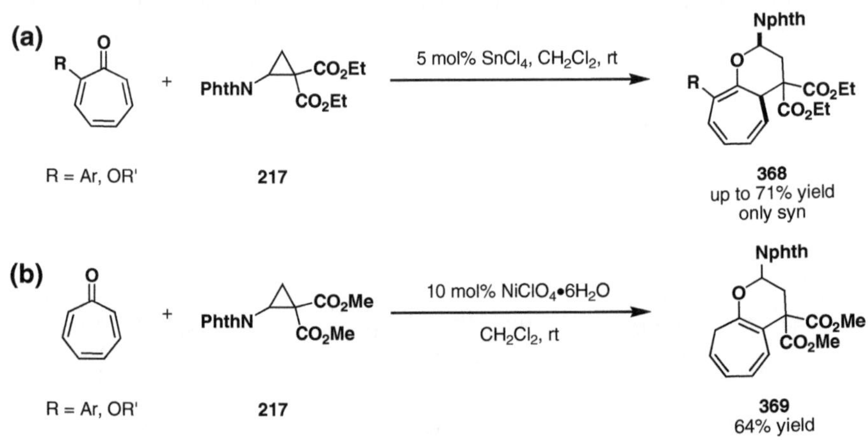

Scheme 4.1 [8 + 3] annulations with aminocyclopropanes

References

1. Racine S, de Nanteuil F, Serrano E, Waser J (in press) Angew Chem Int Ed
2. Rivero AR, Fernández I, Sierra MÁ (2013) Org Lett 15:4928
3. Tejero R, Ponce A, Adrio J, Carretero JC (2013) Chem Comm 49:10406

Chapter 5
Experimental Part

5.1 General Methods

All reactions were carried out in oven-dried glassware under nitrogen or argon atmosphere with magnetic stirring, unless stated otherwise. THF, Et_2O, CH_3CN, toluene, hexane and dichloromethane were dried by passage over activated alumina under nitrogen atmosphere (water content <30 ppm, Karl-Fischer titration) on an Innovative Technology Solvent Delivery System. All chemicals were purchased from Strem, Acros, Aldrich, Fluka, VWR, Aplichem or Merck and used as such unless stated otherwise. Chromatographic purification was performed as flash chromatography using Macherey-Nagel silica 40–63, 60 Å or using aluminium oxide, basic, Brockmann I purchased from Acros, using the solvents indicated as eluent with 0.1–0.5 bar pressure. For flash chromatography, previously distilled technical grade solvents were used. TLC was performed on Merck silica gel 60 F254 TLC glass plates or aluminium plates and visualized with UV light, and by permanganate stain, CAN stain or p-anisaldehyde stain followed by heating. Melting points were measured on a calibrated Büchi B-540 melting point apparatus using open glass capillaries. ^{1}H-NMR spectra were recorded on a Brucker DPX-400 400 MHz spectrometer in $CDCl_3$, DMSO-d_6, CD_2Cl_2 or CD_3OD, all signals are reported in ppm with the internal chloroform signal at 7.26 ppm, the internal DMSO signal at 2.50 ppm, the internal CD_2Cl_2 signal at 5.31 ppm, or the internal MeOD signal at 3.30 ppm as standard. The data are reported as follows: (s = singlet, d = doublet, t = triplet, q = quadruplet, qi = quintet, m = multiplet or unresolved, br = broad signal, coupling constant(s) in Hz, integration, interpretation) or (s = singlet, d = doublet, t = triplet, q = quadruplet, qi = quintet, m = multiplet or unresolved, br = broad signal, integration, coupling constant(s) in Hz, interpretation). ^{13}C-NMR spectra were recorded with ^{1}H-decoupling on a Brucker DPX-400 100 MHz spectrometer in $CDCl_3$, DMSO-d_6, CD_2Cl_2 or CD_3OD, all signals are reported in ppm with the internal chloroform signal at 77.0 ppm, the internal DMSO signal at 39.5 ppm, the internal CD_2Cl_2 signal at 53.5 ppm or the internal MeOD

F. de Nanteuil, *Synthesis and Reactivity of Donor-Acceptor Substituted Aminocyclopropanes and Aminocyclobutanes*, Springer Theses,
DOI 10.1007/978-3-319-23006-1_5

signal at 49.0 ppm as standard. Infrared spectra were recorded on a JASCO FT-IR B4100 spectrophotometer with an ATR PRO410-S and a ZnSe prisma and are reported as cm^{-1} (w = weak, m = medium, s = strong, sh = shoulder). High resolution mass spectrometry measurements were performed by the mass spectrometry service of ISIC at the EPFL on a MICROMASS (ESI) Q-TOF Ultima API. HPLC measurements were done on a JASCO HPLC system with an AS2055 Autosampler, a PU 2089 Pump, a UV 2075 detector using a CHIRALPAK IC, IB, IF or IA column from DAICEL Chemical. Optical rotations were measured on a polarimeter using a 10 cm cell with a Na 589 nm filter. The specific solvents and concentrations (in g/100 mL) are indicated.

5.2 [3 + 2] Annulation with Enols Ethers

Benzyl vinylcarbamate (195)

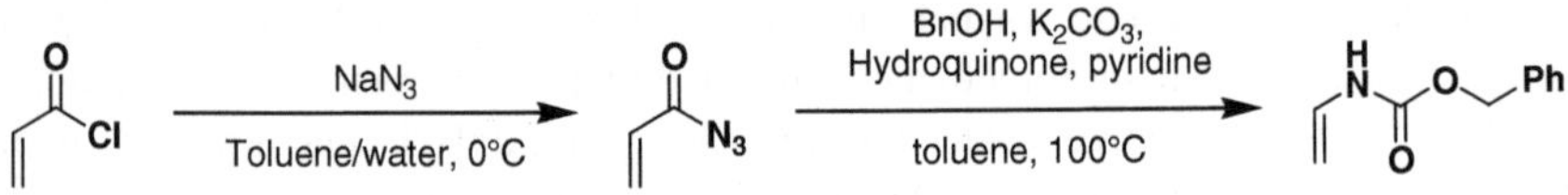

193 194 195

Following a reported procedure [1], a solution of acryloyl chloride **193** (8.97 mL, 110 mmol, 1.00 equiv) and TBAI (2.04 g, 5.52 mmol, 0.05 equiv) in toluene (100 mL) was added dropwise to a solution of sodium azide (8.60 g, 132 mmol, 1.20 equiv) in H_2O (100 mL) at 0 °C. The biphasic reaction was stirred at 0 °C for 5 h then the layers were separated. The organic layer was dried over Na_2SO_4 and filtered. The filtrate was added carefully into a solution of benzyl alcohol (120 g, 1.10 mol, 10 equiv), pyridine (523 mg, 6.63 mmol, 0.06 equiv) and hydroquinone (607 mg, 5.52 mmol, 0.05 equiv) at 85 °C. The reaction was stirred for 1 h at 100 ° C then distilled under reduced pressure. The crude distillate (140 °C, 0.13 mmHg) was recrystallized in cyclohexane to give benzyl-vinylcarbamate (**195**) as colorless crystals (4.82 g, 27.2 mmol, 25 %).

R_f 0.7 (1:9 AcOEt/PET). **^{1}H NMR** (400 MHz, $CDCl_3$) δ 7.39–7.31 (m, 5 H, Ar), 6.78–6.64 (m, 1 H, CH vinyl), 6.52–6.36 (m, 1 H, NH), 5.15 (s, 2 H, CH_2 benzyl), 4.48 (d, 1 H, J = 15.7 Hz, CH_2 vinyl), 4.30 (d, 1 H, J = 8.6 Hz, CH_2 vinyl).

The characterization data for **195** corresponded to the reported values [1].

3-Vinyloxazolidin-2-one (185)

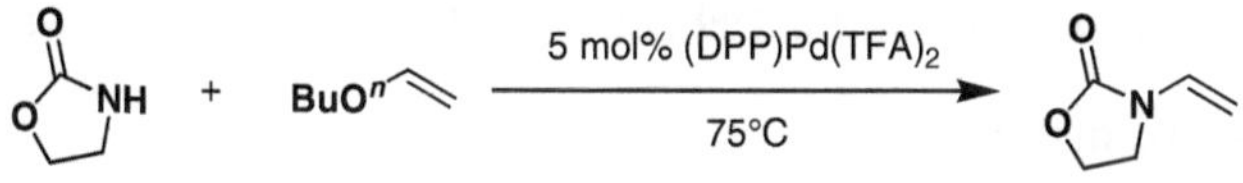

196 197 185

Following a reported procedure [2], oxazolidinone **196** (2.00 g, 23.0 mmol, 1 eq), butyl vinyl ether **197** (30 mL, 10 eq) and palladium trifluoroacetate-phenanthroline complex (609 mg, 1.15 mmol, 0.05 eq) were added in a flask under air. The reaction is heated without condenser at 75 °C for 4 h and concentrated under reduced pressure in order to remove the major part of butyl vinyl ether. Purification by column chromatography (9:1 diethyl ether/petroleum ether) afforded 3-Vinyloxazolidin-2-one (**185**) (2.42 g, 21.4 mmol, 93 %) as a yellow oil.

R_f 0.43 (1:9 PET/Et_2O). **^{1}H NMR** (400 MHz, $CDCl_3$) δ 6.85 (dd, 1 H, J = 15.8, 9.0 Hz, CH vinyl), 4.48–4.39 (m, 3 H, CH_2 + CH_2 vinyl), 4.28 (dd, 1 H, J = 15.8, 1.1 Hz, CH_2 vinyl), 3.73–3.66 (m, 2 H, CH_2).

The characterization data for **185** corresponded to the reported values [3].

Dimethyl 2-diazomalonate (199)

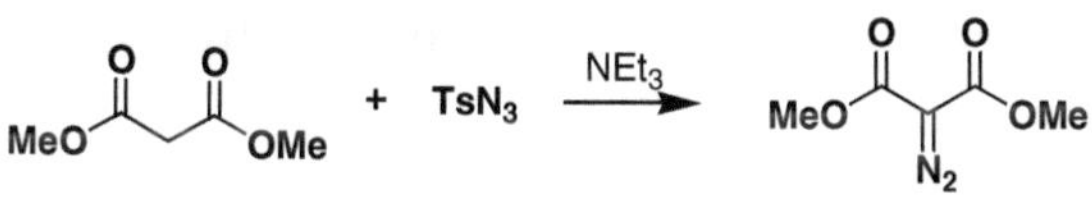

198 370 199

Following a reported procedure [4], dimethylmalonate **198** (7.93 mL, 69.7 mmol, 1 eq), triethylamine (10.6 mL, 76.6 mmol, 1.1 eq) and tosyl azide **370** (15.1 g, 76.6 mmol, 1.1 eq) were dissolved in acetonitrile (100 mL). The solution was stirred at 23 °C for 20 h. The solution was concentrated under reduced pressure and partitioned between dichloromethane and water. The layers were separated and the aqueous layer was extracted with dichloromethane (x1). The organic layers were combined and dried over $MgSO_4$. The crude was first filtered over a plug of silica gel (PET/Et_2O 1/1) to remove most of the tosylamide formed during the reaction. Then, purification by column chromatography (PET/Et_2O 1/1) afforded dimethyl 2-diazomalonate (**199**) (10.4 g, 65.5 mmol, 94 %) as a yellow oil which solidified under storage at 4 °C.

R_f 0.32 (1:1 PET/Et_2O). **^{1}H NMR** (400 MHz, $CDCl_3$) δ 3.87 (s, 1 H). **^{13}C NMR** (101 MHz, $CDCl_3$) δ 161.2, 52.4.[1]

The characterization data for **199** corresponded to the reported values [4].

Diethyl 2-diazomalonate (201)

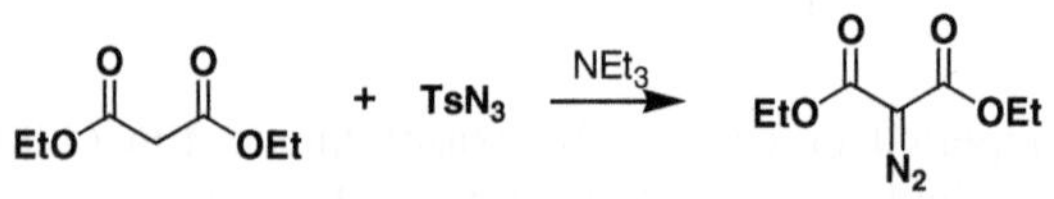

200 **370** **201**

Following a reported procedure [4], diethylmalonate **200** (2.17 mL, 14.3 mmol, 1 eq), triethylamine (2.18 mL, 15.7 mmol, 1.1 eq) and tosyl azide **370** (3.09 g, 15.7 mmol, 1.1 eq) were dissolved in acetonitrile (20 mL). The solution was stirred at 23 °C for 20 h. The solution was concentrated under reduced pressure and partitioned between DCM and water. The layers were separated and the aqueous layer was extracted with dichloromethane (x1). The organic layers were combined and dried over $MgSO_4$. The crude was first filtered over a plug of silica gel (PET/Et_2O 1/1) to remove most of the tosylamide formed during the reaction. Then, purification by column chromatography (PET/Et_2O 1/1) afforded dimethyl 2-diazomalonate (**201**) (2.3 g, 12 mmol, 86 %) as a yellow oil which solidified under storage at 4 °C.

R_f 0.48 (1:1 PET/Et_2O). **^{1}H NMR** (400 MHz, $CDCl_3$) δ 4.27 (q, 4 H, $J = 7.1$ Hz), 1.28 (t, 6 H, $J = 7.1$ Hz). **^{13}C NMR** (101 MHz, $CDCl_3$) δ 161.0, 61.5, 14.3 (See footnote 1).

The characterization data for **201** corresponded to the reported values [5].

General procedure 1 (GP1) for Rhodium catalyzed cyclopropanation

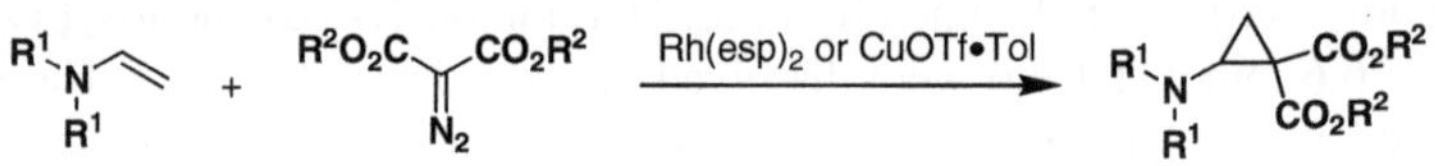

Following a modified procedure [6], bis[rhodium(α,α,α′,α′-tetramethyl-1,3-benzenedipropionic acid)] (0.1 mol%) is weighted in the glovebox. The flask is closed with a septum and put under N_2 atmosphere. A solution of vinyl imide (1 eq) in dry dichloromethane is added and the resulting green suspension is cooled down to 0 °C with an ice/water bath. A solution of diazo compound (1.1 eq)

[1]The diazo carbon could not be detected.

in dichloromethane is added over 5 min. When the addition is complete, the reaction is allowed to warm to room temperature. After 5 h at room temperature, the solvent is removed under reduced pressure and the crude is directly purified by column chromatography.

GP2 for Copper catalyzed cyclopropanation

Copper-(I)-trifluoromethanesulfonate-toluene complex (2.5 mol%) is weighted in the glovebox. The flask is closed with a septum, protected from light and put under N_2 atmosphere. A solution of vinyl imide (1 eq) in dry dichloromethane is added. The diazo compound (4 eq) is added via a syringe pump over 18 h. When the addition is complete, the solvent is removed under reduced pressure and the crude is directly purified by column chromatograpy.

Trans-ethyl 2-((benzyloxy)carbonyl)amino)cyclopropanecarboxylate (203)

Following GP1, using *N*-Vinyl-*O*-benzyl Urethane (50 mg, 0.28 mmol, 1 eq), ethyl-diazoacetate (64 mg, 0.56 mmol, 2 eq) and 2.1 mg (0.0028 mmol, 1 mol%) of bis[rhodium(α,α,α′,α′-tetramethyl-1,3-benzenedipropionic acid)]. After column chromatography (100 % Heptane + 1 % NEt_3 to 9:1 Heptane/Ethyl Acetate + 1 % NEt_3), *trans*-ethyl 2-((benzyloxy)carbonyl)amino)cyclopropanecarboxylate (**203**) (27 mg, 0.10 mmol, 36 % yield) is obtained as a colorless oil.

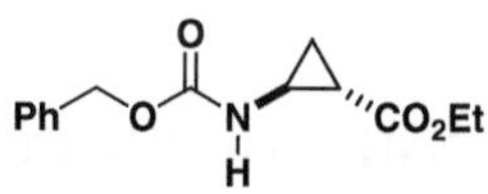

R_f 0.5 (7:3 Petroleum ether/Ethyl Acetate) **^{1}H NMR** (400 MHz, $CDCl_3$) δ 7.42–7.32 (m, 5 H, Ar), 5.13 (m, 2 H, Ar–CH_2), 4.16 (m, 2 H, O–CH_2–CH_3, 3.12 (m, 1 H, N–C–H), 1.79 (m, 1 H, CH_2), 1.45 (m, 1 H, CH_2), 1.31–1.25 (t, 3 H, J = 7.5 Hz, O–CH_2–CH_3), 1.14–1.19 (m, 1 H, CH_2).

The characterization data for **203** corresponded to the reported values [7].

Ethyl 2-(2-oxooxazolidin-3-yl)cyclopropanecarboxylate (204)

Following GP2, using *N*-vinyl-oxazolidinone (2.42 g, 21.4 mmol, 1 eq), ethyl diazoacetate (9.0 mL, 86 mmol, 4 eq) and copper-(I)-trifluoromethanesulfonate-toluene complex (277 mg, 0.534 mmol, 2.5 mol%). After purification by column chromatography (9/1 Petroleum Ether/Ethyl Acetate to 100 Ethyl Acetate), ethyl 2-(2-oxooxazolidin-3-yl)cyclopropanecarboxylate (**204**) (3.21 g, 16.1 mmol, 75 % yield) was obtained as a yellow oil.

1:1 Mixture of *cis* and *trans* non-separable diastereoisomers.

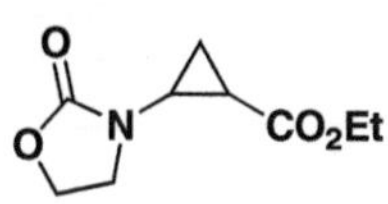

R$_f$ 0.3 (98:2 DCM/MeOH). ^{1}**H NMR** (400 MHz, $CDCl_3$) δ 4.35–4.26 (m, 4 H, CH_2 oxazolidinone), 4.19–4.11 (m, 4 H, O–CH_2–CH_3), 3.74 (dd, 1 H, J = 7.9, 7.9 Hz, N–C–H), 3.60–3.50 (m, 3 H, N–C–H and CH_2 oxazolidinone), 3.07–3.00 (m, 2 H, CH_2 oxazolidinone), 2.02 (m, 1 H, $CHCO_2Et$), 1.90 (m, 1 H, $CHCO_2Et$), 1.55 (q, 1 H, J = 6.0 Hz, CH_2), 1.52–1.43 (m, 2 H, CH_2), 1.38–1.31 (m, 1 H, CH_2), 1.30–1.24 (m, 6 H, O–CH_2–CH_3). ^{13}**C NMR** (101 MHz, $CDCl_3$) δ 171.9, 170.5, 158.5, 157.9, 62.3, 61.9, 61.0, 61.0, 45.7, 45.7, 34.4, 33.0, 20.7, 20.6, 15.0, 14.2, 12.8.[2] **IR** 2986 (w), 2985 (w), 2917 (w), 1751 (s), 1723 (s), 1722 (s), 1426 (m), 1409 (m), 1185 (s), 1040 (m). **HRMS (ESI)** calcd for $C_9H_{14}NO_4^+$ $[M + H]^+$ 200.0917; found 200.0908.

Trans-ethyl 2-(2-oxopyrrolidin-1-yl)cyclopropanecarboxylate (205)

Following GP2, using commercially available *N*-vinyl-pyrrolidone, (250 mg, 2.25 mmol, 1 eq), ethyl diazoacetate (0.95 mL, 8.9 mmol, 4 eq) and copper-(I)-trifluoromethanesulfonate-toluene complex (29 mg, 0.056 mmol, 2.5 mol%). After purification by column chromatography (9/1 Petroleum Ether/Ethyl Acetate to 100 Ethyl Acetate), *trans*-ethyl 2-(2-oxopyrrolidin-1-yl)cyclopropanecarboxylate (**205**) (240 mg, 1.22 mmol, 54 % yield) is obtained as a yellow oil.

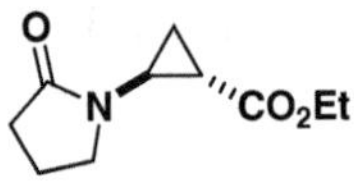

R$_f$ 0.2 (Ethyl acetate). ^{1}**H NMR** (400 MHz, $CDCl_3$) δ 4.15 (qd, 2 H, J = 7.2, 1.5 Hz, O–CH_2–CH_3), 3.31 (t, 2 H, J = 7.2 Hz, CH_2 pyrrolidone), 3.20–3.15 (m, 1 H, N–C–H), 2.39 (t, 2 H, J = 8.0 Hz, CH_2 pyrrolidone), 2.05–1.96 (m, 2 H, CH_2 pyrrolidone), 1.88–1.82 (m, 1 H, $CHCO_2Et$), 1.51–1.38 (m, 2 H, CH_2), 1.27 (t, 3 H, J = 7.1 Hz, O–CH_2–CH_3). ^{13}**C NMR** (101 MHz, $CDCl_3$) δ 175.9, 172.3, 60.9, 47.2, 34.1, 31.7, 19.7, 18.0, 14.2, 14.1.

The characterization data for **205** corresponded to the reported values [8].

Trans-ethyl 2-(1,3-dioxoisoindolin-2-yl)cyclopropanecarboxylate (206)

Following GP2, using *N*-vinyl-phthalimide (500 mg, 2.04 mmol, 1 eq), ethyl diazoacetate (0.860 mL, 8.16 mmol, 4 eq) and copper-(I)-trifluoromethanesulfonate-toluene complex (26.4 mg, 0.0510 mmol, 2.5 mol%). After column chromatograpy (8:2 Petroleum Ether/Ethyl Acetate) a yellow solid is obtained and recrystallized in Petroleum Ether/Ethyl Acetate to give *trans*-ethyl 2-(1,3-dioxoisoindolin-2-yl)cyclopropanecarboxylate (**206**) (205 mg, 0.791 mmol, 40 % yield) as a colorless solid.

[2]One carbon aliphatic of one diastereoisomer could not be resolved.

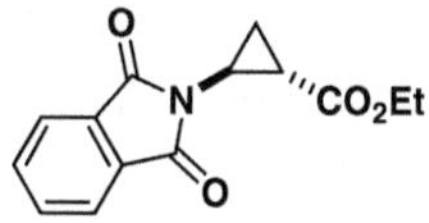

R_f 0.36 (8:2, Petroleum Ether/Ethyl acetate). **^{1}H NMR** (400 MHz, $CDCl_3$) δ 7.87–7.81 (m, 2 H, Phth), 7.77–7.71 (m, 2 H, *Phth*), 4.22 (q, 2 H, J = 7.2 Hz, O–CH_2–CH_3), 3.35–3.28 (m, 1 H, N–C–H), 2.25–2.20 (m, 1 H, C*H*CO_2Et), 1.76 (dt, 1 H, J = 9.3, 5.5 Hz, CH_2), 1.65 (dt, 1 H, J = 8.1, 5.9 Hz, CH_2), 1.32 (t, 3 H, J = 7.2 Hz, O–CH_2–CH_3). **^{13}C NMR** (101 MHz, $CDCl_3$) δ 172.1, 168.0, 134.3, 131.6, 123.4, 61.1, 29.5, 20.0, 14.2, 13.6.

The characterization data for **206** corresponded to the reported values [8].

Diethyl 2-(1,3-dioxoisoindolin-2-yl)cyclopropane-1,1-dicarboxylate (217)

Following GP1, using commercially available *N*-vinyl-phthalimide (3.0 g, 18 mmol, 1 eq) diethyl-2-diazomalonate (4.0 g, 21 mmol, 1.2 eq) and bis[rhodium(α,α,α′,α′-tetramethyl-1,3-benzenedipropionic acid)] (14 mg, 0.018 mmol, 0.1 mol%). After column chromatography (100 % Heptane + 1 % NEt_3 to 9:1 Heptane/Ethyl Acetate + 1 % NEt_3) diethyl 2-(1,3-dioxoisoindolin-2-yl)cyclopropane-1,1-dicarboxylate (**217**) (5.4 g, 16 mmol, 90 % yield) is obtained as a white solid.

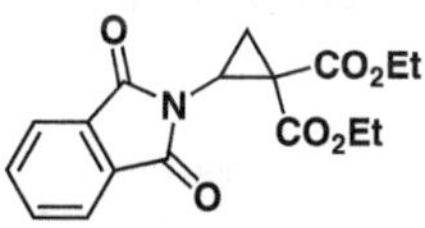

R_f 0.36 (6:4, Hexane/Ethyl acetate) **Mp** 93.1 °C. **^{1}H NMR** (400 MHz, $CDCl_3$) δ 7.85 (m, 2 H, Phth), 7.74 (m, 2 H, Phth), 4.30 (m, 2 H, OCH_2), 4.07 (m, 2 H, OCH_2), 3.71 (dd, 1 H, J = 8.5, 6.6 Hz, N–C–H), 2.74 (t, 1 H, J = 6.5 Hz, CH_2), 2.02 (dd, 1 H, J = 8.5, 6.4 Hz, CH_2), 1.34 (t, 3 H, J = 7.1 Hz, CH_3), 1.12 (t, 3 H, J = 7.1 Hz, CH_3). **^{13}C NMR** (101 MHz, $CDCl_3$) δ 168.2, 167.8, 166.4, 134.3, 131.6, 123.4, 62.0, 61.8, 34.7, 33.5, 19.2, 14.1, 13.8. **IR** 2985 (w), 2938 (w), 2907 (w), 1783 (m), 1719 (s), 1614 (w), 1393 (s), 1321 (m), 1218 (s), 1133 (m), 719 (s). **HRMS (ESI)** calcd for $C_{17}H_{18}NO_6^+$ $[M + H]^+$ 332.1129; found 332.1135.

Dimethyl 2-(1,3-dioxoisoindolin-2-yl)cyclopropane-1,1-dicarboxylate (223)

Following GP1, using commercially available *N*-vinyl-phthalimide (2.5 g, 14 mmol, 1 eq) diethyl-2-diazomalonate (2.5 g, 15 mmol, 1.1 eq) and bis[rhodium (α,α,α′,α′-tetramethyl-1,3-benzenedipropionic acid)] (10.9 mg, 0.014 mmol, 0.1 mol%). After column chromatography (9:1 Hexane/Ethyl Acetate to 7:3 Hexane/Ethyl Acetate) diethyl 2-(1,3-dioxoisoindolin-2-yl) cyclopropane-1,1-dicarboxylate (**223**) (3.40 g, 11.2 mmol, 78 % yield) is obtained as a white solid.

The two enantiomers were separated by HPLC using Chiralpack IA column (0.46 × 25 cm), 95:5 Hexane/Isopropanol, 1 mL/min. tr_1 = 23.9 min, $[\alpha]_D^{25.0}$ 122 (c = 0.69, $CHCl_3$). tr_2 = 27.1 min,

$[\alpha]_D^{25.0}$ − 122 (c = 0.1, $CHCl_3$)

Preparative HPLC: Column IA 20 × 250 mm, 98:2 Hexane/Isopropanol, 16 mL/min. tr_1 = 45 min, tr_2 = 75 min.

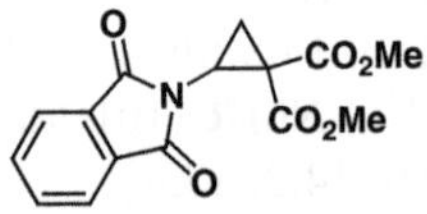

R_f 0.27 (6:4, Hexane/Ethyl acetate). **Mp** 124.6 – 125 °C. **^{1}H NMR** (400 MHz, $CDCl_3$) δ 7.86 (m, 2 H, Phth), 7.75 (m, 2 H, Phth), 3.85 (s, 3 H, OMe), 3.72 (dd, 1 H, *J* = 8.5, 6.6 Hz, N–CH), 3.64 (s, 3 H, OMe), 2.73 (dd, 1 H, *J* = 6.5, 6.5 Hz, CH_2), 2.06 (dd, 1 H, *J* = 8.5, 6.4 Hz, CH_2). **^{13}C NMR** (101 MHz, $CDCl_3$) δ 168.5, 167.8, 166.9, 134.3, 131.4, 123.5, 53.1, 53.0, 34.9, 33.1, 19.6. **IR** 2956 (w), 1783 (w), 1727 (s), 1468 (w), 1439 (w), 1399 (m), 1329 (m), 1294 (m), 1222 (m), 1134 (w), 909 (w), 876 (w), 720 (m). **HRMS (ESI)** calcd for $C_{15}H_{14}NO_6^+$ $[M + H]^+$ 304.0816; found 304.0804.

General procedures for the synthesis of Silyl Enol Ethers

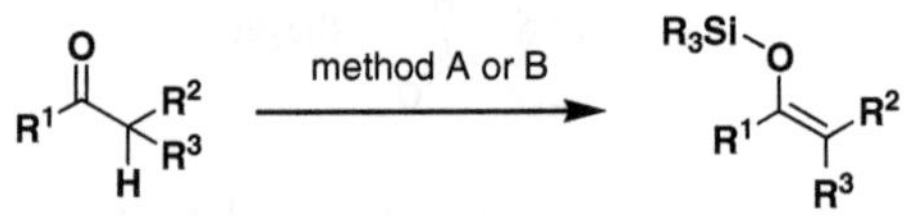

GP3

The ketone (1 eq) in anhydrous THF is added in an oven-dried flask sealed with a septum and under N_2 atmosphere. The solution is cooled down to −78 °C and a 2 M solution of NaHMDS (1.22 eq) is added dropwise. The cold bath is removed and the pale yellow solution is stirred for 1 h at room temperature. The reaction is cooled again at −78 °C and the corresponding silyl chloride (1.2 eq) is added dropwise. The reaction is stirred at room temperature for 5 h after what the solvent is directly removed under reduced pressure. The resulting orange oil is purified by plug or by column chromatography on triethylamine-deactivated silica (100 % Hexane).

GP4

The ketone (1 eq) in anhydrous dichloromethane is added in an oven-dried flask sealed with a septum and under N_2 atmosphere. Triethylamine (1.8 eq) and then triisopropylsilyl-trifluoromethanesulfonate (1.2 eq) are added at room temperature. After 2 h, the reaction is diluted with dichloromethane and washed with a solution of sat. NH_4Cl. The aqueous layer is extracted three times with ethyl acetate. The organic layers are collected, washed with brine, dried over anhydrous sodium sulfate, filtered and concentrated under reduced pressure. The crude oil is purified by plug or column chromatography on triethylamine deactivated silica (100 % Hexane).

Triisopropyl(1-phenylvinyl)oxy)silane (219a)

Following GP3, starting from 580 mg of acetophenone (4.82 mmol), 1.03 g (3.72 mmol, 77 % yield) of a colorless oil was obtained.

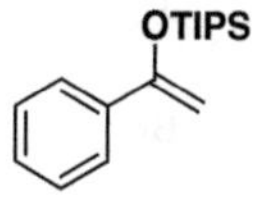

^{1}H NMR (400 MHz, $CDCl_3$) δ 7.69–7.65 (m, 2 H, Ar), 7.38–7.29 (m, 3 H, Ar), 4.85 (d, 1 H, J = 1.8 Hz, C=CH_2), 4.41 (d, 1 H, J = 1.8 Hz, C=CH_2), 1.39–1.27 (m, 3 H, $SiCH(CH_3)_2$), 1.19–1.13 (m, 18 H, $SiCH(CH_3)_2$). **^{13}C NMR** (101 MHz, $CDCl_3$) δ 156.2, 138.0, 128.2, 128.1, 125.4, 90.0, 18.2, 12.9.

The characterization data for **219a** corresponded to the reported values [9].

***Tert*-butyldimethyl(1-phenylvinyl)oxy)silane (219b):**

Following GP3, starting from 580 mg of acetophenone (4.82 mmol), 960 mg (4.10 mmol, 85 % yield) of a colorless oil was obtained.

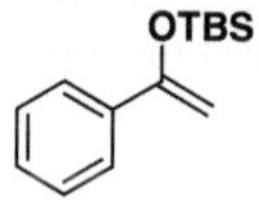

^{1}H NMR (400 MHz, $CDCl_3$) δ 7.65–7.60 (m, 2 H, Ar), 7.39–7.29 (m, 3 H, Ar), 4.89 (d, 1 H, J = 1.7 Hz, C=CH_2), 4.42 (d, 1 H, J = 1.7 Hz, C=CH_2), 1.00 (s, 9 H, Si$(CH_3)_2C(CH_3)_3$), 0.21 (s, 5 H, $Si(CH_3)_2C(CH_3)_3$).**^{13}C NMR** (101 MHz, $CDCl_3$) δ 156.0, 137.8, 128.2, 128.1, 125.3, 90.9, 25.9, 18.4, −4.6.

The characterization data for **219b** corresponded to the reported values [9].

Triethyl(1-phenylvinyl)oxy)silane (219c):

Following GP3, starting from 580 mg of acetophenone (4.82 mmol), 869 mg (3.71 mmol, 77 % yield) of a colorless oil were obtained.

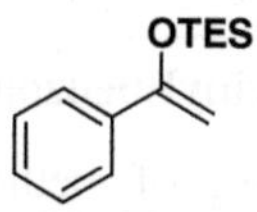

^{1}H NMR (400 MHz, $CDCl_3$) δ 7.62 (m, 2 H, Ar), 7.31 (m, 3 H, Ar), 4.88 (d, 1 H, J = 1.7 Hz, C=CH_2), 4.43 (d, 1 H, J = 1.7 Hz, C=CH_2), 1.01 (m, 9 H, $SiCH_2CH_3$), 0.77 (m, 6 H, $SiCH_2CH_3$). **^{13}C NMR** (101 MHz, $CDCl_3$) δ 156.0, 137.8, 128.2, 128.1, 125.3, 90.4, 6.8, 5.1.

The characterization data for **219c** corresponded to the reported values [9].

(1-(2-Bromophenyl)vinyl)oxy)triisopropylsilane (219d)

Following GP3, starting from 0.50 mL of 2′-bromoacetophenone (3.7 mmol), 1.15 g (3.26 mmol, 88 % yield) of a colorless oil were obtained.

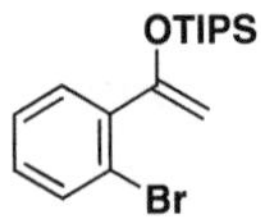

R*f* 0.80 (9:1, Hexane/Ethyl acetate). **^{1}H NMR** (400 MHz, $CDCl_3$) δ 7.60 (ddd, 1 H, *J* = 8.0, 1.2, 0.2 Hz, Ar), 7.46 (dd, 1 H, *J* = 7.6, 1.8 Hz, Ar), 7.29 (td, 1 H, *J* = 7.4, 1.3 Hz, Ar), 7.16 (ddd, 1 H, *J* = 8.0, 7.4, 1.8 Hz, Ar), 4.63 (d, 1 H, *J* = 1.5 Hz, C=CH_2), 4.55 (d, 1 H, *J* = 1.5 Hz, C=CH_2), 1.32–1.21 (m, 3 H, SiCH $(CH_3)_2$), 1.16–1.11 (m, 18 H, SiCH$(CH_3)_2$). **^{13}C NMR** (101 MHz, $CDCl_3$) δ 156.2, 140.5, 133.3, 130.3, 129.1, 126.9, 121.5, 95.4, 18.1, 12.7. **IR** 2945 (w), 2891 (w), 2866 (m), 1705 (w), 1626 (w), 1589 (w), 1466 (m), 1317 (s), 1024 (s), 883 (m). **HRMS (ESI)** calcd for $C_{17}^{79}BrH_{28}OSi^+$ $[M + H]^+$ 355.1087; found 355.1075.

Triisopropyl(1-(naphthalen-1-yl)vinyl)oxy)silane (219e)

Following GP4, starting from 500 mg of 1′-acetonaphthone (2.94 mmol), 865 mg (2.65 mmol, 91 % yield) of a colorless oil were obtained.

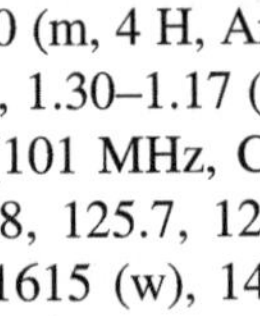

R*f* 0.66 (100 % Hexane). **^{1}H NMR** (400 MHz, $CDCl_3$) δ 8.42–8.37 (m, 1 H, Ar), 7.88–7.80 (m, 2 H, Ar), 7.59–7.40 (m, 4 H, Ar), 4.76 (d, 1 H, *J* = 0.8 Hz, C=CH_2), 4.63 (d, 1 H, *J* = 0.8 Hz, C=CH_2), 1.30–1.17 (m, 3 H, SiCH$(CH_3)_2$), 1.12–1.05 (m, 18 H, SiCH$(CH_3)_2$). **^{13}C NMR** (101 MHz, $CDCl_3$) δ 157.6, 137.7, 133.7, 131.0, 128.4, 128.1, 126.6, 125.9, 125.8, 125.7, 125.0, 95.9, 18.0, 12.7. **IR** 2944 (m), 2892 (w), 2866 (m), 1626 (w), 1615 (w), 1464 (w), 1304 (s), 1016 (s), 778 (s). **HRMS (ESI)** calcd for $C_{21}H_{31}OSi^+$ $[M + H]^+$ 327.2139; found 327.2132.

Methyl 4-(1-(triisopropylsilyl)oxy)vinyl)benzoate (219f)

Following GP4, starting from 500 mg of methyl 4-acetylbenzoate (2.81 mmol), 856 mg (2.56 mmol, 91 % yield) of a colorless oil were obtained.

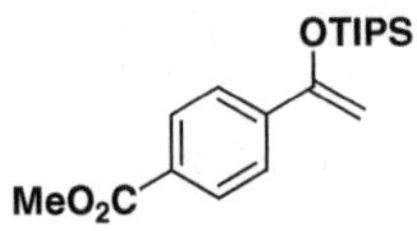

R_f 0.5 (9:1, Hexane/Ethyl acetate). **^{1}H NMR** (400 MHz, $CDCl_3$) δ 8.00 (m, 2 H, Ar), 7.70 (m, 2 H, Ar), 4.96 (d, 1 H, J = 2.0 Hz, C=CH_2), 4.53 (d, 1 H, J = 2.0 Hz, C=CH_2), 3.92 (s, 3 H, CO_2CH_3), 1.37–1.26 (m, 3 H, $SiCH(CH_3)_2$), 1.18–1.12 (m, 18 H, $SiCH(CH_3)_2$). **^{13}C NMR** (101 MHz, $CDCl_3$) δ 166.9, 155.3, 142.3, 129.6, 129.5, 125.2, 92.1, 52.1, 18.1, 12.8. **IR** 3669 (w), 3522 (w), 2959 (m), 2955 (m), 2943 (m), 2867 (m), 1722 (s), 1679 (m), 1282 (s), 1113 (s). **HRMS** (ESI) calcd for $C_{19}H_{31}O_3Si^+$ $[M + H]^+$ 335.2037; found 335.2032.

(1-(4-Fluorophenyl)vinyl)oxy)triisopropylsilane (219g)

Following GP4, starting from 500 mg of 4′-fluoroacetophenone (3.62 mmol), 776 mg (2.64 mmol, 73 % yield) of a colorless oil were obtained.

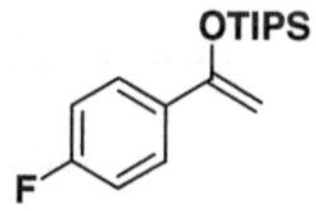

^{1}H NMR (400 MHz, $CDCl_3$) δ 7.61 (m, 2 H, Ar), 7.00 (m, 2 H, Ar), 4.77 (d, 1 H, J = 2.0 Hz, C=CH_2), 4.39 (d, 1 H, J = 1.9 Hz, C=CH_2), 1.43–1.29 (m, 3 H, SiCH $(CH_3)_2$), 1.22–1.14 (m, 18 H, $SiCH(CH_3)_2$). **^{13}C NMR** (101 MHz, $CDCl_3$) δ 162.8 (d, J = 248 Hz), 155.3, 134.1 (d, J = 3 Hz), 127.1 (d, J = 8.1 Hz), 114.9 (d, J = 21.5 Hz), 89.7 (d, J = 2 Hz), 18.1, 12.8.

The characterization data for **219g** corresponded to the reported values [9].

4-(1-(Triisopropylsilyl)oxy)vinyl)benzonitrile (219h)

Following GP4, starting from 0.50 mg of 4-acetylbenzonitrile (3.4 mmol), 1.0 g (3.3 mmol, 97 % yield) of a pale orange oil were obtained.

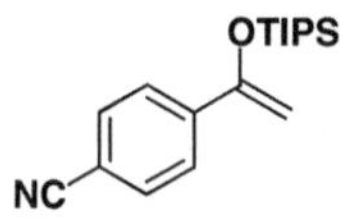

R_f 0.55 (9:1, Hexane/Ethyl acetate). **^{1}H NMR** (400 MHz, $CDCl_3$) δ 7.73 (m, 2 H, Ar), 7.62 (m, 2 H, Ar), 4.96 (d, 1 H, J = 2.3 Hz, C=CH_2), 4.57 (d, 1 H, J = 2.3 Hz, C=CH_2), 1.39–1.26 (m, 3 H, $SiCH(CH_3)_2$), 1.18–1.12 (m, 18 H, SiCH $(CH_3)_2$). **^{13}C NMR** (101 MHz, $CDCl_3$) δ 154.5, 142.2, 132.0, 125.8, 119.0, 111.5, 92.8, 18.1, 12.7. **IR** 2946 (m), 2945 (m), 2892 (w), 2868 (m), 2229 (w), 1614 (w), 1464 (w), 1317 (s), 1302 (m), 1111 (s), 1015 (s). **HRMS (ESI)** calcd for $C_{18}H_{28}NOSi^+$ $[M + H]^+$ 302.1935; found 302.1935.

Triisopropyl(1-(4-methoxyphenyl)vinyl)oxy)silane (219i)

Following GP3starting from 724 mg of 4′-methoxyacetophenone (4.82 mmol), 1.3 g (4.2 mmol, 87 % yield) of a colorless oil were obtained.

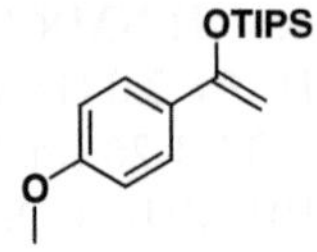

R_f 0.75 (9:1, Hexane/Ethyl acetate). **^{1}H NMR** (400 MHz, $CDCl_3$) δ 7.65 (m, 2 H, Ar), 6.92 (m, 2 H, Ar), 4.81 (d, 1 H, J = 1.8 Hz, C=CH_2), 4.39 (d, 1 H, J = 1.8 Hz, C=CH_2), 3.86 (s, 3 H, OCH_3), 1.43–1.31 (m, 3 H, $SiCH(CH_3)_2$), 1.23–1.17 (m, 18 H, $SiCH(CH_3)_2$). **^{13}C NMR** (101 MHz, $CDCl_3$) δ 159.7, 155.9, 130.7, 126.7, 113.4, 88.4, 55.2, 18.2, 12.9. **IR** 2944 (w), 2891 (w), 2867 (m), 2838 (w), 1608 (m), 1510 (s), 1464 (m), 1249 (s), 1175 (s), 1014 (s), 734 (s). **HRMS (ESI)** calcd for $C_{18}H_{31}O_2Si^+$ $[M + H]^+$ 307.2088; found 307.2076.

(3,4-Dihydronaphthalen-1-yl)oxy)triisopropylsilane (219j)

Following GP3, starting from 540 mg of alpha-tetralone (3.69 mmol), 973 mg (3.22 mmol, 87 % yield) of a colorless oil were obtained.

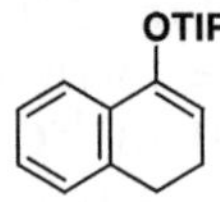

^{1}H NMR (400 MHz, $CDCl_3$) δ 7.58 (dd, J = 7.4, 1.0 Hz, Ar), 7.29–7.12 (m, 3 H, Ar), 5.18 (t, 1 H, J = 4.6 Hz, C=CH), 2.76 (t, 2 H, J = 7.7 Hz, Ar–CH_2), 2.37–2.29 (m, 2 H, Ar–CH_2–CH_2), 1.37–1.26 (m, 3 H, $SiCH(CH_3)_2$), 1.19–1.13 (m, 18 H, $SiCH(CH_3)_2$). **^{13}C NMR** (101 MHz, $CDCl_3$) δ 148.5, 137.2, 133.8, 127.2, 126.9, 126.2, 122.0, 103.8, 28.3, 22.3, 18.2, 12.9.

The characterization data for **219j** corresponded to the reported values [10].

(Cyclohex-1-en-1-yloxy)triisopropylsilane (219k)

Following GP3, starting from 474 mg of cyclohexanone (4.82 mmol), 1.22 g (4.82 mmol, 99 % yield) of a colorless oil were obtained.

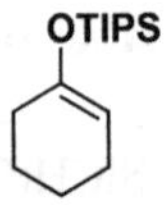

^{1}H NMR (400 MHz, $CDCl_3$) δ 4.88 (m, 1 H, C=CH), 2.11–1.99 (m, 4 H, –CH_2–), 1.72–1.64 (m, 2 H, –CH_2–), 1.57–1.50 (m, 2 H, –CH_2–), 1.22–1.04 (m, 21 H, TIPS). **^{13}C NMR** (101 MHz, $CDCl_3$) δ 150.6, 103.6, 30.0, 23.9, 23.3, 22.4, 18.0, 12.7.

The characterization data for **219k** corresponded to the reported values [11].

(Z)-Triisopropyl(1-phenylprop-1-en-1-yl)oxy)silane (219l)

Following GP3, starting from 500 mg of propiophenone (11.2 mmol), 3.2 g (11 mmol, 98 % yield) of a colorless oil were obtained.

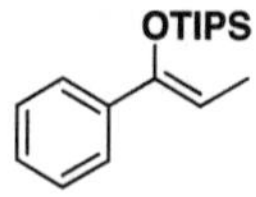

R_f 0.85 (9:1, Hexane/Ethyl acetate). **^{1}H NMR** (400 MHz, $CDCl_3$) δ 7.53–7.47 (m, 2 H, Ar), 7.36–7.25 (m, 3 H, Ar), 5.06 (q, 1 H, $J = 6.9$ Hz, $C{=}CH_2$), 1.85–1.80 (d, 3 H, $J = 6.9$ Hz, CH_3), 1.19–1.05 (m, 21 H, TIPS). **^{13}C NMR** (101 MHz, $CDCl_3$) δ 151.3, 140.4, 127.9, 127.3, 126.0, 105.2, 18.0, 13.6, 11.8. **IR** 3058 (w), 2945 (s), 2918 (m), 2867 (s), 1694 (m), 1653 (w), 1464 (m), 1325 (s), 1080 (s), 1064 (s), 883 (s). **HRMS** (ESI) calcd for $C_{18}H_{31}OSi^+$ $[M + H]^+$ 291.2139; found 291.2145.

Triisopropyl(2-methyl-1-phenylprop-1-en-1-yl)oxy)silane (219m)

Following GP3, starting from 548 mg of propiophenone (3.70 mmol), 1.1 g (3.6 mmol, 98 % yield) of a colorless oil were obtained.

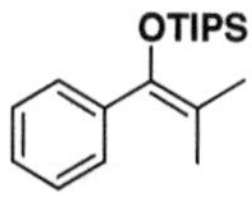

R_f 0.85 (9:1, Hexane/Ethyl acetate). **^{1}H NMR** (400 MHz, $CDCl_3$) δ 7.39–7.25 (m, 5 H, Ar), 1.88 (s, 3 H, CH_3), 1.63 (s, 3 H, CH_3), 1.08–0.96 (m, 21 H, TIPS). **^{13}C NMR** (101 MHz, $CDCl_3$) δ 144.2, 139.4, 129.5, 127.6, 127.2, 111.6, 20.0, 18.2, 17.9, 13.2. **IR** 2966 (m), 2962 (m), 2960 (m), 2945 (m), 2924 (m), 2867 (m), 1671 (w), 1464 (w), 1162 (s), 832 (s). **HRMS** (ESI) calcd for $C_{19}H_{33}OSi^+$ $[M + H]^+$ 305.2295; found 305.2308.

2-(1-(Triisopropylsilyl)oxy)vinyl)pyridine (219n)

Following GP3, starting from 448 mg of 1-(pyridin-2-yl)ethan-1-one (3.70 mmol), 600 mg (2.16 mmol, 58 % yield) of a colorless oil were obtained.

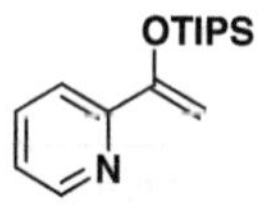

^{1}H NMR (400 MHz, $CDCl_3$) δ 8.57 (m, 1 H, Ar), 7.71 (m, 2 H, Ar), 7.19 (m, 1 H, Ar), 5.67 (m, 1 H, $C{=}CH_2$), 4.58 (m, 1 H, $C{=}CH_2$), 1.35 (m, 3 H, TIPS), 1.16 (m, 18 H, TIPS).

The characterization data for **219n** corresponded to the reported value [12].

Triisopropyl(pent-2-en-3-yloxy)silane (219o)

Following GP3, starting from 318 mg of 3-pentanone (3.69 mmol), 549 mg (2.26 mmol, 61 % yield) of a colorless oil were obtained containing a 8.3:1 mixture of *Z*/*E* isomers.

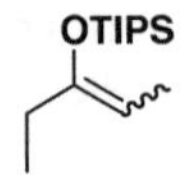

^{1}H NMR (400 MHz, $CDCl_3$) δ 4.56 (q, 0.12 H, J = 6.9 Hz, C=CH E isomer), 4.46 (qt, J = 6.5, 1.0 Hz, C=CH Z isomer), 2.16–2.07 (m, 2.2 H, CH_2 Z + E isomers), 1.64–1.59 (m, 3 H, $CHCH_3$ Z isomer), 1.57–1.54 (d, 0.4 H, J = 6.8 Hz, E isomer), 1.25–1.02 (m, 27 H, TIPS + CH_2CH_3 Z + E isomers). **^{13}C NMR** (101 MHz, $CDCl_3$) (only Z isomer) δ 153.2, 99.4, 29.4, 18.1, 13.3, 11.8, 10.7.

The characterization data for 219o corresponded to the reported values [13].

(Hex-1-en-2-yloxy)triisopropylsilane (219p)

Following GP3, starting from 370 mg of 2-hexanone (3.69 mmol), 873 mg (3.40 mmol, 92 % yield) of a colorless oil were obtained containing a 10:1.3:0.3 mixture of the di- and Z/E trisubstituted isomers.

Major Isomer

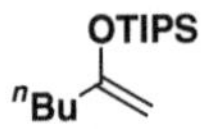

R_f 0.75 (9:1, Hexane/Ethyl acetate). **^{1}H NMR** (400 MHz, CDC13) δ 4.03 (s, 1 H, C=CH_2), 4.00 (s, 1 H, C=CH_2), 2.08 (m, 2 H, C–CH_2), 1.51 (m, 2 H, –CH_2–), 1.37 (m, 2 H, –CH_2–), 1.27–1.16 (m, 3 H, $SiCH(CH_3)_2$), 1.16–1.08 (m, 18 H, SiCH $(CH_3)_2$), 0.93 (t, 3 H, J = 7.3 Hz, –CH_3). **^{13}C NMR** (101 MHz, $CDCl_3$) δ 159.9, 88.6, 36.4, 29.2, 22.3, 18.0, 14.0, 12.7. **IR** 2945 (s), 2895 (w), 2868 (s), 1674 (w), 1657 (w), 1617 (w), 1464 (m), 1272 (s), 1020 (s), 883 (s). **HRMS (ESI)** calcd for $C_{15}H_{33}OSi^+$ $[M + H]^+$ 257.2295; found 257.2308.

GP5 Lewis acid optimization (Table 2.1)

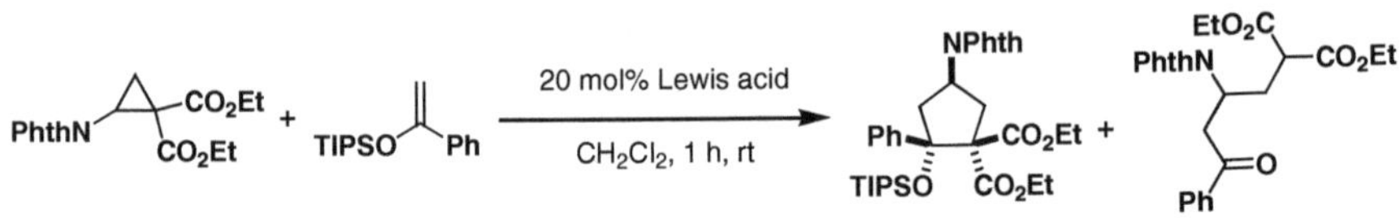

217 219a 220a 221

In a microwave vial, under nitrogen atmosphere are added cyclopropane **217** (13 mg, 0.040 mmol, 1 eq), enol ether **219a** (17 mg, 0.060 mmol, 1.5 eq) and 20 mol% Lewis acid in 0.5 mL dichloromethane at the indicated temperature. The reaction is quenched by filtration on silica or by adding 0.1 mL of triethylamine for Sn catalyzed reaction. Crude mixture is then analyzed by ^{1}H NMR.

GP6 for the $SnCl_4$-catalyzed synthesis of N-phthalimide aminocyclopentane

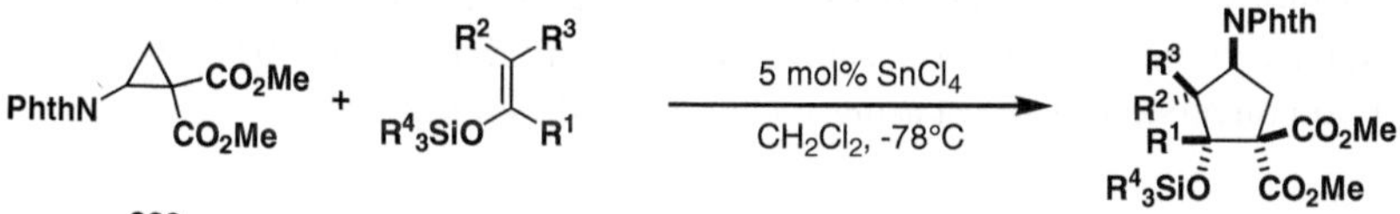

223

In an oven-dried flask sealed with a septum and under N_2 atmosphere is added the N-phthalimide aminocyclopropane (0.3 mmol, 1 eq) and the silyl-enol ether (1.5 eq) in dry dichloromethane (0.15 M). The solution is cooled down to −78 °C and a 0.43 M solution of tin tetrachloride (5 mol%) in dry dichloromethane is added. The reaction is stirred for 1 h at −78 °C. Triethylamine (0.2 mL) is then added in one portion at −78 °C. The reaction is warmed at room temperature and stirred for 15 min. Dichloromethane is removed under reduced pressure and the crude is directly purified by column chromatography (8:2 Hexane/Ethyl Acetate).

Control experiments

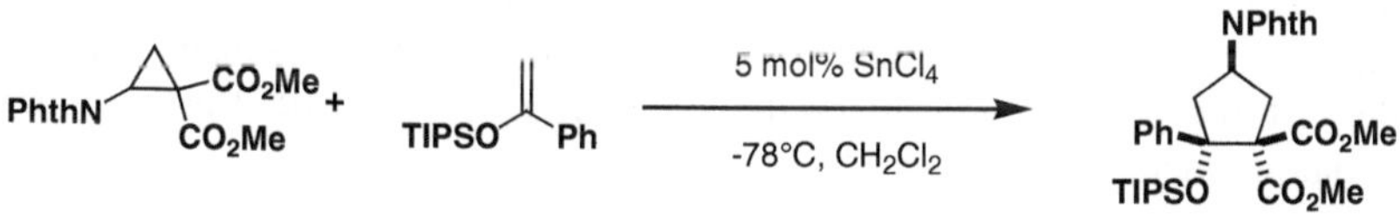

223 219a 220b

To a solution of aminocyclopropane **223** (50 mg, 0.17 mmol, 1 eq) and silyl enol ether **219a** (68 mg, 0.25 mmol, 1.5 eq) in 1 mL of dry dichloromethane under nitrogen, is added a tin tetrachloride solution (0.43 M, 19.2 µl, 0.00826 mmol, 5 mol%) at room temperature. The reaction is stirred 1 h at room temperature after what 0.1 mL of triethylamine is added. The solution is concentrated under reduced pressure and purified by column chromatography (9:1–8:2 Hexane/Ethyl Acetate). 69 mg (0.12 mmol, 72 %) of **220b** and 16.4 (0.04 mmol, 24 %) of **225** are isolated as a solid and an oil respectively.

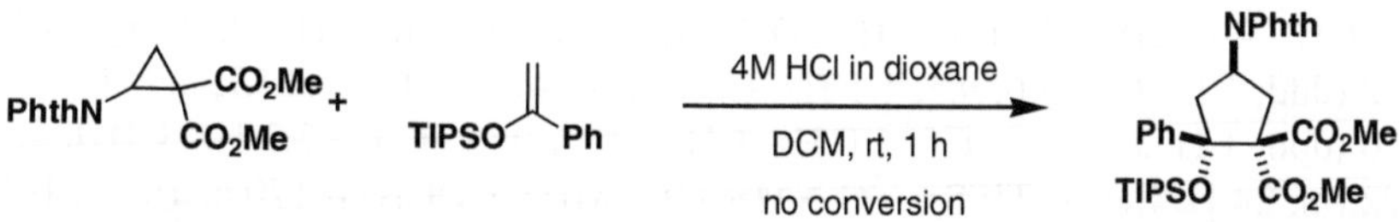

223 219a 220b

To a solution of aminocyclopropane **223** (30 mg, 0.098 mmol, 1 eq) and silyl enol ether **219a** (41 mg, 0.15 mmol, 1.5 eq) in dry dichloromethane (0.30 mL) under nitrogen, is added an HCl solution (4 M, 5.0 µL, 0.019 mmol, 20 mol%) in dioxane at room temperature. The reaction is stirred at room temperature for 12 h and no conversion of the starting material could be observed.

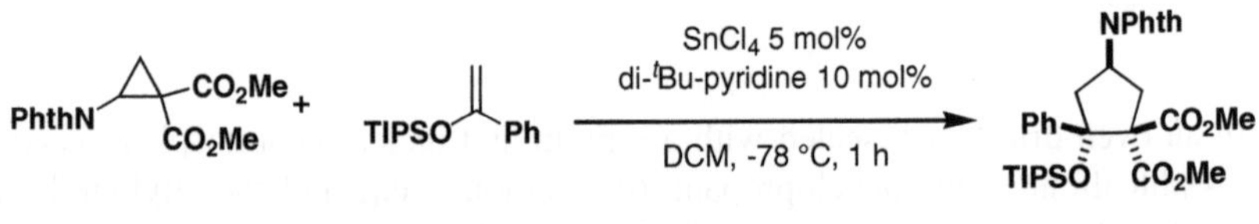

223 **219a** **220b**

In an oven-dried flask sealed with a septum and under N_2 atmosphere is added 50 mg of N-phthalimide aminocyclopropane **223** (0.16 mmol, 1 eq), silyl-enol ether **219a** (68 mg, 0.25 mmol, 1.5 eq) and 2,6-di-*tert*-butylpyridine (3.1 mg, 0.016 mmol, 10 mol%) in dry dichloromethane (0.5 mL). The solution is cooled down to −78 °C and a solution of tin tetrachloride (0.43 M, 19.2 µL, 5 mol%) in dry dichloromethane is added. The reaction is stirred for 1 h at −78 °C. Triethylamine (0.1 mL) is then added in one portion at −78 °C. The reaction is warmed at room temperature and stirred for 15 min. Dichloromethane is removed under reduced pressure and the crude is analyzed by ^{1}H NMR. Ratio of starting material/product = 2.5:1.

***Trans*-diethyl-4-(1,3-dioxoisoindolin-2-yl)-2-phenyl-2-(triisopropylsilyl)oxy) cyclopentane-1,1-dicarboxylate (220a)**

Following GP6, 180 mg (0.296 mmol, 98 %) of a single diastereoisomer as a colorless solid was obtained.

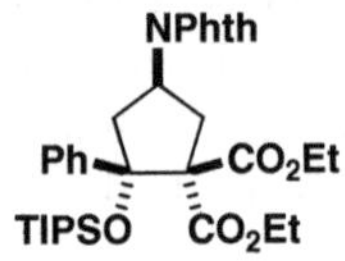

R_f 0.54 (7:3, Hexane/Ethyl acetate). **Mp** 113.4 °C. **^{1}H NMR** (400 MHz, $CDCl_3$) δ 7.90–7.81 (m, 4 H, Phth + Ar), 7.74 (m, 2 H, Phth), 7.33–7.26 (m, 3 H, Ar), 5.30 (m, 1 H, N–C–H), 4.33 (m, 2 H, CO_2CH_2), 4.00–3.81 (m, 3 H, CO_2CH_2 + CH_2), 3.42 (ddd, 1 H, J = 13.6, 9.5, 1.1 Hz, CH_2), 2.89 (dd, 1 H, J = 13.7, 8.9 Hz, CH_2), 2.46 (ddd, 1 H, J = 12.3, 6.0, 1.0 Hz, CH_2), 1.36 (t, 3 H, J = 7.1 Hz, CH_3), 1.03–0.92 (m, 24 H, CH_3 + TIPS). **^{13}C NMR** (101 MHz, $CDCl_3$) δ 170.6, 168.4, 168.3, 142.0, 134.0, 132.0, 128.6, 127.9, 127.1, 123.2, 87.5, 70.1, 61.5, 61.0, 47.9, 41.8,

36.3, 18.2, 18.2, 13.7.[3] **IR** 2944 (w), 2868 (w), 2258 (w), 1775 (w), 1734 (m), 1712 (s), 1377 (s), 1253 (m), 1128 (s), 981 (m). **HRMS (ESI)** calcd for $C_{34}H_{46}NO_7Si^+$ $[M + H]^+$ 608.3038; found 608.3050.

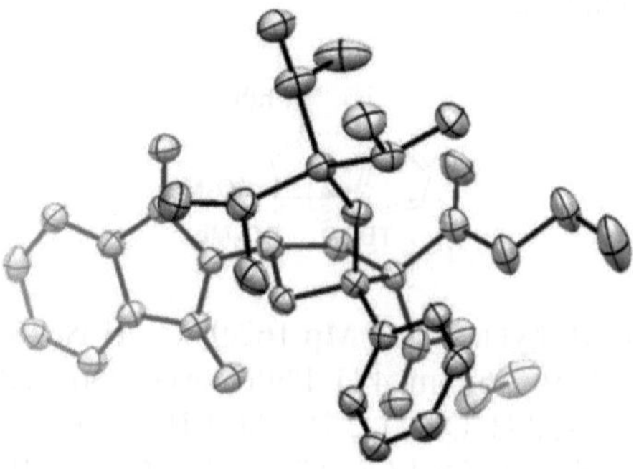

Recrystallized in ethanol. The crystal structure has been deposited at the Cambridge Crystallographic Data Centre and allocated the deposition number: CCDC 842232.

***Trans*-dimethyl-4-(1,3-dioxoisoindolin-2-yl)-2-phenyl-2-(triisopropylsilyl)oxy) cyclopentane-1,1-dicarboxylate (220b)**

Following GP6, 165 mg (0.284 mmol, 95 %) of a single diastereoisomer as a colorless solid was obtained.

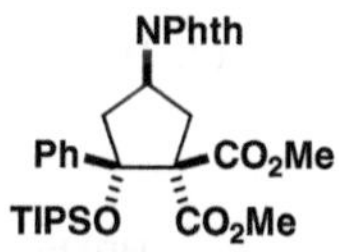

The two enantiomers were separated by HPLC using Chiralpack IA column (0.46 × 25 cm), 95:5 Hexane/Isopropanol, 0.5 mL/min. tr_1 = 20.0 min, $[\alpha]_D^{25.0}$ − 40.3 (c = 1, $CHCl_3$). tr_2 = 21.4 min.

R_f 0.55 (6:4, Hexane/Ethyl acetate). **Mp** 134.5 °C. **^{1}H NMR** (400 MHz, $CDCl_3$) δ 7.87 (m, 2 H, Phth), 7.80–7.72 (m, 4 H, Phth + Ar), 7.34–7.25 (m, 3 H, Ar), 5.29 (m, 1 H, N–C–H), 3.86 (s, 3 H, OMe), 3.81 (t, 1 H, *J* = 12.2 Hz, CH_2), 3.47–3.39 (m, 4 H, OMe + CH_2), 2.91 (dd, 1 H, *J* = 13.7, 8.7 Hz, CH_2), 2.46 (dd, 1 H, *J* = 12.4, 6.2 Hz, CH_2), 1.01–0.92 (m, 21 H, TIPS). **^{13}C NMR** (101 MHz, $CDCl_3$) δ 171.0, 168.7, 168.4, 141.8, 134.1, 132.0, 128.4, 128.0, 127.1, 123.2, 87.6, 70.1, 52.4, 52.1, 47.8, 41.7, 36.2, 18.2, 18.2, 13.7. **IR** 2952 (w), 2868 (w), 2259 (w), 1738 (m), 1712 (s), 1378 (m), 1129 (s), 981 (m). **HRMS (ESI)** calcd for $C_{32}H_{42}NO_7Si^+$ $[M + H]^+$ 580.2725; found 580.2717.

[3]The CH_3 carbons of TIPS are splitting.

Trans-dimethyl-2-(tert-butyldimethylsilyl)oxy)-4-(1,3-dioxoisoindolin-2-yl)-2-phenylcyclopentane-1,1-dicarboxylate (220c)

Following GP6, 157 mg (0.293 mmol, 98 %) of a single diastereoisomer as a colorless solid was obtained.

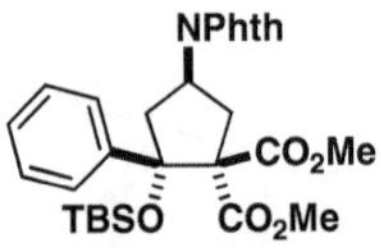

R_f 0.38 (7:3, Hexane/Ethyl acetate). **Mp** 162.9 °C. **^{1}H NMR** (400 MHz, $CDCl_3$) δ 7.86 (m, 2 H, Phth), 7.79–7.68 (m, 4 H, Phth + Ar), 7.36–7.26 (m, 3 H, Ar), 5.27 (m, 1 H, N–C–H), 3.87 (s, 3 H, OMe), 3.78 (dd, 1 H, *J* = 12.2, 12.2 Hz, CH_2), 3.47 (s, 3 H, OMe), 3.41 (ddd, 1 H, *J* = 13.5, 9.3, 1.2 Hz, CH_2), 2.93 (dd, 1 H, *J* = 13.6, 8.9 Hz, CH_2), 2.36 (dd, 1 H, *J* = 12.0, 5.0 Hz, CH_2), 0.95 (s, 9 H, TBS), 0.04 (s, 3 H, TBS), −0.47 (s, 3 H, TBS). **^{13}C NMR** (101 MHz, $CDCl_3$) δ 170.9, 168.7, 168.3, 141.3, 134.1, 132.0, 128.4, 127.9, 127.2, 123.2, 87.1, 69.8, 52.4, 52.1, 47.7, 41.7, 36.2, 25.8, 18.5, −2.5, −3.5. **IR** 2953 (w), 2932 (w), 2887 (w), 2857 (w), 2258 (w), 1737 (m), 1712 (s), 1378 (m), 1128 (m), 909 (s). **HRMS (ESI)** calcd for $C_{29}H_{36}NO_7Si^+$ [M + Na]$^+$ 560.2075; found 560.2012.

Trans-dimethyl-4-(1,3-dioxoisoindolin-2-yl)-2-phenyl-2-(triethylsilyl)oxy) cyclopentane-1,1-dicarboxylate (220d)

Following GP6, 160 mg (0.298 mmol, 99 %) of a single diastereoisomer as a colorless solid was obtained.

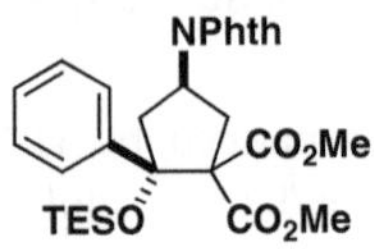

R_f 0.42 (7:3, Hexane/Ethyl acetate). Mp 109.4 °C. **^{1}H NMR** (400 MHz, $CDCl_3$) δ 7.87 (m, 2 H, Phth), 7.75 (m, 2 H, Phth), 7.70 (m, 2 H, Ar), 7.31 (m, 3 H, Ar), 5.24 (m, 1 H, N–C–H), 3.86 (s, 3 H, OMe), 3.78 (t, 1 H, *J* = 12.2 Hz, CH_2), 3.49 (m, 3 H, OMe), 3.37 (ddd, 1 H, *J* = 13.5, 9.2, 0.9 Hz, CH_2), 2.92 (dd, 1 H, *J* = 13.6, 8.9 Hz, CH_2), 2.31 (ddd, 1 H, *J* = 12.0, 6.0, 1.0 Hz, CH_2), 0.87 (m, 9 H, TES), 0.44 (m, 6 H, TES). **^{13}C NMR** (101 MHz, $CDCl_3$) δ 170.8, 168.8, 168.4, 141.6, 134.1, 132.0, 128.1, 127.9, 127.2, 123.2, 87.0, 69.8, 52.2, 52.1, 47.8, 41.8, 36.0, 6.9, 6.1. **IR** 2956 (w), 2927 (w), 2915 (w), 2878 (w), 2855 (w), 2842 (w), 2365 (w), 1739 (s), 1714 (s), 1379 (m), 1131 (m). **HRMS (ESI)** calcd for $C_{29}H_{36}NNaO_7Si^+$ [M + Na]$^+$ 560.2075; found 560.3220.

***Trans*-dimethyl-4-(1,3-dioxoisoindolin-2-yl)-2-phenyl-2-(trimethylsilyl)oxy) cyclopentane-1,1-dicarboxylate (220e)**

Following GP6, 141 mg (0.285 mmol, 95 %) of a single diastereoisomer as a colorless solid was obtained.

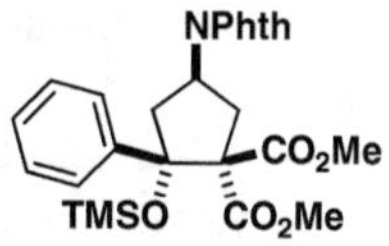

R*f* 0.35 (7:3, Hexane/Ethyl acetate). **Mp** 139.5 °C. **^{1}H NMR** (400 MHz, $CDCl_3$) δ 7.87 (m, 2 H, Phth), 7.76 (m, 2 H, Phth), 7.69–7.65 (m, 2 H, Ar), 7.35–7.27 (m, 3 H, Ar), 5.23 (m, 1 H, N–C–H), 3.85 (s, 3 H, OMe), 3.80 (t, 1 H, J = 12.2 Hz, CH_2), 3.49 (s, 3 H, OMe), 3.36 (ddd, 1 H, J = 13.5, 9.3, 1.4 Hz, CH_2), 2.93 (dd, 1 H, J = 13.6, 9.0 Hz, CH_2), 2.28 (ddd, 1 H, J = 12.2, 5.9, 1.2 Hz, CH_2), −0.03 (s, 9 H, TMS). **^{13}C NMR** (101 MHz, $CDCl_3$) δ 170.7, 168.9, 168.4, 141.3, 134.1, 132.0, 128.2, 127.8, 127.2, 123.2, 87.2, 69.8, 52.2, 52.1, 47.9, 41.6, 35.8, 1.6. **IR** 2953 (w), 2924 (w), 2850 (w), 1773 (w), 1738 (s), 1712 (s), 1379 (m), 1253 (m), 1131 (m), 985 (m), 845 (s). **HRMS (ESI)** calcd for $C_{26}H_{30}NO_7Si^+$ $[M + H]^+$ 496.1786; found 496.1800.

***Trans*-dimethyl-4-(1,3-dioxoisoindolin-2-yl)-2-(4-(methoxycarbonyl)phenyl)-2-(triisopropyl-silyl)oxy) cyclopentane-1,1-dicarboxylate (220f)**

Following GP6, 181 mg (0.284 mmol, 95 %) of a single diastereoisomer as a colorless solid was obtained.

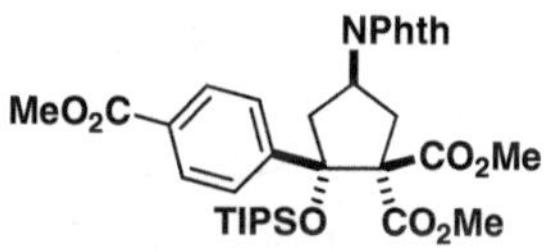

R*f* 0.5 (6:4, Hexane/Ethyl acetate). **Mp** 151.5 °C. **^{1}H NMR** (400 MHz, $CDCl_3$) δ 7.99 (m, 2 H, Ar), 7.88–7.82 (m, 4 H, Ar + Phth), 7.75 (m, 2 H, Phth), 5.28 (m, 1 H, N–C–H), 3.92 (s, 3 H, OMe), 3.85 (s, 3 H, OMe), 3.81 (dd, 1 H, J = 12.4 Hz, CH_2), 3.52–3.42 (m, 4 H, OMe + CH_2), 2.88 (dd, 1 H, J = 13.9, 8.1 Hz, CH_2), 2.47 (dd, 1 H, J = 12.4, 6.2 Hz, CH_2), 1.03–0.91 (m, 21 H, TIPS). **^{13}C NMR** (101 MHz, $CDCl_3$) δ 170.7, 168.5, 168.3, 166.9, 146.7, 134.1, 131.9, 129.5, 128.7, 128.3, 123.3, 87.2, 70.0, 52.5, 52.3, 52.1, 47.4, 41.7, 36.1, 18.2, 18.1, 13.7 (See footnote 3). **IR** 2953 (w), 2870 (w), 2256 (w), 1712 (m), 1283 (w), 908 (s), 731 (s). **HRMS (ESI)** calcd for $C_{34}H_{43}NNaO_9Si^+$ $[M + Na]^+$ 660.2599; found 660.2604.

Trans-dimethyl-4-(1,3-dioxoisoindolin-2-yl)-2-(4-fluorophenyl)-2-(triisopropylsilyl)oxy)cyclo-pentane-1,1-dicarboxylate (220g)

Following GP6, 163 mg (0.273 mmol, 91 %) of a single diastereoisomer as a colorless solid was obtained.

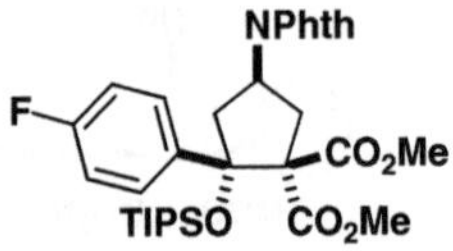

R$_f$ 0.6 (6:4, Hexane/Ethyl acetate). **Mp** 152.5 °C. **^{1}H NMR** (400 MHz, $CDCl_3$) δ 7.86 (m, 2 H, Phth), 7.79–7.73 (m, 4 H, Phth + Ar), 7.00 (t, 2 H, J = 8.7 Hz, Ar), 5.27 (m, 1 H, N–C–H), 3.86 (s, 3 H, OMe), 3.79 (dd, 1 H, J = 12.2, 12.2 Hz, CH_2), 3.51 (s, 3 H, OMe), 3.44 (ddd, 1 H, J = 13.8, 9.7, 0.9 Hz, CH_2), 2.89 (dd, 1 H, J = 13.9, 8.4 Hz), CH_2, 2.44 (dd, 1 H, J = 12.5, 6.3 Hz, CH_2), 1.03–0.93 (m, 21 H, TIPS). **^{13}C NMR** (101 MHz, $CDCl_3$) δ 170.9, 168.6, 168.3, 162.4 (d, J = 247 Hz), 137.7 (d, J = 4 Hz), 134.1, 132.0, 130.4 (d, J = 8 Hz), 123.2, 113.9 (d, J = 21 Hz), 87.1, 70.0, 52.4, 52.3, 47.6, 41.9, 36.1, 18.2, 18.1, 13.7 (See footnote 3). **IR** 2951 (w), 2894 (w), 2869 (w), 1775 (w), 1739 (m), 1714 (s), 1514 (w), 1379 (m), 1129 (m). **HRMS (ESI)** calcd for $C_{32}H_{41}FNO_7Si$ $[M + H]^+$ 598.2631; found 598.2708.

Trans-dimethyl-2-(4-cyanophenyl)-4-(1,3-dioxoisoindolin-2-yl)-2-(triisopropylsilyl)oxy)cyclo-pentane-1,1-dicarboxylate (220h)

Following GP6, 127 mg (0.210 mmol, 70 %) of a single diastereoisomer as a colorless solid was obtained.

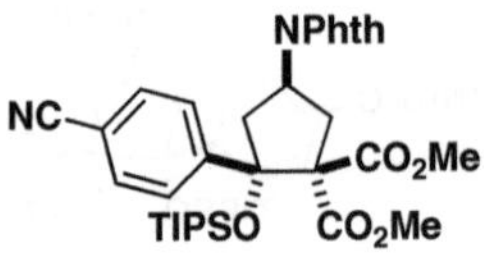

R$_f$ 0.42 (6:4, Hexane/Ethyl acetate). **Mp** 116.5 °C. **^{1}H NMR** (400 MHz, $CDCl_3$) δ 7.93–7.84 (m, 4 H, Ar + Phth), 7.76 (m, 2 H, Phth), 7.63 (m, 2 H, Ar), 5.27 (m, 1 H, N–C–H), 3.85 (s, 3 H, OMe), 3.77 (dd, 1 H, 12.2, 12.2 Hz, CH_2), 3.55 (s, 3 H, OMe), 3.50 (dd, 1 H, J = 14.1, 10.2 Hz, CH_2), 2.86 (dd, 1 H, J = 14.1, 7.8 Hz, CH_2), 2.46 (dd, 1 H, J = 12.5, 6.5 Hz, CH_2), 1.04–0.91 (m, 21 H, TIPS). **^{13}C NMR** (101 MHz, $CDCl_3$) δ 170.6, 168.3, 168.2, 146.9, 134.2, 131.9, 130.8, 129.5, 123.3, 118.9, 111.8, 86.9, 69.9, 52.6, 52.5, 47.0, 41.5, 36.0, 18.2, 18.1, 13.8 (See footnote 3). **IR** 2951 (w), 2869 (w), 2230 (w), 1739 (m), 1712 (s), 1380 (m), 1136 (m), 1124 (m), 983 (m), 912 (m), 735 (s), 727 (s). **HRMS (ESI)** calcd for $C_{33}H_{40}N_2NaO_7Si^+$ $[M + Na]^+$ 627.2497; found 627.2499.

Trans-dimethyl4-(1,3-dioxoisoindolin-2-yl)-2-(4-methoxyphenyl)-2-(triisopropylsilyl)-oxy)cyclo-pentane-1,1-dicarboxylate (220i)

Following GP6, 182 mg (0.298 mmol, 99 %) of a single diastereoisomer as a colorless solid was obtained.

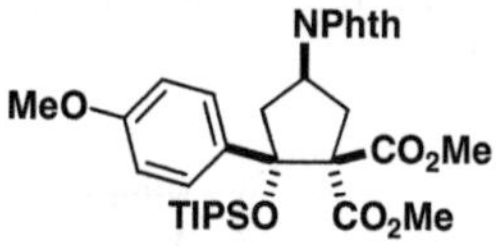

R_f 0.48 (6:4, Hexane/Ethyl acetate). **Mp** 142.3 °C. **^{1}H NMR** (400 MHz, $CDCl_3$) δ 7.86 (m, 2 H, Phth), 7.74 (m, 2 H, Phth), 7.69 (d, J = 9.0 Hz, 2 H, Ar), 6.83 (d, J = 9.0 Hz, 2 H, Ar), 5.27 (m, 1 H, N–C–H), 3.86 (s, 3 H, OMe), 3.83 (s, 3 H, OMe), 3.78 (dd, 1 H, J = 12.2, 12.2 Hz, CH_2), 3.51 (s, 3 H, OMe), 3.42 (ddd, 1 H, J = 13.6, 9.7, 0.7 Hz, CH_2), 2.89 (dd, 1 H, J = 13.8, 8.6 Hz, CH_2), 2.43 (dd, 1 H, J = 12.3, 6.2 Hz, CH_2), 1.04–0.93 (m, 21 H, TIPS). **^{13}C NMR** (101 MHz, $CDCl_3$) δ 171.1, 168.8, 168.3, 159.1, 134.0, 133.99, 132.0, 129.7, 123.2, 112.3, 87.4, 70.0, 55.2, 52.3, 52.2, 47.8, 41.9, 36.2, 18.2, 18.2, 13.7 (See footnote 3). **IR** 2949 (w), 2868 (w), 1776 (w), 1737 (m), 1712 (s), 1378 (m), 1255 (m), 1125 (m), 981 (m), 722 (s). **HRMS (ESI)** calcd for $C_{33}H_{43}NNaO_8Si^+$ $[M + Na]^+$ 632.2650; found 632.2651.

Trans-dimethyl-2-(2-bromophenyl)-4-(1,3-dioxoisoindolin-2-yl)-2-(triisopropylsilyl)oxy)cyclo-pentane-1,1-dicarboxylate (220j)

Following GP6, 183 mg (0.277 mmol, 92 %) of a single diastereoisomer as a colorless solid was obtained.

The two enantiomers were separated by HPLC using Chiralpack IC column (0.46 × 25 cm), 95:5 Hexane/Isopropanol, 1 mL/min. tr_1 = 20.9 min. tr_2 = 24.9 min, $[\alpha]_D^{25.0}$ − 43.0 (c = 1, $CHCl_3$).

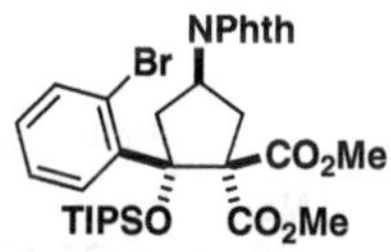

R_f 0.46 (6:4, Hexane/Ethyl acetate). **Mp** 155.7 °C. **^{1}H NMR** (400 MHz, $CDCl_3$) δ 7.88–7.81 (m, 3 H, Phth + Ar), 7.71 (m, 2 H, Phth), 7.57 (dd, 1 H, J = 8.0, 1.4 Hz, Ar), 7.24 (m, 1 H, Ar), 7.06 (m, 1 H, Ar), 5.22 (m, 1 H, N–C–H), 4.40 (dd, 1 H, J = 12.8, 12.8 Hz, CH_2), 3.79 (s, 3 H, OMe), 3.60 (s, 3 H, OMe), 3.35 (dd, 1 H, J = 13.5, 10.0 Hz, CH_2), 2.78 (dd, 1 H, J = 13.7, 7.4 Hz, CH_2), 2.60 (dd, 1 H, J = 12.9, 6.0 Hz, CH_2), 1.04–0.94 (m, 21 H, TIPS). **^{13}C NMR** (101 MHz, $CDCl_3$) δ 174.4, 172.0, 171.6, 143.1, 137.1, 135.8, 135.2, 133.7, 130.8, 127.3, 124.6, 123.2, 89.0, 70.3, 50.4, 50.4, 45.7, 40.8, 33.5, 14.9, 14.7, 10.3 (See footnote 3). **IR** 2949 (w), 2868 (w), 2259 (w), 2255 (w), 1774 (w), 1758 (m), 1737 (s), 1712 (s), 1378 (m), 913 (m), 734 (s), 722 (s). **HRMS (ESI)** calcd for $C_{32}^{79}BrH_{41}NO_7Si^+$ $[M + H]^+$ 658.1830; found 658.1835.

***Trans*-dimethyl4-(1,3-dioxoisoindolin-2-yl)-2-(naphthalen-2-yl)-2-(triisopropylsilyl)oxy)cyclo-pentane-1,1-dicarboxylate (220k)**

Following GP6, 171 mg (0.272 mmol, 91 %) of a single diastereoisomer as a colorless solid were obtained.

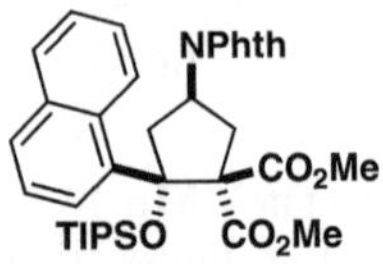

R_f 0.6 (6:4, Hexane/Ethyl acetate). **Mp** 193.8 °C. **^{1}H NMR** (400 MHz, $CDCl_3$) δ 8.39 (d, 1 H, $J = 8.4$ Hz, Ar), 7.90–7.79 (m, 5 H, Phth + Ar), 7.75 (m, 2 H, Phth), 7.45 (t, 1 H, $J = 7.8$ Hz, Ar), 7.42–7.33 (m, 2 H, Ar), 5.42 (m, 1 H, N–C–H), 3.90 (dd, 1 H, $J = 12.5$, 1.5 Hz, CH_2), 3.76 (dd, 1 H, $J = 14.3$, 11.7 Hz, CH_2), 3.63 (s, 3 H, OMe), 3.43 (s, 3 H, OMe), 2.95 (dd, 1 H, $J = 14.4$, 4.7 Hz, CH_2), 2.67 (dd, 1 H, $J = 12.5$, 7.3 Hz, CH_2), 1.08–1.03 (m, 12 H, TIPS), 0.73–0.68 (m, 9 H, TIPS). **^{13}C NMR** (101 MHz, $CDCl_3$) δ 171.0, 169.8, 168.3, 136.4, 134.5, 134.0, 132.4, 132.0, 129.7, 128.8, 128.5, 124.7, 124.0, 123.6, 123.3, 90.9, 68.6, 52.4, 52.2, 45.5, 43.7, 37.8, 18.5, 17.9, 14.3 (See footnote 3). **IR** 3050 (w), 2949 (w), 2894 (w), 2868 (w), 2258 (w), 1775 (w), 1762 (w), 1734 (s), 1712 (s), 1378 (m), 1123 (m), 970 (m), 910 (m), 732 (s). **HRMS (ESI)** calcd for $C_{36}H_{44}NO_7Si^+$ $[M + H]^+$ 630.2882; found 630.2876.

Dimethyl 4-(1,3-dioxoisoindolin-2-yl)-3-methyl-2-phenyl-2-(triisopropylsilyl) oxy)cyclo-pentane-1,1-dicarboxylate (220l)

Following GP6, 164 mg (0.275 mmol, 92 %) of two diastereoisomers (20:1) as a colorless solid was obtained.

Major diastereoisomer:

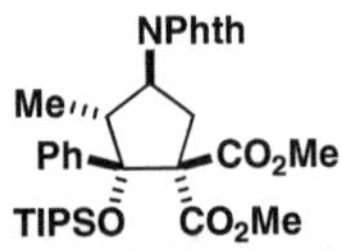

The two enantiomers were separated by HPLC using Chiralpack IA column (0.46 × 25 cm), 93:7 Hexane/Isopropanol, 1 mL/min. $tr_1 = 7.5$ min, $[\alpha]_D^{25.0}$ 48.4 (c = 1, $CHCl_3$). $tr_2 = 12.8$ min.

R_f 0.52 (6:4, Hexane/Ethyl acetate). **Mp** 120 °C. **^{1}H NMR** (400 MHz, $CDCl_3$) δ 7.84 (m, 2 H, Phth), 7.73 (m, 2 H, Phth), 7.69–7.65 (m, 2 H, Ar), 7.36–7.27 (m, 3 H, Ar), 4.90 (td, 1 H, $J = 11.1$, 6.9 Hz, N–C–H), 3.99 (m, 1 H, $CHCH_3$), 3.78 (s, 3 H, OMe), 3.53 (s, 3 H, OMe), 3.45 (dd, 1 H, $J = 14.3$, 11.0 Hz, CH_2), 2.72 (dd, $J = 14.3$, 6.9 Hz, 1 H, CH_2), 1.25–1.07 (m, 15 H, TIPS + Me), 0.99–0.95 (m, 9 H, *TIPS*). **^{13}C NMR**

(101 MHz, $CDCl_3$) δ 170.8, 168.7, 168.5, 139.6, 134.0, 131.9, 129.6, 127.8, 126.7, 123.2, 90.6, 70.1, 53.1, 52.3, 43.4, 34.1, 18.8, 18.7, 15.1, 11.9 (See footnote 3). **IR** 2950 (w), 2869 (w), 1758 (w), 1739 (s), 1739 (s), 1717 (s), 1381 (m), 1126 (m), 735 (m), 724 (s). **HRMS (ESI)** calcd for $C_{33}H_{43}NNaO_7Si^+$ $[M + Na]^+$ 616.2701; found 616.2717.

Dimethyl-3-(1,3-dioxoisoindolin-2-yl)-9b-(triisopropylsilyl)oxy)-3,3a,4,5-tetrahydro-1H-cyclopenta[a]naphthalene-1,1(2 H,9bH)-dicarboxylate (220m)

Following GP6, 166 mg (0.274 mmol, 91 %) of two diastereoisomers (10:1) as a colorless solid were obtained.

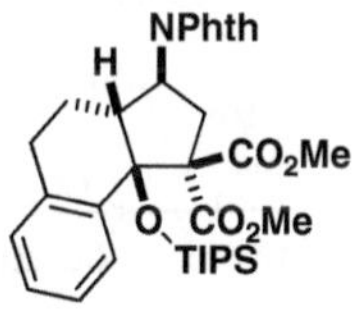

R_f 0.50 (6:4, Hexane/Ethyl acetate). **Mp** 179.1 °C. **^{1}H NMR** (400 MHz, $CDCl_3$) δ 8.20 (d, 1 H, J = 7.0 Hz, Ar), 7.88 (m, 2 H, Phth), 7.76 (m, 2 H, Phth), 7.22 (dd, 1 H, J = 7.2, 7.2 Hz, Ar), 7.16 (dd, 1 H, J = 7.4, 7.0 Hz, Ar), 7.07 (d, 1 H, J = 7.6 Hz, Ar), 5.17 (m, 1 H, N–C–H), 3.98 (dd, 1 H, J = 12.9, 12.9 Hz, CH_2–$C(CO_2Me)_2$), 3.88 (s, 3 H, OMe), 3.32 (m, 1 H, –CH–), 3.26 (s, 3 H, OMe), 3.04 (m, 1 H, Ar–CH_2), 2.63 (m, 1 H, Ar–CH_2), 2.16 (dd, 1 H, J = 12.6, 5.6 Hz, CH_2–$C(CO_2Me)_2$), 1.87 (m, 1 H, Ar–CH_2–CH_2), 1.66 (m, 1 H, Ar–CH_2–CH_2), 1.10–0.87 (m, 21 H, TIPS). **^{13}C NMR** (101 MHz, $CDCl_3$) δ 170.8, 168.5, 168.3, 140.2, 138.2, 134.1, 131.9, 129.2, 127.5, 127.3, 126.1, 123.3, 83.4, 71.0, 52.5, 52.1, 50.3, 49.8, 34.3, 25.3, 21.2, 18.4, 18.0, 14.0 (See footnote 3). **IR** 2950 (w), 2868 (w), 2256 (w), 1773 (w), 1715 (s), 1381 (m), 1128 (m), 908 (s), 731 (s). **HRMS (ESI)** calcd for $C_{34}H_{43}NNaO_7Si^+$ $[M + Na]^+$ 628.2701; found 628.2722.

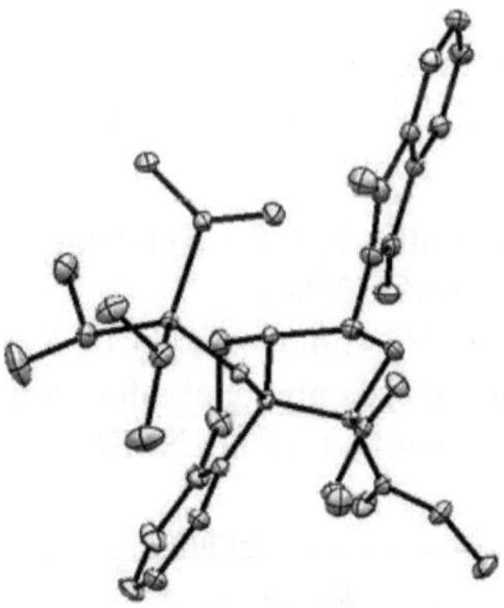

The major diastereoisomer was recrystallized from isopropanol. The crystal structure has been deposited at the Cambridge Crystallographic Data Centre and allocated the deposition number: CCDC 842233.

Dimethyl-4-(1,3-dioxoisoindolin-2-yl)-3,3-dimethyl-2-phenyl-2-(triisopropylsilyl)oxy)cyclopen-tane-1,1-dicarboxylate (220n)

Following GP6, 147 mg (0.242 mmol, 81 %) of two non-separable diastereoisomers (4:1) as a colorless solid were obtained.

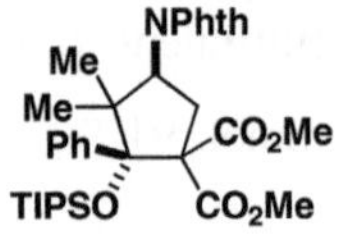

Isomer 1

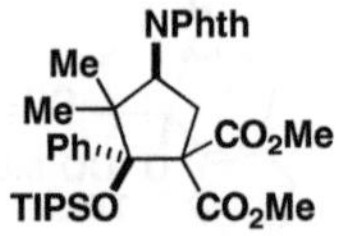

Isomer 2

$\mathbf{R_f}$ 0.52 (6:4, Hexane/Ethyl acetate) **^{1}H NMR** (400 MHz, $CDCl_3$) δ 7.93 (m, 2 H, Ar, Isomer 1 & 2), 7.84 (m, 2.5 H, Phth, Isomer 1 & 2), 7.79–7.72 (m, 3 H, Phth + Ar, Isomer 1 & 2), 7.51 (m, 0.5 H, Ar, Isomer 2), 7.34–7.24 (m, 4 H, Ar, Isomer 1 & 2), 5.12 (dd, 1 H, J = 13.0, 7.3 Hz, N–C–H, Isomer 1), 4.77 (m, 0.25 H, N–C–H, Isomer 2), 4.48 (dd, 0.25 H, J = 13.4, 13.4 Hz, CH_2, Isomer 2), 3.86 (s, 3 H, OMe, Isomer 1), 3.78 (s, 0.75 H, OMe, Isomer 2), 3.69–3.61 (m, 1.75 H, OMe, Isomer 2, CH_2, Isomer 1), 3.55 (s, 3 H, OMe, Isomer 1), 3.02 (dd, 1 H, J = 13.4, 7.4 Hz, CH_2, Isomer 1), 2.40 (dd, 0.25 H, J = 13.3, 7.4 Hz, CH_2, Isomer 2), 1.52 (s, 3 H, Me, Isomer 1), 1.41 (s, 0.75 H, Me, Isomer 2), 1.31 (s, 3 H, Me, Isomer 1), 1.04 (s, 0.75 H, Me, Isomer 2), 1.02–0.98 (m, 12 H, TIPS, Isomer 1 & 2), 0.97–0.92 (m, 10 H, TIPS, Isomer 1 & 2), 0.90–0.80 (m, 4 H, TIPS, Isomer 1 & 2). **^{13}C NMR** (101 MHz, $CDCl_3$) δ 172.2, 170.2, 169.7, 169.3, 141.8, 141.4, 134.1, 134.0, 131.8, 129.5, 128.5, 127.7, 127.4, 127.2, 126.9, 123.2, 93.3, 69.6, 58.8, 57.8, 53.3, 52.9, 52.7, 52.5, 52.4, 32.9, 32.5, 29.0, 25.2, 23.6, 21.4, 19.0, 18.8, 18.8, 15.0, 14.6 (See footnote 3).[4] **IR** 2949 (w), 2869 (w), 1777 (w), 1735 (m), 1716 (s), 1373 (m), 1034 (m), 911 (m), 731 (s). **HRMS (ESI)** calcd for $C_{34}H_{46}NO_7Si^+$ $[M + H]^+$ 608.3038; found 608.3010.

Dimethyl-4-(1,3-dioxoisoindolin-2-yl)-2-ethyl-3-methyl-2-(triisopropylsilyl)oxy) cyclopentane-1,1-dicarboxylate (220o)

Following GP6, 157 mg (0.288 mmol, 96 %) of a colorless oil were obtained. Starting from a 1:8 mixture of *Z:E* silyl enol ether, a mixture of 3 diastereoisomer with a ratio of 11:4:1 was identified by 2D NMR.

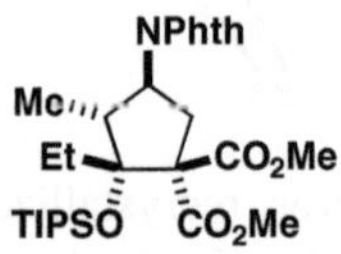

[4]Not all the signals of the minor diastereoisomer were resolved by ^{13}C.

R_f 0.7 (6:4, Hexane/Ethyl acetate). **^{1}H NMR** (400 MHz, $CDCl_3$) δ 7.88–7.81 (m, 2 H), 7.69–7.76 (m, 3 H), 4.86–4.73 (m, 1 H), 4.55–4.43 (m, 0.28 H), 3.81–3.69 (m, 8 H), 3.54–3.40 (m, 1 H), 3.39–3.31 (m, 1 H), 3.08 (dd, 1 H, J = 14.3, 10.6 Hz), 2.72–2.58 (m, 2 H), 2.36–2.20 (m, 1 H), 1.99–1.88 (m, 1 H), 1.88–1.80 (m, 0.26 H), 1.23 (t, 1 H, J = 7.0 Hz), 1.20–1.09 (m, 29 H), 1.09–0.96 (m, 9 H). ^{13}C NMR (101 MHz, $CDCl_3$) δ 171.6, 171.3, 169.6, 169.5, 168.6, 168.3, 133.9, 132.0, 131.8, 123.2, 123.1, 89.9, 87.8, 67.6, 67.2, 54.0, 52.6, 52.4, 52.4, 52.3, 48.5, 42.6, 35.1, 34.8, 30.5, 29.7, 29.4, 18.7, 18.6, 18.5, 18.4, 18.0, 14.6, 14.1, 13.4, 13.3, 11.3, 9.9, 9.7 (See footnote 3). **IR** 2951 (w), 2869 (w), 1737 (m), 1715 (s), 1378 (m), 1266 (w), 1249 (w), 1127 (m). **HRMS (ESI)** calcd for $C_{29}H_{44}NO_7Si$ + $[M + H]^+$ 546.2882; found 546.2870.

***Trans*-dimethyl2-butyl-4-(1,3-dioxoisoindolin-2-yl)-2-(triisopropylsilyl)oxy) cyclopentane-1,1-dicarboxylate (220p)**

Following GP6, 164 mg (0.293 mmol, 98 %) of a colorless oil were obtained. A mixture of 4 diatereo- and regioisomers in the crude NMR with a ratio of 16.6:5:5:1 were identified by 2D NMR.

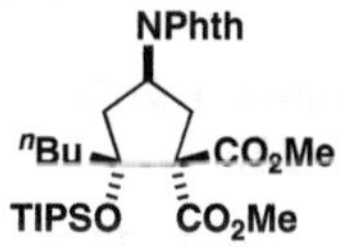

R_f 0.25 (9:1, Hexane/Ethyl acetate). **^{1}H NMR** (400 MHz, $CDCl_3$) δ 7.88–7.80 (m, 3 H), 7.76–7.69 (m, 3 H), 5.34 (q, 0.1 H, J = 9.8 Hz), 5.12 (m, 1 H), 4.76 (m, 0.25 H), 4.46 (m, 0.06 H), 3.85–3.82 (m, 3.5 H), 3.80–3.74 (m, 4.6 H), 3.39 (t, 0.05 H, J = 12.7 Hz), 3.38–3.26 (m, 1.30 H), 2.95 (dd, 1 H, J = 12.7, 11.3 Hz), 2.75 (m, 0.06 H), 2.64 (dd, 1 H, J = 14.5, 7.0 Hz), 2.57–2.47 (m, 0.9 H), 2.47–2.40 (dd, 0.2 H, J = 14.6, 5.7 Hz), 2.10 (dd, 1 H, J = 12.8, 7.3 Hz), 1.91 (m, 1 H), 1.57 (m, 0.3 H), 1.53–1.26 (m, 8 H), 1.06–1.21 (m, 30 H), 0.90 (t, 4 H, J = 7.3 Hz), 0.75 (t, 0.9 H, J = 7.2 Hz), 0.67 (t, 0.3 H, J = 7.3 Hz). **^{13}C NMR** (101 MHz, $CDCl_3$) δ 171.5, 169.3, 168.3, 133.9, 132.0, 123.2, 123.1, 88.6, 68.9, 52.5, 52.4, 50.8, 47.4, 41.2, 36.6, 35.1, 34.6, 29.6, 28.1, 23.4, 22.3, 21.3, 18.6, 18.5, 18.4, 18.3, 14.4, 14.1, 13.9, 13.8 (See footnote 3). **IR** 2952 (w), 2895 (w), 2868 (w), 1776 (w), 1741 (s), 1714 (s), 1377 (m), 1129 (m). **HRMS (ESI)** calcd for $C_{29}H_{44}NO_7Si^+$ $[M + H]^+$ 560.3038; found 560.3030.

Dimethyl-3-(1,3-dioxoisoindolin-2-yl)-7a-(triisopropylsilyl)oxy) octahydro-1H-indene-1,1-dicarboxylate (220q)

Following GP6, 159 mg (0.285 mmol, 90 %) of two diastereoisomers (4:1, determined by integration of peaks at 4.40 and 4.80 ppm in crude ^{1}H-NMR) as a colorless oil were obtained. The major diastereoisomer was separated from others by column chromatography.

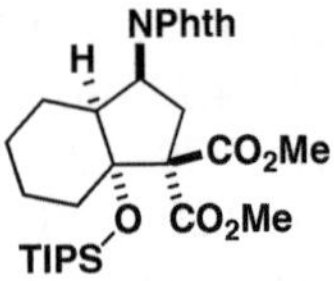

The two enantiomers were separated by HPLC using Chiralpack IB column (0.46 × 25 cm), 98:2 Hexane/Isopropanol, 0.6 mL/min. tr_1 = 20.0 min, $[\alpha]_D^{25.0}$ 20.1 (c = 0.5, $CHCl_3$). tr_2 = 24.4 min.

R_f 0.70 (6:4, Hexane/Ethyl acetate). **^{1}H NMR** (400 MHz, $CDCl_3$) δ 7.82 (m, 2 H, Phth), 7.71 (m, 2 H, Phth), 5.25 (ddd, 1 H, *J* = 10.1, 10.1, 7.1 Hz, N–C–H), 4.04 (dd, 1 H, *J* = 14.4, 10.5 Hz, CH_2 cyclopentane), 3.78 (s, 3 H, OMe), 3.74 (s, 3 H, OMe), 3.03 (dd, 1 H, *J* = 14.4, 9.8 Hz, CH_2 cyclopentane), 2.91 (m, 1 H, CH_2), 2.38 (m, 1 H, –CH–), 1.80–1.54 (m, 4 H, CH_2), 1.54–1.41 (m, 1 H, CH_2), 1.39–1.28 (m, 1 H, CH_2), 1.23–1.09 (m, 22 H, TIPS + CH_2). **^{13}C NMR** (101 MHz, $CDCl_3$) δ 170.2, 169.3, 169.2, 133.9, 131.9, 123.1, 87.1, 68.0, 52.8, 52.8, 52.6, 52.5, 34.4, 29.9, 23.9, 23.3, 23.2, 18.4, 18.6, 13.7 (See footnote 3). **IR** 2949 (w), 2867 (w), 1774 (w), 1739 (m), 1715 (s), 1373 (w), 1133 (w). **HRMS (ESI)** calcd for $C_{30}H_{44}NO_7Si$ + [M + H] + 558.2882; found 558.2892.

Trans-dimethyl-4-(1,3-dioxoisoindolin-2-yl)-2-(trimethylsilyl)oxy) cyclopentane-1,1-dicarboxylate (220r)

Following GP6, 97.0 mg (0.231 mmol, 77 %) of a single diastereoisomer as a colorless solid was obtained.

The two enantiomers were separated by HPLC using Chiralpack IA column (0.46 × 25 cm), 98:2 Hexane/Isopropanol, 1 mL/min. tr_1 = 19.8 min. tr_2 = 24.0 min, $[\alpha]_D^{25.0}$ − 51 (c = 0.68, $CHCl_3$).

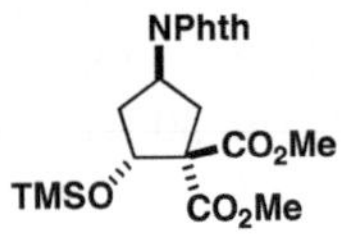

R_f 0.40 (6:4, Hexane/Ethyl acetate). **Mp** 120.6 °C. **^{1}H NMR** (400 MHz, $CDCl_3$) δ 7.80 (m, 2 H, Phth), 7.69 (m, 2 H, Phth), 5.11 (m, 1 H, N–C–H), 4.99 (dd, 1 H, *J* = 4.1, 1.8 Hz, O–C–H), 3.82 (s, 3 H, OMe), 3.72 (s, 3 H, OMe), 3.14 (dd, 1 H, *J* = 14.4, 10.7 Hz, CH_2), 2.55 (m, 1 H, CH_2), 2.46 (dd, 1 H, *J* = 14.5, 6.6 Hz, CH_2), 1.98 (ddd, 1 H, *J* = 12.9, 7.9, 1.9 Hz, CH_2), 0.11 (m, 9 H, TMS). **^{13}C NMR** (101 MHz, $CDCl_3$) δ 171.3, 169.0, 168.0, 134.0, 131.9, 123.2, 76.4, 66.0, 52.8, 52.5, 47.6, 38.7, 33.8, 0.0. **IR** 2955 (w), 1775 (w), 1737 (s), 1710 (s), 1378 (m), 1251 (m), 1124 (m), 844 (s), 720 (s). **HRMS (ESI)** calcd for $C_{20}H_{26}NO_7Si^+$ $[M + H]^+$ 420.1473; found 420.1483.

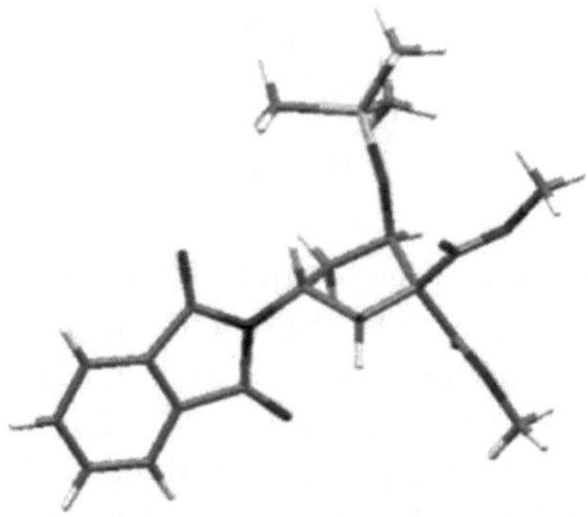

The crystal structure has been deposited at the Cambridge Crystallographic Data Centre and allocated the deposition number: CCDC 842234.

Trans-dimethyl2-butoxy-4-(1,3-dioxoisoindolin-2-yl) cyclopentane-1,1-dicarboxylate (220s)

Following GP6, 119 mg (0.296 mmol, 99 %) of two unseparable diastereoisomers (20:1) as colorless oil were obtained.

Major diastereoisomer:

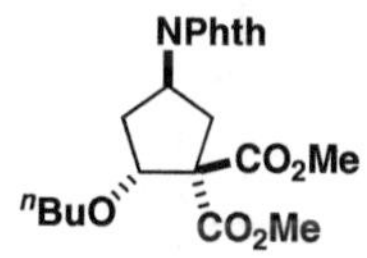

R_f 0.5 (6:4 Hexane/Ethyl acetate) **^{1}H NMR** (400 MHz, $CDCl_3$) δ 7.84 (m, 2 H, Phth), 7.72 (m, 2 H, phth), 5.06 (m, 1 H, N–C–H), 4.64 (dd, 1 H, $J = 4.2, 2.1$ Hz, –CH–O), 3.87 (s, 3 H, OMe), 3.77 (s, 3 H, OMe), 3.57 (dt, 1 H, $J = 9.2, 6.1$ Hz, OCH_2 ^{n}butyl), 3.38 (dt, 1 H, $J = 9.2, 6.5$ Hz, OCH_2 ^{n}butyl), 3.14 (dd, 1 H, $J = 14.5, 10.7$ Hz, CH_2–$C(CO_2Me)_2$), 2.54–2.42 (m, 2 H, CH_2–CH–O–), 2.23 (ddd, 1 H, $J = 13.2, 8.1, 2.1$ Hz, CH_2–$C(CO_2Me)_2$), 1.59–1.45 (m, 2 H, CH_2 ^{n}butyl), 1.42–1.25 (m, 2 H, CH_2 ^{n}butyl), 0.92 (t, 4 H, $J = 7.4$ Hz, CH_3). **^{13}C NMR** (101 MHz, $CDCl_3$) δ 171.3, 169.0, 168.0, 134.0, 131.9, 123.2, 83.1, 69.6, 64.9, 52.9, 52.6, 47.4, 34.8, 34.1, 31.8, 19.3, 13.9. **HRMS (ESI)** calcd for $C_{21}H_{26}NO_7^+$ $[M + H]^+$ 404.1704; found 404.1720.

Dimethyl 5-(1,3-dioxoisoindolin-2-yl)hexahydrocyclopenta[b]pyran-7,7(7aH) dicarboxylate (220t)

Following GP6, 115 mg (0.297 mmol, 99 %) of two diastereoisomers (2:1) as a colorless solid was obtained. The two isomers were separated by column chromatography.

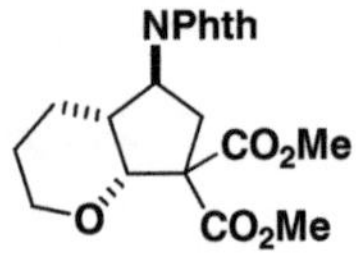

R_f 0.13 (6:4, Hexane/Ethyl acetate). **Mp** 169.2 °C. **^{1}H NMR** (400 MHz, $CDCl_3$) δ 7.82 (m, 2 H, Phth), 7.72 (m, 2 H, Phth), 4.87 (m, 1 H, N–C–H), 4.29 (d, 1 H, *J* = 11.8 Hz, O–C–H), 4.05 (dd, 1 H, *J* = 11.5, 4.5 Hz, CH_2–O), 3.89 (s, 3 H, OMe), 3.79 (s, 3 H, OMe), 3.51 (dd, 1 H, *J* = 12.1, 12.1, 3.0 Hz, CH_2–O), 3.04 (dd, 1 H, *J* = 15.2, 9.6 Hz, CH_2 cyclopentane), 2.70 (dd, 1 H, *J* = 15.2, 2.7 Hz, CH_2 cyclopentane), 2.25 (m, 1 H, CH cyclopentane), 1.87 (m, 1 H, CH_2 pyran), 1.68–1.49 (m, 2 H, CH_2 pyran), 1.14 (qd, 1 H, *J* = 12.2, 4.3 Hz, CH_2 pyran). **^{13}C NMR** (101 MHz, $CDCl_3$) δ 171.8, 171.0, 168.8, 134.2, 131.5, 123.4, 84.2, 69.2, 61.8, 53.1, 53.0, 49.2, 45.0, 35.0, 25.2, 25.0. **IR** 2951 (w), 2853 (w), 2255 (w), 1779 (w), 1749 (w), 1731 (m), 1708 (m), 1370 (m), 1276 (m), 723 (s), 648 (s), 634 (s). **HRMS (ESI)** calcd for $C_{20}H_{22}NO_7^+$ $[M + H]^+$ 388.1391; found 388.1404.

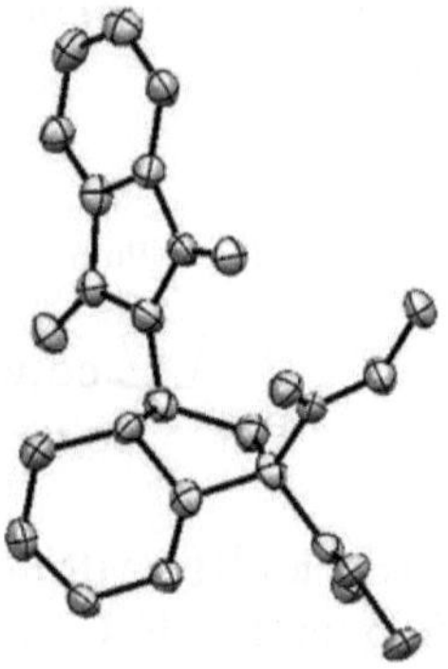

Recrystallized in isopropanol. The crystal structure has been deposited at the Cambridge Crystallographic Data Centre and allocated the deposition number: CCDC 842235.

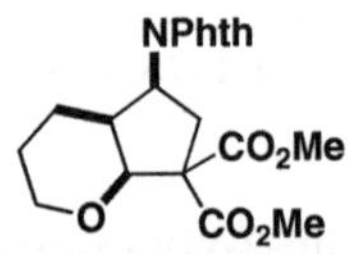

R_f 0.31 (6:4, Hexane/Ethyl acetate). **Mp** 173.5 °C. **^{1}H NMR** (400 MHz, $CDCl_3$) δ 7.83 (m, 2 H, Phth), 7.72 (m, 2 H, Phth), 5.05 (ddd, 1 H, *J* = 10.9, 10.9, 6.6 Hz, N–C–H), 4.47 (d, 1 H, *J* = 3.0 Hz, O–C–H), 4.00 (dd, 1 H, *J* = 11.4, 4.4 Hz, CH_2–O), 3.88 (s, 3 H, OMe), 3.77 (s, 3 H, OMe), 3.40 (dd, 1 H, *J* = 12.2, 2.1 Hz, CH_2–O),

3.27 (dd, 1 H, J = 14.6, 11.0 Hz, CH_2 cyclopentane), 2.87 (m, 1 H, CH cyclopentane), 2.64 (dd, 1 H, J = 14.6, 6.5 Hz, CH_2 cyclopentane), 1.99 (m, 1 H, CH_2 pyran), 1.74 (m, 1 H, CH_2 pyran), 1.59 (m, 1 H, CH_2 pyran), 1.40 (m, 1 H, CH_2 pyran). **^{13}C NMR** (101 MHz, $CDCl_3$) δ 171.0, 168.8, 168.2, 134.0, 131.9, 123.2, 82.1, 67.9, 63.9, 52.9, 52.9, 49.0, 40.9, 33.0, 21.0, 20.0. **IR** 2954 (w), 2858 (w), 1736 (s), 1710 (s), 1383 (m), 1108 (m), 912 (m), 719 (s) **HRMS (ESI)** calcd for $C_{20}H_{22}NO_7^+$ $[M + H]^+$ 388.1391; found 388.1395.

Diethyl 2-(2-(1,3-dioxoisoindolin-2-yl)-4-oxo-4-phenylbutyl)malonate (221)

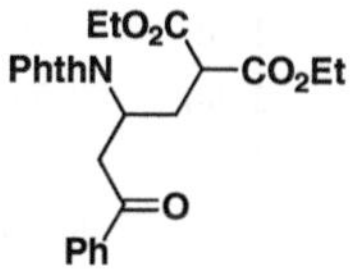

R_f 0.32 (6:4 Hexane/Ethyl Acetate).**^{1}H NMR** (400 MHz, $CDCl_3$) δ 7.93 (m, 2 H, Ar), 7.87–7.80 (m, 2 H, Phth), 7.75–7.68 (m, 2 H, Phth), 7.59–7.53 (m, 1 H, Ar), 7.48–7.41 (m, 2 H, Ar), 5.02–4.94 (m, 1 H, N–C–H), 4.29–4.20 (m, 2 H, $CO_2CH_2CH_3$), 4.09–3.92 (m, 3 H, $CO_2CH_2CH_3$ + $CH(CO_2Et)_2$), 3.51 (dd, 1 H, J = 17.7, 5.3 Hz, CH_2–CO–Ph), 3.40 (dd, 1 H, J = 8.3, 6.2 Hz, CH_2–CO–Ph), 2.79 (ddd, 1 H, J = 14.3, 11.0, 6.2 Hz, CH_2), 2.47 (ddd, 1 H, J = 14.3, 8.3, 4.1 Hz, CH_2), 1.29 (t, 3 H, J = 7.1 Hz, $CO_2CH_2CH_3$), 1.17 (t, 3 H, J = 7.1 Hz, $CO_2CH_2CH_3$). **^{13}C NMR** (101 MHz, $CDCl_3$) δ 196.7, 168.7, 168.4, 168.2, 136.5, 134.0, 133.4, 131.8, 128.7, 128.1, 123.3, 61.8, 61.6, 49.6, 45.6, 40.7, 31.6, 14.0, 13.9. **IR** 2983 (w), 1776 (w), 1748 (m), 1731 (s), 1711 (s), 1687 (w), 1393 (w), 1373 (m). **HRMS (ESI)** calcd for $C_{25}H_{26}NO_7^+$ $[M + H]^+$ 452.1704; found 452.1708.

Dimethyl 2-(2-(1,3-dioxoisoindolin-2-yl)-4-oxo-4-phenylbutyl)malonate (225)

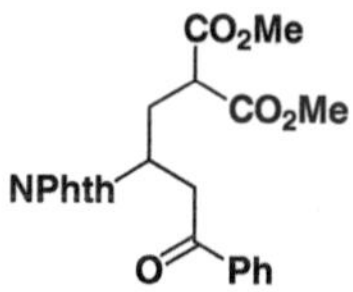

The two enantiomers were separated by HPLC using Chiralpack IA column (0.46 × 25 cm), 75:25 Hexane/Isopropanol, 1 mL/min.

87:13 er tr_1 = 26.5 min $[\alpha]_D^{25.0}$ − 7.3 (c = 1, $CHCl_3$). tr_2 = 28.2 min.

99:1 er tr_1 = 26.5 min. tr_2 = 28.2 min. $[\alpha]_D^{25.0}$ + 41 (c = 0.6, $CHCl_3$).

R_f 0.45 (6:4 Hexane/Ethyl Acetate). **^{1}H NMR** (400 MHz, $CDCl_3$) δ 7.96–7.90 (m, 2 H, Ar), 7.87–7.81 (m, 2 H, Phth), 7.76–7.70 (m, 2 H, Phth), 7.59–7.54 (m, 1 H, Ar), 7.49–7.43 (m, 2 H, Ar), 5.02–4.92 (m, 1 H, N–C–H), 4.02 (dd, 1 H, J = 17.9, 8.7 Hz, $CH(CO_2Et)_2$), 3.80 (s, 3 H, CO_2CH_3), 3.57 (s, 3 H, CO_2CH_3), 3.51 (dd, 1 H, J = 17.8, 5.3 Hz, CH_2–CO–Ph), 3.45 (dd, 1 H, J = 8.5, 6.1 Hz, CH_2–CO–Ph), 2.80 (ddd, 1 H, J = 14.3, 11.2, 6.1 Hz, CH_2), 2.48 (ddd, 1 H, J = 14.3, 8.6,

Table 5.1 Enantiospecific reaction of aminocyclopropanes

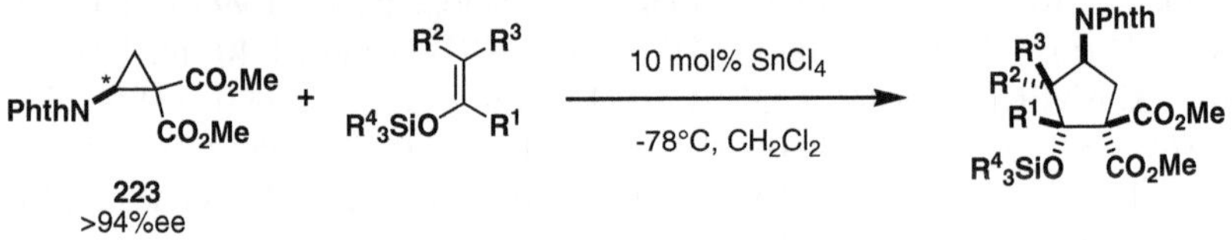

Entry	er **223**	**219**	**220**	Yield %[a]	*dr*[a]	er **220**
1	98.5:1.5	**219a**	**220b**	82	>20:1	99:1
2	99.5:0.5	**219d**	**220j**	73	>20:1	90.5:9.5
3	97.5:2.5	**219l**	**220l**	76	3:1	97:3
4	99.5:0.5	**219k**	**220q**	77.5	3:1:1	97:3
5	99:1	**219r**	**220r**	74	>20:1	98:2

[a]Due to the small scale for these experiments, product of retroaldol was detected, affecting the yield and *dr*

3.9 Hz, CH_2). **^{13}C NMR** (101 MHz, $CDCl_3$) δ 196.7, 169.1, 168.8, 168.2, 136.4, 134.1, 133.5, 131.7, 128.7, 128.1, 123.4, 52.9, 52.7, 49.2, 45.5, 40.7, 31.7. **IR** 2956 (w), 1774 (w), 1753 (m), 1734 (s), 1709 (s), 1687 (w), 1373 (m), 1372 (m), 724 (m). **HRMS (ESI)** calcd for C23H22NO7 + [M + H] + 424.1391; found 424.1386.

Enantiospecific reactions

In an oven-dried flask sealed with a septum and under N_2 atmosphere is added the chiral N-phthalimide aminocyclopropane **223** (16–26 mg, 0.053–0.062 mmol, 1 eq) and the silyl-enol ether (1.5 eq) in dry dichloromethane (0.15 M). The solution is cooled down to −78 °C and a 0.43 M solution of tin tetrachloride (10 mol%) in dry dichloromethane is added. The reaction is stirred for 1 h at −78 °C. Triethylamine (0.1 mL) is then added in one portion at −78 °C. The reaction is warmed at room temperature and stirred for 15 min. Dichloromethane is removed under reduced pressure and the crude is directly purified by column chromatography (8:2 Hexane/Ethyl Acetate) (Table 5.1).

Access to linear building blocks

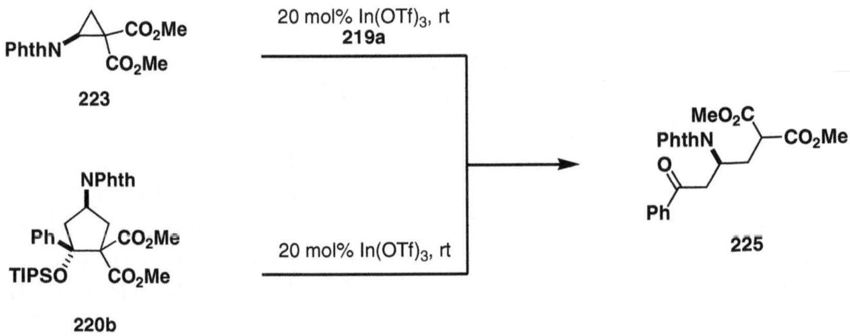

Indium (III) trifluoromethanesulfonate (17.0 mg, 0.026 mmol, 20 mol%) is weighted in the glovebox. The flask is closed with a septum and put under N_2 atmosphere. A solution of aminocyclopropane **223** (50 mg, 0.15 mmol, 1 eq) and silyl enol ether **219a** (50 mg, 0.18 mmol, 1.2 eq) in 1 mL of dry dichloromethane is added. The reaction is stirred overnight at room temperature. The reaction is filtered through a plug of silica in order to remove the catalyst and concentrated under reduced pressure. Purification by column chromatograpy (6:4 Hexane/Ethyl Acetate) furnished 58 mg (0.13 mmol, 85 % yield) of the desired product **225** as a colorless oil.

er **220b**	Yield %	er **225**
97:3	85	87:13

Indium (III) trifluoromethanesulfonate (3.5 mg, 0.0062 mmol, 20 mol%) is weighted in the glovebox. The flask is closed with a septum and put under N_2 atmosphere. A solution of **220b** (18.1 mg, 0.031 mmol, 1 eq) in 0.3 mL of dry dichloromethane (0.15 M) is added. The reaction is stirred overnight at room temperature. The reaction is filtered through a plug of silica in order to remove the catalyst and concentrated under reduced pressure. Purification by column chromatography (6:4 Hexane/Ethyl Acetate) furnished 7.2 mg (0.017 mmol, 55 % yield) of the desired product **225** as a colorless oil.

er **220b**	Yield %	er **225**
99:1	55	99:1

Dimethyl 2-(2-(1,3-dioxoisoindolin-2-yl)-3-methyl-4-oxo-4-phenylbutyl)malonate (226)

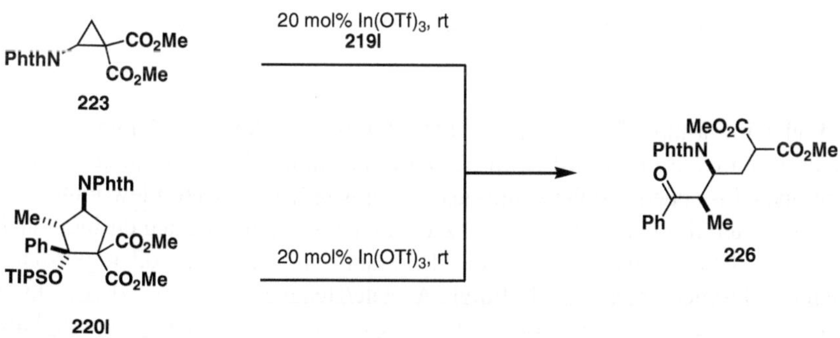

Following the same procedure described above, using 50 mg (0.17 mmol, 1 eq) of aminocyclopropane **223**, 72 mg (0.25 mmol, 1.5 eq) of silyl enol ether **219l** and 18.5 mg (0.033 mmol, 20 mol%) of indium (III) trifluoromethanesulfonate, 49 mg (0.11 mmol, 67 % yield) of a colorless oil is obtained after purification by column chromatography (9:1 Hexane/Ethyl Acetate to 8:2 Hexane/Ethyl Acetate). 4:1 Mixture of syn/anti diastereoisomers of **226.**

Indium (III) trifluoromethanesulfonate (5 mg, 0.009 mmol, 20 mol%) is weighted in the glovebox. The flask is closed with a septum and put under N_2 atmosphere. A solution of **220l** (26 mg, 0.044 mmol, 1 eq) in 0.5 mL of dry dichloromethane is added. The reaction is stirred overnight at room temperature. The reaction is filtered through a plug of silica in order to remove the catalyst and concentrated under reduced pressure. Purification by column chromatography (6:4 Hexane/Ethyl Acetate) furnished 15 mg (0.043 mmol, 78 % yield) of the desired product **226** as a colorless oil.

dr > 20:1 syn/anti $\mathbf{R_f}$ 0.29 (6:4 Hexane/Ethyl Acetate). ^{1}**H NMR** (400 MHz, $CDCl_3$) δ 8.07 (d, 2 H, J = 7.4 Hz, Ar), 7.91 (m, 2 H, Phth), 7.79 (m, 2 H, Phth), 7.64 (m, 1 H, Ar), 7.54 (t, 2 H, J = 7.8 Hz, Ar), 4.84 (ddd, 1 H, J = 10.9, 10.9, 3.1 Hz, N–C–H), 4.48 (m, 1 H, Me–CH–CO–Ph), 3.76 (s, 3 H, CO_2CH_3), 3.47 (s, 3 H, CO_2CH_3), 3.39 (dd, 1 H, J = 7.2, 7.2 Hz, $CH(CO_2Et)_2$), 2.77 (ddd, 1 H, J = 14.4, 11.4, 6.8 Hz, CH_2), 2.24 (ddd, 1 H, J = 14.4, 7.9, 3.2 Hz, CH_2), 1.10 (d, 3 H, J = 7.1 Hz,, CH_3). ^{13}**C NMR** (101 MHz, $CDCl_3$) δ 201.2, 169.1, 168.6, 168.4, 136.1, 134.3, 133.6, 131.5, 128.9, 128.6, 123.5, 52.8, 52.5, 51.9, 49.8, 42.3, 30.0, 16.0. **IR** 2955 (w), 1754 (m), 1736 (s), 1712 (s), 1388 (m), 1368 (m), 723 (s). **HRMS (ESI)** calcd for $C_{24}H_{24}NO_7^+$ $[M + H]^+$ 438.1547; found 438.1505.

***Trans*-dimethyl4-amino-2-phenyl-2-(triisopropylsilyl)oxy)cyclopentane-1,1-dicarboxylate (227)**

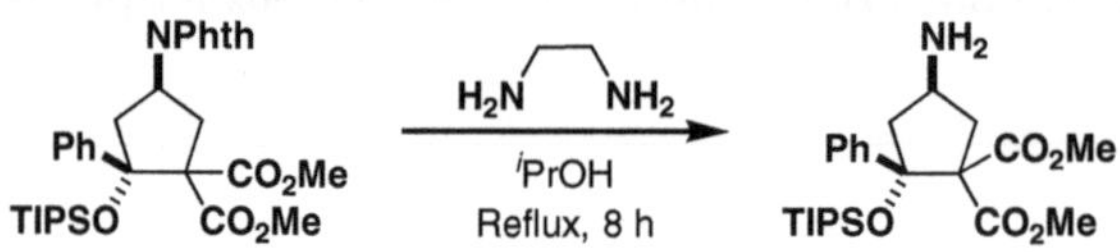

220b **227**

Following a modified procedure [14], 100 mg of **220b** (0.172 mmol, 1 eq) is added in an oven dried flask with 52 mg (0.86 mmol, 5 eq) of ethylenediamine. Isopropanol is added and the white suspension is refluxed. After a few minutes, the starting material is completely solubilized and the reaction is heated during 8 h. The solvent is then removed under reduced pressure and the crude oil is purified by column chromatography (9:1 Ethyl Acetate/Hexane + 1 % NEt_3). 55 mg (0.12 mmol, 72 %) of a colorless oil corresponding to the free cyclopentylamine **227** is isolated.

R_f 0.2 (100 % Ethyl Acetate). **Mp** 200 °C decomposition. **^{1}H NMR** (400 MHz, $CDCl_3$) δ 7.78–7.74 (m, 2 H, Ar), 7.32–7.24 (m, 3 H, Ar), 3.82–3.82 (m, 1 H, N–C–H), 3.77 (s, 3 H, CO_2CH_3), 3.31 (s, 3 H, CO_2CH_3), 3.17 (dd, 1 H, J = 13.9, 8.7 H, CH_2), 2.73 (dd, 1 H, J = 13.5, 9.3 Hz, CH_2), 2.61 (dd, 1 H, J = 13.4, 6.9 Hz, CH_2), 1.99 (dd, 1 H, J = 14.0, 6.2 Hz, CH_2), 1.76 (br s, 2 H, NH_2), 0.98–0.91 (m, 21 H, TIPS). **^{13}C NMR** (101 MHz, $CDCl_3$) δ 172.1, 169.2, 142.0, 128.0, 127.8, 127.2, 89.0, 71.3, 52.1, 49.4, 49.0, 44.6, 29.7, 18.2, 13.8. **IR** 2949 (w), 2949 (w), 2949 (w), 2949 (w), 2868 (w), 1752 (s), 1736 (s), 1717 (s), 1448 (w), 1263 (m), 1106 (s), 1077 (s). **HRMS (ESI)** calcd for $C_{24}H_{40}NO_5Si^+$ $[M + H]^+$ 450.2670; found 450.2660.

Methyl 4-(1,3-dioxoisoindolin-2-yl)-2-phenylcyclopent-1-enecarboxylate (228)

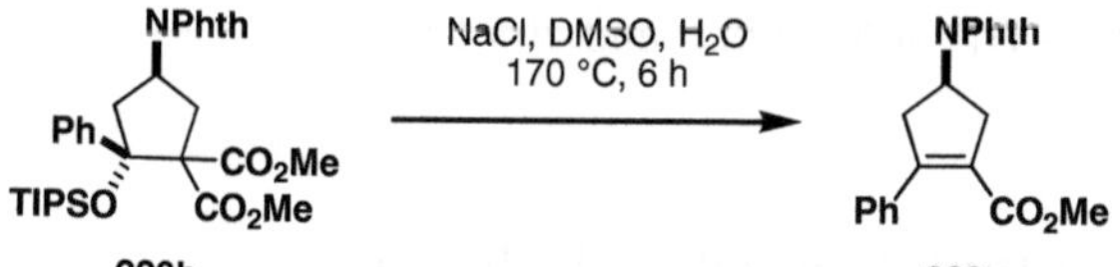

220b 228

Following a modified procedure [15], **220b** (90 mg, 0.15 mmol, 1 eq) is dissolved in 2 mL dimethylsulfoxide and 5 µl (0.3 mmol, 2 eq) of water. The solution is stirred for 6 h at 170 °C. After cooling to room temperature, the reaction mixture is extracted three times with 5 mL of diethyl ether, washed three times with 5 mL of water, three times with 5 mL brine dried over magnesium sulfate and filtered through a cotton plug. The solvents are evaporated under reduced pressure and the crude is purified by column chromatography (8:2 Hexane/Ethyl Acetate). 47 mg (0.14 mmol, 88 % yield) of **228** as a white solid are isolated.

R_f 0.51 (6:4 Hexane/Ethyl Acetate). **Mp** 165 °C. **^{1}H NMR**[5] (400 MHz, $CDCl_3$) δ 7.88 (m, 2 H, Phth), 7.75 (m, 2 H, Phth), 7.44–7.31 (m, 5 H, Ar), 5.14 (m, 1 H, N–C–H), 3.66 (s, 3 H, CO_2CH_3), 3.50 (m, 1 H, CH_2), 3.35 (m, 1 H, CH_2), 3.19 (m, 2 H, CH_2). **^{13}C NMR** (101 MHz, $CDCl_3$) δ 171.5, 168.6, 154.4, 138.0, 135.9, 133.7, 129.8, 129.4, 129.3, 128.1, 124.6, 49.3, 44.5, 41.3, 36.5. **IR** 2949 (w), 1773 (w), 1710 (s), 1393 (m), 1378 (m), 1236 (m), 722 (m). **HRMS (ESI)** calcd for $C_{21}H_{17}NNaO_4^+$ $[M + Na]^+$ 370.1050; found 370.1063.

Dimethyl 4-(1,3-dioxoisoindolin-2-yl)-2-phenylcyclopent-2-ene-1,1-dicarboxylate (229)

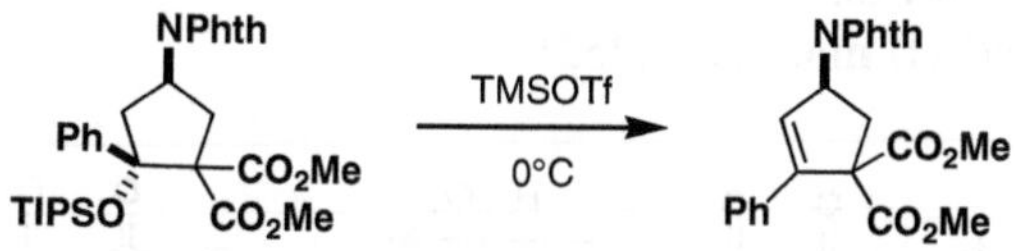

220b 229

[5]Peaks are splitting due to conformers of methyl ester.

220b (80 mg, 0.14 mmol, 1 eq) is added in an oven dried flask, under nitrogen. 1 mL of dry dichloromethane is added and the solution is cooled to 0 °C with an ice/water bath. TMSOTf (28 μl, 0.15 mmol, 1.1 eq) is added and the reaction is stirred for 10 min at 0 °C. The solvent is evaporated under reduced pressure and the crude is purified by column chromatography (8:2 Hexane/Ethyl Acetate). **229** is isolated as a colorless solid (30 mg, 0.074 mmol, 53 % yield).

R_f 0.43 (6:4 Hexane/Ethyl Acetate). **Mp** 132 °C. **^{1}H NMR** (400 MHz, $CDCl_3$) δ 7.90–7.83 (m, 2 H, Phth), 7.79–7.72 (m, 2 H, Phth), 7.51–7.46 (m, 2 H, Ar), 7.37–7.29 (m, 3 H, Ar), 6.25 (d, 1 H, *J* = 2.3 Hz, C=CH), 5.65–5.58 (m, 1 H, N–C–H), 3.82 (s, 3 H, CO_2CH_3), 3.72 (s, 3 H, CO_2CH_3), 3.15 (m, 2 H, CH_2). **^{13}C NMR** (101 MHz, $CDCl_3$) δ 171.6, 170.5, 167.9, 143.9, 134.6, 134.1, 131.9, 131.3, 128.0, 128.0, 127.7, 123.4, 53.5, 53.1, 52.8, 40.3, 29.7. **IR** 1773 (w), 1732 (m), 1715 (s), 1391 (w), 1368 (w), 1273 (w), 1124 (w), 720 (m). **HRMS (ESI)** calcd for $C_{23}H_{20}NO_6^+$ $[M + H]^+$ 406.1285; found 406.1295.

5.3 [3 + 2] Annulation with Aldehydes

The experiments of this project were carried out by Dr. Fides Benfatti and therefore the experimental part will not be included in this thesis. It can be found in the supporting information of our publication: Benfatti, F. de Nanteuil, F. Waser, J. *Org. Lett.* **2011**, *14*, 386, which will be added in the appendix of this thesis.

5.4 [3 + 2] Annulation with Ketones

The experiments of this project were carried out by Dr. Fides Benfatti and therefore the experimental part will not be included in this thesis. It can be found in the supporting information of our publication: Benfatti, F. de Nanteuil, F. Waser, J. *Chem. Eur. J.* **2012**, *18*, 4844, which will be added in the appendix of this thesis.

5.5 Friedel-Craft Alkylation with Indoles

Starting material synthesis
Bis(2,2,2-trifluoroethyl) malonate (252)

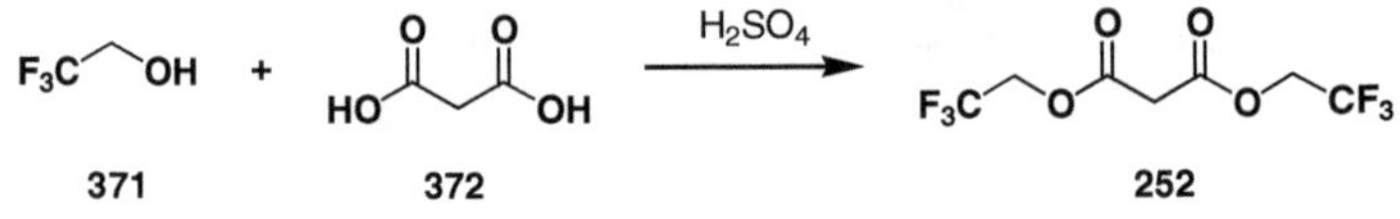

Following a reported procedure [16], a 250 mL round-bottom flask equipped with a condenser was charged with malonic acid (**372**) (8.00 g, 77.0 mmol, 1 equiv), 2,2,2-trifluoroethanol (**371**) (29.6 mL, 412 mmol, 5.4 equiv), benzene (40.0 mL) and sulfuric acid (1.00 mL, 19.0 mmol, 0.25 equiv). The flask was flushed with nitrogen and the reaction mixture refluxed overnight under nitrogen atmosphere. Afterwards, the reaction mixture was allowed to cool to room temperature, diluted with additional benzene (80 mL), and washed sequentially with 10 % sodium carbonate (3 × 80 mL), water (80 mL), and brine (80 mL). The organic layer was dried over anhydrous sodium sulfate and concentrated under reduced pressure to afford bis(2,2,2-trifluoroethyl) malonate (**252**) (7.19 g, 26.8 mmol, 35 %) as a colorless oil.

^{1}**H NMR** (400 MHz, $CDCl_3$) δ 4.54 (q, 4 H, J = 8.3 Hz, CH_2–CF_3), 3.60 (s, 2 H, CH_2–$(CO)_2$). ^{13}**C NMR** (101 MHz, $CDCl_3$) δ 164.1, 122.5 (q, $J_{C\text{-}F}$ = 277 Hz), 61.1 (q, $J_{C\text{-}F}$ = 37 Hz), 40.0. ^{19}**F NMR** (376 MHz, $CDCl_3$) δ −74.11 (t, J = 8.3 Hz).

The characterization data correspond to the reported values [16].

Dibenzyl 2-diazomalonate (253)

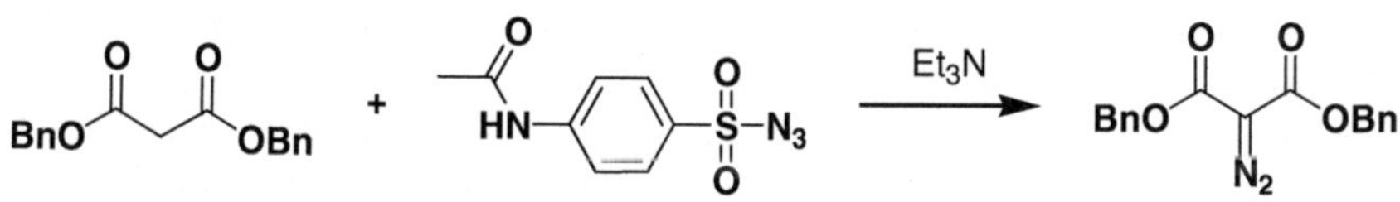

Dibenzyl malonate (**251**) (2.00 g, 7.03 mmol, 1 equiv), triethylamine (1.10 mL, 7.70 mmol, 1.1 equiv) and 4-acetamidobenzenesulfonyl azide (**373**) (1.81 g, 7.53 mmol, 1.1 equiv) were dissolved in acetonitrile (70.0 mL) at room temperature. After stirring the resulting mixture overnight, the suspension was filtered through a plug of cotton wool and the solvent was removed under reduced pressure. The crude was partitioned between dichloromethane (100 mL) and water (100 mL), and filtered through a plug of cotton wool. The two layers were separated and the aqueous layer was extracted with dichloromethane (100 mL). The combined organic layers were dried over $MgSO_4$, concentrated under reduced pressure and dried under vacuum. Filtration over a plug of silica (Diethyl ether/hexane 1/1, 150 mL) afforded dibenzyl 2-diazomalonate (**253**) (2.03 g, 6.53 mmol, 93 % yield) as a yellow solid.

$\mathbf{R_f}$ 0.69 (6:4 Hexane/AcOEt). **Mp** 48.2–51.1 °C. ^{1}**H NMR** (400 MHz, $CDCl_3$) δ 7.38–7.32 (m, 10 H, Ar), 5.28 (s, 4 H, CH_2). ^{13}**C NMR** (101 MHz, $CDCl_3$) δ 160.9, 135.4, 128.8, 128.6, 128.4, 67.2, 41.7. **IR** 3064 (w), 3035 (w), 2950 (w), 2888 (w),

2140 (s), 1756 (s), 1732 (s), 1692 (m), 1456 (w), 1385 (m), 1315 (s), 1268 (m), 1087 (s). **HRMS** (ESI) calcd for $C_{17}H_{15}N_2O_4^+$ $[M + H]^+$ 311.1026; found 311.1016.

Bis(2,2,2-trifluoroethyl) 2-diazomalonate (254)

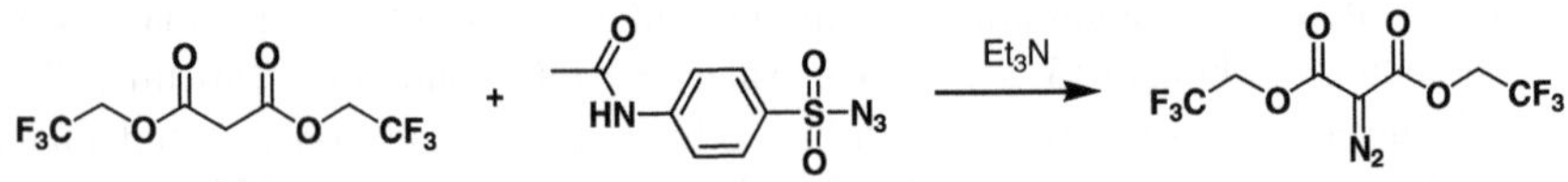

252 373 254

Bis(2,2,2-trifluoroethyl) malonate (**252**) (5.00 g, 18.7 mmol, 1 equiv), triethylamine (2.84 mL, 20.5 mmol, 1.1 equiv) and 4-acetamidobenzenesulfonyl azide (**373**) (4.93 g, 20.5 mmol, 1.1 equiv) were dissolved in acetonitrile (180 mL) at room temperature. After stirring the resulting mixture overnight, the suspension was filtered through a plug of cotton wool and the solvent was removed under reduced pressure. The crude was dissolved in dichloromethane (200 mL), filtered through a plug of cotton wool, and partitioned between dichloromethane and water (200 mL). The two layers were separated and the aqueous layer was extracted with dichloromethane (2 × 150 mL). The combined organic layers were dried over $MgSO_4$, concentrated under reduced pressure and dried under vacuum. Filtration through a plug of silica (AcOEt/hexane 1/1 + 1 % NEt_3, 500 mL) afforded bis (2,2,2-trifluoroethyl) 2-diazomalonate (**254**) (5.44 g, 18.5 mmol, 99 % yield) as a yellow oil which was not further purified.

$\mathbf{R}_f$ 0.67 (6:4 hexane/ethyl acetate). **^{1}H NMR** (400 MHz, $CDCl_3$) δ 4.62 (q, 4 H, $J = 8.2$ Hz, CH_2). **^{13}C NMR** (101 MHz, $CDCl_3$) δ 158.7, 122.6 (q, $J_{C\text{-}F} = 277$ Hz), 60.9 (q, $J_{C\text{-}F} = 37$ Hz).[6] **IR** 2985 (w), 2154 (m), 1777 (s), 1715 (m), 1417 (m), 1357 (m), 1280 (s), 1168 (s), 1102 (s). **HRMS** (ESI) calcd for $C_7H_4F_6N_2NaO_4^+$ $[M + Na]^+$ 316.9967; found 316.9972.

Dimethyl 2-(5-bromo-1,3-dioxoisoindolin-2-yl)cyclopropane-1,1-dicarboxylate (245)

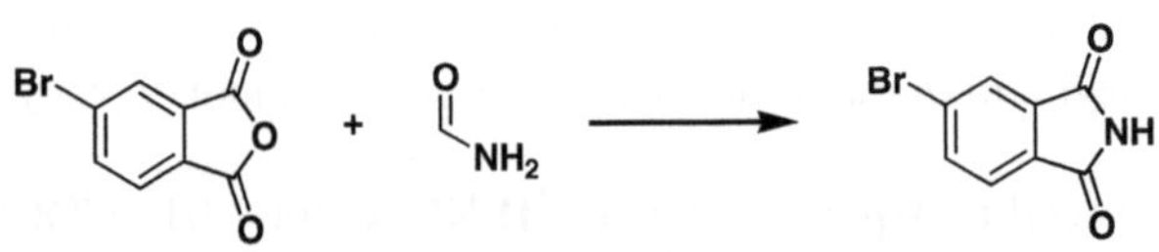

374 244 245

Following a modified procedure [17], 5-bromoisobenzofuran-1,3-dione (**374**) (500 mg, 2.20 mmol, 1 equiv) and formamide (**244**) (8.82 mL, 220 mmol, 100

[6]The diazo carbon could not be detected.

equiv) were added in a 20 mL microwave vial and sealed with a microwave cap. The mixture was stirred until the product was completely dissolved. The mixture was heated twice at 200 °C for 30 with 10 s pre-stirring, using Biotage Initiator 2.0 microwave reactor. The product crystallized spontaneously as colorless needles in the pale yellow solution. The mixture was cooled to 0 °C and cold water (10 mL) was added into the tube. The solid was filtrated over a filter paper, washed with water (15 mL) and hexane (20 mL) and dried under reduced pressure to afford 5-bromoisoindoline-1,3-dione (**245**) (394 mg, 1.75 mmol, 79 % yield) as a colorless solid which was used without further purification.

^{1}H NMR (400 MHz, DMSO) δ 11.4 (s, 1 H, NH), 8.04–7.99 (m, 2 H, Ar), 7.76 (d, 1 H, J = 7.7 Hz, Ar).

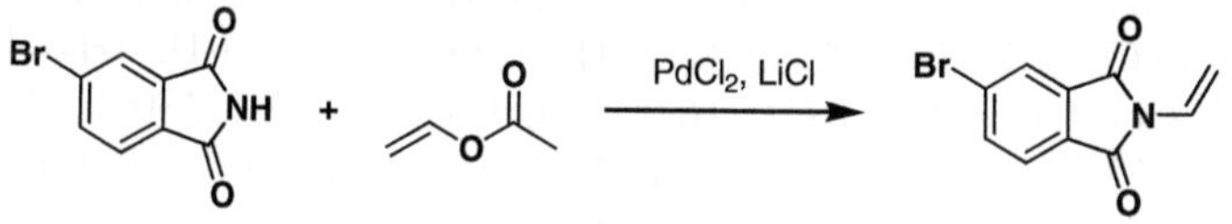

245 **224** **247**

Following a modified procedure [18], 5-bromoisoindoline-1,3-dione (**245**) (1.50 g, 6.64 mmol, 1 equiv), palladium(II) chloride (118 mg, 0.664 mmol, 0.1 equiv), lithium chloride (28.0 mg, 0.660 mmol, 0.1 equiv) and vinyl acetate (**224**) (16.5 mL, 178 mmol, 27 equiv) were added in a microwave tube sealed with a microwave cap. The mixture was stirred for 28 h at 80 °C, and then cooled down to room temperature. The solvent was removed under reduced pressure. Purification by silica gel chromatography (hexane/ethyl acetate 20/1 to 15/1) afforded 5-bromo-2-vinylisoindoline-1,3-dione (**247**) (1.66 g, 6.59 mmol, 99 % yield) as a yellow solid.

R_f 0.47 (9:1 hexane/ethyl acetate). **Mp** 74.9–75.6 °C. **^{1}H NMR** (400 MHz, DMSO) δ 8.11 (d, 1 H, J = 1.4 Hz, Ar), 8.07 (dd, 1 H, J = 8.0, 1.8 Hz, Ar), 7.84 (d, 1 H, J = 7.9 Hz, Ar), 6.81 (dd, 1 H, J = 16.3, 9.8 Hz, =CH), 5.93 (d, 1 H, J = 16.3 Hz, =CH), 5.08 (d, 1 H, J = 9.8 Hz, =CH). **^{13}C NMR** (101 MHz, $CDCl_3$) δ 165.7, 165.1, 138.6, 133.3, 130.1, 129.5, 127.0, 125.0, 123.7, 105.1. **IR** 2927 (w), 2854 (w), 2361 (m), 1728 (s), 1640 (w), 1377 (s), 1303 (w), 1169 (m). **HRMS** (ESI) calcd for $C_{10}^{79}BrH_7NO_2^+$ $[M + H]^+$ 251.9655; found 251.9656.

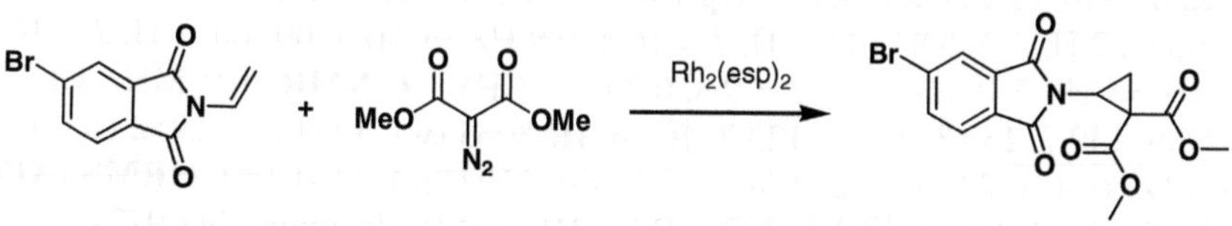

247 **199** **255**

Following a modified procedure [6], a solution of dimethyl 2-diazomalonate (**199**) (0.05 g, 0.3 mmol, 1.5 equiv) in dichloromethane (0.4 mL) was added dropwise over 5 min to a solution of 5-bromo-2-vinylisoindoline-1,3-dione (**247**) (0.05 g, 0.2 mmol, 1 equiv) and bis[rhodium(α,α,α′,α′-tetramethyl-1,3-benzenedipropionic acid)] (0.3 mg, 0.4 μmol, 0.2 mol%) in dichloromethane (0.6 mL) at 0 °C. After stirring the resulting mixture overnight at room temperature, the solution was concentrated under reduced pressure. Purification by silica gel chromatography (hexane/ethyl acetate 75/25–70/30) afforded dimethyl 2-(5-bromo-1,3-dioxoisoindolin-2-yl) cyclopropane-1,1-dicarboxylate (**255**) (0.08 g, 0.2 mmol, 100 %) as a colorless oil.

R_f 0.30 (hexane/ethyl acetate 3/1). **Mp** 114.8–117.5 °C. **^{1}H NMR** (400 MHz, $CDCl_3$) δ 7.99–7.95 (m, 1 H, Ar), 7.86 (dd, 1 H, J = 7.9, 1.6 Hz, Ar), 7.70 (d, 1 H, J = 7.9 Hz, Ar), 3.82 (s, 3 H, Me–O), 3.70–3.64 (m, 1 H, CH–NPhth), 3.62 (s, 3 H, Me–O), 2.66 (dd, 1 H, J = 6.5, 6.5 Hz, CH_2), 2.06–2.01 (m, 1 H, CH_2). **^{13}C NMR** (101 MHz, $CDCl_3$) δ 168.4, 167.0, 167.0, 166.5, 137.4, 133.1, 130.0, 129.4, 126.9, 124.9, 53.2, 53.0, 34.9, 33.1, 19.6. **IR** 2956 (w), 2855 (w), 2361 (w), 1782 (w), 1730 (s), 1605 (w), 1439 (w), 1397 (m), 1331 (m), 1295 (w), 1223 (m), 1134 (w). **HRMS** (ESI) calcd for $C_{15}^{79}BrH_{13}NO_6^+$ $[M + H]^+$ 381.9921; found 381.9920.

Dimethyl 2-(5,6-dichloro-1,3-dioxoisoindolin-2-yl)cyclopropane-1,1-dicarbo xylate (256)

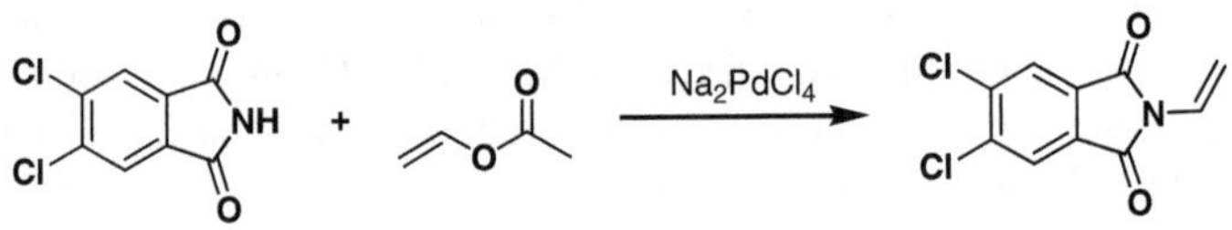

Following a modified procedure [18], Na_2PdCl_4 (27.0 mg, 0.0930 mmol, 2 mol %) was added to a stirred solution of 4,5-dichlorophthalimide (**375**) (1.00 g, 4.63 mmol, 1.00 equiv) in vinyl acetate (**224**) (11.5 mL, 124 mmol, 26.8 equiv), and the mixture was heated under reflux for 48 h. After solvent evaporation, the crude was purified by Biotage (SNAP Cartridge KP-Sil 25 g, 8:2 Hexane/AcOEt) to obtain5,6-dichloro-2-vinylisoindoline-1,3-dione (**250**) (1.12 g, 4.63 mmol, 46 % yield) as a yellow solid.

R_f 0.53 (6:4 Hexane/AcOEt). **Mp** 164.7–166.6 °C. **^{1}H NMR** (400 MHz, $CDCl_3$) δ 7.96 (s, 2 H, Ar), 6.84 (dd, 1 H, J = 16.4, 9.8 Hz, =CH), 6.09 (dd, 1 H, J = 16.4, 0.3 Hz, =CH), 5.10 (dd, 1 H, J = 9.8, 0.3 Hz, =CH). **^{13}C NMR** (101 MHz, $CDCl_3$) δ 164.6, 139.7, 130.8, 125.8, 123.7, 105.6. **IR** 3065 (w), 3034 (w), 2952 (w), 1726 (s), 1392 (m), 1322 (m), 1288 (m), 1217 (m), 1194 (m), 1131 (m). **HRMS (APPI ionization)** calcd for $C_{10}^{35}Cl_2H_6NO_2^+$ $[M + H]^+$ 241.9770; found 241.9771.

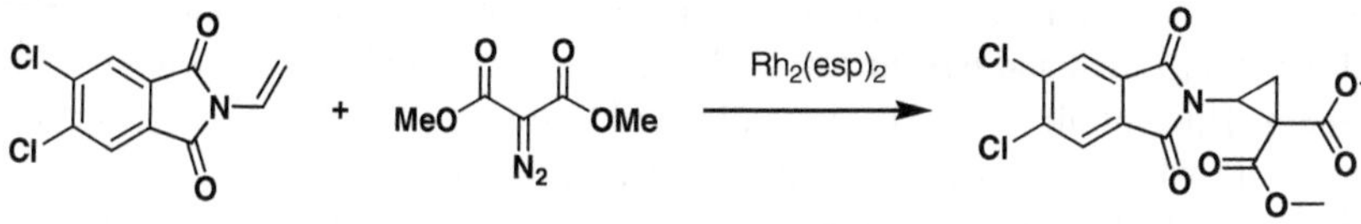

250 199 256

Following a modified procedure [6], a solution of dimethyl 2-diazomalonate (**199**) (0.70 g, 4.4 mmol, 1.5 equiv) in dichloromethane (8.0 mL) was added dropwise over 5 min to a solution of 5,6-dichloro-2-vinylisoindoline-1,3-dione (**250**) (72 mg, 3.0 mmol, 1 equiv) and bis[rhodium(α,α,α′,α′-tetramethyl-1,3-benzenedipropionic acid)] (4.5 mg, 5.9 μmol, 0.2 mol%) in dichloromethane (4.0 mL) at 0 °C. After stirring the resulting mixture overnight at room temperature, the solution was concentrated under reduced pressure. Purification by Biotage (SNAP cartridge KP-Sil 25 g, hexane/AcOEt 95/5 to 60/40) afforded dimethyl 2-(5,6-dichloro-1,3-dioxoisoindolin-2-yl) cyclopropane-1,1-dicarboxylate **(256)** (0.81 g, 2.2 mmol, 77 % yield) as a colorless solid.

R_f 0.27 (8:2 hexane/ethyl acetate). **Mp** 145.9–148.1 °C. **^{1}H NMR** (400 MHz, $CDCl_3$) δ 7.92 (s, 2 H, Ar), 3.82 (s, 3 H, Me–O), 3.66 (dd, 1 H, J = 8.5, 6.7 Hz, CH–Phth), 3.63 (s, 3 H, Me–O), 2.64 (dd, 1 H, J = 6.5. 6.5 Hz, CH_2), 2.07–2.01 (m, 1 H, CH_2). **^{13}C NMR** (101 MHz, $CDCl_3$) δ 168.4, 167.1, 166.0, 139.5, 130.6, 125.7, 53.3, 53.2, 35.0, 33.1, 19.8. **IR** 3096 (w), 3033 (w), 2955 (w), 2851 (w), 1787 (m), 1725 (s), 1438 (m), 1398 (s), 1308 (m), 1222 (m), 1134 (m). **HRMS** (ESI) calcd for $C_{15}{}^{35}Cl_2H_{12}NO_6{}^+$ $[M + H]^+$ 372.0036; found 372.0022.

Dimethyl 2-(1,3-dioxo-1H-benzo[*f*]isoindol-2(3H)-yl)cyclopropane-1,1-dicarboxylate (257)

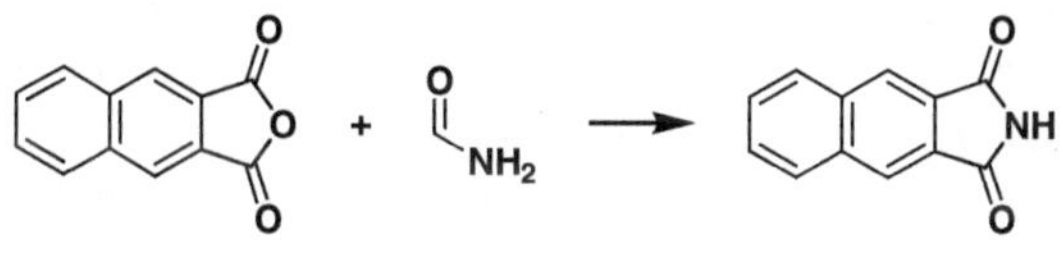

376 244 246

Following a modified procedure [17], naphtho[2,3-*c*]furan-1,3-dione (**376**) (500 mg, 2.52 mmol, 1 equiv) and formamide (**244**) (10.0 mL, 252 mmol, 100 equiv) were added in a 20 mL microwave vial and sealed with a microwave cap. The mixture was stirred until the product was completely dissolved. The mixture was heated twice at 200 °C for 30 s with 10 s pre-stirring, using Biotage Initiator 2.0 microwave reactor. The mixture was cooled to 0 °C and cold water

(10 mL) was added into the tube. The solid was filtrated over a filter paper, washed with water (15 mL) and hexane (20 mL) and dried under reduced pressure to afford 1*H*-benzo[*f*]isoindole-1,3(2 H)-dione (**246**) (432 mg, 2.19 mmol, 87 % yield) as a beige solid which was used without further purification.

R*f* 0.44 (6:4Hexane/AcOEt). **Mp** 267°C decomp. **^{1}H NMR** (400 MHz, DMSO) δ 11.5 (s, 1 H, NH), 8.45 (s, 2 H, Ar), 8.26 (dd, 2 H, J = 6.1, 3.3 Hz, Ar), 7.76 (dd, 2 H, J = 6.6, 3.3 Hz, Ar). **^{13}C NMR** (101 MHz, DMSO) δ 168.9, 135.1, 130.2, 129.1, 128.7, 124.2. **IR** 3224 (w), 3071 (w), 2925 (w), 2852 (w), 1707 (s), 1447 (w), 1316 (m), 1113 (m), 1012 (w), 905 (w). **HRMS** (ESI) calcd for $C_{12}H_7NNaO_2^+$ $[M + Na]^+$ 220.0369; found 220.0380.

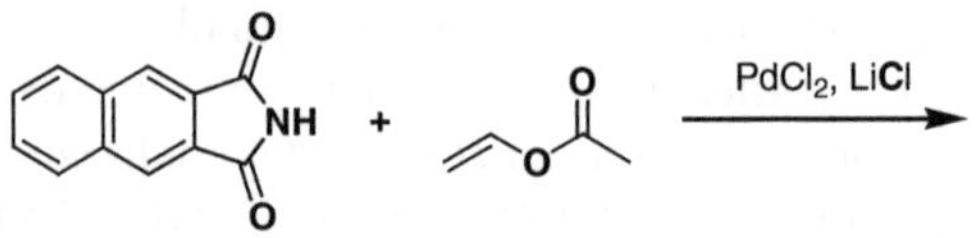

246 **224** **248**

Following a modified procedure [18], 1*H*-benzo[*f*]isoindole-1,3(2 H)-dione (**246**) (1.70 g, 8.62 mmol, 1 equiv), palladium(II) chloride (0.150 g, 0.860 mmol, 0.1 equiv), lithium chloride (0.0370 g, 0.860 mmol, 0.1 equiv) and vinyl acetate (**224**) (21.4 mL, 231 mmol, 27 equiv) were added in a microwave tube sealed with a microwave cap. After stirring for 31 h at 80 °C, the resulting mixture was cooled down to room temperature. Purification by silica gel chromatography (hexane/ethyl acetate 17/1 to 10/1) afforded 2-vinyl-1*H*-benzo[*f*]isoindole-1,3(2 H)-dione (**248**) (1.26 g, 5.66 mmol, 66 % yield) as a colorless solid.

R*f* 0.51 (8:2 hexane/ethyl acetate). **Mp** 201.9–202.8 °C. **^{1}H NMR** (400 MHz, $CDCl_3$) δ 8.37 (s, 2 H, Ar), 8.07 (dd, 2 H, J = 6.1, 3.4 Hz, Ar), 7.72 (dd, 2 H, J = 6.3, 3.3 Hz, Ar), 6.97 (dd, 1 H, J = 16.4, 9.7 Hz, =CH), 6.20 (d, 1 H, J = 16.4 Hz, =CH), 5.12 (d, 1 H, J = 9.9 Hz, =CH). **^{13}C NMR** (101 MHz, $CDCl_3$) δ 166.4, 135.9, 130.5, 129.6, 127.3, 125.4, 124.3, 105.3. **IR** 3029 (w), 2949 (w), 1704 (s), 1379 (w), 1303 (w), 1218 (w), 1139 (m), 1012 (w), 975 (w), 882 (w). **HRMS** (ESI) calcd for $C_{14}H_{10}NO_2^+$ $[M + H]^+$ 224.0706; found 224.0710.

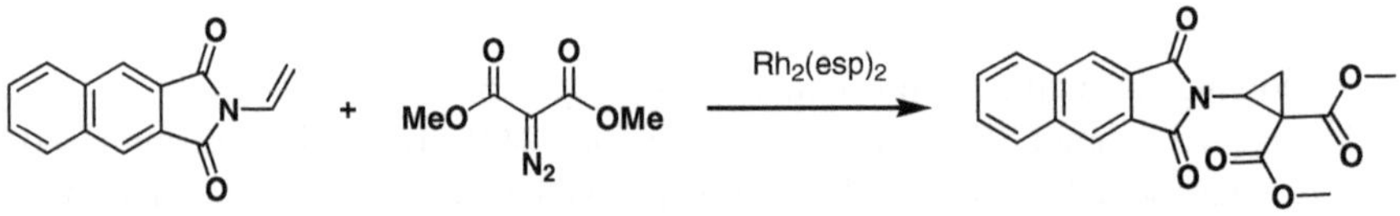

248 **199** **257**

Following a modified procedure [6], a solution of dimethyl 2-diazomalonate (**199**) (0.20 g, 1.3 mmol, 1.5 equiv) in dichloromethane (2.0 mL) was added dropwise over 5 min to a solution of 2-vinyl-1*H*-benzo[*f*]isoindole-1,3(2 H)-dione (**248**) (0.19 g, 0.85 mmol, 1 equiv) and bis[rhodium(α,α,α′,α′-tetramethyl-1,3-benzenedipropionic acid)] (1.3 mg, 1.7 μmol, 0.2 mol%) in dichloromethane (3.0 mL) at 0 °C. After stirring the resulting mixture for 26 h at room temperature the solution was concentrated under reduced pressure. Purification by silica gel chromatography (8:2 hexane/ethyl acetate to 6/4) afforded dimethyl 2-(1,3-dioxo-1*H*-benzo[*f*]isoindol-2(3 H)-yl) cyclopropane-1,1-dicarboxylate (**257**) (0.28 g, 0.80 mmol, 94 % yield) as a colorless solid.

R*f* 0.39 (6:4 hexane/ethyl acetate). **Mp** 165.6–167.5 °C. **^{1}H NMR** (400 MHz, $CDCl_3$) δ 8.34 (s, 2 H, Ar), 8.06 (dd, 2 H, *J* = 6.0, 3.3 Hz, Ar), 7.70 (dd, 2 H, *J* = 6.3, 3.3 Hz, Ar), 3.84 (s, 3 H, Me–O), 3.77 (dd, 1 H, *J* = 8.5, 6.7 Hz, CH–N), 3.60 (s, 3 H, Me–O), 2.78 (dd, 1 H, *J* = 6.5, 6.5 Hz, CH_2), 2.08 (dd, 1 H, *J* = 8.5, 6.4, CH_2). **^{13}C NMR** (101 MHz, $CDCl_3$) δ 168.7, 167.7, 167.0, 135.7, 130.4, 129.4, 127.2, 125.1, 53.2, 53.1, 35.3, 33.3, 19.8. **IR** 3034 (w), 2955 (w), 1720 (s), 1439 (m), 1393 (m), 1332 (m), 1292 (m), 1222 (m), 1134 (m), 912 (m). **HRMS** (ESI) calcd for $C_{19}H_{16}NO_6^+$ $[M + H]^+$ 354.0972; found 354.0968.

Dimethyl 2-(2,5-dioxo-2,5-dihydro-1H-pyrrol-1-yl)cyclopropane-1,1-dicarboxylate (258)

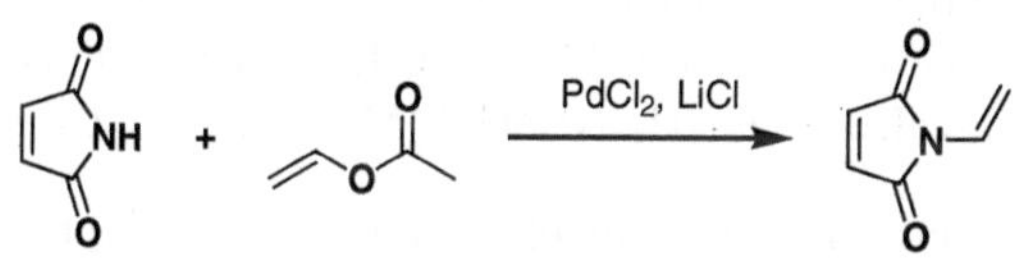

Following a modified procedure, maleimide (**377**) (1.30 g, 13.4 mmol, 1 equiv), palladium (II) chloride (0.237 g, 1.34 mmol, 0.1 equiv), lithium chloride (57.0 mg, 1.34 mmol, 0.1 equiv) and vinyl acetate (**224**) (33.2 mL, 359 mmol, 27 equiv) were added in a microwave tube sealed with a microwave cap. After stirring at 80 °C for 23 h, the resulting mixture was cooled down to room temperature. Purification by Biotage (SNAP cartridge KP-Sil 50 g, hexane/AcOEt 93/7–40/60) afforded 1-vinyl-1*H*-pyrrole-2,5-dione (**249**) (1.74 g, 14.1 mmol, quantitative) as a bright yellow oil.

R*f* 0.54 (7:3 hexane/ethyl acetate). **^{1}H NMR** (400 MHz, $CDCl_3$) δ 6.74 (s, 2 H, CH–C=O), 6.67 (dd, 1 H, *J* = 16.4, 9.8 Hz, CH–N), 5.87 (d, 1 H, *J* = 16.3 Hz, =CH_2), 4.94 (d, 1 H, *J* = 9.8 Hz, =CH_2). **^{13}C NMR** (101 MHz, $CDCl_3$) δ 168.7, 134.5, 123.1, 103.4. **IR** 3087 (w), 2359 (w), 2113 (w), 1716 (s), 1641 (m), 1384 (s), 1307 (w), 1221 (w), 1130 (w), 896 (w), 845 (m). **HRMS** (APPI) calcd for $C_6H_5NO_2$ $[M^+]$ 123.0320; found 123.0323.

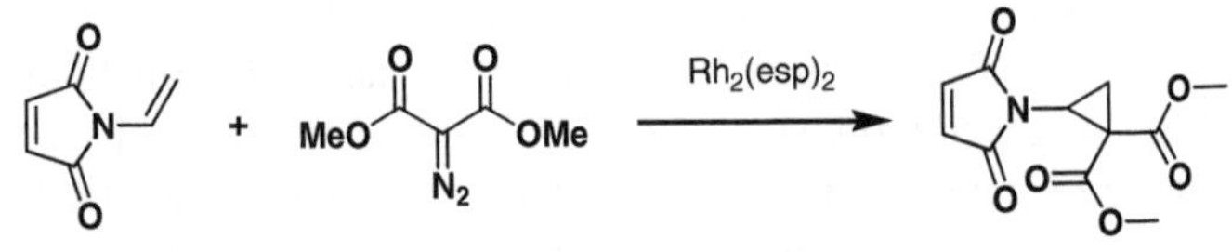

249 199 258

Following a modified procedure [6], a solution of dimethyl 2-diazomalonate (**199**) (96 mg, 0.61 mmol, 1.5 equiv) in dichloromethane (1.0 mL) was added dropwise over 5 min to a solution of 1-vinyl-1*H*-pyrrole-2,5-dione (**249**) (50 mg, 0.41 mmol, 1 equiv) and bis[rhodium(α,α,α′,α′-tetramethyl-1,3-benzenedipropionic acid)] (0.7 mg, 0.9 μmol, 0.2 mol%) in dichloromethane (2.0 mL) at 0 °C. The resulting mixture was stirred for 5 h at room temperature and finally concentrated under reduced pressure. Purification by Biotage (SNAP cartridge KP-Sil 10 g, hexane/AcOEt 95/5 to 70/30) afforded dimethyl 2-(2,5-dioxo-2,5-dihydro-1*H*-pyrrol-1-yl)cyclopropane-1,1-dicarboxylate (**258**) (66.9 mg, 0.264 mmol, 65 % yield) as a colorless oil.

R_f 0.38 (6:4 hexane/ethyl acetate). **Mp** 78.4–80.7 °C. **^{1}H NMR** (400 MHz, $CDCl_3$) δ 6.67 (s, 2 H, CH–C=O), 3.79 (s, 3 H, Me–O), 3.66 (s, 3 H, Me–O), 3.56–3.51 (m, 1 H, CH–N), 2.56 (dd, 1 H, J = 6.4, 6.5 Hz, CH_2), 1.96–1.91 (m, 1 H, CH_2). **^{13}C NMR** (101 MHz, $CDCl_3$) δ 170.0, 168.4, 167.0, 134.1, 53.1, 53.0, 34.3, 32.9, 19.3. **IR** 2363 (w), 1727 (s), 1437 (w), 1332 (w), 1296 (w), 1220 (w), 1135 (w). **HRMS** (ESI) calcd for $C_{11}H_{11}NNaO_6^+$ $[M + Na]^+$ 276.0479; found 276.0485.

Bis(2,2,2-trifluoroethyl) 2-(1,3-dioxoisoindolin-2-yl)cyclopropane-1,1-dicarbox ylate (259)

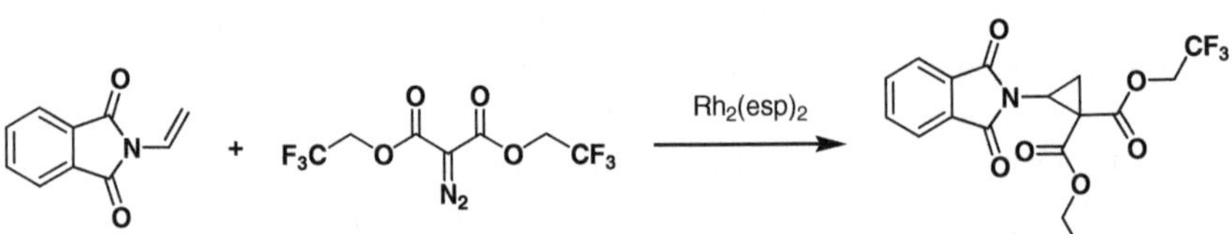

183 254 259

Following a modified procedure [6], a solution of bis(2,2,2-trifluoroethyl) 2-diazomalonate (**254**) (1.0 g, 3.4 mmol, 1.1 equiv) in dichloromethane (4.0 mL) was added dropwise over 5 min to a solution of 2-vinylisoindoline-1,3-dione (**183**) (540 mg, 3.10 mmol, 1 equiv) and bis[rhodium(α,α,α′,α′-tetramethyl-1,3-benzenedipropionic acid)] (4.7 mg, 6.2 μmol, 0.2 mol%) in dichloromethane (6.0 mL) at 0 °C. After stirring the resulting mixture overnight at room temperature, the solution was concentrated under reduced pressure. Purification by silica gel chromatography (hexane/AcOEt 95/5–75/25) afforded bis (2,2,2-trifluoroethyl) 2-(1,3-dioxoisoindolin-2-yl)cyclopropane-1,1-dicarboxylate (**259**) (1.3 g, 3.0 mmol, 97 % yield) as a colorless solid.

R$_f$ 0.61 (6:4 hexane/ethyl acetate). **Mp** 76.3–78.0 °C. **^{1}H NMR** (400 MHz, $CDCl_3$) δ 7.87–7.82 (m, 2 H, Phth), 7.77–7.72 (m, 2 H, Phth), 4.61 (q, 2 H, J = 8.2 Hz, CH_2–CF_3), 4.50–4.30 (m, 2 H, CH_2–CF_3), 3.83 (dd, 1 H, J = 8.6, 6.9 Hz, CH–Phth), 2.90 (dd, 1 H, J = 6.8, 6.8 Hz, CH_2), 2.19 (dd, 1 H, J = 8.6, 6.6 Hz, CH_2). **^{13}C NMR** (101 MHz, $CDCl_3$) δ 167.7, 166.4, 164.5, 134.7, 131.4, 123.8, 122.7 (q, $J_{C\text{-}F}$ = 277 Hz), 122.5 (q, $J_{C\text{-}F}$ = 277 Hz), 61.7 (q, $J_{C\text{-}F}$ = 37 Hz), 61.5 (q, $J_{C\text{-}F}$ = 37 Hz), 36.4, 32.7, 21.0. **IR** 3495 (w), 3121 (w), 3037 (w), 2981 (w), 1724 (s), 1397 (m), 1277 (s), 1163 (s), 1116 (s), 975 (m). **HRMS** (ESI) calcd for $C_{17}F_6H_{12}NO_6^+$ $[M + H]^+$ 440.0563; found 440.0555.

Dibenzyl 2-(1,3-dioxoisoindolin-2-yl)cyclopropane-1,1-dicarboxylate (260)

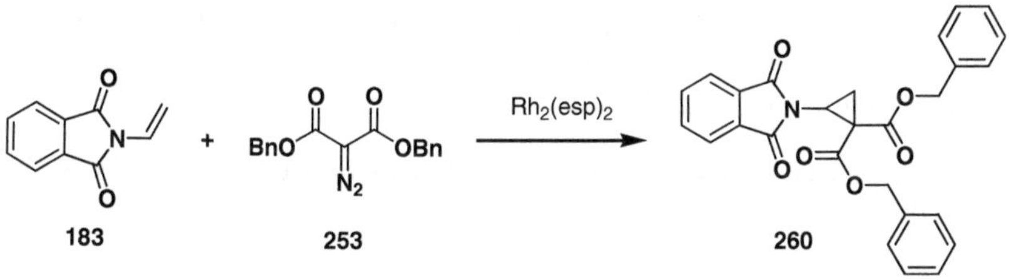

Following a modified procedure [6], a solution of dibenzyl 2-diazomalonate (**253**) (2.0 g, 6.5 mmol, 1.10 equiv) in dichloromethane (12.0 mL) was added dropwise over 5 min to a solution of 2-vinylisoindoline-1,3-dione (**183**) (1.0 g, 5.9 mmol, 1.00 equiv) and bis[rhodium(α,α,α′,α′-tetramethyl-1,3-benzenedipropionic acid)] (9.0 mg, 0.012 mmol, 0.2 mol%) in dichloromethane (10.0 mL) at 0 °C. After stirring the resulting mixture overnight at room temperature, the solution was concentrated under reduced pressure. Purification by Biotage (SNAP cartridge KP-Sil 50 g, hexane/AcOEt 90/10 to 20/80) afforded dibenzyl 2-(1,3-dioxoisoindolin-2-yl) cyclopropane-1,1-dicarboxylate (**260**) (0.91 g, 2.0 mmol, 34 % yield) as a colorless oil.

R$_f$ 0.60 (6:4 hexane/ethyl acetate). **^{1}H NMR** (400 MHz, $CDCl_3$) δ 7.78–7.73 (m, 2 H, Phth), 7.72–7.67 (m, 2 H, Phth), 7.36–7.28 (m, 5 H, Ar), 7.20–7.13 (m, 5 H, Ar), 5.29–5.19 (m, 2 H, CH_2–Ph), 5.04–4.95 (m, 2 H, CH_2–Ph), 3.73 (dd, 1 H, J = 8.5, 6.7 Hz, CH–Phth), 2.79 (dd, 1 H, J = 6.5, 6.5 Hz, CH_2), 2.02 (m, 1 H, CH_2). **^{13}C NMR** (101 MHz, $CDCl_3$) δ 168.1, 167.8, 166.2, 135.4, 135.1, 134.3, 131.5, 128.7, 128.4, 128.4, 128.3, 128.1, 123.6, 67.8, 67.7, 35.3, 33.5, 19.8.[7] **IR** 3063 (w), 3034 (w), 2954 (w), 1722 (s), 1388 (m), 1319 (m), 1285 (m), 1215 (m), 1128 (m), 980 (w). **HRMS** (ESI) calcd for $C_{27}H_{21}NNaO_6^+$ $[M + Na]^+$ 478.1261; found 478.1262.

[7]One of the aromatic carbons is overlapping.

GP7: Optimization of FC alkylation Table 2.11

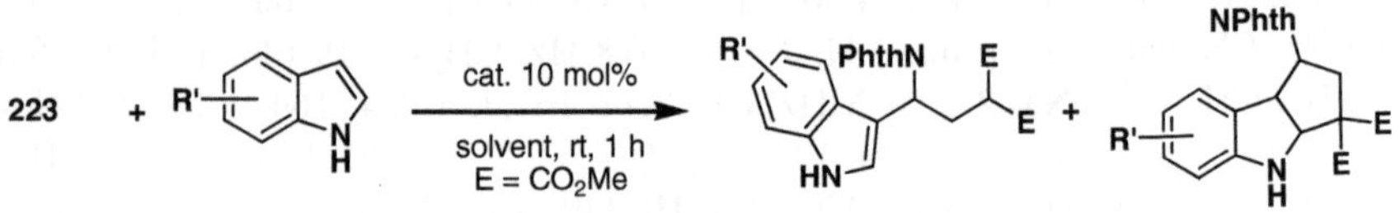

Dimethyl 2-(1,3-dioxoisoindolin-2-yl)cyclopropane-1,1-dicarboxylate (**223**) (10 mg, 0.034 mmol, 1 equiv) and indole (0.051 mmol, 1.5 equiv) were dissolved in solvent (0.20 mL) and added to the Lewis acid (3.4 μmol, 0.1 equiv) in a sealed tube closed with a septum. The mixture was stirred at room temperature for 1 h. The mixture was then filtered through a short plug of silica gel (AcOEt/hexane 1/1, deactivated with 2 % NEt_3, 1 mL) and concentrated under reduced pressure to remove volatiles. The by ^{1}H NMR.

Dimethyl 2-(2-(1,3-dioxoisoindolin-2-yl)-2-(1-methyl-1H-indol-3-yl)ethyl)malonate (240a)

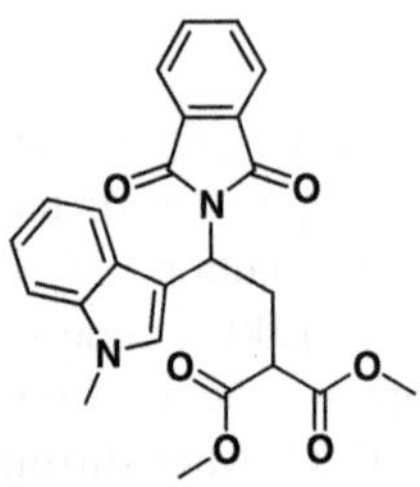

R*f* 0.40 (6:4 Hexane/AcOEt). **^{1}H NMR** (400 MHz, $CDCl_3$) δ 7.82–7.77 (m, 2 H, Phth), 7.75–7.71 (m, 1 H, ArH), 7.71–7.65 (m, 2 H, Phth), 7.41 (s, 1 H, ArH), 7.32–7.27 (m, 1 H, ArH), 7.26–7.19 (m, 1 H, ArH), 7.16–7.09 (m, 1 H, ArH), 5.79 (dd, 1 H, J = 9.7, 6.5 Hz, N–C–H), 3.80 (s, 3 H, Me), 3.76 (s, 3 H, Me), 3.68 (s, 3 H, Me), 3.53–3.47 (m, 1 H, CH), 3.30–3.20 (m, 1 H, CH_2), 3.09–3.00 (m, 1 H, CH_2). **^{13}C NMR** (101 MHz, $CDCl_3$) δ 169.2, 169.2, 168.1, 136.5, 133.9, 131.9, 128.7, 126.9, 123.2, 121.9, 119.6, 119.0, 111.6, 109.3, 52.8, 52.7, 49.6, 44.5, 33.0, 30.9. **IR** 2953 (w), 2365 (w), 1753 (m), 1736 (s), 1711 (s), 1614 (w), 1543 (w), 1469 (w), 1436 (w), 1383 (m), 1355 (w), 1329 (m), 1275 (w), 1158 (w), 913 (w), 739 (m), 728 (s). **HRMS (ESI)** calcd for $C_{24}H_{23}N_2O_6^+$ $[M + H]^+$ 435.1551; found 435.1548.

Dimethyl-1-(1,3-dioxoisoindolin-2-yl)-4-methyl-1,3a,4,8b-tetrahydrocyclopenta [b]indole-3,3(2H)-dicarboxylate (241a)

Diastereomer 1:

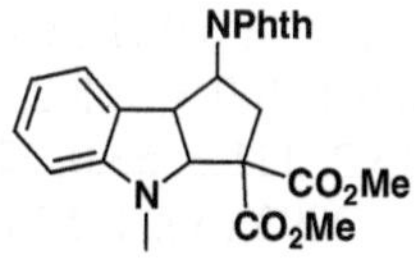

R_f 0.50 (6:4 Hexane/AcOEt). **^{1}H NMR** (400 MHz, $CDCl_3$) δ 7.90–7.85 (m, 2 H, Phth), 7.78–7.72 (m, 2 H, Phth), 7.10–7.04 (m, 1 H, ArH), 6.87–6.83 (m, 1 H, ArH), 6.59 (td, 1 H, $J = 7.4, 0.9$ Hz, ArH), 6.41 (d, 1 H, $J = 7.9$ Hz, ArH), 5.26 (dt, 1 H, $J = 11.3, 7.4$ Hz, CH), 4.96 (d, 1 H, $J = 11.2$ Hz, CH), 4.44 (dd, 1 H, $J = 11.0, 8.1$ Hz, CH), 3.80 (s, 3 H, CH_3), 3.58 (s, 3 H, CH_3), 2.92 (s, 3 H, CH_3), 2.79 (dd, 1 H, $J = 12.9, 11.3$ Hz, CH_2), 2.57 (dd, 1 H, $J = 12.9, 7.0$ Hz, CH_2). **^{13}C NMR** (101 MHz, CDCl3) δ 171.5, 170.1, 168.2, 152.0, 134.3, 131.9, 129.9, 128.5, 123.5, 123.3, 118.3, 107.5, 74.8, 65.4, 56.9, 53.0, 52.5, 49.1, 36.2, 35.9. **IR** 2954 (w), 2257 (w), 1731 (s), 1710 (s), 1378 (m), 1256 (m), 719 (s). **HRMS** (ESI) calcd for $C_{24}H_{23}N_2O_6^+$ $[M + H]^+$ 435.1551; found 435.1567.

Diastereomer 2:

R_f 0.50 (6:4 Hexane/AcOEt). **^{1}H NMR** (400 MHz, $CDCl_3$) δ 7.92–7.56 (m, 4 H, Phth), 7.08–7.01 (m, 1 H, ArH), 6.52 (d, 1 H, $J = 7.9$ Hz, ArH), 6.45–6.41 (m, 1 H, ArH), 6.32 (td, 1 H, $J = 7.4, 0.7$ Hz, ArH), 4.76-4.66 (m, 1 H, CH), 4.64 (dd, 1 H, $J = 8.1, 0.8$ Hz, CH), 4.27 (t, 1 H, $J = 8.9$ Hz, CH), 3.98 (t, 1 H, $J = 13.4$ Hz, CH_2), 3.84 (s, 3 H, CH_3), 3.83 (s, 3 H, CH_3), 2.76 (s, 3 H, CH_3), 2.34 (dd, 1 H, $J = 13.0, 6.4$ Hz, CH_2). **^{13}C NMR** (101 MHz, $CDCl_3$) δ 171.2, 168.8, 155.4, 134.3 (br), 128.9, 127.5, 124.2, 123.3, 118.6, 109.4, 76.1, 64.0, 53.7, 53.3, 52.7, 48.5, 38.8, 30.8.[8] **IR** 2955 (w), 2924 (w), 1732 (s), 1714 (s), 1375 (m), 1268 (w), 1130 (w), 719 (s). **HRMS** (ESI) calcd for $C_{24}H_{23}N_2O_6^+$ $[M + H]^+$ 435.1551; found 435.1533.

Dimethyl 2-(2-(1,3-dioxoisoindolin-2-yl)-2-(1H-indol-3-yl)ethyl)malonate (240b)

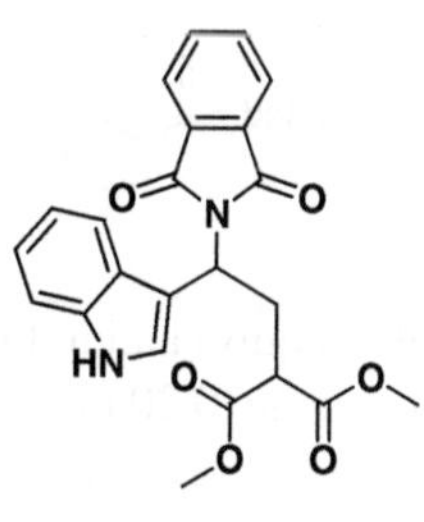

[8] Some phthalimide peaks could not be resolved.

R_f 0.19 (6:4 Hexane/AcOEt). **Mp** 78.2–82.9 °C. **^{1}H NMR** (400 MHz, $CDCl_3$) δ 8.41 (s, 1 H, N*H*), 7.80–7.73 (m, 2 H, Phth), 7.71 (d, 1 H, *J* = 7.9 Hz, ArH), 7.68–7.62 (m, 2 H, Phth), 7.50 (d, 1 H, *J* = 1.9 Hz, ArH), 7.34 (d, 1 H, *J* = 8.0 Hz, ArH), 7.19–7.14 (m, 1 H, ArH), 7.13–7.08 (m, 1 H, ArH), 5.79 (dd, 1 H, *J* = 9.8, 6.4 Hz, PhthN–CH), 3.74 (s, 3 H, CH_3), 3.65 (s, 3 H, CH_3), 3.50 (t, 1 H, *J* = 7.4 Hz, CH–$(CO)_2$), 3.26 (ddd, 1 H, *J* = 14.2, 9.9, 6.9 Hz, CH_2), 3.09–3.00 (m, 1 H, CH_2). **^{13}C NMR** (101 MHz, $CDCl_3$) δ 169.4, 169.3, 168.2, 135.8, 134.1, 131.8, 126.4, 124.3, 123.3, 122.5, 120.1, 118.9, 113.2, 111.3, 53.0, 52.9, 49.7, 44.6, 30.8. **IR** 3403 (w), 2954 (w), 2865 (w), 2363 (w), 2093 (w), 1753 (m), 1733 (s), 1710 (s), 1358 (m). **HRMS** (ESI) calcd for $C_{23}H_{20}N_2NaO_6^+$ $[M + Na]^+$ 443.1214; found 443.1216.

Dimethyl 2-(2,2-di(1H-indol-3-yl)ethyl)malonate (243)

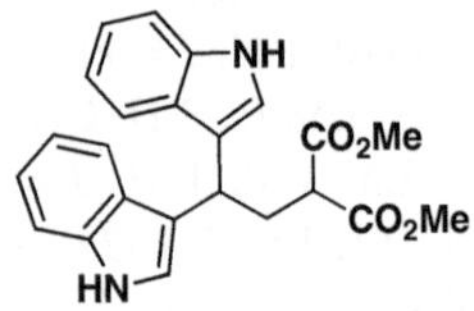

R_f 0.54 (1:1 Hexane/AcOEt). **^{1}H NMR** (400 MHz, $CDCl_3$)[9] δ 7.97 (s, 2 H, NH), 7.63 (d, 2 H, *J* = 7.9 Hz, ArH), 7.36–7.32 (m, 2 H, ArH), 7.19–7.14 (m, 2 H, ArH), 7.07–7.03 (m, 2 H, ArH), 7.02 (d, 2 H, *J* = 1.7 Hz, ArH), 4.55 (t, 1 H, *J* = 7.8 Hz, CH), 3.67 (s, 6 H, CH_3), 3.53 (t, 1 H, *J* = 7.4 Hz, CH), 2.85 (t, 2 H, *J* = 7.5 Hz, CH_2).[10] **^{13}C NMR** (101 MHz, $CDCl_3$) δ 170.1, 136.7, 126.9, 122.0, 121.8, 119.7, 119.3, 118.5, 111.1, 52.5, 50.4, 34.5, 32.3. **IR** 3414 (m), 3056 (w), 2952 (w), 2865 (w), 2360 (w), 1731 (s), 1458 (m), 1437 (m), 1342 (m), 1269 (m), 1229 (m), 1159 (m). **HRMS** (ESI) calcd for $C_{23}H_{22}N_2NaO_4^+$ $[M + Na]^+$ 413.1472; found 413.1468.

GP8: Screening of aminocyclopropanes (Table 2.12)

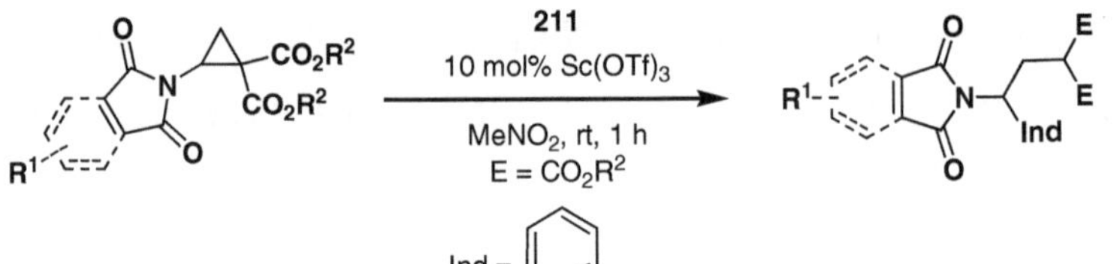

255-260

261-266

Cyclopropane (0.034 mmol, 1 equiv) and indole (**211**) (6.0 mg, 0.051 mmol, 1.5 equiv) were dissolved in nitromethane (0.20 mL) and added to the catalyst (3.4 μmol,

[9]Measured at 268 K.

[10]An impurity is present in approx. 5 % in ^{1}H NMR spectrum.

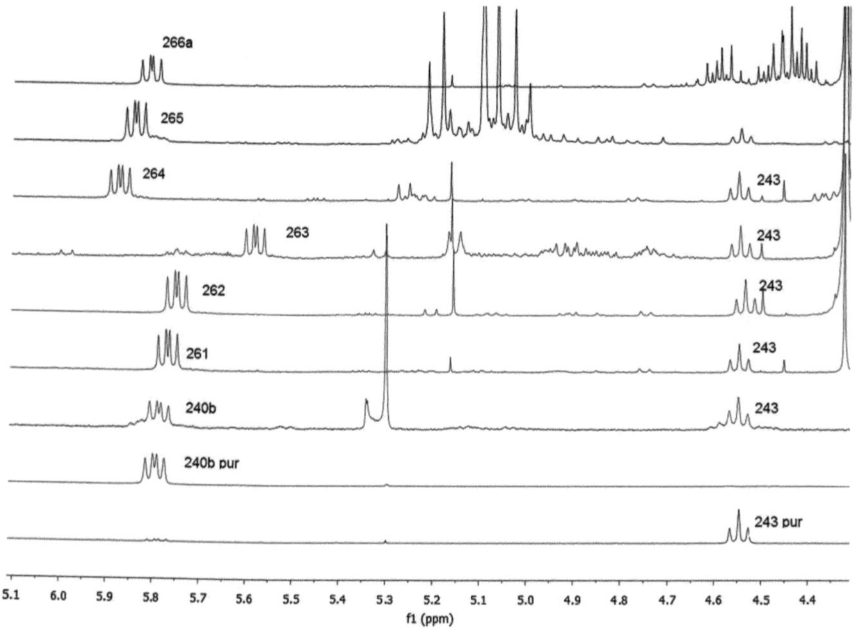

Fig. 5.1 Crude ^{1}H NMR of entries 1–6, Table 2.10

0.100 equiv) in a tube sealed with a septum. The mixture was stirred at room temperature for 1 h. The mixture was then filtered through a short plug of silica gel (AcOEt/hexane 1/1, deactivated with 2 % NEt_3, 1 mL) and concentrated under reduced pressure to remove volatiles and the crude was analyzed by ^{1}H NMR. The products **261-265** were not isolated and solely identified in the ^{1}H crude NMR

Ratios of Friedel-Crafts products and side products were obtained by integrating the doublet of doublet at 5.7–5.9 ppm for mono-addition and the triplet at 4.55 ppm for **243**. For entry 6 of the Table 2.12 with **266a**, absence of bis-addition side product was confirmed by checking the crude at 2.75 ppm (Figs. 5.1 and 5.2).

GP9: Last optimization (not described in main part).

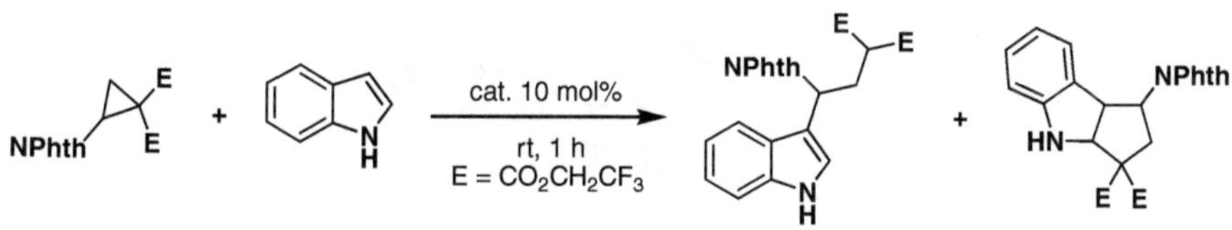

259 **32** **266a** **266a'**

Bis(2,2,2-trifluoroethyl) 2-(1,3-dioxoisoindolin-2-yl) cyclopropane-1,1-dicarboxylate **259** (15.0 mg, 0.0340 mmol, 1.00 equiv) and 1H-

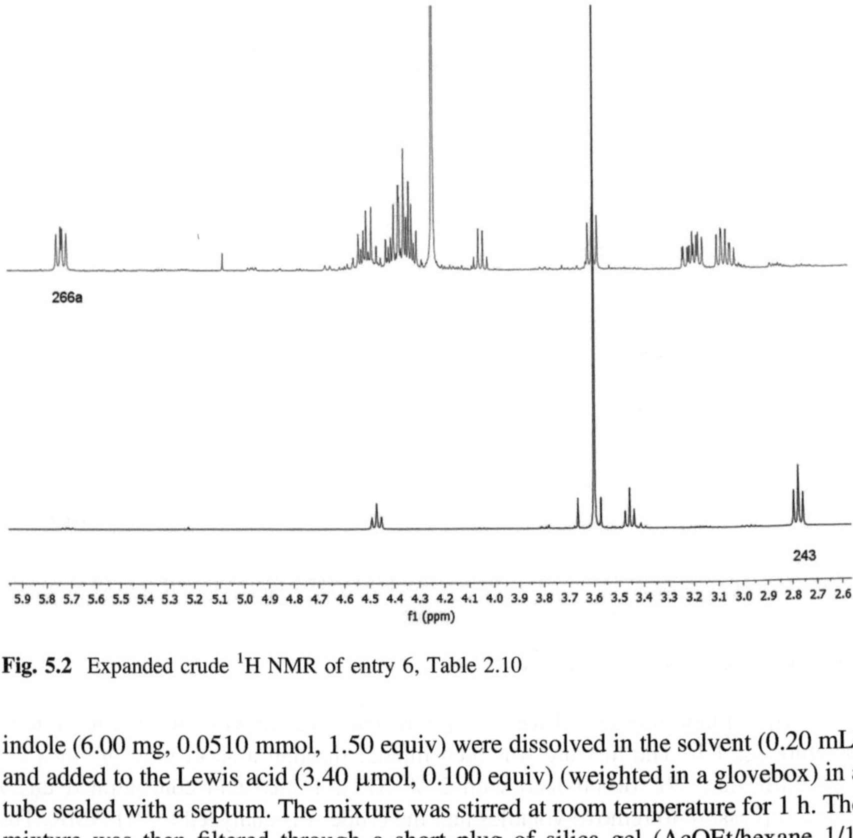

Fig. 5.2 Expanded crude ^{1}H NMR of entry 6, Table 2.10

indole (6.00 mg, 0.0510 mmol, 1.50 equiv) were dissolved in the solvent (0.20 mL) and added to the Lewis acid (3.40 μmol, 0.100 equiv) (weighted in a glovebox) in a tube sealed with a septum. The mixture was stirred at room temperature for 1 h. The mixture was then filtered through a short plug of silica gel (AcOEt/hexane 1/1, deactivated with 2 % NEt_3, 1 mL) and concentrated under reduced pressure to remove volatiles and the crude was analyzed by ^{1}H NMR (Table 5.2).

Bis(2,2,2-trifluoroethyl)-1-(1,3-dioxoisoindolin-2-yl)-1,3a,4,8b-tetrahydrocyclopenta-[b]indole-3,3(2 H)-dicarboxylate (266a′)

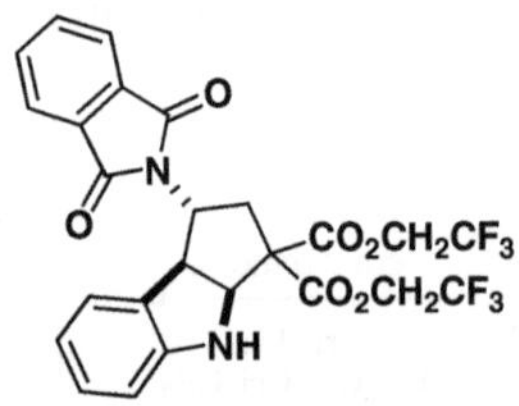

R$_f$ 0.56 (6:4 Hexane/AcOEt). **Mp** 59.2–63.5 °C. **^{1}H NMR** (400 MHz, $CDCl_3$)[11] δ 7.88 (d, 1 H, J = 7.3 Hz, Phth), 7.70 (dt, 2 H, J = 19.7, 7.3 Hz, Phth), 7.59 (d, 1 H,

[11]Measured at 268 K.

Table 5.2 Optimization of the Friedel-Crafts

Entry	Catalyst	Solvent	266a/266a′	dr	Conv. (%)
1	$Cu(OTf)_2$	ACN	>20:1	nd	100
2	$Cu(OTf)_2$	Toluene	1:5.4	1:2.2	100
3	$Cu(OTf)_2$	DCM	1:3.2	1:1.7	100
4	$Cu(OTf)_2$	Et_2O	>20:1	nd	100
5	$Cu(OTf)_2$	THF	1:0.45	<20:1	15
6	$Sc(OTf)_3$	ACN	>20:1	nd	100
7	$Sc(OTf)_3$	Toluene	1:0.3	<1:20	100
8	$Sc(OTf)_3$	DCM	1:0.6	13:1	100
9	$Sc(OTf)_3$	Et_2O	>20:1	nd	100
10	$Sc(OTf)_3$	THF	1:0.4	<1:20	40

J = 7.1 Hz, Phth), 7.00 (t, 1 H, J = 7.6 Hz, ArH), 6.65 (d, 1 H, J = 7.8 Hz, ArH), 6.50 (d, 1 H, J = 7.4 Hz, ArH), 6.38 (t, 1 H, J = 7.4 Hz, ArH), 5.07–4.95 (m, 2 H, CH–NH + PhthN–CH), 4.87–4.43 (m, 5 H, CF_3–CH_2 + NH), 4.30 (t, 1 H, J = 9.0 Hz, Ar–CH), 4.08 (t, 1 H, J = 13.3 Hz, CH_2), 2.41 (dd, 1 H, J = 13.3, 6.9 Hz, CH_2). **^{13}C NMR** (101 MHz, $CDCl_3$)[12] δ 168.9, 168.3, 167.9, 166.5, 151.3, 134.4, 134.0, 132.0, 130.6, 128.9, 127.4, 124.7, 123.4, 122.7 (q, $J_{C\text{-}F}$ = 277 Hz), 122.4 (q, $J_{C\text{-}F}$ = 278 Hz), 119.5, 110.2, 67.6, 64.6, 61.2 (q, $J_{C\text{-}F}$ = 37 Hz), 61.2 (q, $J_{C\text{-}F}$ = 37 Hz), 53.4, 48.9, 30.3. **IR** 3375 (w), 3057 (w), 1754 (m), 1713 (s), 1376 (m), 1284 (s), 1163 (s), 1125 (s). **HRMS** (ESI) calcd for $C_{25}F_6H_{19}N_2O_6^+$ $[M + H]^+$ 557.1142; found 557.1147.

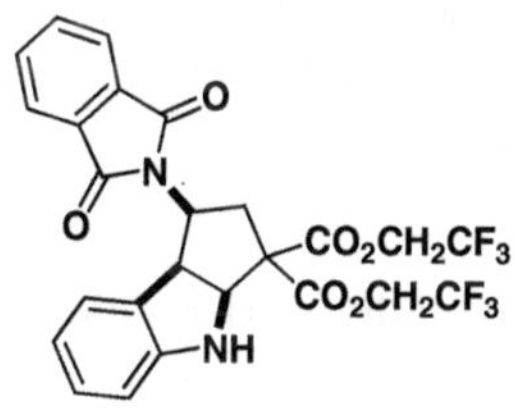

^{1}H NMR (400 MHz, $CDCl_3$) (3:1 mixture with the trans diastereomer) δ 7.91–7.86 (m, 2 H, Phth, cis diastereomer), 7.79–7.74 (m, 2 H, Phth, cis diastereomer), 7.09–7.04 (m, 1 H, ArH, cis diastereomer), 7.01–6.96 (m, 1 H, ArH, trans diastereomer) 6.99 (d, 1 H, J = 7.3 Hz, ArH, cis diastereomer), 6.72 (td, 1 H, J = 7.5, 0.6 Hz, ArH, cis diastereomer), 6.65 (d, 1 H, J = 7.8 Hz, ArH, trans diastereomer) 6.61 (d, 1 H, J = 7.8 Hz, ArH, cis diastereomer), 6.50 (d, 1 H, J = 7.8 Hz, ArH, trans diastereomer), 6.39–6.34 (m, 1 H, ArH, trans diastereomer), 5.32 (d, 1 H, J = 9.8 Hz, CH–NH, cis diastereomer), 5.15–5.08 (m, 1 H, PhthN–CH, cis diastereomer), 5.07–4.98 (m, 2 H, CH–NH + PhthN–CH, trans diastereomer), 4.84–

[12]Measured at 268 K. The carbon atoms of the phthalimide group are not equivalent.

4.73 (m, 1 H, CF_3–CH_2, trans diastereomer), 4.69–4.45 (m, 6 H, CF_3–CH_2, cis + trans diastereomers), 4.40 (dd, 1 H, J = 9.7, 7.3 Hz, Ar–CH, cis diastereomer), 4.34–4.23 (m, 2 H, CF_3–CH_2 (cis diastereomer) + Ar–CH (trans), 4.10 (t, 1 H, J = 13.3 Hz, CH_2 trans diastereomer) 2.91–2.81 (m, 2 H, CH_2, cis diastereomer), 2.44 (dd, 1 H, J = 13.3, 7.2 Hz, CH_2, trans diastereomer).[13] **^{13}C NMR** (101 MHz, $CDCl_3$) (from the mixture with the other diastereomer) δ 168.6, 168.1, 167.0, 148.9, 134.4, 131.9, 129.5, 128.7, 124.1, 123.6, 122.7 (q, $J_{C\text{-}F}$ = 277 Hz), 122.7 (q, $J_{C\text{-}F}$ = 277 Hz), 120.2, 110.2, 68.5, 65.1, 61.6 (q, $J_{C\text{-}F}$ = 37 Hz), 61.5 (q, $J_{C\text{-}F}$ = 37 Hz), 56.6, 50.9, 35.9. **IR** (2.5:1 mixture with the other diastereomer) 1758 (m), 1716 (s), 1613 (w), 1381 (m), 1288 (m), 1239 (m), 1174 (s).

Tetrakis(2,2,2-trifluoroethyl) 2,2′-(oxybis(2-(1,3-dioxoisoindolin-2-yl)ethane-2,1-diyl)dimalonate (267)

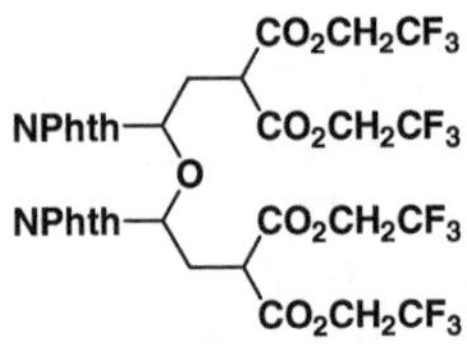

R_f 0.31 (6:4 Hexane/AcOEt). **^{1}H NMR** (400 MHz, $CDCl_3$) δ 7.65–7.60 (m, 4 H, Phth), 7.58–7.53 (m, 4 H, Phth), 5.65 (dd, 2 H, J = 8.3, 5.7 Hz, PhthN–CH), 4.69–4.49 (m, 8 H, CF_3–CH_2), 3.82 (m, 2 H, CH–$(CO)_2$), 3.16–3.08 (m, 2 H, CH_2), 2.66–2.57 (m, 2 H, CH_2). **^{13}C NMR** (101 MHz, $CDCl_3$) δ 166.7, 166.3, 166.1, 134.4, 131.2, 123.4, 122.5 (q, $J_{C\text{-}F}$ = 277 Hz), 78.2, 61.5 (q, $J_{C\text{-}F}$ = 37 Hz), 47.5, 31.1. **IR** 2984 (w), 2957 (w), 2927 (w), 2854 (w), 1776 (m), 1758 (s), 1730 (s), 1375 (m), 1286 (s), 1172 (s). **HRMS** (ESI) calcd for $C_{34}F_{12}H_{25}N_2O_{13}^+$ $[M + H]^+$ 897.1160; found 897.1148.

Synthesis of indoles

1-(2-(Tri*iso*propylsilyl)oxy)ethyl)-1H-indole (242 g)

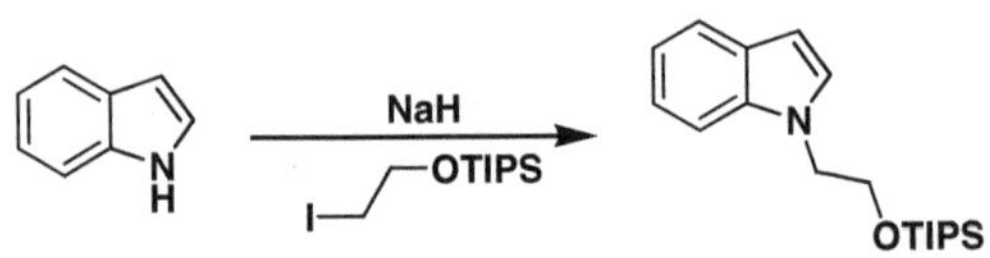

32 242g

1H-indole (**32**) (0.843 g, 7.20 mmol, 1.2 equiv) was dissolved in N, N-dimethylformamide (6 mL) and NaH (60 % in mineral oil, 0.360 g, 9.00 mmol, 1.33 equiv, 1.25 equiv compared to indole) was added at rt under strong stirring and the reaction mixture was stirred for 1 h. N,N-Dimethylformamide (18 mL) was

[13]The phthalimide peaks of the trans diastereomer could not be resolved due to NMR exchange.

added to dissolve the white precipitate and to give a greenish solution. The reaction was cooled to 0 °C and (2-iodoethoxy)triisopropylsilane (1.97 g, 6.00 mmol, 1 equiv) was added dropwise. The reaction was stirred overnight and let to slowly warm up to rt. The reaction was then quenched with water (20 mL) and the reaction mixture was extracted with EtOAc (3 × 25 mL). The combined organic layers were washed with water (10 mL), brine (3 × 10 mL) and dried over $MgSO_4$. The solvent was evaporated and the crude product was dried under vacuum with stirring. The crude NMR did not show the presence of the alkylating agent. TLC (10:1 hexanes: EtOAc, **R_f** prod.: 0.7). Purification by flash chromatography (SiO_2, 1–10 % EtOAc in hexane) gave 1-(2-(triisopropylsilyl)oxy)ethyl)-1H-indole (**242 g**) (1.56 g, 4.91 mmol, 82 % yield) as a colorless oil.

R_f 0.65 (10:1 Hexane/AcOEt). **^{1}H NMR** (400 MHz, $CDCl_3$) δ 7.66 (m, 1 H, ArH), 7.38 (dd, 1 H, J = 8.2, 0.8 Hz, ArH), 7.25–7.19 (m, 2 H, ArH) 7.13 (m, 1 H, ArH), 6.52 (dd, 1 H, J = 3.1, 0.8 Hz, ArH), 4.30 (t, 2 H, J = 6.0 Hz, CH_2), 4.04 (t, 2 H, J = 5.8 Hz, CH_2), 1.17–0.85 (m, 21 H, TIPS). **^{13}C NMR** (101 MHz, $CDCl_3$) δ 136.1, 128.7, 128.6, 121.3, 120.9, 119.2, 109.3, 101.0, 62.8, 48.8, 17.9, 11.9. **IR** 3056 (w), 2942 (m), 2891 (m), 2865 (s), 1514 (w), 1464 (s), 1439 (w), 1400 (w), 1387 (w), 1360 (w), 1334 (w), 1317 (m), 1250 (w), 1200 (w), 1115 (s), 1077 (m), 1013 (m), 997 (w), 923 (m), 883 (s), 819 (w). **HRMS (ESI)** calcd for $C_{19}H_{32}NOSi^+$ $[M + H]^+$ 318.2248; found 318.2236.

The characterization data correspond to the reported values [19].

1-Methyl-2-(triisopropylsilyl)ethynyl)-1H-indole (242j)

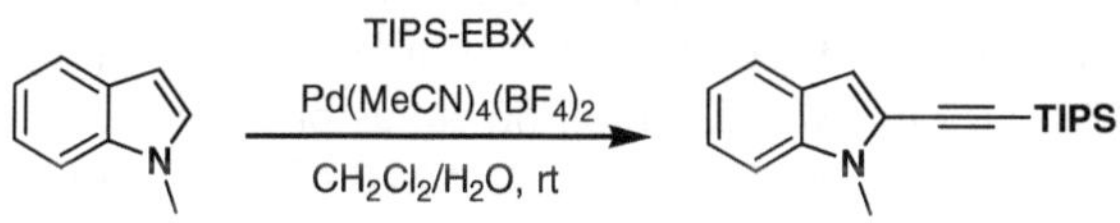

211 **242j**

In a 10 mL round bottom-flask, 1-methyl-1H-indole (**211**) (64 μl, 66 mg, 0.50 mmol, 1 equiv) and 1-[(triisopropylsilyl)ethynyl]-1,2-benziodoxol-3(1*H*)-one (TIPS-EBX) (0.64 g, 1.5 mmol, 3 equiv) were dissolved in DCM (5 mL) under air, then water was added (0.10 mL). Lastly $Pd(MeCN)_4(BF_4)_2$ (4.4 mg, 10 μmol, 2 %) was added with strong stirring. The flask was closed and the reaction mixture was stirred overnight, when it became brownish. The solvent was evaporated under reduced pressure. EtOAc (25 mL) was added to the crude product, and the solution was washed with $NaOH_{aq}$ (0.1 M, 25 mL), conc. $NaHCO_3$ (2 × 25 mL) and brine (25 mL). The organic layer was dried over $MgSO_4$, filtered and the solvent was evaporated under reduced pressure. Purification by column chromatography (SiO_2, hexane to hexane/DCM 90/10) gave 1-methyl-2-(triisopropylsilyl)ethynyl)-1H-indole (**242j**) (102 mg, 0.33 mmol, 66 %) as a pale yellow oil.

R_f 0.75 (10:1 Hexanes:EtOAc). **^{1}H NMR** (400 MHz, $CDCl_3$) δ 7.63 (dt, 1 H, J = 8.0, 0.9 Hz, ArH), 7.33–7.28 (m, 2 H, ArH), 7.16 (q, 1 H, J = 4 Hz, ArH), 6.86 (s, 1 H, ArH), 3.78 (s, 3 H, Me), 1.30–1.09 (m, 21 H, TIPS). **^{13}C NMR**

(101 MHz, $CDCl_3$) δ 137.1, 127.1, 123.1, 122.3, 121.1, 120.1, 109.4, 107.7, 98.2, 97.8, 30.6, 18.8, 11.4. **IR** 3058 (w), 2942 (s), 2891 (m), 2864 (s), 2150 (s), 1463 (s), 1429 (w), 1383 (m), 1364 (m), 1339 (s), 1317 (m), 1238 (m), 1170 (w), 1152 (w), 1073 (w), 1012 (m), 997 (m), 920 (m), 883 (s), 854 (m). **HRMS (ESI)** calcd. for $C_{20}H_{30}NSi^+$ $[M + H]^+$ 312.2142; found 312.2147.

The characterization data correspond to the reported values [19].

3-(Tri*iso*-propylsilyl)ethynyl)-1H-indole (242l)

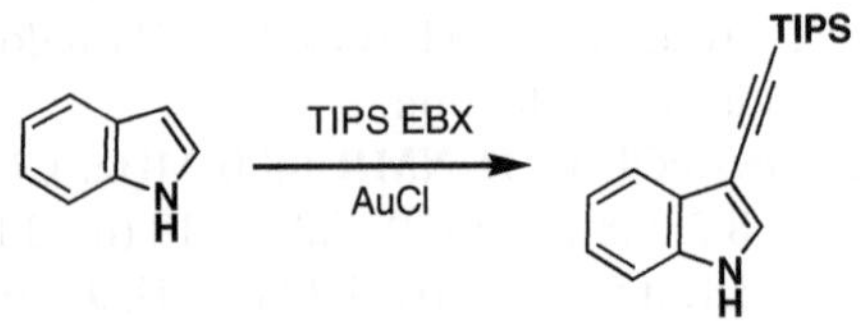

32 242l

1-[(Triisopropylsilyl)ethynyl]-1,2-benziodoxol-3(1*H*)-one (206 mg, 0.480 mmol, 1.2 equiv) was added to a stirring solution of AuCl (4.6 mg, 0.020 mmol, 0.05 equiv) and N–H indole (**32**) (0.400 mmol, 1.0 equiv) in Et_2O (8 mL) under air. The reaction was sealed and stirred at room temperature for 12 h. Et_2O (10 mL) was added, the organic layer was washed twice with 0.1 M NaOH (15 mL). The aqueous layers were combined and extracted with Et_2O (20 mL). The organic layers were combined, washed with saturated $NaHCO_3$ (20 mL), brine (20 mL), dried with $MgSO_4$ and concentrated under reduced pressure. Purification by flash chromatography (PET/Et_2O 8/2) afforded 3-(Tri*iso*-propylsilyl)ethynyl)-*1H*-indole (**242l**) (102 mg, 0.342 mmol, 86 %) as brown solid.

$\mathbf{R}_f$ 0.4 (7:3 PET/Et_2O). Mp 55–58 °C. **^{1}H NMR** ($CDCl_3$, 400 MHz) δ 8.11 (br s, 1 H NH), 7.79 (m, 1 H. ArH), 7.40 (d, J = 2.7 Hz, 1 H ArH), 7.36 (m, 1 H, ArH), 7.26 (m, 2 H ArH), 1.22 (m, 21 H TIPS). **^{13}C NMR** ($CDCl_3$, 100 MHz) δ 135.1, 128.9, 128.3, 123.1, 120.8, 120.1, 111.4, 100.4, 99.3, 92.19, 18.8, 11.5. **IR** ν 3407 (m), 3062 (w), 2942 (s), 2891 (m), 2864 (s), 2152 (s), 1620 (w), 1532 (w), 1457 (s), 1416 (m), 1383 (w), 1341 (w), 1325 (m), 1239 (s), 1128 (m), 1071 (m), 996 (m), 910 (m), 883 (s), 774 (s), 742 (s), 676 (s), 658 (s), 628 (s). **HRMS(ESI)** calcd for $C_{19}H_{28}NSi^+$ $[M + H]^+$ 298.1991, found 298.2001.

The characterization data correspond to the reported values [20].

3-(2-(Trimethylsilyl)oxy)ethyl)-1H-indole (242m)

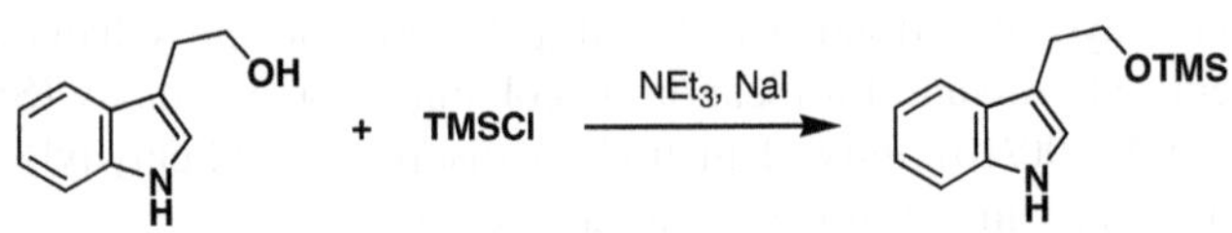

378 379 242m

Sodium iodide (581 mg, 3.88 mmol, 1.25 equiv), 2-(1H-indol-3-yl)ethanol (**378**) (500 mg, 3.10 mmol, 1 equiv), acetonitrile (5.00 mL) and hexane (5.00 mL) were mixed in a 25 mL flask. Triethylamine (0.537 mL, 3.88 mmol, 1.25 equiv) followed by chlorotrimethylsilane (**379**) (0.496 mL, 3.88 mmol, 1.25 equiv) were added in one portion with efficient stirring of the bi-phasic mixture. The mixture was then stirred for 1 h at room temperature. The two immiscible layers were separated with a separatory funnel. The organic layer was filtered through a plug of silica gel (hexane/AcOEt 8/2, deactivated with 2 % NEt_3, 30 mL) and concentrated under reduced pressure to afford 3-(2-(trimethylsilyl)oxy)ethyl)-1H-indole (**242m**) (131 mg, 0.560 mmol, 18 % yield) as a colorless solid. The product was used without further purification.

R_f 0.49 (8:2 hexane/ethyl acetate). **^{1}H NMR** (400 MHz, $CDCl_3$) δ 7.99 (br s, 1 H, N*H*), 7.63 (dd, 1 H, *J* = 7.9, 0.6 Hz, ArH), 7.38–7.34 (m, 1 H, ArH), 7.23–7.18 (m, 1 H, ArH), 7.16–7.11 (m, 1 H, ArH), 7.05–7.02 (m, 1 H, ArH), 3.87 (t, 2 H, *J* = 7.4 Hz, CH_2), 3.03 (td, 2 H, *J* = 7.7, 0.6 Hz, CH_2), 0.12 (s, 9 H, TMS). **^{13}C NMR** (101 MHz, $CDCl_3$) δ 136.3, 127.8, 122.1, 122.0, 119.4, 119.0, 113.1, 111.2, 63.4, 29.1, −0.3. **HRMS** (APPI) calcd for $C_{13}H_{20}NOSi^+$ $[M + H]^+$ 234.1309; found 234.1300.

Synthesis of bis(2,2,2-trifluoroethyl) 2-(2-(1,3-dioxoisoindolin-2-yl)-2-(1H-indol-3-yl)ethyl)malonates

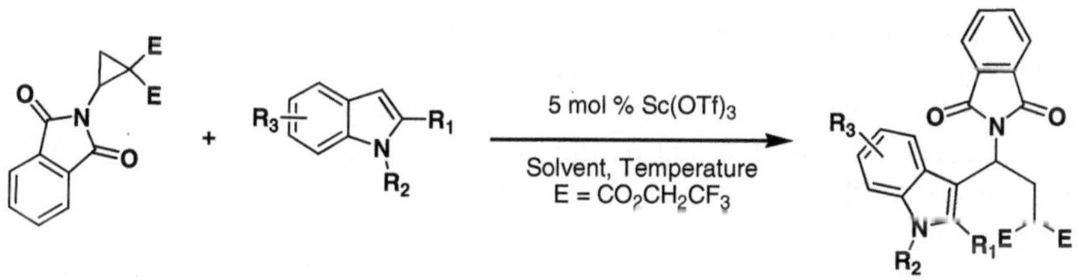

GP10

Bis(2,2,2-trifluoroethyl) 2-(1,3-dioxoisoindolin-2-yl) cyclopropane-1,1-dicarboxylate (**259**) (0.088 g, 0.20 mmol, 1 equiv) and the corresponding indole (0.22 mmol, 1.1 equiv) were dissolved in diethyl ether (1.2 mL) and added to $Sc(OTf)_3$ (4.9 mg, 10 μmol, 5.0 %) in a tube sealed with a septum. The mixture was stirred at room temperature until full conversion was observed by TLC (6:4 Hexane/EtOAc, Anisaldehyde). The mixture was then filtered over a short plug of silica gel (AcOEt/hexane 1/1, 5 mL). The crude was concentrated under reduced pressure to remove volatiles. Purification by Biotage (SNAP cartridge KP-Sil 10 g) afforded the desired product.

GP11

Bis(2,2,2-trifluoroethyl) 2-(1,3-dioxoisoindolin-2-yl)cyclopropane-1,1-dicarboxyl ate (**259**) (0.088 g, 0.20 mmol, 1 equiv) and the corresponding indole

(0.22 mmol, 1.1 equiv) were dissolved in diethyl ether (1.2 mL) and added to Sc $(OTf)_3$ (4.9 mg, 10 μmol, 5.0 %) in a tube sealed with a septum. The mixture was stirred at 35 °C until full conversion was observed by TLC (6:4 Hexane/EtOAc, Anisaldehyde). The mixture was then filtered over a short plug of silica gel (AcOEt/hexane 1/1, 5 mL). The crude was concentrated under reduced pressure to remove volatiles. Purification by Biotage (SNAP cartridge KP-Sil 10 g, hexane/AcOEt) afforded the desired product.

GP12

Bis(2,2,2-trifluoroethyl) 2-(1,3-dioxoisoindolin-2-yl)cyclopropane-1,1-dicarboxylate (**259**) (0.088 g, 0.20 mmol, 1 equiv) and the corresponding indole (0.22 mmol, 1.1 equiv) were dissolved in toluene (1.2 mL) and added to $Sc(OTf)_3$ (4.9 mg, 10 μmol, 5.0 %) in a tube sealed with a septum. The mixture was stirred at 60 °C until full conversion was observed by TLC (6:4 Hexane/EtOAc, Anisaldehyde). The mixture was then filtered over a short plug of silica gel (AcOEt/hexane 1/1, 5 mL). The crude was concentrated under reduced pressure to remove volatiles. Purification by Biotage (SNAP cartridge KP-Sil) afforded the desired product.

Bis(2,2,2-trifluoroethyl) 2-(2-(1,3-dioxoisoindolin-2-yl)-2-(1H-indol-3-yl)ethyl) malonate (266a)

Following GP10 and starting from 1H-indole (**32**) (26 mg, 0.22 mmol, 1.1 equiv), bis(2,2,2-trifluoroethyl) 2-(2-(1,3-dioxoisoindolin-2-yl)-2-(1H-indol-3-yl)ethyl) malonate (**266a**) (95 mg, 0.17 mmol, 85 % yield) was obtained as a colorless solid after a reaction time of 1 h and purification by Biotage (SNAP cartridge KP-Sil 10 g, hexane/AcOEt 98/2 to 75/25 + 1 % AcOH).

Large scale procedure:

Bis(2,2,2-trifluoroethyl) 2-(1,3-dioxoisoindolin-2-yl)cyclopropane-1,1-dicarboxylate (**259**) (1.40 g, 3.19 mmol, 1 equiv) and indole (**32**) (411 mg, 3.51 mmol, 1.1 equiv) were dissolved in diethyl ether (19 mL) and added to $Sc(OTf)_3$ (78.0 mg, 159 μmol, 5.0 mol%) in a one neck flask sealed with a septum under nitrogen atmosphere. The mixture was stirred at room temperature for 50 min, when full conversion was observed by TLC (6:4 Hexane/EtOAc, Anisaldehyde). The mixture was then filtered over a short plug of silica gel (AcOEt/hexane 1/1, 5 mL). The crude was concentrated under reduced pressure to remove volatiles. Purification by column chromatography (silica gel, hexane:AcOEt 95:5–70:30) afforded bis (2,2,2-trifluoroethyl) 2-(2-(1,3-dioxoisoindolin-2-yl)-2-(1H-indol-3-yl)ethyl)malonate (**266a**) (1.54 g, 2.77 mmol, 87 % yield).

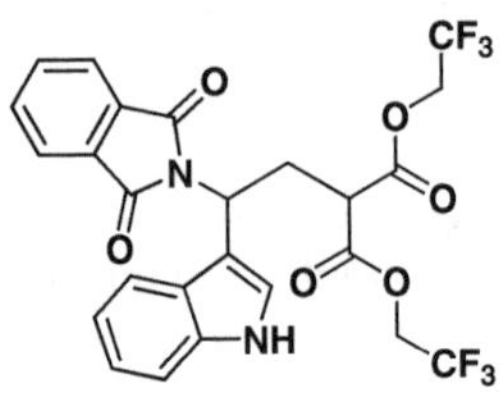

R_f 0.49 (1:1 hexane/ethyl acetate). **Mp** 69.3–72.6 °C. **^{1}H NMR** (400 MHz, $CDCl_3$) δ 8.26 (s, 1 H, NH), 7.81–7.75 (m, 2 H, Phth), 7.72 (d, 1 H, J = 8.0 Hz, ArH), 7.70–7.65 (m, 2 H, Phth), 7.52 (d, 1 H, J = 2.5 Hz, ArH), 7.37–7.33 (m, 1 H, ArH), 7.21–7.16 (m, 1 H, ArH), 7.15–7.10 (m, 1 H, ArH), 5.81 (dd, 1 H, J = 9.4, 6.9 Hz, PhthN–CH), 4.65–4.36 (m, 4 H, CF_3–CH_2), 3.71–3.66 (m, 1 H, CH–$(CO)_2$), 3.32–3.23 (m, 1 H, CH_2), 3.18–3.10 (m, 1 H, CH_2). **^{13}C NMR** (101 MHz, $CDCl_3$) δ 168.2, 166.7, 166.6, 135.8, 134.2, 131.8, 126.3, 124.3, 123.4, 122.7, 122.6 (q, $J_{C\text{-}F}$ = 277 Hz), 122.5 (q, $J_{C\text{-}F}$ = 277 Hz), 120.4, 118.9, 112.8, 111.4, 61.5 (q, $J_{C\text{-}F}$ = 37 Hz), 61.4 (q, $J_{C\text{-}F}$ = 37 Hz), 49.0, 44.3, 30.7. **IR** 3412 (w), 3061 (w), 2979 (w), 1757 (m), 1708 (s), 1381 (m), 1357 (m), 1330 (m), 1283 (s), 1166 (s), 975 (m). **HRMS** (ESI) calcd for $C_{25}H_{18}F_6N_2NaO_6^+$ $[M + Na]^+$ 579.0961; found 579.0964.

Bis(2,2,2-trifluoroethyl) 2-(2-(5-chloro-1H-indol-3-yl)-2-(1,3-dioxoisoindolin-2-yl)ethyl)malonate (266b)

Following GP10 and starting from 5-chloro-1H-indole (**242a**) (33 mg, 0.22 mmol, 1.1 equiv), bis(2,2,2-trifluoroethyl) 2-(2-(5-chloro-1H-indol-3-yl)-2-(1,3-dioxoisoin dolin-2-yl)ethyl)malonate (**266b**) (97 mg, 0.16 mmol, 82 % yield) was obtained as a colorless solid after a reaction time of 1.5 h and purification by Biotage (SNAP cartridge KP-Sil 10 g, hexane/AcOEt 98/2 to 75/25 + 1 % AcOH).

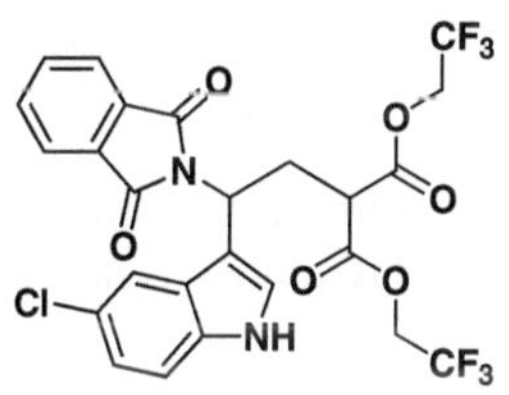

R_f 0.40 (6:4 hexane/ethyl acetate). **Mp** 98.7–102.0 °C. **^{1}H NMR** (400 MHz, $CDCl_3$) δ 8.29 (s, 1 H, NH), 7.83–7.78 (m, 2 H, Phth), 7.72–7.67 (m, 2 H, Phth), 7.67 (d, 1 H, J = 2.0 Hz, ArH), 7.55 (d, 1 H, J = 2.6 Hz, ArH), 7.26 (dd, 1 H, J = 8.7, 0.4 Hz, ArH), 7.14 (dd, 1 H, J = 8.5, 1.9 Hz, ArH), 5.72 (dd, 1 H, J = 9.6, 6.7 Hz, PhthN–CH), 4.66-4.37 (m, 4 H, CF_3–CH_2), 3.67–3.62 (m, 1 H, CH–$(CO)_2$), 3.30–3.21 (m, 1 H, CH_2), 3.12–3.03 (m, 1 H, CH_2). **^{13}C NMR** (101 MHz, $CDCl_3$) δ 168.1, 166.6, 166.5, 134.4, 134.1, 131.8, 127.4, 126.3, 125.7, 123.6, 123.2, 122.6 (q, $J_{C\text{-}F}$ = 277 Hz), 122.5 (q, $J_{C\text{-}F}$ = 277 Hz), 118.4, 112.7, 112.4, 61.6 (q, J = 37 Hz), 61.4 (q, J = 37 Hz), 48.9, 44.0, 30.7. **IR** 3419 (w), 1757 (m), 1709 (s), 1464 (w), 1384 (m), 1286 (s), 1170 (s), 973 (w), 724 (m). **HRMS** (ESI) calcd for $C_{25}^{35}ClF_6H_{18}N_2O_6^+$ $[M + H]^+$ 591.0752; found 591.0761.

Bis(2,2,2-trifluoroethyl) 2-(2-(5-methoxy-1H-indol-3-yl)-2-(1,3-dioxoisoindolin-2-yl)ethyl)malonate (266c)

Following GP10 and starting from 5-methoxy-1H-indole (**242b**) (32 mg, 0.22 mmol, 1.1 equiv), bis(2,2,2-trifluoroethyl) 2-(2-(5-methoxy-1H-indol-3-yl)-2-

(1,3-dioxoisoindolin-2-yl)ethyl)malonate (**266c**) (79 mg, 0.14 mmol, 68 % yield) was obtained as a colorless solid after a reaction time of 0.5 h and purification by Biotage (SNAP cartridge KP-Sil 10 g, hexane/AcOEt 97/3 to 75/25 + 1 % AcOH).

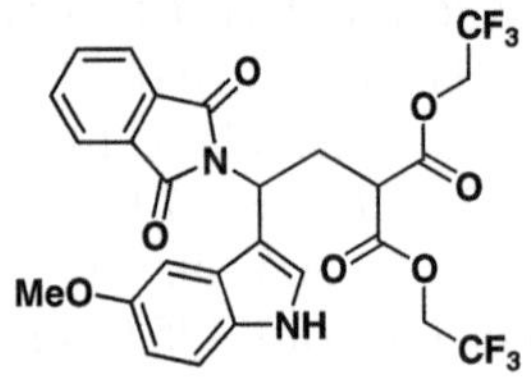

R*f* 0.41 (6:4 hexane/ethyl acetate). **Mp** 92.6–96.8 °C. **^{1}H NMR** (400 MHz, $CDCl_3$) δ 8.17 (s, 1 H, NH), 7.81–7.76 (m, 2 H, Phth), 7.70–7.65 (m, 2 H, Phth), 7.48 (d, 1 H, J = 2.6 Hz, ArH), 7.23 (d, 1 H, J = 8.8 Hz, ArH), 7.18 (d, 1 H, J = 2.4 Hz, ArH), 6.84 (dd, 1 H, J = 8.8, 2.4 Hz, ArH), 5.76 (dd, 1 H, J = 9.4, 6.9 Hz, PhthN–CH), 4.65–4.38 (m, 4 H, CF_3–CH_2), 3.84 (s, 3 H, OMe), 3.71–3.66 (m, 1 H, CH–$(CO)_2$), 3.31–3.23 (m, 1 H, CH_2), 3.17–3.09 (m, 1 H, CH_2). **^{13}C NMR** (101 MHz, $CDCl_3$) δ 168.2, 166.7, 166.6, 154.6, 134.3, 131.8, 130.8, 126.8, 124.9, 123.4, 122.6 (q, $J_{C\text{-}F}$ = 277 Hz), 122.5 (q, $J_{C\text{-}F}$ = 277 Hz), 113.2, 112.6, 112.1, 100.4, 61.5 (q, $J_{C\text{-}F}$ = 37 Hz), 61.4 (q, $J_{C\text{-}F}$ = 37 Hz), 55.9, 49.0, 44.4, 30.6. **IR** 3414 (w), 2949 (w), 2840 (w), 1758 (m), 1708 (s), 1490 (w), 1284 (s), 1170 (s), 975 (m), 721 (m). **HRMS** (ESI) calcd for $C_{26}F_6H_{21}N_2O_7^+$ $[M + H]^+$ 587.1247; found 587.1252.

Bis(2,2,2-trifluoroethyl) 2-(2-(1,3-dioxoisoindolin-2-yl)-2-(5-(4,4,5,5-tetramet hyl-1,3,2-dioxaborolan-2-yl)-1H-indol-3-yl)ethyl)malonate (266d)

Following GP10 and starting from 5-(4,4,5,5-tetramethyl-1,3,2-dioxaborolan-2-yl)-1H-indole (**242c**) (54 mg, 0.22 mmol, 1.1 equiv), bis(2,2,2-trifluoroethyl) 2-(2-(1,3-dioxoisoindolin-2-yl)-2-(5-(4,4,5,5-tetramethyl-1,3,2-dioxaborolan-2-yl)-1H-indol-3-yl)ethyl)malonate (**266d**) (86 mg, 0.13 mmol, 63 % yield) was obtained as an colorless solid after a reaction time of 1 h and purification by Biotage (SNAP cartridge KP-Sil 10 g, hexane/AcOEt 97/3–75/25).

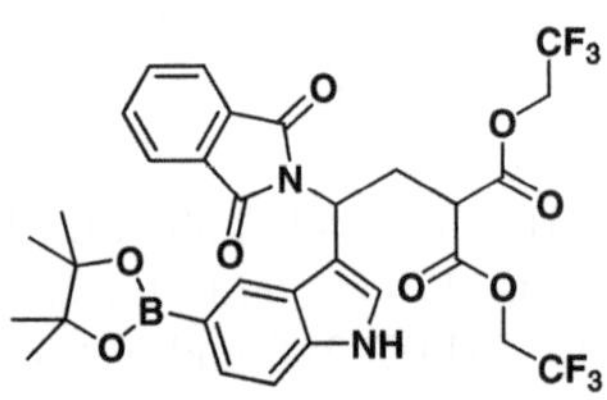

R*f* 0.47 (6:4 hexane/ethyl acetate). **Mp** 83.6–87.2 °C. **^{1}H NMR** (400 MHz, $CDCl_3$) δ 8.31 (s, 1 H, NH), 8.16 (s, 1 H, ArH), 7.82–7.77 (m, 2 H, Phth), 7.70–7.65 (m, 2 H, Phth), 7.63 (d, 1 H, J = 8.2 Hz, ArH), 7.57 (d, 1 H, J = 2.5 Hz, ArH), 7.34 (d, 1 H, J = 8.2 Hz, ArH), 5.89–5.84 (m, 1 H, PhthN–CH), 4.65–4.36 (m, 4 H,

CF_3–CH_2), 3.63 (t, 1 H, J = 7.4 Hz, CH–$(CO)_2$), 3.24–3.06 (m, 2 H, CH_2), 1.35 (s, 6 H, CH_3), 1.35 (s, 6 H, CH_3). **^{13}C NMR** (101 MHz, $CDCl_3$) δ 168.1, 166.7, 166.6, 137.6, 134.2, 131.9, 128.9, 126.1, 124.6, 123.5, 122.6 (q, $J_{C\text{-}F}$ = 277 Hz), 122.6 (q, $J_{C\text{-}F}$ = 277 Hz), 120.3 (br), 113.4, 110.8, 83.7, 61.5 (q, $J_{C\text{-}F}$ = 37 Hz), 61.4 (q, $J_{C\text{-}F}$ = 37 Hz), 49.0, 44.2, 31.6, 25.1, 24.9. **IR** 3405 (w), 2982 (w), 2934 (w), 1759 (m), 1712 (s), 1357 (s), 1286 (s), 1170 (s), 969 (w), 723 (m). **HRMS** (ESI) calcd for $C_{31}H_{29}^{11}BF_6N_2NaO_8^+$ $[M + Na]^+$ 705.1813; found 705.1801.

Bis(2,2,2-trifluoroethyl) 2-(2-(7-bromo-1H-indol-3-yl)-2-(1,3-dioxoisoindolin-2-yl)ethyl)malonate (266e)

Following GP10 and starting from 7-bromo-1H-indole (**242d**) (43 mg, 0.22 mmol, 1.1 equiv), bis(2,2,2-trifluoroethyl) 2-(2-(7-bromo-1H-indol-3-yl)-2-(1,3-dioxoiso indolin-2-yl)ethyl)malonate (**266e**) (89 mg, 0.14 mmol, 70 % yield) was obtained as a pale yellow solid after a reaction time of 50 min and purification by Biotage (SNAP cartridge KP-Sil 25 g, hexane/AcOEt 90/10–45/55).

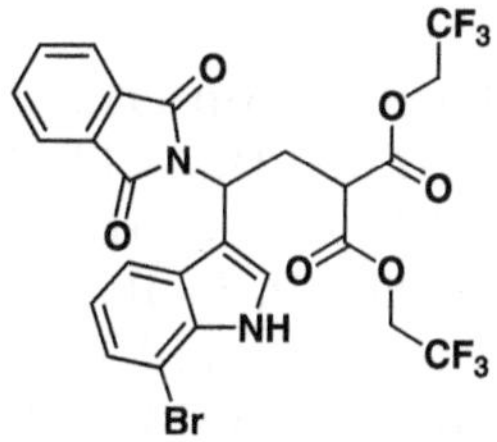

R_f 0.52 (6:4 hexane/ethyl acetate). **Mp** 182.9–184.2 °C. **^{1}H NMR** (400 MHz, $CDCl_3$) δ 8.32 (s, 1 H, NH), 7.75–7.67 (m, 2 H, Phth), 7.65–7.60 (m, 2 H, Phth), 7.58 (d, 1 H, J = 8.0 Hz, ArH), 7.52 (d, 1 H, J = 2.4 Hz, ArH), 7.34 (d, 1 H, J = 7.6 Hz, ArH), 7.01 (t, 1 H, J = 7.8 Hz, ArH), 5.69 (dd, 1 H, J = 9.5, 6.8 Hz, PhthN–CH), 4.61–4.48 (m, 1 H, CF_3–CH_2), 4.47–4.28 (m, 3 H, CF_3–CH_2), 3.62–3.55 (m, 1 H, CH–$(CO)_2$), 3.26–3.15 (m, 1 H, CH_2), 3.09–2.99 (m, 1 H, CH_2). **^{13}C NMR** (101 MHz, $CDCl_3$) δ 168.1, 166.63, 166.61, 134.5, 134.4, 131.8, 127.4, 125.1, 124.8, 123.5, 122.6 (q, $J_{C\text{-}F}$ = 277 Hz), 122.5 (q, $J_{C\text{-}F}$ = 277 Hz), 121.6, 118.2, 114.2, 105.0, 61.5 (q, $J_{C\text{-}F}$ = 37 Hz), 61.4 (q, $J_{C\text{-}F}$ = 37 Hz), 48.9, 44.3, 30.6. **IR** 3387 (w), 1774 (m), 1759 (m), 1712 (s), 1386 (w), 1333 (m), 1286 (m), 1173 (s). **HRMS** (ESI) calcd for $C_{25}^{79}BrF_6H_{18}N_2O_6^+$ $[M + H]^+$ 635.0247; found 635.0240.

Dimethyl 2-(2-(1,3-dioxoisoindolin-2-yl)-2-(6-nitro-1H-indol-3-yl)ethyl)malonate (266f)

GP10 was followed using 6-nitro-1H-indole (**242e**) (36 mg, 0.22 mmol, 1.1 equiv). The mixture was stirred at room temperature for 30 min and then filtered over a short plug of silica gel (AcOEt/hexane 1/1). The crude was concentrated under reduced pressure to remove volatiles. Potassium carbonate (2.77 mg, 0.0200 mmol, 0.100 equiv) was added to a solution of the crude in MeOH (2.00 mL) at 0 °C. The resulting suspension was stirred for 40 min. The mixture was then partitioned between dichloromethane (5 mL) and saturated aqueous ammonium chloride-brine (1:2) (5 mL). The aqueous layer was extracted two times with dichloromethane (5 mL) and the combined organic layers were washed with brine (5 mL) and dried over anhydrous magnesium sulfate. Volatiles were removed under reduced pressure. The crude was then purified by Biotage (SNAP cartridge KP-Sil 10 g, pentane/AcOEt 87/13–1/99) to afford dimethyl 2-(2-(1,3-dioxoisoindolin-2-yl)-2-(3-(2-hydroxyethyl)-1H-indol-2-yl)ethyl)malonate (**266f**) (53.7 mg, 0.115 mmol, 58 % yield) as a yellow solid.

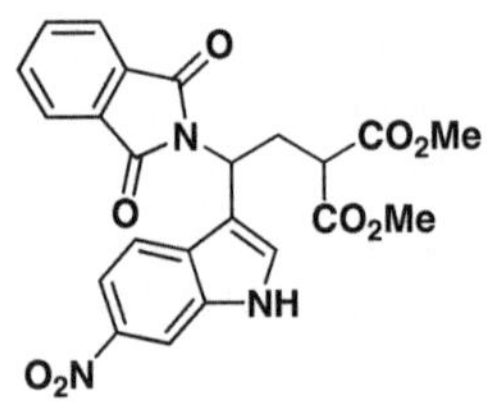

R$_f$ 0.21 (6:4 hexane/ethyl acetate). **Mp** 171.3–174.2 °C. **^{1}H NMR** (400 MHz, $CDCl_3$) δ 8.84 (s, 1 H, NH), 8.32 (d, 1 H, J = 1.9 Hz, ArH), 8.00 (dd, 1 H, J = 8.9, 2.0 Hz, ArH), 7.84–7.78 (m, 3 H, Phth + ArH), 7.75 (d, 1 H, J = 8.9 Hz, ArH), 7.73–7.67 (m, 2 H, Phth), 5.78 (dd, 1 H, J = 10.1, 6.0 Hz, PhthN–CH), 3.75 (s, 3 H, OMe), 3.65 (s, 3 H, OMe), 3.49–3.44 (m, 1 H, CH–$(CO)_2$), 3.31–3.22 (m, 1 H, CH_2), 3.04–2.96 (m, 1 H, CH_2). **^{13}C NMR** (101 MHz, $CDCl_3$) δ 169.2, 169.1, 168.1, 143.8, 134.4, 134.2, 131.7, 131.0, 129.9, 123.6, 119.1, 115.8, 114.6, 108.3, 53.1, 53.0, 49.5, 44.1, 30.7. **IR** 3367 (w), 2359 (w), 1734 (s), 1712 (s), 1513 (m), 1338 (s). **HRMS** (ESI) calcd for $C_{23}H_{19}N_3NaO_8^+$ $[M + Na]^+$ 488.1064; found 488.1059.

Bis(2,2,2-trifluoroethyl) 2-(2-(1,3-dioxoisoindolin-2-yl)-2-(1-methyl-1H-indol-3-yl)ethyl)malonate (266g)

Following GP10 and starting from 1-methyl-1H-indole (**43**) (29 mg, 0.22 mmol, 1.1 equiv), bis(2,2,2-trifluoroethyl) 2-(2-(1,3-dioxoisoindolin-2-yl)-2-(1-methyl-1H-indol-3-yl)ethyl)malonate (**266g**) (108 mg, 0.188 mmol, 94 % yield) was obtained as a colorless solid after a reaction time of 0.5 h and purification by Biotage (SNAP cartridge KP-Sil 10 g, hexane/AcOEt 97/3–85/15 + 1 % AcOH).

Larger scale procedure:

Bis(2,2,2-trifluoroethyl) 2-(1,3-dioxoisoindolin-2-yl)cyclopropane-1,1-dicarboxylate (**259**) (1.25 g, 2.85 mmol, 1 equiv) and 1-methyl-1H-indole (**43**) (411 mg, 3.13 mmol, 1.1 equiv) were dissolved in diethyl ether (17 mL) and added to Sc $(OTf)_3$ (70.0 mg, 142 μmol, 5.0 mol%) in a one neck flask sealed with a septum under nitrogen atmosphere. The mixture was stirred at room temperature for 50 min, when full conversion was observed by TLC (6:4 Hexane/EtOAc, Anisaldehyde). The mixture was then filtered over a short plug of silica gel (AcOEt/hexane 1/1, 5 mL). The crude was concentrated under reduced pressure to remove volatiles. Purification by column chromatography (silica gel, hexane:AcOEt 95:5–70:30) afforded bis(2,2,2-trifluoroethyl) 2-(2-(1,3-dioxoisoindolin-2-yl)-2-(1H-indol-3-yl)ethyl)malonate (**266g**) (1.34 g, 2.35 mmol, 83 % yield).

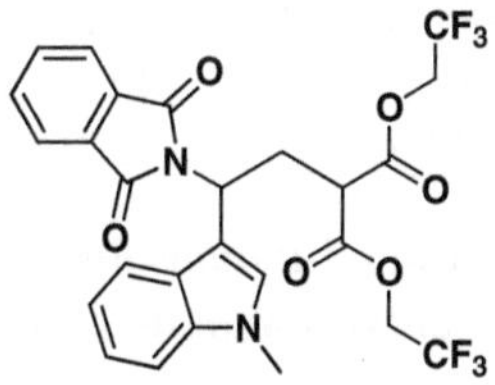

R$_f$ 0.49 (6:4 hexane/ethyl acetate). **Mp** 47.8–49.5 °C. **^{1}H NMR** (400 MHz, $CDCl_3$) δ 8.06–8.01 (m, 1 H, ArH), 7.94–7.89 (m, 1 H, ArH), 7.81–7.74 (m, 2 H, Phth), 7.69–7.63 (m, 2 H, Phth), 7.26–7.22 (m, 1 H, ArH), 7.16–7.09 (m, 2 H, ArH), 5.80 (dd, 1 H, $J = 9.3$, 7.1 Hz, PhthN–CH), 4.65–4.37 (m, 4 H, CF_3–CH_2), 3.78 (s, 3 H, CH_3), 3.70–3.65 (m, 1 H, CH–$(CO)_2$), 3.30–3.21 (m, 1 H, CH_2), 3.18–3.09 (m, 1 H, CH_2). **^{13}C NMR** (101 MHz, $CDCl_3$) δ 168.1, 166.7, 166.6, 136.6, 134.2, 131.9, 128.9, 126.9, 123.4, 122.6 (q, $J_{C\text{-}F} = 277$ Hz), 122.5 (q, $J_{C\text{-}F} = 277$ Hz), 122.2, 119.9, 119.0, 111.1, 109.5, 61.5 (q, $J_{C\text{-}F} = 37$ Hz), 61.4 (q, $J_{C\text{-}F} = 37$ Hz), 49.0, 44.3, 33.1, 30.9. **IR** 3059 (w), 2921 (w), 1758 (m), 1712 (s), 1382 (m), 1327 (m), 1284 (s), 1169 (s), 977 (m), 724 (m). **HRMS** (ESI) calcd for $C_{26}H_{20}F_6N_2NaO_6^+$ $[M + Na]^+$ 593.1118; found 593.1136.

Bis(2,2,2-trifluoroethyl) 2-(2-(1,3-dioxoisoindolin-2-yl)-2-(6-(methoxycarbonyl)-1-methyl-1H-indol-3-yl)ethyl)malonate (266h)

Following GP10 and starting from methyl 1-methyl-1*H*-indole-6-carboxylate (**242f**) (38 mg, 0.22 mmol, 1.1 equiv), bis(2,2,2-trifluoroethyl) 2-(2-(1,3-dioxoisoindolin-2-yl)-2-(6-(methoxycarbonyl)-1-methyl-1H-indol-3-yl)ethyl) malonate (**266 h**) (0.11 g, 0.17 mmol, 86 % yield) was obtained as a colorless solid after a reaction time of 1.5 h and purification by Biotage (SNAP cartridge KP-Sil 10 g, hexane/AcOEt 94/6–68/32).

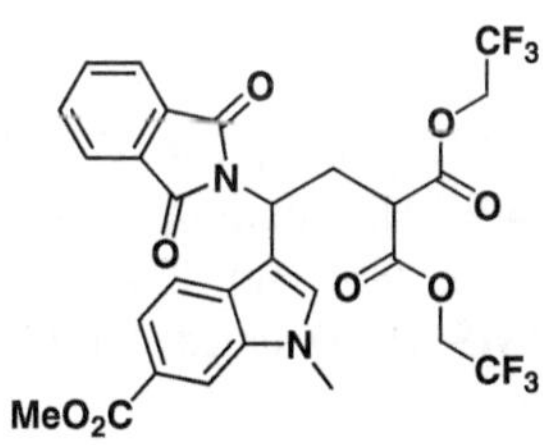

R$_f$ 0.26 (6:4 hexane/ethyl acetate). **Mp** 69.2–73.0 °C. **^{1}H NMR** (400 MHz, $CDCl_3$) δ 8.05 (d, 1 H, $J = 0.6$ Hz, ArH), 7.81–7.76 (m, 3 H, Phth + ArH), 7.72–7.65 (m, 3 H, Phth + ArH), 7.55 (s, 1 H, ArH), 5.79 (dd, $J = 9.5$, 6.9 Hz, PhthN–CH), 4.66–4.36 (m, 4 H, CF_3–CH_2), 3.92 (s, 3 H, CH_3), 3.84 (s, 3 H, CH_3), 3.68–

3.62 (m, 1 H, CH–(CO)$_2$), 3.31–3.21 (m, 1 H, CH$_2$), 3.14–3.04 (m, 1 H, CH$_2$). **^{13}C NMR** (101 MHz, CDCl$_3$) δ 168.1, 168.0, 166.6, 166.5, 136.0, 134.3, 132.1, 131.7, 130.3, 123.9, 123.5, 122.6 (q, $J_{C\text{-}F}$ = 277 Hz), 122.5 (q, $J_{C\text{-}F}$ = 277 Hz), 121.0, 118.6, 112.0, 111.7, 61.5 (q, $J_{C\text{-}F}$ = 37 Hz), 61.4 (q, $J_{C\text{-}F}$ = 37 Hz), 52.1, 48.9, 44.0, 33.3, 30.8. **IR** 3406 (w), 2954 (w), 2867 (w), 2094 (w), 1774 (m), 1713 (s), 1474 (w), 1439 (w), 1385 (m), 1278 (s), 1174 (s). **HRMS** (ESI) calcd for $C_{28}F_6H_{23}N_2O_8^+$ [M + H]$^+$ 629.1353; found 629.1347.

Bis(2,2,2-trifluoroethyl) 2-(2-(1,3-dioxoisoindolin-2-yl)-2-(1-(2-(triisopropylsilyl)oxy)ethyl)-1H-indol-3-yl)ethyl)malonate (266i)

Following GP10 and starting from 1-(2-(triisopropylsilyl)oxy)ethyl)-1H-indole (**242g**) (64 mg, 0.22 mmol, 1.1 equiv), bis(2,2,2-trifluoroethyl) 2-(2-(1,3-dioxoisoindolin-2-yl)-2-(1-(2-(triisopropylsilyl)oxy)ethyl)-1H-indol-3-yl) ethyl)malonate (**266i**) (0.12 g, 0.16 mmol, 80 % yield) was obtained as an colorless solid after a reaction time of 45 min and purification by Biotage (SNAP cartridge KP-Sil 10 g, hexane/AcOEt 95/5–78/22).

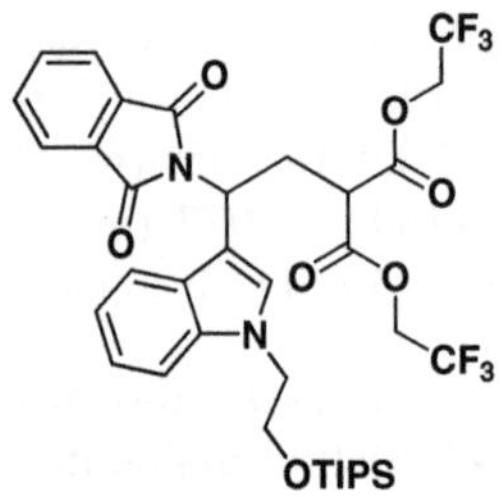

R_f 0.68 (6:4 hexane/ethyl acetate). **Mp** 68.8–72.3 °C. **^{1}H NMR** (400 MHz, CDCl$_3$) δ 7.81–7.75 (m, 2 H, Phth), 7.71 (d, 1 H, J = 8.1 Hz, ArH), 7.69–7.65 (m, 2 H, Phth), 7.48 (s, 1 H, Ar*H*), 7.32–7.29 (m, 1 H, ArH), 7.18 (t, 1 H, J = 7.2 Hz, ArH), 7.10 (t, 1 H, J = 7.4 Hz, ArH), 5.78 (dd, 1 H, J = 9.8, 6.7 Hz, PhthN–CH), 4.67–4.55 (m, 1 H, CF$_3$–CH$_2$), 4.54–4.36 (m, 3 H, CF$_3$–CH$_2$), 4.30–4.18 (m, 2 H, CH$_2$), 3.99 (t, 2 H, J = 5.6 Hz, CH$_2$), 3.68–3.63 (m, 1 H, CH–CO)$_2$), 3.34–24 (m, 1 H, CH$_2$), 3.12–3.03 (m, 1 H, CH$_2$), 1.06–0.85 (m, 21 H, TIPS). **^{13}C NMR** (101 MHz, CDCl$_3$) δ 168.1, 166.7, 166.6, 136.1, 134.2, 131.9, 128.6, 127.0, 123.4, 122.6 (q, $J_{C\text{-}F}$ = 277 Hz), 122.5 (q, $J_{C\text{-}F}$ = 277 Hz), 122.0, 119.8, 119.0, 111.4, 109.7, 62.7, 61.5 (q, $J_{C\text{-}F}$ = 37 Hz), 61.3 (q, $J_{C\text{-}F}$ = 37 Hz), 49.1, 49.0, 44.4, 30.9, 17.9, 17.9, 11.9. **IR** 2946 (w), 2868 (w), 1775 (m), 1760 (m), 1714 (s), 1469 (w), 1286 (m), 1173 (s). **HRMS** (ESI) calcd for $C_{36}H_{42}F_6N_2NaO_7Si^+$ [M + Na]$^+$ 779.2563; found 779.2548.

Bis(2,2,2-trifluoroethyl) 2-(2-(1,3-dioxoisoindolin-2-yl)-2-(2-methyl-1H-indol-3-yl)ethyl)malonate (266j)

Following GP10 and starting from 2-methyl-1H-indole (**242h**) (29 mg, 0.22 mmol, 1.1 equiv), bis(2,2,2-trifluoroethyl) 2-(2-(1,3-dioxoisoindolin-2-yl)-2-(2-methyl-1H-indol-3-yl)ethyl)malonate (**266j**) (0.11 g, 0.19 mmol, 97 % yield) was obtained as an colorless solid after a reaction time of 0.5 h and purification by Biotage (SNAP cartridge KP-Sil 10 g, hexane/AcOEt 97/3–85/15 + 1 % AcOH).

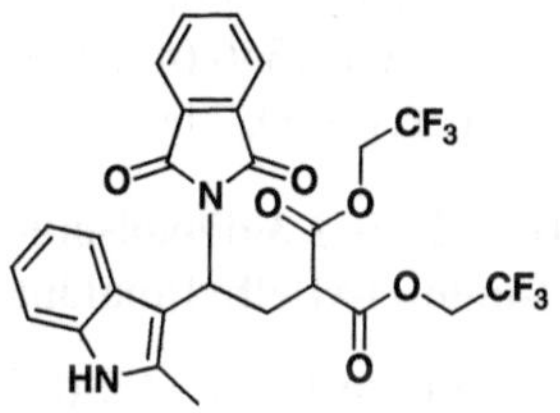

$\mathbf{R_f}$ 0.52 (6:4 hexane/ethyl acetate). **Mp** 49.4–52.3 °C. **^{1}H NMR** (400 MHz, $CDCl_3$) δ 7.80–7.75 (m, 2 H, Phth), 7.72 (dt, J = 8.0, 1.0 Hz, 1 H, ArH), 7.69–7.64 (m, 2 H, Phth), 7.40 (s, 1 H, NH), 7.28 (dt, J = 8.3, 1.0 Hz, 1 H, ArH), 7.22 (ddd, J = 8.2, 6.9, 1.2 Hz, 1 H, ArH), 7.12 (ddd, J = 8.0, 6.9, 1.2 Hz, 1 H, ArH), 5.80 (dd, J = 9.3, 7.1 Hz, 1 H, NCH), 4.67–4.53 (m, 1 H, CH_2CF_3), 4.53–4.36 (m, 3 H, CH_2CF_3), 3.78 (s, 3 H, Me), 3.68 (dd, J = 7.9, 6.9 Hz, 1 H, CH), 3.26 (ddd, J = 14.0, 9.3, 6.9 Hz, 1 H, CH_2), 3.13 (ddd, J = 14.1, 8.0, 7.1 Hz, 1 H, CH_2). **^{13}C NMR** (101 MHz, $CDCl_3$) δ 168.0, 166.5, 166.5, 136.5, 134.1, 131.8, 128.8, 126.7, 123.3, 122.6 (q, $J_{C\text{-}F}$ = 277 Hz), 122.5 (q, $J_{C\text{-}F}$ = 277 Hz), 122.1, 119.8, 118.9, 111.0, 109.4, 61.5 (q, $J_{C\text{-}F}$ = 37 Hz), 61.4 (q, $J_{C\text{-}F}$ = 37 Hz), 48.9, 44.2, 33.0, 30.7. **IR** 3396 (w), 3060 (w), 2974 (w), 2949 (w), 2872 (w), 1759 (m), 1711 (s), 1463 (w), 1356 (m), 1287 (m), 1174 (s). **HRMS** (ESI) calcd for $C_{26}F_6H_{21}N_2O_6^+$ $[M + H]^+$ 571.1298; found 571.1303.

Bis(2,2,2-trifluoroethyl) 2-(2-(1,3-dioxoisoindolin-2-yl)-2-(2-phenyl-1H-indol-3-yl)ethyl)malonate (266k)

Following GP10 and starting from 2-methyl-1H-indole (**242i**) (42 mg, 0.22 mmol, 1.1 equiv), bis(2,2,2-trifluoroethyl)2-(2-(1,3-dioxoisoindolin-2-yl)-2-(2-methyl-1H-indol-3-yl)ethyl)-malonate (**266k**) (118 mg, 0.187 mmol, 93 % yield) was obtained as a colorless solid after a reaction time of 45 min and purification by Biotage (SNAP cartridge KP-Sil 10 g, hexane/AcOEt 98/2–70/30).

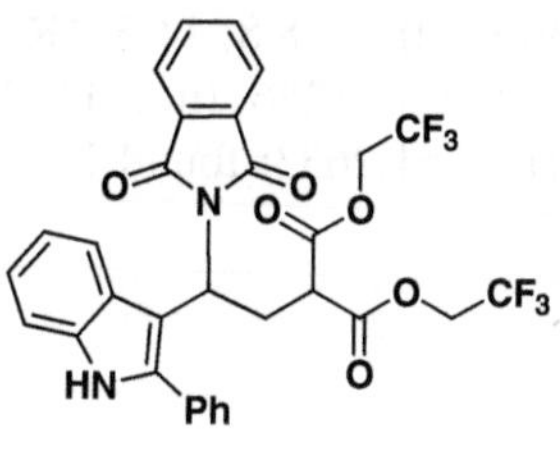

R_f 0.56 (6:4 hexane/ethyl acetate). **Mp** 81.6–84.9 °C. **^{1}H NMR** (400 MHz, $CDCl_3$) δ 8.18 (s, 1 H, NH), 8.14–8.10 (m, 1 H, ArH), 7.79–7.74 (m, 2 H, Phth), 7.69–7.62 (m, 4 H, Phth + ArH), 7.52–7.43 (m, 3 H, ArH), 7.37–7.33 (m, 1 H, ArH), 7.25–7.16 (m, 2 H, ArH), 5.80–5.73 (m, 1 H, PhthN–CH), 4.48–4.36 (m, 1 H, CF_3–CH_2), 4.35–4.19 (m, 3 H, CF_3–CH_2), 3.58–3.53 (m, 1 H, CH–$(CO)_2$), 3.40–3.31 (m, 1 H, CH_2), 3.27–3.17 (m, 1 H, CH_2). **^{13}C NMR** (101 MHz, $CDCl_3$) δ 168.6, 166.6, 166.5, 138.0, 135.8, 134.2, 132.4, 131.9, 129.8, 128.9, 128.8, 126.6, 123.4, 122.7, 122.5 (q, $J_{C\text{-}F}$ = 277 Hz), 121.2, 120.6, 111.0, 109.6, 61.2 (q, $J_{C\text{-}F}$ = 37 Hz), 61.2 (q, $J_{C\text{-}F}$ = 37 Hz), 49.0, 47.1, 30.7. **IR** 3395 (w), 3063 (w), 2977 (w), 1757 (m), 1709 (s), 1454 (w), 1356 (m), 1284 (s), 1168 (s), 977 (m), 720 (m). **HRMS** (ESI) calcd for $C_{31}F_6H_{23}N_2O_6^+$ $[M + H]^+$ 633.1455; found 633.1476.

Bis(2,2,2-trifluoroethyl) 2-(2-(1,3-dioxoisoindolin-2-yl)-2-(1-methyl-2-(triisopropylsilyl)ethynyl)-1H-indol-3-yl)ethyl)malonate (266l)

Following GP10 and starting from 1-methyl-2-(triisopropylsilyl)ethynyl)-1H-indole (**242j**) (62 mg, 0.22 mmol, 1.1 equiv), bis(2,2,2-trifluoroethyl) 2-(2-(1,3-dioxoisoindolin-2-yl)-2-(1-methyl-2-(triisopropylsilyl)ethynyl)-1H-indol-3-yl) ethyl)malonate (**266l**) (78 mg, 0.10 mmol, 52 % yield) was obtained as a colorless solid after a reaction time of 1 h and purification by Biotage (SNAP cartridge KP-Sil 10 g, hexane/AcOEt 95/5–75/25).

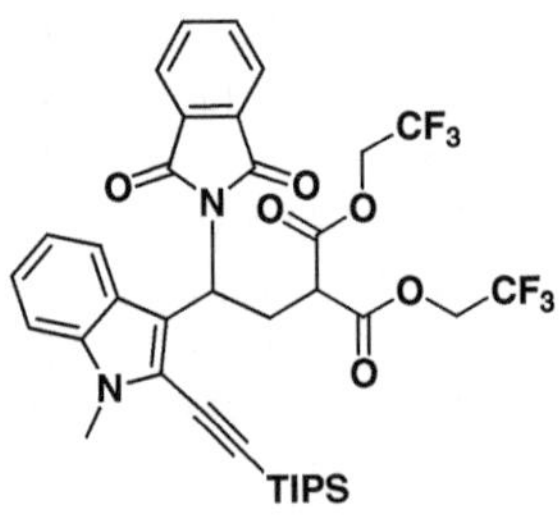

R_f 0.56 (7:3 hexane/ethyl acetate). **Mp** 71.6–74.0 °C. **^{1}H NMR** (400 MHz, $CDCl_3$) δ 7.93 (d, 1 H, J = 8.1 Hz, ArH), 7.80–7.75 (m, 2 H, Phth), 7.68–7.63 (m, 2 H, Phth), 7.28–7.21 (m, 2 H, ArH), 7.17–7.12 (m, 1 H, ArH), 5.97–5.92 (m, 1 H, PhthN–CH), 4.54–4.31 (m, 4 H, CF_3–CH_2), 3.79 (s, 3 H, CH_3), 3.70–3.64 (m, 1 H, CH–$(CO)_2$), 3.54–3.45 (m, 1 H, CH_2), 3.40–3.31 (m, 1 H, CH_2), 1.32–1.13 (m, 21 H, TIPS). **^{13}C NMR** (101 MHz, $CDCl_3$) δ 167.8, 166.6, 166.5, 137.0, 134.0, 132.0, 125.6, 123.6, 123.4, 122.7, 122.6 (q, $J_{C\text{-}F}$ = 277 Hz), 122.5 (q, $J_{C\text{-}F}$ = 277 Hz), 120.9, 120.6, 115.5, 109.6, 103.3, 96.6, 61.3 (q, $J_{C\text{-}F}$ = 37 Hz), 61.3 (q, $J_{C\text{-}F}$ = 37 Hz), 49.2, 46.2, 30.8, 30.1, 18.8, 11.5. **IR** 2944 (w), 2867 (w), 2153 (w), 1776 (m), 1761 (m), 1718 (s), 1286 (m), 1173 (s). **HRMS** (ESI) calcd for $C_{37}F_6H_{41}N_2O_6Si^+$ $[M + H]^+$ 751.2633; found 751.2633.

Bis(2,2,2-trifluoroethyl) 2-(2-(1,3-dioxoisoindolin-2-yl)-2-(3-methyl-1H-indol-2-yl)ethyl)malonate (266m)

Following GP10 and starting from 3-methyl-1H-indole (**242k**) (29 mg, 0.22 mmol, 1.1 equiv), bis(2,2,2-trifluoroethyl) 2-(2-(1,3-dioxoisoindolin-2-yl)-2-(3-methyl-1H-indol-2-yl)ethyl)malonate (**266m**) (68 mg, 0.12 mmol, 60 % yield) was obtained as a colorless solid after a reaction time of 1.5 h and purification by Biotage (SNAP cartridge KP-Sil 10 g, hexane/dichloromethane 80/20–2/98).

Following GP11 and starting from 3-methyl-1H-indole (**242k**) (29 mg, 0.22 mmol, 1.1 equiv), bis(2,2,2-trifluoroethyl) 2-(2-(1,3-dioxoisoindolin-2-yl)-2-(3-methyl-1H-indol-2-yl)ethyl)malonate (**266m**) (94 mg, 0.16 mmol, 82 % yield) was obtained as a colorless solid after a reaction time of 1.5 h and purification by Biotage (SNAP cartridge KP-Sil 10 g, hexane/dichloromethane 80/20–2/98).

Following GP12 and starting from 3-methyl-1H-indole (**242k**) (29 mg, 0.22 mmol, 1.1 equiv), bis(2,2,2-trifluoroethyl) 2-(2-(1,3-dioxoisoindolin-2-yl)-2-(3-methyl-1H-indol-2-yl)ethyl)malonate (**266m**) (76 mg, 0.13 mmol, 67 % yield) was obtained as a colorless solid after a reaction time of 1.5 h and purification by Biotage (SNAP cartridge KP-Sil 10 g, hexane/dichloromethane 80/20–2/98).

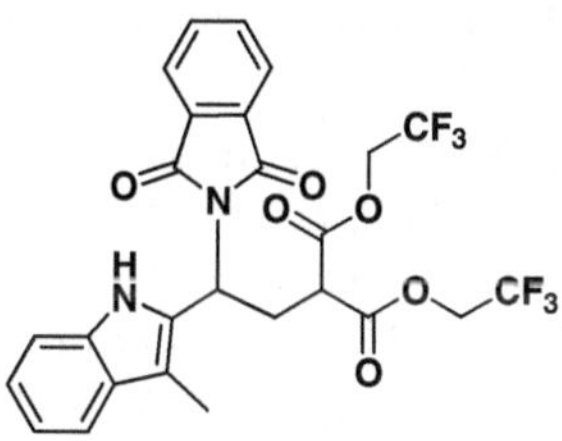

R_f 0.50 (7:3 hexane/ethyl acetate). **Mp** 59.9–64.0 °C. **^{1}H NMR** (400 MHz, $CDCl_3$) δ 9.08 (s, 1 H, NH), 7.88–7.83 (m, 2 H, Phth), 7.77–7.71 (m, 2 H, Phth), 7.52 (d, 1 H, $J = 7.9$ Hz, ArH), 7.36 (d, 1 H, $J = 8.2$ Hz, ArH), 7.21 (td, 1 H, $J = 7.1$, 0.8 Hz, ArH), 7.13–7.08 (m, 1 H, ArH), 5.85 (dd, 1 H, $J = 9.5$, 6.5 Hz, PhthN–CH), 4.57–4.36 (m, 3 H, CF_3–CH_2), 4.25–4.15 (m, 1 H, CF_3–CH_2), 3.49 (t, 1 H, $J = 7.3$ Hz, CH–$(CO)_2$), 3.26–3.17 (m, 1 H, CH_2), 2.92–2.83 (m, 1 H, CH_2), 2.34 (s, 3 H, CH_3). **^{13}C NMR** (101 MHz, $CDCl_3$) δ 168.6, 166.4, 166.2, 136.1, 134.6, 131.6, 129.5, 127.9, 123.8, 123.2, 122.5 (q, $J_{C\text{-}F} = 277$ Hz), 119.7, 119.3, 111.4, 61.5 (q, $J_{C\text{-}F} = 37$ Hz), 48.6, 44.0, 31.2, 8.4.[14] **IR** 1773 (m), 1713 (s), 1288 (m), 1176 (s), 978 (w). **HRMS** (ESI) calcd for $C_{26}F_6H_{21}N_2O_6^+$ $[M + H]^+$ 571.1298; found 571.1307.

Bis(2,2,2-trifluoroethyl) 2-(2-(1,3-dioxoisoindolin-2-yl)-2-(3-(triisopropylsilyl) ethynyl)-1H-indol-2-yl)ethyl)malonate (266n)

Following GP12 and starting from 3-(triisopropylsilyl)ethynyl)-1H-indole (**242l**) (66 mg, 0.22 mmol, 1.1 equiv), bis(2,2,2-trifluoroethyl) 2-(2-(1,3-dioxoisoind

[14]Only one CF_3 peak could be detected.

olin-2-yl)-2-(3-(triisopropylsilyl)ethynyl)-1H-indol-2-yl)ethyl)malonate (**266n**) (71 mg, 0.097 mmol, 49 % yield) was obtained as a colorless solid, after a reaction time of 6 h and purification by Biotage (SNAP cartridge KP-Sil 10 g, hexane/toluene 6/4–9/1).

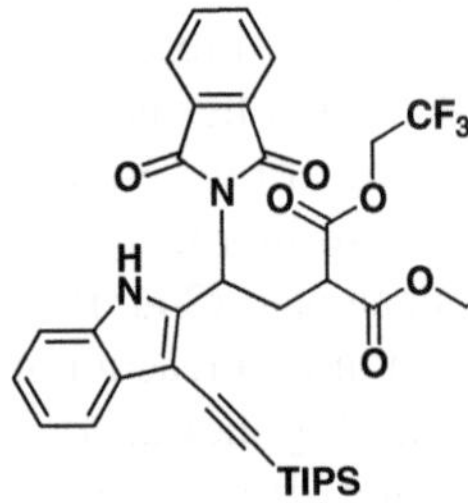

$\mathbf{R_f}$ 0.60 (6:4 hexane/ethyl acetate). **Mp** 49.2–51.8 °C. **^{1}H NMR** (400 MHz, $CDCl_3$) δ 9.51 (s, 1 H, NH), 7.91–7.86 (m, 2 H, Phth), 7.78–7.73 (m, 2 H, Phth), 7.64 (d, 1 H, J = 7.8 Hz, Ar*H*), 7.40–7.36 (m, 1 H, ArH), 7.27–7.22 (m, 1 H, ArH), 7.19–7.14 (m, 1 H, ArH), 6.13–6.08 (m, 1 H, PhthN–CH), 4.53–4.28 (m, 3 H, CF_3–CH_2), 3.97–3.87 (m, 1 H, CF_3–CH_2), 3.63 (dd, 1 H, J = 8.2, 2.3 Hz, CH–$(CO)_2$), 3.14 (dt, 1 H, J = 14.1, 8.8 Hz, CH_2), 2.87 (ddd, 1 H, J = 13.9, 7.8, 5.9 Hz, CH_2), 1.22 (s, 21 H, TIPS). **^{13}C NMR** (101 MHz, $CDCl_3$) δ 168.3, 166.2, 165.8, 137.1, 135.5, 134.7, 131.6, 128.1, 124.1, 123.9, 122.5 (q, $J_{C\text{-}F}$ = 275 Hz), 121.1, 120.4, 111.8, 100.1, 98.3, 96.0, 61.5 (q, $J_{C\text{-}F}$ = 37 Hz), 61.3 (q, $J_{C\text{-}F}$ = 37 Hz), 48.5, 45.3, 31.5, 18.9, 11.5. **IR** 3394 (w), 2151 (w), 1776 (m), 1716 (m), 1277 (s), 1173 (s). **HRMS** (ESI) calcd for $C_{36}F_6H_{39}N_2O_6Si^+$ $[M + H]^+$ 737.2476; found 737.2480.

Dimethyl 2-(2-(1,3-dioxoisoindolin-2-yl)-2-(3-(2-hydroxyethyl)-1H-indol-2-yl) ethyl)malonate (266o)

GP12 was followed using 3-(2-(trimethylsilyl)oxy)ethyl)-1H-indole (**242m**) (51 mg, 0.22 mmol, 1.1 equiv). The mixture was stirred at 60 °C for 25 min and then filtered over a short plug of silica gel (AcOEt/hexane 1/1). The crude was concentrated under reduced pressure to remove volatiles. Potassium carbonate (2.77 mg, 0.0200 mmol, 0.100 equiv) was added to a solution of the crude in MeOH (2.00 mL) at 0 °C. The resulting suspension was stirred for 30 min. The mixture was then partitioned between dichloromethane (5 mL) and saturated aqueous ammonium chloride-brine (1:2) (5 mL). The aqueous layer was extracted two times with dichloromethane (5 mL) and the combined organic layers were washed with brine (5 mL) and dried over anhydrous magnesium sulfate. Volatiles were removed under reduced pressure. The crude was then purified by Biotage (SNAP cartridge KP-Sil 10 g, pentane/AcOEt 85/15–50/50) to afford dimethyl 2-(2-(1,3-dioxoisoindolin-2-yl)-2-(3-(2-hydroxyethyl)-1H-indol-2-yl)ethyl)malonate (**266o**) (67 mg, 0.14 mmol, 72 % yield) as a pale yellow oil.

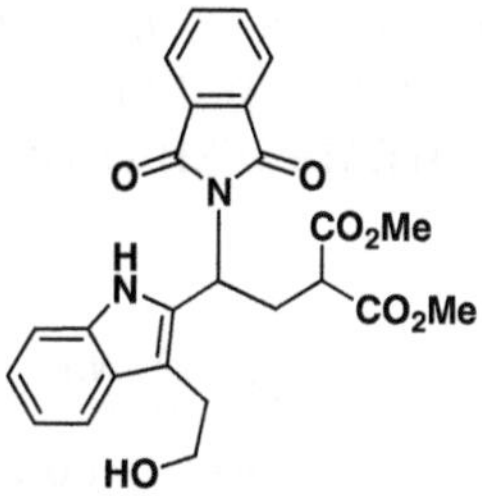

R_f 0.13 (6:4 hexane/ethyl acetate). **^{1}H NMR** (400 MHz, $CDCl_3$) δ 9.18 (s, 1 H, NH), 7.88–7.81 (m, 2 H, Phth), 7.77–7.70 (m, 2 H, Phth), 7.56 (d, 1 H, $J = 8.0$ Hz, ArH), 7.36 (d, 1 H, $J = 8.1$ Hz, ArH), 7.22–7.17 (m, 1 H, ArH), 7.11–7.06 (m, 1 H, ArH), 5.85 (t, 1 H, $J = 8.2$ Hz, PhthN–CH), 3.90–3.85 (m, 2 H, CH_2), 3.69 (s, 3 H, OMe), 3.63 (s, 3 H, OMe), 3.37 (t, 1 H, $J = 7.3$ Hz, CH–$(CO)_2$), 3.18–2.90 (m, 4 H, CH_2). **^{13}C NMR** (101 MHz, $CDCl_3$) δ 168.9, 168.6, 136.0, 134.5, 131.7, 131.6, 127.3, 123.7, 123.0, 119.7, 119.2, 111.6, 111.4, 62.9, 52.9, 49.1, 44.3, 31.3, 27.9. **IR** 3550 (w), 3406 (w), 3058 (w), 2954 (w), 2884 (w), 1735 (s), 1708 (s), 1440 (m), 1385 (m), 1333 (m), 1165 (m). **HRMS** (ESI) calcd for $C_{25}H_{25}N_2O_7^+$ $[M + H]^+$ 465.1656; found 465.1661.

Bis(2,2,2-trifluoroethyl) 2-(2-(1,3-dioxoisoindolin-2-yl)-2-(1H-pyrrol-2-yl)ethyl) malonate (272a)

Bis(2,2,2-trifluoroethyl) 2-(2-(1,3-dioxoisoindolin-2-yl)-2-(1H-pyrrol-3-yl)ethyl) malonate (272b)

Following GP10 and using 1*H*-pyrrole (**268**) (0.015 mL, 0.22 mmol, 1.1 equiv), bis (2,2,2-trifluoroethyl) 2-(2-(1,3-dioxoisoindolin-2-yl)-2-(1*H*-pyrrol-2-yl)ethyl)malonate (**272a**) (62 mg, 0.12 mmol, 61 % yield) and bis(2,2,2-trifluoroethyl) 2-(2-(1,3-dioxoisoindolin-2-yl)-2-(1*H*-pyrrol-3-yl)ethyl)malonate (**272b**) (31 mg, 0.059 mmol, 30 % yield) were obtained as colorless solids, after a reaction time of 1.5 h and purification by Biotage (SNAP cartridge KP-Sil 10 g, Hex/AcOEt 96/4–75/25). The structures were assigned by 2D NMR experiments.

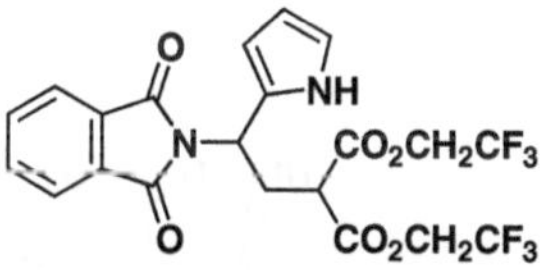

R_f 0.55 (6:4 hexane/ethyl acetate). **Mp** 122.2–123.8 °C. **^{1}H NMR** (400 MHz, $CDCl_3$) δ 9.28 (s, 1 H, NH), 7.86–7.80 (m, 2 H, Phth), 7.75–7.70 (m, 2 H, Phth), 6.80–6.77 (m, 1 H, ArH), 6.21–6.17 (m, 1 H, ArH), 6.10–6.07 (m, 1 H, ArH), 5.49 (dd, 1 H, $J = 9.2$, 7.7 Hz, PhthN–CH), 4.56–4.45 (m, 3 H, CF_3–CH_2), 4.42–4.31 (m, 1 H, CF_3–CH_2), 3.47 (t, 1 H, $J = 7.3$ Hz, CH–$(CO)_2$), 3.04 (ddd, 1 H, $J = 14.0$, 9.2, 7.4 Hz, CH_2), 2.82–2.74 (m, 1 H, CH_2). **^{13}C NMR** (101 MHz, $CDCl_3$) δ 168.4, 166.4, 166.3, 134.6, 131.7, 127.0, 123.7, 122.6 (q, $J_{C\text{-}F} = 277$ Hz), 122.5 (q, $J_{C\text{-}}$

$_F$ = 277 Hz), 119.7, 109.6, 108.0, 61.5 (q, $J_{C\text{-}F}$ = 37 Hz), 61.4 (q, $J_{C\text{-}F}$ = 37 Hz), 48.8, 46.4, 31.7. **IR** 3418 (w), 1760 (m), 1708 (s), 1386 (m), 1359 (m), 1283 (s), 1165 (s), 974 (m), 724 (m). **HRMS** (ESI) calcd for $C_{21}F_6H_{17}N_2O_6^+$ $[M + H]^+$ 507.0985; found 507.0986.

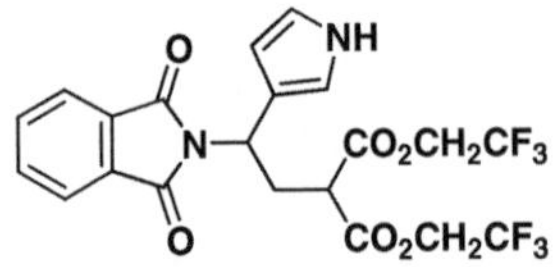

R$_f$ 0.48 (6:4 hexane/ethyl acetate). **Mp** 81.7–84.0 °C. **^{1}H NMR** (400 MHz, $CDCl_3$) δ 8.19 (s, 1 H, NH), 7.83–7.78 (m, 2 H, Phth), 7.72–7.67 (m, 2 H, Phth), 6.92–6.89 (m, 1 H, ArH), 6.73–6.69 (q, 1 H, J = 2.6 Hz, ArH), 6.38–6.34 (m, 1 H, ArH), 5.38 (t, 1 H, J = 8.2 Hz, PhthN–CH), 4.62–4.38 (m, 4 H, CF_3–CH_2), 3.58 (t, 1 H, J = 7.6 Hz, CH–$(CO)_2$), 3.04 (t, 2 H, J = 8.1 Hz, CH_2). **^{13}C NMR** (101 MHz, $CDCl_3$) δ 168.1, 166.7, 134.2, 132.0, 123.4, 122.6 (q, $J_{C\text{-}F}$ = 277 Hz), 122.6 (q, $J_{C\text{-}F}$ = 277 Hz), 120.8, 118.4, 117.3, 108.5, 61.4 (q, $J_{C\text{-}F}$ = 37 Hz), 61.3 (q, $J_{C\text{-}F}$ = 37 Hz), 49.0, 46.3, 31.4.[15] **IR** 3405 (w), 2359 (w), 2335 (w), 1774 (w), 1758 (w), 1712 (m), 1384 (w), 1278 (m), 1176 (m), 751 (s). **HRMS** (ESI) calcd for $C_{21}F_6H_{17}N_2O_6^+$ $[M + H]^+$ 507.0985; found 507.0980.

Bis(2,2,2-trifluoroethyl) 2-(2-(1,3-dioxoisoindolin-2-yl)-2-(1-(triisopropylsilyl)-1H-pyrrol-2-yl)ethyl) malonate (273a)

Following GP10 and starting from 1-(triisopropylsilyl)-1*H*-pyrrole (**269**) (49 mg, 0.22 mmol, 1.1 equiv), a 7:1 mixture of two regioisomers (**273a**) was obtained as a colorless oil (0.12 g, 0.18 mmol, 88 % yield) after a reaction time of 25 min and purification by Biotage (SNAP cartridge KP-Sil 10 g, hexane/AcOEt 93/7–55/45). Peaks at 5.47 and 5.36 ppm in the ^{1}H NMR spectrum were integrated to determine the ratio of regioisomer. The structure of the regioisomers was assigned through 2D NMR experiments. The major isomer was isolated in pure form after a second purification by column chromatography for characterization.

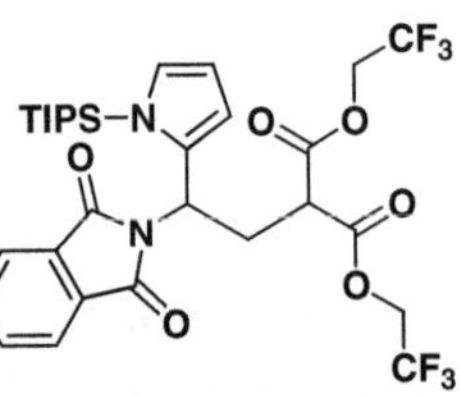

R$_f$ 0.38 (8:2 hexane/ethyl acetate). **^{1}H NMR** (400 MHz, $CDCl_3$) δ 7.84–7.78 (m, 2 H, Phth), 7.73–7.66 (m, 2 H, Phth), 6.83 (t, 1 H, J = 1.7 Hz, ArH), 6.67 (t, 1 H,

[15]Two carbonyls peaks are not overlapping.

J = 2.5 Hz, ArH), 6.42 (dd, 1 H, J = 2.7, 1.4 Hz, ArH), 5.36 (t, 1 H, J = 8.3 Hz, PhthN–CH), 4.62–4.39 (m, 4 H, CF_3–CH_2), 3.53 (t, 1 H, J = 7.7 Hz, CH–$(CO)_2$), 3.08–2.95 (m, 2 H, CH_2), 1.41 (hept, 3 H, J = 7.5 Hz, Si–CH), 1.07 (d, 9 H, J = 2.8 Hz, CH_3), 1.05 (d, 9 H, J = 2.8 Hz, CH_3). **^{13}C NMR** (101 MHz, $CDCl_3$) δ 168.1, 166.7, 134.1, 132.0, 124.8, 123.6, 123.4, 122.9 (q, $J_{C\text{-}F}$ = 277 Hz), 122.9 (q, $J_{C\text{-}F}$ = 277 Hz), 122.4, 110.4, 61.4 (q, $J_{C\text{-}F}$ = 37 Hz), 61.3 (q, $J_{C\text{-}F}$ = 37 Hz), 49.0, 46.4, 31.5, 17.9, 11.7 (See footnote 15). **IR** 2951 (w), 2869 (w), 1775 (m), 1760 (m), 1715 (s), 1287 (m), 1173 (s). **HRMS** (ESI) calcd for $C_{30}F_6H_{37}N_2O_6Si^+$ $[M + H]^+$ 663.2320; found 663.2317.

Bis(2,2,2-trifluoroethyl) 2-(2-(1,3-dioxoisoindolin-2-yl)-2-(2-methoxyphenyl)ethyl)malonate (274a)

Bis(2,2,2-trifluoroethyl) 2-(2-(1,3-dioxoisoindolin-2-yl)-2-(4-methoxyphenyl)ethyl)malonate (274b)

Following GP10 and starting from anisole (**270**) (0.024 mL, 0.22 mmol, 1.1 equiv), an inseparable mixture of bis(2,2,2-trifluoroethyl) 2-(2-(1,3-dioxoisoindolin-2-yl)-2-(2-methoxyphenyl)ethyl)malonate (**274a**) and bis(2,2,2-trifluoroethyl) 2-(2-(1,3-dioxoisoindolin-2-yl)-2-(4-methoxyphenyl)ethyl)malonate (**274b**) (66 mg, 0.12 mmol, 60 % yield) was obtained as a colorless solid after a reaction time of 45 min and purification by Biotage (SNAP cartridge KP-Sil 10 g, Hex/AcOEt 93/7–70/30). The integration of the peaks at 5.81 and 5.33 ppm in the ^{1}H NMR spectrum were used to determine the ratio of regioisomers.

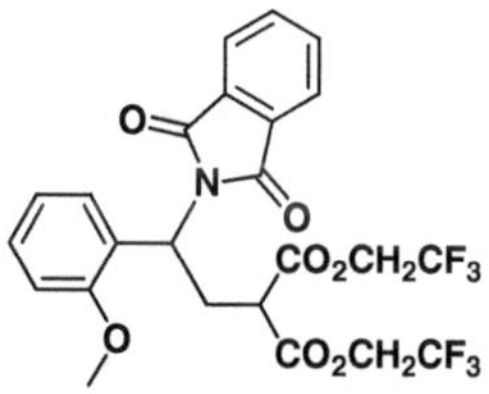

Isomer 1

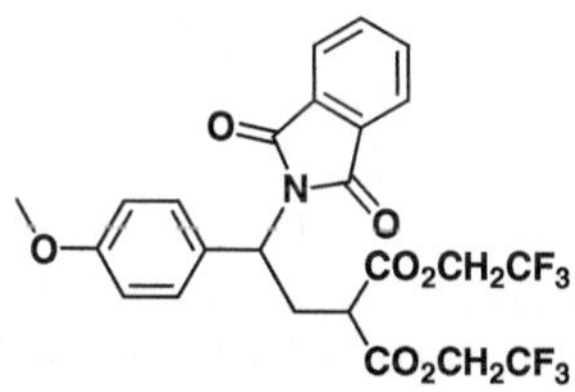

Isomer 2

R_f 0.60 (6:4 hexane/ethyl acetate). **^{1}H NMR** (400 MHz, $CDCl_3$) δ 7.83–7.78 (m, 4 H, Phth, Isomer 1 & 2), 7.73–7.68 (m, 4 H, Phth, Isomer 1 & 2), 7.62 (dd, 1 H, J = 7.7, 1.3 Hz, ArH, Isomer 1), 7.47 (d, 2 H, J = 8.7 Hz, ArH, Isomer 2), 7.30–7.25 (m, 1 H, Ar*H*, *Isomer 1*), 6.97 (dt, 1 H, J = 7.5, 0.9 Hz, ArH, Isomer 1), 6.88–6.82 (m, 3 H, ArH, Isomer 1 & 2), 5.81 (dd, 1 H, J = 10.1, 5.9 Hz, PhthN–CH, Isomer 1),

5.33 (dd, 1 H, J = 9.7, 7.0 Hz, PhthN–CH, Isomer 2), 4.66–4.54 (m, 2 H, CF_3–CH_2, Isomer 1 & 2), 4.53–4.35 (m, 6 H, CF_3–CH_2, Isomer 1 & 2), 3.78 (s, 3 H, OMe), 3.77 (s, 3 H, OMe), 3.63 (dd, 1 H, J = 8.5, 6.3 Hz, CH–$(CO)_2$, Isomer 1), 3.55 (dd, 1 H, J = 8.5, 6.8 Hz, CH–$(CO)_2$, Isomer 2), 3.26–3.13 (m, 2 H, CH_2, Isomer 1 & 2), 3.07–2.98 (m, 1 H, CH_2, Isomer 2), 2.97–2.89 (m, 1 H, CH_2, Isomer 1). **^{13}C NMR** (101 MHz, $CDCl_3$) δ 168.2, 168.2, 166.7, 166.6, 166.5, 159.7, 156.9, 134.3, 134.2, 131.8, 131.8, 130.0, 129.6, 128.6, 126.0, 123.6, 123.5, 122.7 (q, $J_{C\text{-}F}$ = 277 Hz), 122.6 (q, $J_{C\text{-}F}$ = 277 Hz), 122.6 (q, $J_{C\text{-}F}$ = 277 Hz), 120.6, 114.3, 110.8, 61.5 (q, $J_{C\text{-}F}$ = 37 Hz), 61.4 (q, $J_{C\text{-}F}$ = 37 Hz), 55.7, 55.4, 52.0, 48.9, 46.3, 30.3, 30.1.[16] **IR** 2937 (w), 2846 (w), 1761 (m), 1715 (s), 1361 (m), 1286 (s), 1172 (s). **HRMS** (ESI) calcd for $C_{24}F_6H_{20}NO_7^+$ $[M + H]^+$ 548.1138; found 548.1119.

Bis(2,2,2-trifluoroethyl) 2-(2-(1,3-dioxoisoindolin-2-yl)-2-(2-hydroxyphenyl) ethyl)malonate (275a)

Bis(2,2,2-trifluoroethyl) 2-(2-(1,3-dioxoisoindolin-2-yl)-2-phenoxyethyl)malonate (275b)

Following GP10 and starting from phenol (**271**) (21 mg, 0.22 mmol, 1.1 equiv), bis (2,2,2-trifluoroethyl) 2-(2-(1,3-dioxoisoindolin-2-yl)-2-(2-hydroxyphenyl)ethyl) malonate (**275a**) (13 mg, 0.024 mmol, 12 % yield) and bis(2,2,2-trifluoroethyl) 2-(2-(1,3-dioxoisoindolin-2-yl)-2-phenoxyethyl)malonate (**275b**) (79 mg, 0.15 mmol, 74 % yield) were obtained as colorless solids after a reaction time of 40 min and purification by Biotage (SNAP cartridge KP-Sil 10 g, Hex/AcOEt 95/5–70/30).

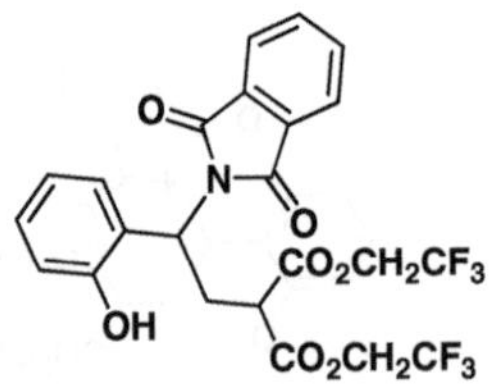

R_f 0.46 (6:4 hexane/ethyl acetate). **Mp** 103.5–106.0 °C. **^{1}H NMR** (400 MHz, $CDCl_3$) δ 7.86–7.81 (m, 2 H, Phth), 7.76–7.70 (m, 2 H, Phth), 7.52 (d, 1 H, J = 7.7 Hz, ArH), 7.22 (t, 1 H, J = 7.5 Hz, ArH), 6.94 (t, 1 H, J = 7.5 Hz. ArH), 6.89 (d, 1 H, J = 8.1 Hz, ArH), 6.76 (s, 1 H, OH), 5.62 (t, 1 H, J = 8.1 Hz, PhthN–CH), 4.62–4.38 (m, 4 H, CF_3–CH_2), 3.58 (t, 1 H, J = 7.4 Hz, CH–$(CO)_2$), 3.26–3.09 (m, 2 H, CH_2). **^{13}C NMR** (101 MHz, $CDCl_3$) δ 168.7, 166.2, 154.0, 134.5, 131.3, 130.2, 129.5, 123.6, 123.3, 122.6 (q, $J_{C\text{-}F}$ = 277 Hz), 122.5 (q, $J_{C\text{-}F}$ = 277 Hz), 121.1, 117.9, 61.5 (q, $J_{C\text{-}F}$ = 37 Hz), 61.4 (q, $J_{C\text{-}F}$ = 38 Hz), 48.6, 47.4, 29.0. **IR** 3452 (w), 3062 (w), 2932 (w), 1766 (m), 1715 (s), 1378 (m), 1287 (s), 1172 (s). **HRMS** (ESI) calcd for $C_{23}F_6H_{18}NO_7^+$ $[M + H]^+$ 534.0982; found 534.0989.

[16]All carbons couldn't be resolved.

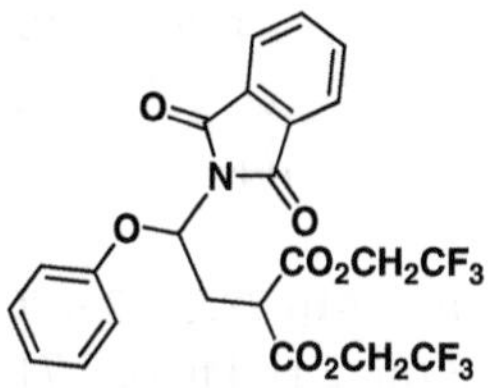

R_f 0.63 (6:4 hexane/ethyl acetate). **Mp** 110.2–111.9 °C. **^{1}H NMR** (400 MHz, $CDCl_3$) δ 7.87–7.81 (m, 2 H, Phth), 7.76–7.70 (m, 2 H, Phth), 7.26–7.20 (m, 2 H, ArH), 7.02–6.95 (m, 3 H, ArH), 6.32 (dd, 1 H, J = 8.6, 5.1 Hz, PhthN–CH), 4.65–4.43 (m, 4 H, CF_3–CH_2), 3.99–3.94 (m, 1 H, CH–$(CO)_2$), 3.43–3.33 (m, 1 H, CH_2), 2.80–2.71 (m, 1 H, CH_2). **^{13}C NMR** (101 MHz, $CDCl_3$) δ 167.1, 166.4, 155.5, 134.7, 131.4, 129.9, 124.0, 123.2, 122.6 (q, $J_{C\text{-}F}$ = 277 Hz), 122.5 (q, $J_{C\text{-}F}$ = 277 Hz), 116.8, 76.6, 61.5 (q, $J_{C\text{-}F}$ = 37 Hz), 61.5 (q, $J_{C\text{-}F}$ = 37 Hz), 47.7, 31.6. **IR** 3068 (w), 2940 (w), 1765 (m), 1723 (s), 1361 (m), 1289 (s), 1174 (s). **HRMS** (ESI) calcd for $C_{23}H_{17}F_6NNaO_7^+$ $[M + Na]^+$ 556.0801; found 556.0804.

Product modifications

Methyl 5-(1H-indol-3-yl)-2-oxopyrrolidine-3-carboxylate (277)

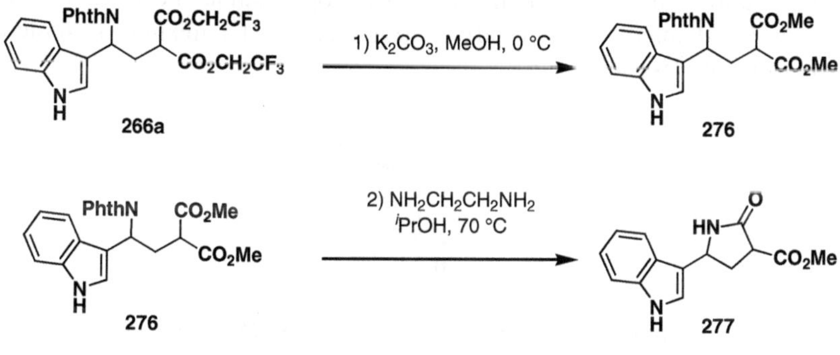

To a solution of bis(2,2,2-trifluoroethyl) 2-(2-(1,3-dioxoisoindolin-2-yl)-2-(1H-indol-3-yl)ethyl)malonate (**266a**) (556 mg, 1.00 mmol, 1 equiv) in MeOH (10.0 mL) was added potassium carbonate (13.8 mg, 0.100 mmol, 0.1 equiv) at 0 °C. The resulting suspension was stirred for 30 min and was partitioned between dichloromethane (20 mL) and saturated aqueous ammonium chloride-brine (1:2) (20 mL). The aqueous layer was extracted with dichloromethane (20 mL) twice and the combined organic layers were dried over anhydrous magnesium sulfate. The solvent was removed under reduced pressure to afford the crude as a yellow solid, which was purified by Biotage (SNAP cartridge KP-Sil 25 g, pentane/AcOEt 85/15–60/40) to afford dimethyl 2-(2-(1,3-dioxoisoindolin-2-yl)-2-(1H-indol-3-yl) ethyl)malonate (**276**) (302 mg, 0.717 mmol, 72 % yield) as an colorless solid.

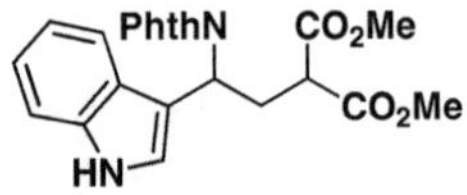

R_f 0.19 (6:4 Hexane/AcOEt). **Mp** 78.2–82.9 °C. **^{1}H NMR** (400 MHz, $CDCl_3$) δ 8.41 (s, 1 H, NH), 7.80–7.73 (m, 2 H, Phth), 7.71 (d, 1 H, J = 7.9 Hz, ArH), 7.68–7.62 (m, 2 H, Phth), 7.50 (d, 1 H, J = 1.9 Hz, ArH), 7.34 (d, 1 H, J = 8.0 Hz, ArH), 7.19–7.14 (m, 1 H, ArH), 7.13–7.08 (m, 1 H, ArH), 5.79 (dd, 1 H, J = 9.8, 6.4 Hz, PhthN–CH), 3.74 (s, 3 H, CH_3), 3.65 (s, 3 H, CH_3), 3.50 (t, 1 H, J = 7.4 Hz, CH–$(CO)_2$), 3.26 (ddd, 1 H, J = 14.2, 9.9, 6.9 Hz, CH_2), 3.09–3.00 (m, 1 H, CH_2). **^{13}C NMR** (101 MHz, $CDCl_3$) δ 169.4, 169.3, 168.2, 135.8, 134.1, 131.8, 126.4, 124.3, 123.3, 122.5, 120.1, 118.9, 113.2, 111.3, 53.0, 52.9, 49.7, 44.6, 30.8. **IR** 3403 (w), 2954 (w), 2865 (w), 2363 (w), 2093 (w), 1753 (m), 1733 (s), 1710 (s), 1358 (m). **HRMS** (ESI) calcd for $C_{23}H_{20}N_2NaO_6^+$ $[M + Na]^+$ 443.1214; found 443.1216.

Following a modified procedure [21], dimethyl 2-(2-(1,3-dioxoisoindolin-2-yl)-2-(1H-indol-3-yl)ethyl)malonate (**276**) (84.0 mg, 0.200 mmol, 1 equiv) was added in a tube sealed with a septum with ethane-1,2-diamine (0.0410 mL, 1.00 mmol, 5 equiv). Isopropanol (1.30 mL) was added and the resulting mixture was stirred at 70 °C for 30 min. The solution was concentrated under reduced pressure and the crude was purified by Biotage (SNAP cartridge KP-Sil 25 g, pentane/AcOEt 75/25–0/100) to afford methyl 5-(1H-indol-3-yl)-2-oxopyrrolidine-3-carboxylate (**277**) (39.1 mg, 0.151 mmol, 76 % yield) as an inseparable 1:1 mixture of diastereoisomers as a colorless solid.

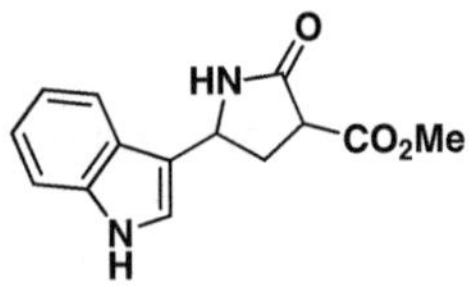

R_f 0.40/0.32 (ethyl acetate). **Mp** 51.2–53.6 °C. **^{1}H NMR** (400 MHz, $CDCl_3$) δ 8.21–8.11 (m, 2 H, NH, Isomer 1 & 2), 7.74 (d, 1 H, J = 7.9 Hz, ArH, Isomer 2), 7.61 (d, 1 H, J = 7.9 Hz, ArH, Isomer 1), 7.42 (s, 1 H, ArH), 7.40 (s, 1 H, ArH), 7.28–7.22 (m, 3 H, ArH), 7.20–7.13 (m, 3 H, ArH), 5.97 (s, 1 H, NH, Isomer 1), 5.90 (s, 1 H, NH, Isomer 2), 5.26 (t, 1 H, J = 6.9 Hz, CH–NH, Isomer 1), 5.05 (t, 1 H, J = 8.0 Hz, CH–NH, Isomer 2), 3.83 (s, 3 H, CH_3), 3.83 (s, 3 H, CH_3), 3.69–3.60 (m, 2 H, CH–$(CO)_2$, Isomer 1 & 2), 2.98–2.90 (m, 1 H, CH_2, Isomer 1), 2.89–2.73 (m, 2 H, CH_2, Isomer 2), 2.57–2.47 (m, 1 H, CH_2, Isomer 1). **^{13}C NMR** (101 MHz, $CDCl_3$) δ 172.8, 172.6, 170.6, 137.0, 136.9, 125.2, 125.0, 122.7, 122.6, 122.3, 121.6, 120.0, 119.9, 119.1, 118.8, 116.2, 115.3, 111.9, 111.8, 52.9, 52.8, 50.5, 50.1, 48.9, 48.1, 33.5, 33.5.(See footnote 16) **IR** 3274 (w), 2955 (w), 2361 (w), 1738 (s), 1696 (s), 1438 (w), 1268 (w). **HRMS** (ESI) calcd for $C_{14}H_{15}N_2O_3^+$ $[M + H]^+$ 259.1077; found 259.1086.

Methyl 5-(1-methyl-1H-indol-3-yl)-2-oxopyrrolidine-3-carboxylate (279)

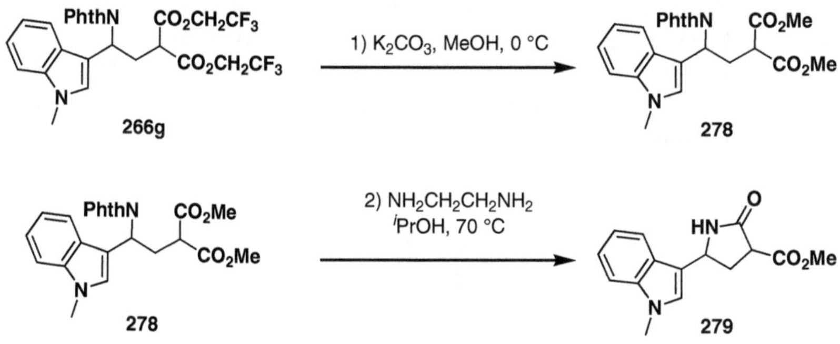

To a solution of bis(2,2,2-trifluoroethyl) 2-(2-(1,3-dioxoisoindolin-2-yl)-2-(1-methyl-1H-indol-3-yl)ethyl)malonate (**266 g**) (570 mg, 1.00 mmol, 1 equiv) in MeOH (10.0 mL) was added potassium carbonate (13.8 mg, 0.100 mmol, 0.1 equiv) at 0 °C. The resulting suspension was stirred for 30 min. Then, the mixture was partitioned between dichloromethane (20 mL) and saturated aqueous ammonium chloride-brine (1:2) (20 mL). The aqueous layer was extracted with dichloromethane (20 mL) twice and the combined organic layers were dried over anhydrous magnesium sulfate. The solvent was removed under reduced pressure to afford the crude as a yellow solid, which was purified by Biotage (SNAP cartridge KP-Sil 25 g, pentane/AcOEt 85/15 to 60/40) to afford dimethyl 2-(2-(1,3-dioxoisoindolin-2-yl)-2-(1-methyl-1H-indol-3-yl)ethyl)malonate (**278**) (299 mg, 0.688 mmol, 69 % yield) as a colorless solid.

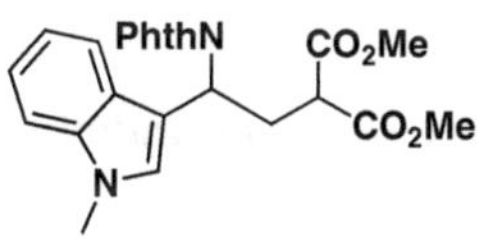

R_f 0.40 (6:4 Hexane/AcOEt). **^{1}H NMR** (400 MHz, $CDCl_3$) δ 7.82–7.77 (m, 2 H, Phth), 7.75–7.71 (m, 1 H, ArH), 7.71–7.65 (m, 2 H, Phth), 7.41 (s, 1 H, ArH), 7.32–7.27 (m, 1 H, ArH), 7.26–7.19 (m, 1 H, ArH), 7.16–7.09 (m, 1 H, ArH), 5.79 (dd, 1 H, J = 9.7, 6.5 Hz, N–C–H), 3.80 (s, 3 H, Me), 3.76 (s, 3 H, Me), 3.68 (s, 3 H, Me), 3.53–3.47 (m, 1 H, CH), 3.30–3.20 (m, 1 H, CH_2), 3.09–3.00 (m, 1 H, CH_2). **^{13}C NMR** (101 MHz, $CDCl_3$) δ 169.2, 169.2, 168.1, 136.5, 133.9, 131.9, 128.7, 126.9, 123.2, 121.9, 119.6, 119.0, 111.6, 109.3, 52.8, 52.7, 49.6, 44.5, 33.0, 30.9. **IR** 2953 (w), 2365 (w), 1753 (m), 1736 (s), 1711 (s), 1614 (w), 1543 (w), 1469 (w), 1436 (w), 1383 (m), 1355 (w), 1329 (m), 1275 (w), 1158 (w), 913 (w), 739 (m), 728 (s). **HRMS (ESI)** calcd for $C_{24}H_{23}N_2O_6^+$ $[M + H]^+$ 435.1551; found 435.1548.

Following a modified procedure [21], dimethyl 2-(2-(1,3-dioxoisoindolin-2-yl)-2-(1-methyl-1H-indol-3-yl)ethyl)malonate (**278**) (87.0 mg, 0.200 mmol, 1 equiv) was added into a tube sealed with a septum with ethane-1,2-diamine (0.0410 mL, 1.00 mmol, 5 equiv). Isopropanol (1.30 mL) was added and the resulting mixture was stirred at 70 °C for 30 min. The solution was concentrated under reduced pressure and the crude was purified by Biotage (SNAP cartridge KP-Sil 25 g, pentane/AcOEt 75/25–0/100) to afford Methyl 5-(1-methyl-1H-indol-3-yl)-2-oxopyrrolidine-3-carboxylate (**279**) (43.7 mg, 0.160 mmol, 80 % yield) as an inseparable 1:1 mixture of diastereoisomers as a colorless solid.

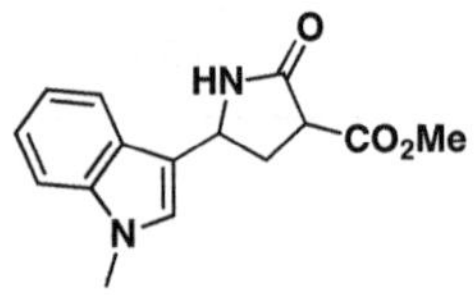

R_f 0.42/0.28 (ethyl acetate). **^{1}H NMR** (400 MHz, $CDCl_3$) δ 7.73 (d, 1 H, J = 8.0 Hz, ArH, Isomer 2), 7.59 (d, 1 H, J = 7.9 Hz, ArH, Isomer 1), 7.36–7.31 (m, 2 H, ArH), 7.30–7.27 (m, 2 H, ArH), 7.19–7.12 (m, 2 H, ArH), 7.09 (s, 1 H, ArH, Isomer 2), 7.00 (s, 1 H, ArH, Isomer 1), 6.00 (br s, 1 H, NH, Isomer 1), 5.92 (br s, 1 H, NH, Isomer 2), 5.24 (t, 1 H, J = 6.9 Hz, CH–NH, Isomer 1), 5.03 (t, 1 H, J = 7.9 Hz, CH–NH, Isomer 2), 3.83 (s, 3 H, CH_3), 3.82 (s, 3 H, CH_3), 3.78 (s, 3 H, CH_3), 3.77 (s, 3 H, CH_3), 3.68–3.59 (m, 2 H, CH–$(CO)_2$, Isomer 1 & 2), 2.96–2.88 (m, 1 H, CH_2, Isomer 1), 2.87–2.70 (m, 2 H, CH_2, Isomer 2), 2.55–2.46 (m, 1 H, CH_2, Isomer 1). **^{13}C NMR** (101 MHz, $CDCl_3$) δ 172.3, 172.1, 170.6, 170.5, 137.7, 137.6, 126.9, 126.2, 125.7, 125.5, 122.5, 122.5, 119.8, 119.7, 119.3, 119.0, 114.8, 113.8, 109.9, 109.8, 53.0, 52.9, 50.4, 49.8, 48.8, 48.1, 33.9, 33.7, 33.0 (See footnote 16). **IR** 3235 (w), 2953 (w), 1741 (s), 1703 (s), 1475 (w), 1334 (w), 1264 (w), 1168 (w). **HRMS** (ESI) calcd for $C_{15}H_{16}N_2NaO_3^+$ $[M + Na]^+$ 295.1053; found 295.1059.

2,2,2-Trifluoroethyl 4-(1,3-dioxoisoindolin-2-yl)-4-(1-methyl-1H-indol-3-yl) butanoate (280)

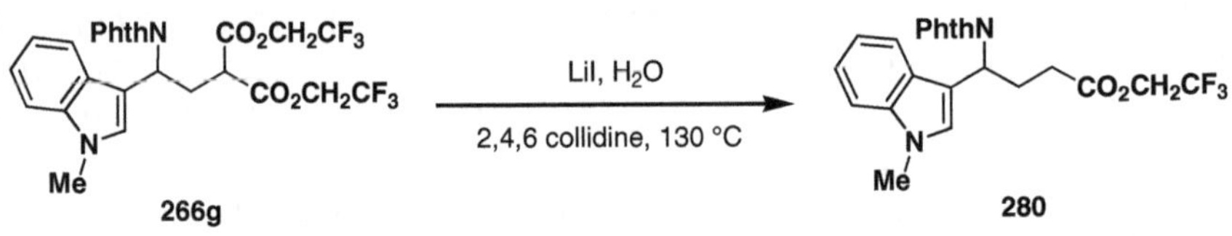

Following a modified procedure [22], a mixture of bis(2,2,2-trifluoroethyl) 2-(2-(1,3-dioxoisoindolin-2-yl)-2-(1-methyl-1H-indol-3-yl)ethyl)malonate (**266 g**) (114 mg, 0.200 mmol, 1 equiv), lithium iodide (40.2 mg, 0.300 mmol, 1.5 equiv) and water (7.20 μl, 0.400 mmol, 2 equiv) in 2,4,6-collidine (2.00 mL) in a microwave vial sealed with a microwave cap was heated at 130 °C for 3 h. After cooling down to room temperature, the mixture was diluted with AcOEt (5 mL), washed with 1 M HCl (5 mL) four times and sat. $NaHCO_3$-brine (1:3) (5 mL), dried over $MgSO_4$ and concentrated under reduced pressure to afford 2,2,2-Trifluoroethyl 4-(1,3-dioxoisoindolin-2-yl)-4-(1-methyl-1H-indol-3-yl)butanoate (**280**) as a pale yellow solid (90.3 mg, 0.200 mmol, 99 % yield).

R_f 0.58 (6:4 hexane/AcOEt). **Mp** 54.6–57.0 °C. **^{1}H NMR** (400 MHz, $CDCl_3$) δ 7.81–7.76 (m, 2 H, Phth), 7.74 (d, 1 H, J = 8.1 Hz, ArH), 7.69–7.63 (m, 2 H, Phth), 7.39 (s, 1 H, ArH), 7.31–7.26 (m, 1 H, ArH), 7.22 (t, 1 H, J = 7.5 Hz, ArH), 7.15–7.09 (m, 1 H, ArH), 5.77 (dd, 1 H, J = 9.1, 7.1 Hz, PhthN–CH), 4.52–4.38 (m, 2 H, CF_3–CH_2), 3.79 (s, 3 H, CH_3), 2.98–2.76 (m, 2 H, CH_2), 2.63–2.49 (m, 2 H, CH_2). **^{13}C NMR** (101 MHz, $CDCl_3$) δ 171.2, 168.3, 136.6, 134.0, 132.0, 128.8, 127.0, 123.3, 123.0 (q, $J_{C\text{-}F}$ = 277 Hz), 122.0, 119.7, 119.1, 111.9, 109.4, 60.5 (q, $J_{C\text{-}F}$ = 37 Hz), 45.9, 33.1, 31.4, 26.9. **IR** 3061 (w), 2941 (w), 1761 (m), 1711 (s), 1473 (w), 1384 (m), 1330 (m), 1285 (m), 1171 (s). **HRMS** (ESI) calcd for $C_{23}F_3H_{20}N_2O_4^+$ $[M + H]^+$ 445.1370; found 445.1356.

2-(10-Methyl-6-oxo-6,7,8,9-tetrahydropyrido[1,2-*a*]indol-9-yl) isoindoline-1,3-dione (281)

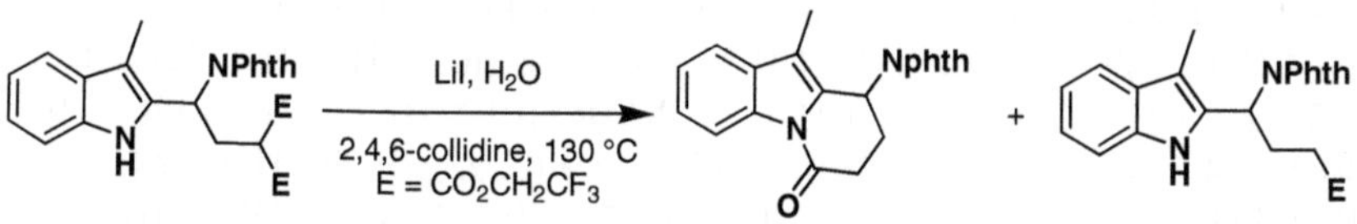

266j 281 282

A suspension of bis(2,2,2-trifluoroethyl) 2-(2-(1,3-dioxoisoindolin-2-yl)-2-(3-methyl-1H-indol-2-yl)ethyl)malonate **266j** (122 mg, 0.214 mmol, 1equiv), lithium iodide (42.9 mg, 0.321 mmol, 1.5 equiv) and water (7.71 μl, 0.428 mmol, 2 equiv) in 2,4,6-collidine (2.10 mL) in a microwave vial sealed with a microwave cap was heated at 130 °C for 18.5 h and then at 140 °C for 5.5 h. After cooling down to room temperature, the mixture was diluted with AcOEt (5 mL), washed with 1 M HCl (5 mL) four times and sat. $NaHCO_3$-brine (1:3) (5 mL) twice, dried over $MgSO_4$, and concentrated under reduced pressure to afford the crude as a light brown solid. The crude was purified by Biotage (SNAP cartridge KP-Sil 25 g, pentane/AcOEt 90/10–61/39) to afford 2-(10-methyl-6-oxo-6,7,8,9-tetrahydropyrido[1,2-*a*]indol-9-yl)isoindoline-1,3-dione (**281**) (53.4 mg, 0.155 mmol, 73 % yield) as a colorless solid and 2,2,2-trifluoroethyl 4-(1,3-dioxoisoindolin-2-yl)-4-(3-methyl-1H-indol-2-yl)butanoate (**282**) (23.2 mg, 0.0520 mmol, 24 % yield).

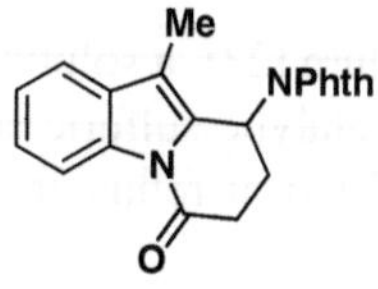

R$_f$ 0.40 (6:4 hexane/AcOEt). **Mp** 177.5–180.8 °C. **^{1}H NMR** (400 MHz, $CDCl_3$) δ 8.54 (d, 1 H, *J* = 8.2 Hz, Ar*H*), 7.87–7.81 (m, 2 H, Phth), 7.78–7.71 (m, 2 H, Phth), 7.43 (d, 1 H, *J* = 7.7 Hz, ArH), 7.40–7.34 (m, 1 H, ArH), 7.31–7.27 (m, 1 H, ArH), 5.85 (t, 1 H, *J* = 4.9 Hz, PhthN–CH), 3.12–3.01 (m, 1 H, CH_2), 2.83 (dt, 1 H, *J* = 17.4, 5.1 Hz, CH_2), 2.54–2.37 (m, 2 H, CH_2),2.09 (s, 3 H). **^{13}C NMR** (101 MHz, $CDCl_3$) δ 168.5, 167.8, 134.9, 134.5, 131.6, 130.4, 129.2, 125.6, 123.9, 123.7, 118.5, 116.8, 115.4, 42.0, 31.6, 27.3, 8.5. **IR** 3068 (w), 1773 (w), 1713 (s), 1458 (m), 1388 (m), 1366 (m), 1326 (m). **HRMS** (ESI) calcd for $C_{21}H_{16}N_2NaO_3^+$ $[M + Na]^+$ 367.1053; found 367.1037.

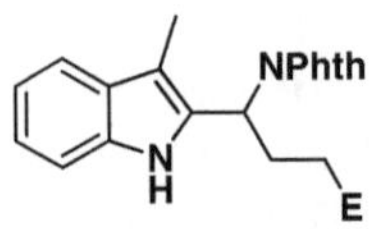

R$_f$ 0.65 (6:4 hexane/AcOEt). **Mp** 47.7–48.8 °C. **^{1}H NMR** (400 MHz, $CDCl_3$) δ 9.08 (br s, 1 H), 7.88–7.81 (m, 2 H, Phth), 7.77–7.69 (m, 2 H, Phth), 7.52 (d, 1 H, *J* = 7.9 Hz, ArH), 7.35 (d, 1 H, *J* = 8.2 Hz, ArH), 7.19 (t, 1 H, *J* = 7.3 Hz, ArH), 7.09 (t, 1 H, *J* = 7.5 Hz, ArH), 5.82–5.76 (m, 1 H, PhthN–CH), 4.48–4.28 (m, 2 H, CF_3–CH_2), 2.89–2.77 (m, 1 H, CH_2), 2.64–2.52 (m, 1 H, CH_2), 2.42 (t, 2 H, *J* = 7.3 Hz, CH_2), 2.36 (s, 3 H, CH_3). **^{13}C NMR** (101 MHz, $CDCl_3$) δ 170.8, 168.8, 136.0, 134.5, 131.7, 130.6, 128.0, 123.7, 122.9 (q, $J_{C\text{-}F}$ = 277 Hz), 122.9, 119.5, 119.2, 111.3, 110.7, 60.6 (q, $J_{C\text{-}F}$ = 37 Hz), 45.5, 30.8, 27.4, 8.5. **IR** 3416 (w), 2987 (w), 2361 (w), 1761 (m), 1710 (s), 1461 (w), 1389 (m), 1332 (w), 1283 (w), 1171 (m). **HRMS** (ESI) calcd for $C_{23}F_3H_{20}N_2O_4^+$ $[M + H]^+$ 445.1370; found 445.1370.

5.6 Dynamic [3 + 2] Kinetic Asymmetric Annulation of DA Aminocyclopropanes

Synthesis of N-vinyl-imides
5-Methoxyisobenzofuran-1,3-dione (288)

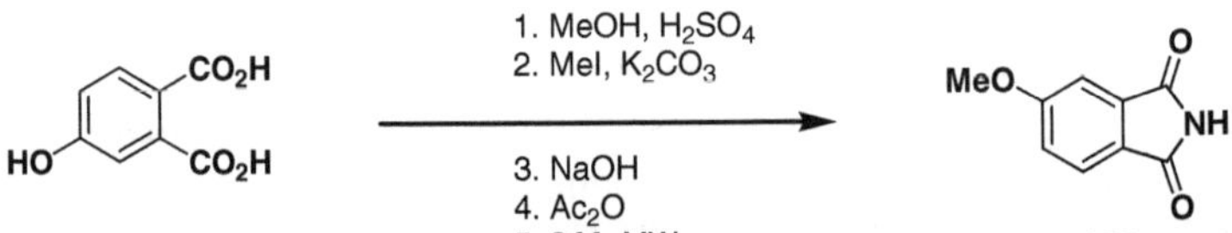

287 **288**

Following a modified procedure [23], a solution of 4-hydroxyphthalic acid (**287**) (2.00 g, 11.0 mmol, 1.00 eq), catalytic sulfuric acid (0.10 mL, 1.9 mmol, 0.17 eq) and MeOH (20.0 mL), was stirred at reflux for 7 h. under air. The solvent was removed under reduced pressure to afford crude dimethyl 4-hydroxyphthalate. The crude diester was dissolved in acetone (70 mL) and reacted with potassium carbonate (7.40 g, 53.5 mmol, 5.00 eq) at 50 °C for 20 min. Iodomethane (1.47 mL, 23.6 mmol, 2.20 eq) was added, and the mixture was stirred at reflux overnight. K_2CO_3 was removed by filtration and the solvent was removed under reduced pressure to afford a colorless oil.

The crude was dissolved in acetone (16.0 mL) and a 11 M solution of sodium hydroxide, (6.00 mL, 66.0 mmol, 6.20 eq) was added, and the solution was stirred for 6 h. under air at rt. The solution was then acidified with 2 M HCl to pH 3, and concentrated under reduced pressure. Then, the crude 4-methoxyphthalic acid was dissolved into acetone (50 mL) and dried over $MgSO_4$, filtered through a plug of cotton wool, and the solvent was removed in vacuo. The crude diacid was partitioned between 2 M NaOH (50 mL) and DCM (50 mL). The organic layer was extracted with NaOH 2 M (50 mL). The combined aqueous phase was cooled down to 0 °C and acidified with 37 % HCl % to pH 3. The aqueous layer was then extracted five times with AcOEt (50 mL). The combined organic layers were dried over $MgSO_4$ and concentrated under reduced pressure to afford the crude diacid as a light brown solid (1.82 g).

A solution of crude 4-methoxyphthalic acid (1.82 g, 9.28 mmol, 1.00 eq) in acetic anhydride (25.0 mL, 266 mmol, 28.7 eq) was stirred at reflux for 21 h. Volatiles were removed in vacuo to afford a dark brown solid. The crude was dissolved in DCM (50 mL) and filtered through fritted glass to remove solid impurities. The solution was concentrated under reduced pressure and dried in vacuo to afford 5-methoxyisobenzofuran-1,3-dione as a light brown solid (1.62 g, 9.08 mmol, 83 % yield over 4 steps)

^{1}H NMR (400 MHz, $CDCl_3$) δ 7.90 (dd, 1 H, J = 8.5, 0.4 Hz, Ar), 7.41 (d, 1 H, J = 2.2 Hz, Ar), 7.35 (dd, 1 H, J = 8.5, 2.3 Hz, Ar), 3.98 (s, 3 H, OMe). **HRMS (ESI)** calcd for $C_9H_7O_4^+$ $[M + H]^+$ 179.0339; found 179.0349.

The data for 5-methoxyisobenzofuran-1,3-dione correspond to the reported values [24].

Following a modified procedure [25], 5-methoxyisobenzofuran-1,3-dione (1.58 g, 8.84 mmol, 1.00 eq) and formamide (**244**) (35.0 mL, 880 mmol, 100 eq) were divided between four 20 mL microwave vials sealed with a microwave cap. The mixture was stirred at rt until the product was completely dissolved, then heated 2 times at 200 °C for 30 s with 10 s pre-stirring, using Biotage Initiator 2.0 microwave reactor. The mixture was cooled to 0 °C to induce crystallization and cold water (10 mL) was added into each vial. The obtained solid was filtrated over filter paper, washed with water (15 mL) and hexanes (20 mL) and dried under reduced pressure to afford 5-methoxyisoindoline-1,3-dione **(288)** as a beige solid (982 mg, 5.54 mmol, 63 % yield) which was used without further purification.

^{1}H NMR (400 MHz, $CDCl_3$) δ 7.77 (dd, 1 H, J = 8.3, 0.4 Hz, Ar), 7.59 (br s, 1 H, NH), 7.33 (d, 1 H, J = 2.2 Hz, Ar), 7.20 (dd, 1 H, J = 8.3, 2.3 Hz, Ar), 3.94 (s, 3 H, OMe). **^{13}C NMR** (101 MHz, $CDCl_3$) δ 167.8, 167.7, 165.0, 135.2, 125.4, 124.5, 120.4, 108.1, 56.2. **HRMS (ESI)** calcd for $C_9H_8NO_3^+$ $[M + H]^+$ 178.0499; found 178.0497.

5-Nitro-2-vinylisoindoline-1,3-dione (283)

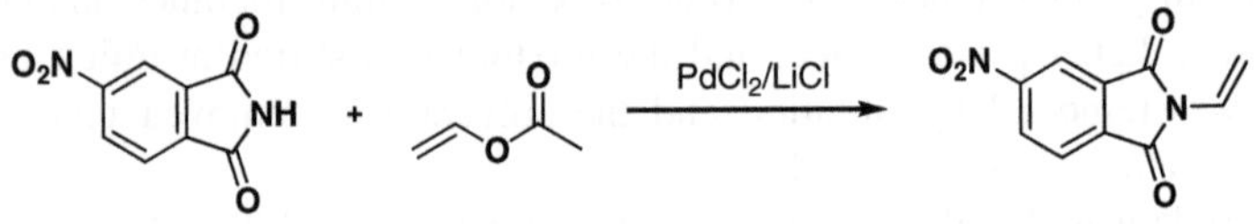

Following a modified procedure [26], 5-nitrosoindoline-1,3-dione (**378**) (1.00 g, 5.20 mmol, 1.00 eq), $PdCl_2$ (92.0 mg, 0.520 mmol, 0.100 eq), LiCl (0.221 mg, 5.20 mmol, 1.00 eq, weighted in a glovebox) and vinyl acetate (**224**) (12.9 mL, 139 mmol, 26.8 eq) were heated under reflux for 20 h. The mixture was cooled down to room temperature and the solvent was evaporated under reduced pressure. The crude was purified by column chromatography using silica gel (Hexane/AcOEt 8:2–5:5) to afford 5-nitro-2-vinylisoindoline-1,3-dione (**283**) as a bright yellow solid (1.14 g, 5.23 mmol, quantitative yield).

$\mathbf{R_f}$ 0.32 (9:1 Pentane/AcOEt). **Mp** 144.3–148.6 °C. **^{1}H NMR** (400 MHz, $CDCl_3$) δ 8.68 (dd, 1 H, J = 2.0, 0.5 Hz, Ar), 8.63 (dd, 1 H, J = 8.1, 2.0 Hz, Ar), 8.08 (m, 1 H, Ar), 6.88 (dd, 1 H, J = 16.4, 9.8 Hz, CH–N), 6.14 (dd, 1 H, J = 16.4, 0.5 Hz, = CH_2), 5.16 (dd, 1 H, J = 9.8, 0.4 Hz, =CH_2). **^{13}C NMR** (101 MHz, $CDCl_3$) δ 164.5, 164.2, 152.1, 136.1, 133.1, 129.8, 125.0, 123.6, 119.2, 106.3. **IR** 3101 (w), 3074 (w), 2924 (w), 1709 (s), 1533 (s), 1383 (s), 1341 (s), 1307 (s), 1062 (m), 1024 (s), 915 (s). **HRMS (ESI)** calcd for $C_{10}H_6N_2O_4$ $[M + H]^+$ 218.0328; found 218.0355.

1-Vinylpyrrolidine-2,5-dione (284)

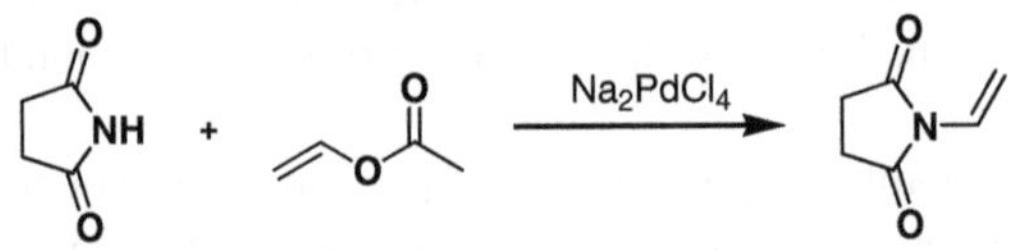

Following a modified procedure [26], succinimide (**379**) (1.00 g, 10.1 mmol, 1.00 eq), vinyl acetate (**224**) (25.0 mL, 270 mmol, 26.8 eq) and Na_2PdCl_4 (59.0 mg, 0.202 mmol, 2.00 mol%) were heated under reflux for 72 h. After solvent evaporation, the crude was purified by Biotage (SNAP Cartridge KP-Sil 50 g, 7:3 Hexane/AcOEt) to obtain 1-Vinylpyrrolidine-2,5-dione (**284**) as a yellow solid (1.22 g, 9.78 mmol, 97 % yield).

$\mathbf{R_f}$ 0.17 (8:2 Hexane/AcOEt). **Mp** 47.6–48.9 °C. **^{1}H NMR** (400 MHz, $CDCl_3$) δ 6.68 (dd, 1 H, J = 16.4, 9.9 Hz, =CH), 6.08 (d, 1 H, J = 16.4 Hz, =CH), 5.06 (d, 1 H, J = 9.9 Hz, =CH), 2.72 (s, 4 H, CH_2). **^{13}C NMR** (101 MHz, $CDCl_3$) δ 175.4, 124.3, 106.6, 27.8. **IR** 2946 (w), 1707 (s), 1382 (s), 1307 (m), 1222 (s), 1113 (s),

974 (m), 906 (m), 821 (w). **HRMS (ESI)** calcd for $C_6H_8NO_2^+$ $[M + H]^+$ 126.0550; found 126.0621.

5-Methoxy-2-vinylisoindoline-1,3-dione (285)

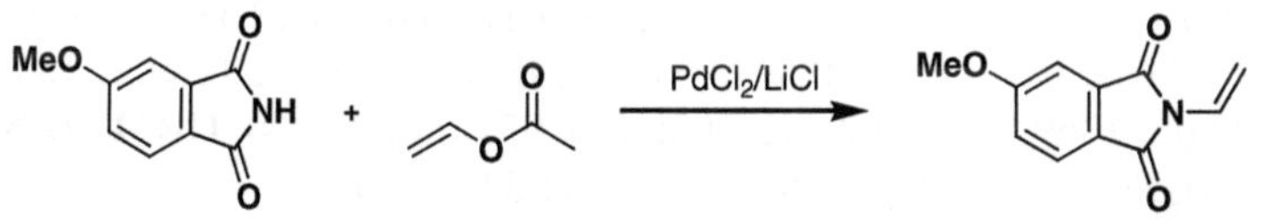

288 **224** **285**

Following a modified procedure, 5-methoxyisoindoline-1,3-dione (**288**) (980 mg, 5.53 mmol, 1.00 eq), $PdCl_2$ (98.0 mg, 0.553 mmol, 0.100 eq), LiCl (235 mg, 5.53 mmol, 1.00 eq, weighted in a glovebox) and vinyl acetate (**224**) (13.7 mL, 148 mmol, 26.8 eq) were heated under reflux for 24 h. The mixture was cooled down to room temperature and diluted with DCM/MeOH 4:1 (20 mL). Activated charcoal was added and the resulting suspension was filtered through a pad of Celite (DCM/MeOH 4:1 100 mL) and concentrated under reduced pressure. Purification by silica gel chromatography (pentane/AcOEt 90:10–75:25) afforded 5-methoxy-2-vinylisoindoline-1,3-dione (**285**) as a colorless solid (828 mg, 4.08 mmol, 74 % yield).

$\mathbf{R}_f$ 0.56 (6:4 Hexane/AcOEt). **Mp** 102.2–105.1 °C. ^{1}**H NMR** (400 MHz, $CDCl_3$) δ 7.76 (d, 1 H, $J = 8.3$ Hz, Ar), 7.32 (d, 1 H, $J = 2.2$ Hz, Ar), 7.17 (dd, 1 H, $J = 8.3$, 2.2 Hz, Ar), 6.83 (dd, 1 H, $J = 16.4$, 9.9 Hz, =CH), 6.03 (d, 1 H, $J = 16.4$ Hz, =CH), 4.99 (d, 1 H, $J = 9.9$ Hz, =CH), 3.93 (s, 3 H, OMe). ^{13}**C NMR** (101 MHz, $CDCl_3$) δ 166.5, 166.3, 165.1, 134.4, 125.5, 124.0, 123.5, 120.6, 108.2, 104.0, 56.3. **IR** 1779 (w), 1720 (s), 1639 (w), 1619 (w), 1493 (w), 1386 (s), 1307 (w), 1295 (w), 1021 (w). **HRMS (ESI)** calcd for $C_{11}H_{10}NO_3^+$ $[M + H]^+$ 204.0655; found 204.0662.

4,5,6,7-Tetrafluoro-2-vinylisoindoline-1,3-dione (286)

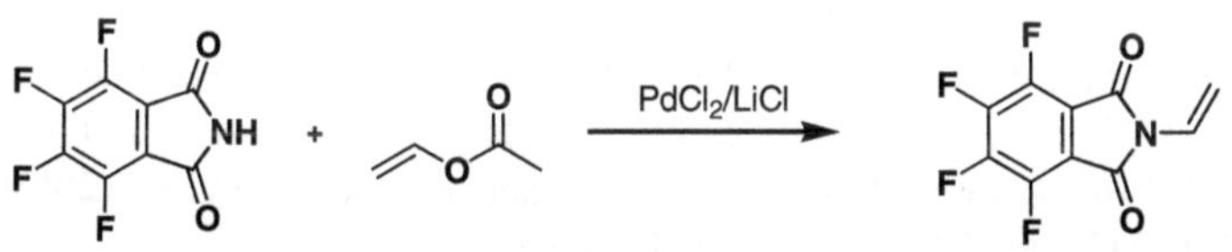

380 **224** **286**

Following a modified procedure, 4,5,6,7-tetrafluoroisoindoline-1,3-dione (500 mg, 2.82 mmol, 1.00 eq), $PdCl_2$ (40.0 mg, 0.228 mmol, 0.100 eq), LiCl (97.0 mg, 2.28 mmol, 1.00 eq, weighted in a glovebox) and vinyl acetate (**224**) (5.70 mL, 61.2 mmol, 26.8 eq) were heated under reflux for 48 h. The mixture was cooled down to room temperature and the solvent was evaporated under reduced pressure. The crude was purified by Biotage (SNAP Cartridge KP-Sil 25 g, 8:2 Hexane/AcOEt) to obtain 4,5,6,7-Tetrafluoro-2-vinylisoindoline-1,3-dione (**286**) as a colorless solid (302 mg, 1.23 mmol, 79 % b.r.s.m, 54 % yield) and (160 mg, 0.730 mmol, 32 % reisolated yield) of the starting material.

R_f 0.65 (9:1 Pentane/AcOEt). **Mp** 144.4–146.0 °C. **^{1}H NMR** (400 MHz, $CDCl_3$) δ 6.78 (dd, 1 H, J = 16.4, 9.8 Hz, N–CH), 6.07 (dd, 1 H, J = 16.4, 0.5 Hz, =CH_2), 5.13 (dd, 1 H, J = 9.8, 0.5 Hz, =CH_2). **IR** 1732 (s), 1639 (w), 1515 (s), 1500 (m), 1402 (s), 1369 (m), 1307 (w), 1038 (w), 951 (m), 908 (s). **HRMS (ESI)** calcd for $C_{10}F_4H_4NO_2^+$ $[M + H]^+$ 246.0173; found 246.0174.

Synthesis of Aminocyclopropanes
GP13 for the of Synthesis of Aminocyclopropanes

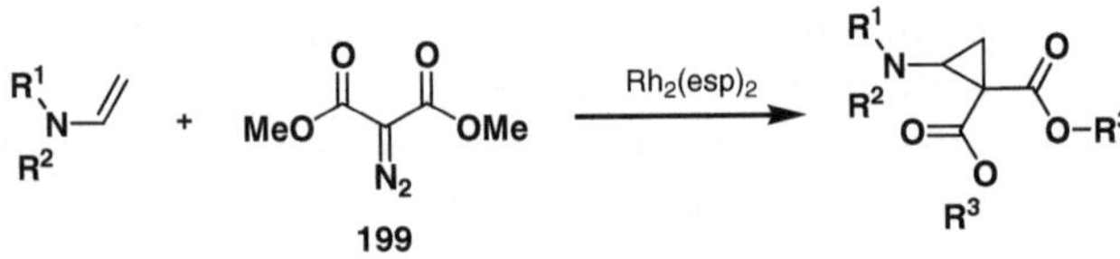

199

Following a modified procedure [6], the corresponding N-vinyl-imide (1.00 eq) was dissolved in dry dichloromethane (10.0 mL) and the solution was cooled down to 0 °C with an ice/water bath. Then, bis[rhodium(α,α,α′, α′-tetramethyl-1,3-benzenedipropionic acid)] (0.1 mol%) was added in one portion. A solution in dichloromethane (2.0 mL) of dimethyldiazomalonate (**199**) (1.20 eq) was added dropwise over 5 min. After the addition, the mixture was allowed to warm to room temperature and stirred overnight. The solvent is then removed under reduced pressure and the crude is directly purified by column chromatography.

Dimethyl 2-(5-nitro-1,3-dioxoisoindolin-2-yl)cyclopropane-1,1-dicarboxylate (289)

GP13 was followed, starting from 5-nitro-2-vinylisoindoline-1,3-dione (**283**) (0.500 g, 2.29 mmol, 1.00 eq), dimethyl diazomalonate (**199**) (0.544 g, 2.75 mmol, 1.20 eq) and bis[rhodium(α,α,α′,α′-tetramethyl-1,3-benzenedipropionic acid)] (1.7 mg, 2.3 μmol, 0.10 mol%). After solvent evaporation, the crude was purified by Biotage (SNAP Cartridge KP-Sil 50 g, 7:3 Hexane/AcOEt), to obtain Dimethyl 2-(5-nitro-1,3-dioxoisoindolin-2-yl)cyclopropane-1,1-dicarboxylate (**289**) as a colorless solid (712 mg, 2.04 mmol, 89 % yield).

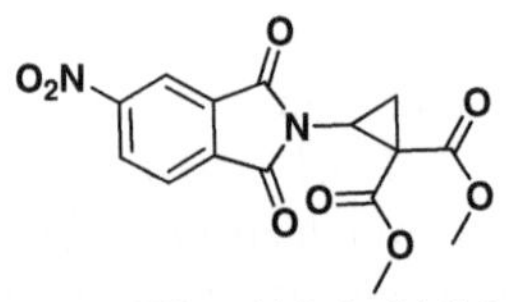

R_f 0.19 (8:2 Pentane/AcOEt). **Mp** 113.0–115.8 °C. **^{1}H NMR** (400 MHz, $CDCl_3$) δ 8.61 (m, 2 H, Ar), 8.03 (d, 1 H, J = 8.1 Hz, Ar), 3.83 (s, 3 H, OMe), 3.70

(m, 1 H, CH–N), 3.62 (s, 3 H, OMe), 2.63 (m, 1 H, CH_2), 2.07 (m, 1 H, CH_2). **^{13}C NMR** (101 MHz, $CDCl_3$) δ 168.2, 167.1, 165.9, 165.6, 152.0, 135.9, 132.9, 129.6, 124.9, 119.0, 53.3, 53.2, 35.0, 33.1, 19.7. **IR** 3110 (w), 2956 (w), 2926 (w), 2853 (w), 1726 (s), 1541 (m), 1400 (m), 1344 (s), 1222 (s), 1130 (m). **HRMS (ESI)** calcd for $C_{15}H_{13}N_2O_8^+$ $[M + H]^+$ 349.0666; found 349.0664.

Dimethyl 2-(4,5,6,7-tetrafluoro-1,3-dioxoisoindolin-2-yl)cyclopropane-1,1-dicarboxylate (290)

GP13 was followed, starting from 4,5,6,7-tetrafluoro-2-vinylisoindoline-1,3-dione (**286**) (0.260 g, 1.06 mmol, 1.00 eq), dimethyl 2-diazomalonate (**199**) (0.252 g, 1.27 mmol, 1.20 eq) and bis[rhodium(α,α,α′,α′-tetramethyl-1,3-benzenedipropionic acid)] (0.80 mg, 1.0 μmol, 0.10 mol%). After solvent evaporation, the crude was purified by Biotage (SNAP Cartridge KP-Sil 10 g, 7:3 Hexane/AcOEt), to obtain Dimethyl 2-(4,5,6,7-tetrafluoro-1,3-dioxoisoindolin-2-yl)cyclopropane-1,1-dicarboxy late (**290**) as a colorless solid (0.300 g, 0.801 mmol, 75 % yield).

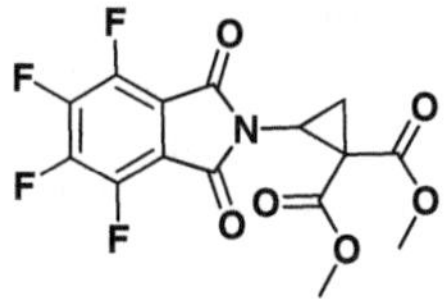

R_f 0.42 (8:2 Pentane/AcOEt). **Mp** 86.2–88.7 °C. **^{1}H NMR** (400 MHz, $CDCl_3$) δ 3.80 (s, 3 H, OMe), 3.64 (s, 3 H, OMe), 3.55 (dd, 1 H, $J = 8.5, 6.5$ Hz, N–CH), 2.50 (t, 1 H, $J = 6.6$ Hz, CH_2), 2.03 (dd, 1 H, $J = 8.5, 6.6$ Hz, CH_2). **^{13}C NMR** (101 MHz, $CDCl_3$) δ 168.1, 167.3, 162.2, 145.4 (m), 143.6 (m), 113.4 (m), 53.4, 53.3, 34.7, 32.8, 19.8. **IR** 2957 (w), 1728 (s), 1515 (s), 1501 (s), 1410 (s), 1219 (m), 945 (s). **HRMS (ESI)** calcd for $C_{15}F_4H_{10}NO_6^+$ $[M + H]^+$ 376.0439; found 376.0436.

Dimethyl 2-(2,5-dioxopyrrolidin-1-yl)cyclopropane-1,1-dicarboxylate (291)

GP13 was followed, starting from N-vinyl-succinimide (**284**) (500 mg, 4.00 mmol, 1.00 eq), dimethyldiazomalonate (**199**) (300 mg, 4.80 mmol, 1.20 eq) and bis [rhodium(α,α,α′, α′-tetramethyl-1,3-benzenedipropionic acid)] (3.0 mg, 4.0 μmol, 0.10 mol%). After solvent evaporation, the residue was purified by Biotage (SNAP Cartridge KP-Sil 50 g, 5:5 Hexane/AcOEt), to obtain dimethyl 2-(2,5-dioxopyrrolidin-1-yl)cyclopropane-1,1-dicarboxylate (**291**) as a yellow solid (801 mg, 3.14 mmol, 79 % yield).

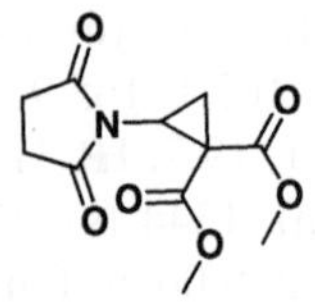

R$_f$ 0.39 (5:5 Hexane/AcOEt). **Mp** 81.9–85.3 °C. **^{1}H NMR** (400 MHz, $CDCl_3$) δ 3.78 (s, 3 H, OMe), 3.68 (s, 3 H, OMe), 3.45 (dd, 1 H, J = 8.5, 6.5 Hz, N–CH), 2.73–2.58 (m, 4 H, O=C–CH_2), 2.45 (t, 1 H, J = 6.5 Hz, CH_2), 1.93 (dd, 1 H, J = 8.5, 6.5 Hz, CH_2). **^{13}C NMR** (101 MHz, $CDCl_3$) δ 176.9, 168.4, 167.2, 53.2, 53.1, 35.1, 32.7, 28.1, 19.7. **IR** 2955 (w), 1717 (s), 1439 (w), 1406 (m), 1332 (m), 1296 (m), 1216 (s), 1132 (m), 1079 (w), 910 (s). **HRMS (ESI)** calcd for $C_{11}H_{14}NO_6^+$ $[M + H]^+$ 256.0816; found 256.0822.

Dicarboxylate dimethyl 2-(5-methoxy-1,3-dioxoisoindolin-2-yl)cyclopropane-1,1-dicarboxylate (292)

GP13 was followed, starting from 5-methoxy-2-vinylisoindoline-1,3-dione (**285**) (0.130 g, 0.640 mmol, 1.00 eq), dimethyldiazomalonate (**199**) (0.121 g, 0.768 mmol, 1.20 eq) and bis[rhodium(α,α,α′,α′-tetramethyl-1,3-benzenedipropionic acid)] (0.5 mg, 0.6 μmol, 0.1 mol%). After solvent evaporation, the crude was purified by Biotage (SNAP Cartridge KP-Sil 10 g, 6:4 Hexane/AcOEt), to obtain dicarboxylate dimethyl 2-(5-methoxy-1,3-dioxoisoindolin-2-yl)cyclopropane-1,1-dicarboxylate (**292**) as a colorless solid (176 mg, 0.528 mmol, 83 % yield).

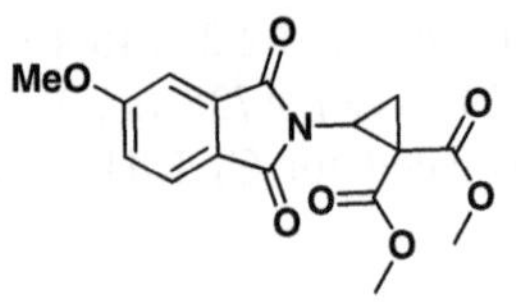

R$_f$ 0.15 (8:2 Pentane/AcOEt). **Mp** 113.5–117.8 °C. **^{1}H NMR** (400 MHz, $CDCl_3$) δ 7.71 (d, 1 H, J = 8.3 Hz, Phth), 7.27 (d, 1 H, J = 2.2 Hz, Phth), 7.14 (dd, 1 H, J = 8.3, 2.3 Hz, Phth), 3.90 (s, 3 H, OMe), 3.80 (s, 3 H, OMe–C=O), 3.66 (dd, 1 H, J = 8.5, 6.6 Hz, N–CH), 3.59 (s, 3 H, OMe–C=O), 2.68 (t, 1 H, J = 6.5 Hz, CH_2), 1.99 (dd, 1 H, J = 8.5, 6.4 Hz, CH_2). **^{13}C NMR** (101 MHz, $CDCl_3$) δ 168.7, 167.8, 167.6, 167.0, 165.0, 134.1, 125.3, 123.4, 120.4, 108.1, 56.2, 53.2, 53.0, 35.0, 33.2, 19.7. **IR** 2955 (w), 1720 (s), 1492 (m), 1437 (m), 1397 (s), 1288 (s), 1133 (m), 1018 (w). **HRMS (ESI)** calcd for $C_{16}H_{16}NO_7^+$ $[M + H]^+$ 334.0921; found 334.0915.

Synthesis of ligands
Synthesis of 2,2′-(cyclopropane-1,1-diyl)bis(8,8a-dihydro-3aH-indeno[1,2-d] oxazole) (L23)

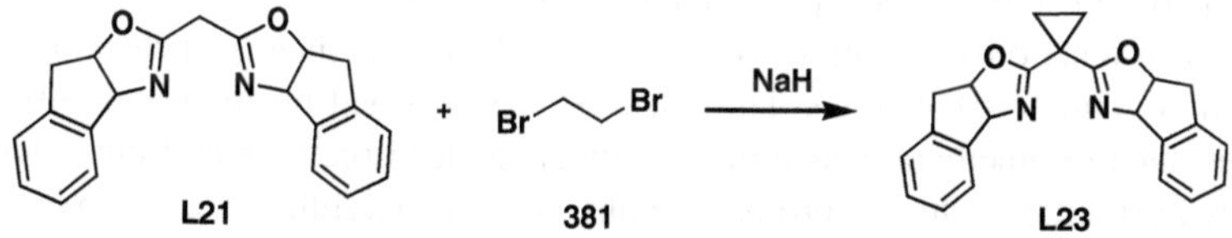

Following a modified procedure [27], inda-BOX (160 mg, 0.484 mmol, 1.00 eq) was dissolved in THF (3.0 mL) in a round bottom vial and the mixture was cooled down to 0 °C with an ice/water bath. Then, sodium hydride (58.1 mg, 1.45 mmol, 3.00 eq) was added carefully in small portions. The solution was stirred for 15 min at 0 °C then warmed up to rt and stirred for 10 more minutes, until no more H_2 evolution was observed. Subsequently, 1.2-dibromoethane (63 μL, 0.73 mmol, 1.5 eq) was added dropwise without letting the mixture rise over 30 °C. The vial was sealed and the mixture was slowly warmed to 52 °C and heated for 4 h. The reaction was quenched with NH_4Cl (5 mL) and extracted with CH_2Cl_2 (3 × 10 mL), the organic layers were combined, washed with brine, dried over $MgSO_4$ and concentrated under reduced pressure. The product was washed with hexanes to obtain a pale beige solid (168 mg, 0.471 mmol, 97 % yield).

$\mathbf{R_f}$ 0.46 (5:5 Hexane/AcOEt). **^{1}H NMR** (400 MHz, $CDCl_3$) δ 7.46–7.43 (m, 4 H, Ar), (See footnote 17) 7.24–7.19 (m, 4 H, Ar),[17] 5.52 (d, 2 H, J = 8.0 Hz, CN–CH), 5.34–5.30 (m, 2 H, CO–CH), 3.38 (dd, 2 H, J = 17.9, 7.0 Hz, CH_2–Ph), 3.19 (dd, 2 H, J = 17.9, 1.6 Hz, CH_2–Ph), 1.36–1.24 (m, 4 H, CH_2–Cyclopropane). **^{13}C NMR** (101 MHz, CDCl3) δ 166.0, 141.9, 139.9, 128.5, 127.5, 125.7, 125.3, 83.5, 76.5, 39.8, 18.5, 15.9. **IR** 3070 (w), 3031 (w), 2954 (w), 2923 (w), 2142 (w), 1756 (w), 1731 (w), 1657 (s), 1367 (m), 1311 (m), 1162 (m), 1111 (m), 1005 (m), 853 (w). **HRMS (ESI)** calcd for $C_{23}H_{21}N_2O_2^+$ $[M + H]^+$ 357.1598; found 357.1591.

The NMR characterization data correspond to the reported values [27].

GP14 for the Racemic Synthesis of Cyclopentylamines Starting from Aminocyclopropanes and Silyl Enol Ethers

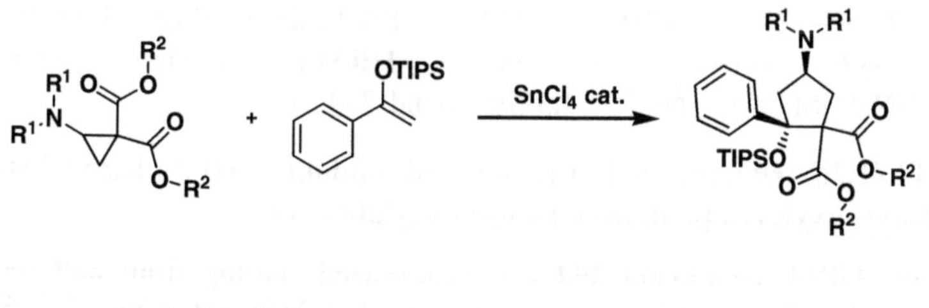

[17]The integration differs from the reported values but the overall number of protons is correct.

Following a reported procedure [28], in an oven-dried flask sealed with a septum and under N_2 atmosphere, the corresponding aminocyclopropane (1.00 eq) and triisopropyl(1-phenylvinyl)oxy)silane (**36**) (1.50 eq) were dissolved in dry CH_2Cl_2 (2.00 mL). The solution was then cooled down to −78 °C and 23.0 μL of a 0.43 M solution of tin tetrachloride (5.0 mol% or stated otherwise) in dry CH_2Cl_2 was added. The reaction was stirred for 1–3 h at −78 °C and then quenched at –78 °C with triethylamine (0.30 mL). The reaction was warmed up to rt and stirred for 15 min. Dichloromethane was removed under reduced pressure and the crude was directly purified by column chromatography (8:2 Hexane/Ethyl Acetate, 3 % Et_3N).

***Trans*-dibenzyl-4-(1,3-dioxoisoindolin-2-yl)-2-phenyl-2-(triisopropylsilyl)oxy) cyclopentane-1,1-dicarboxylate (293)**

Following GP14, compound **293** was synthesized starting from **260** (45.0 mg, 99.0 μmol, 1.00 eq), triisopropylsilyl-enol ether **219a** (41.0 mg, 0.148 mmol, 1.50 eq) and $SnCl_4$ (12 μL 0.43 M, 5.2 μmol, 5.2 mol%) as white solid (60.0 mg, 82.0 μmol, 83 % yield).

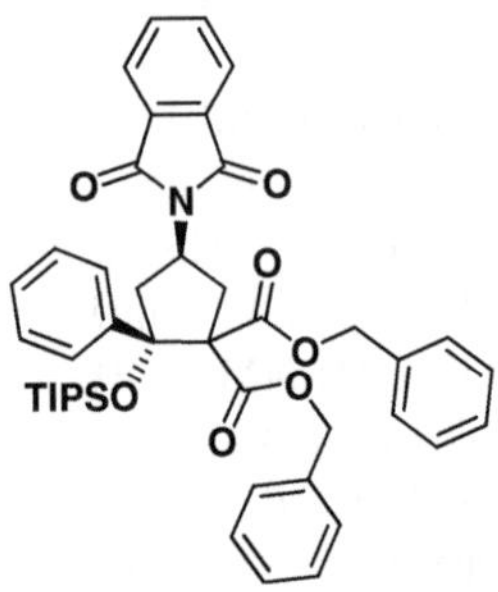

R_f 0.72 (6:4 Hexane/AcOEt). **Mp** 135.6–140.5 °C. **^{1}H NMR** (400 MHz, $CDCl_3$) δ 7.82 (dd, 2 H, *J* = 5.5, 3.1 Hz, Phth), 7.70–7.67 (m, 4 H, Phth + Ar), 7.27–7.04 (m, 13 H, Ar), 5.22–5.18 (m, 3 H, CH_2–Ar + N–CH), 4.86 (s, 2 H, CH_2–Ar), 3.81 (t, 1 H, *J* = 12.2 Hz, CH_2), 3.47 (dd, 1 H, *J* = 13.5, 10.3 Hz, CH_2), 2.92 (dd, 1 H, *J* = 14.0, 8.2 Hz, CH_2), 2.43 (dd, 1 H, *J* = 12.4, 6.2 Hz, CH_2), 0.95–0.89 (m, 21 H, TIPS). **^{13}C NMR** (101 MHz, $CDCl_3$) δ 170.3, 168.4, 168.2, 141.4, 135.5, 135.4, 134.1, 132.1, 128.8, 128.5, 128.4, 128.4, 128.3, 128.2, 128.0, 127.1, 123.3, 87.8, 70.2, 67.6, 67.0, 47.6, 41.9, 36.6, 18.4, 18.3, 13.9. **IR** 3061 (w), 3035 (w), 2947 (w), 2867 (w), 1736 (s), 1715 (s), 1463 (w), 1379 (m), 1259 (m), 1166 (w), 1128 (m), 982 (m), 886 (w). **HRMS** (ESI) calcd for $C_{44}H_{49}NNaO_7Si^+$ $[M + Na]^+$ 754.3170; found 754.3158.

***Trans*-bis(2,2,2-trifluoroethyl)-4-(1,3-dioxoisoindolin-2-yl)-2-phenyl-2-(triiso-propylsilyl)-oxy)cyclopentane-1,1-dicarboxylate (294)**

Following GP14, compound **294** was synthesized starting from **259** (60.0 mg, 0.137 mmol, 1.00 eq), triisopropylsilyl-enol ether **219a** (56.8 mg, 0.205 mmol,

1.50 eq) and $SnCl_4$ (64 μL 0.21 M, 14 μmol, 10 mol%) as a white solid (80.7 mg, 113 μmol, 82 % yield).

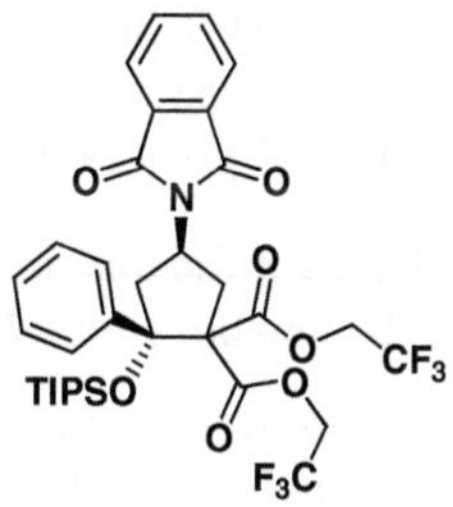

R_f 0.32 (9:1 Pentane/AcOEt). **Mp** 122.5–124.7 °C. **^{1}H NMR** (400 MHz, $CDCl_3$) δ 7.85 (dd, 2 H, J = 5.4, 3.1 Hz, Phth), 7.73 (dd, 2 H, J = 5.4, 3.0 Hz, Phth), 7.69 (m, 2 H, Ar), 7.30 (m, 3 H, Ar), 5.31 (m, 1 H, N–CH), 4.59 (m, 2 H, CH_2–CF_3), 4.25 (m, 2 H, CH_2–CF_3), 3.75 (t, 1 H, J = 12.3 Hz, CH_2), 3.47 (dd, 1 H, J = 14.1, 10.1 Hz, CH_2), 2.96 (dd, 1 H, J = 14.2, 7.8 Hz, CH_2), 2.50 (dd, 1 H, J = 12.7, 6.4 Hz, CH_2), 0.97–0.92 (m, 21 H, TIPS). **^{13}C NMR** (101 MHz, $CDCl_3$) δ 168.3, 168.0, 166.5, 140.5, 134.3, 132.0, 128.6, 128.4, 127.4, 123.4, 122.9 (d, J_{C-F} = 277 Hz), 122.7 (d, J_{C-F} = 277 Hz), 88.0, 69.7, 61.4 (dq, J_{C-F} = 49, 37 Hz), 61.4 (dd, J_{C-F} = 49, 37 Hz), 47.1, 41.7, 36.5, 18.3, 18.2, 13.9. **IR** 2949 (w), 2870 (w), 1757 (m), 1714 (s), 1379 (m), 1285 (m), 1168 (s), 1129 (s), 981 (m). **HRMS** (ESI) calcd for $C_{34}F_6H_{40}NO_7Si^+$ $[M + H]^+$ 716.2473; found 716.2474.

***Trans*-dimethyl4-(5-bromo-1,3-dioxoisoindolin-2-yl)-2-phenyl-2-(triisopropylsilyl)-oxy)cyclopentane-1,1-dicarboxylate (295)**

Following GP14, compound **295** was synthesized starting from **255** (54.0 mg, 0.142 mmol, 1.00 eq), triisopropylsilyl-enol ether **219a** (58.8 mg, 0.213 mmol, 1.50 eq) and $SnCl_4$ (66 μL 0.43 M, 28 μmol, 20 mol%) as a white solid (64.0 mg, 97.0 μmol, 69 % yield).

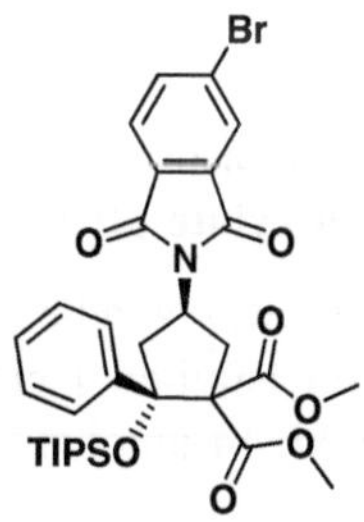

R_f 0.25 (9:1 Pentane/AcOEt). **Mp** 63.6–66.7 °C. **^{1}H NMR** (400 MHz, $CDCl_3$) δ 7.96 (d, 1 H, J = 1.5 Hz, Phth), 7.86 (dd, 1 H, J = 7.9, 1.7 Hz, Phth), 7.73 (m, 2 H, Ar), 7.70 (d, 1 H, J = 7.9 Hz, Phth), 7.31–7.27 (m, 3 H, Ar), 5.24 (m, 1 H, N–CH), 3.84 (s, 3 H, OMe), 3.75 (t, 1 H, J = 12.2 Hz, CH_2), 3.43 (s, 3 H, OMe), 3.43–3.36

(m, 1 H, CH_2), 2.85 (dd, 1 H, J = 13.8, 8.6 Hz, CH_2), 2.43 (dd, 1 H, J = 12.2, 5.9 Hz, CH_2), 0.96–0.91 (m, 21 H, TIPS). **^{13}C NMR** (101 MHz, $CDCl_3$) δ 171.0, 168.8, 167.6, 167.0, 141.7, 137.2, 133.7, 130.6, 129.1, 128.5, 128.1, 127.3, 126.7, 124.7, 87.7, 70.1, 52.5, 52.3, 48.1, 41.8, 36.3, 18.3, 18.3, 13.8. **IR** 2948 (w), 2867 (w), 1738 (m), 1715 (s), 1371 (m), 1253 (w), 1129 (m), 984 (m), 885 (w). **HRMS (ESI)** calcd for $C_{32}H_{40}^{79}BrNNaO_7Si^+$ $[M + Na]^+$ 680.1650; found 680.1649.

***Trans*-dimethyl-4-(5,6-dichloro-1,3-dioxoisoindolin-2-yl)-2-phenyl-2-(triisopropylsilyl)oxy)cyclopentane-1,1-dicarboxylate (296)**

Following GP14, compound **296** was synthesized starting from **256** (74.0 mg, 0.200 mmol, 1.00 eq), triisopropylsilyl-enol ether **219a** (83.0 mg, 0.300 mmol, 1.50 eq) and $SnCl_4$ (23 µL 0.43 M, 1.0 µmol, 5.0 mol%) as a white solid (64.0 mg, 99.0 µmol, 49 % yield).

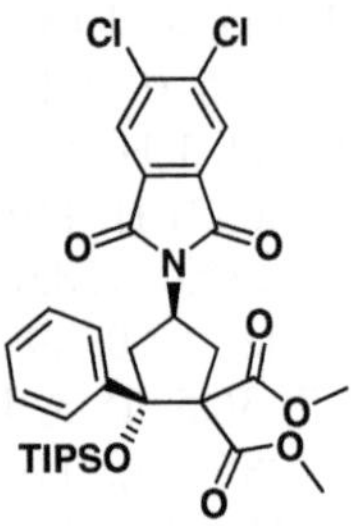

R_f 0.50 (8:2 Hexane/AcOEt). Mp 57.6–61.3 °C. **^{1}H NMR** (400 MHz, $CDCl_3$) δ 7.89 (s, 2 H, Phth), 7.73–7.71 (m, 2 H, Ar), 7.30–7.23 (m, 3 H, Ar), 5.25–5.16 (m, 1 H, N–CH), 3.82 (s, 3 H, OMe), 3.73 (t, 1 H, J = 12.3 Hz, CH_2), 3.41 (s, 3 H, OMe) 3.41–3.35 (m, 2 H, CH_2), 2.82 (dd, 1 H, J = 13.8, 8.6 Hz, CH_2), 2.42 (dd, 1 H, J = 12.3, 5.9 Hz, CH_2), 0.95–0.89 (m, 21 H, TIPS). **^{13}C NMR** (101 MHz, $CDCl_3$) δ 171.0, 168.7, 166.4, 141.6, 139.1, 131.2, 128.5, 128.2, 127.3, 125.4, 87.7, 70.1, 52.5, 52.3, 48.4, 41.8, 36.3, 18.3, 18.2, 13.8. **IR** 2948 (w), 2867 (w), 1719 (s), 1462 (w), 1369 (s), 1256 (w), 1130 (m), 990 (w), 908 (w), 885 (w). **HRMS (ESI)** calcd for $C_{32}H_{39}^{35}Cl_2NNaO_7Si^+$ $[M + Na]^+$ 670.1765; found 670.1743.

***Trans*-dimethyl-4-(5-nitro-1,3-dioxoisoindolin-2-yl)-2-phenyl-2-(triisopropylsilyl)oxy)cyclopentane-1,1-dicarboxylate (297)**

Following GP14, compound **297** was synthesized starting from **289** (56.0 mg, 0.161 mmol, 1.00 eq), triisopropylsilyl-enol ether **219a** (66.9 mg, 0.242 mmol, 1.50 eq) and $SnCl_4$ (67 µL 0.21 M, 14 µmol, 14 mol%) as a white solid (79.0 mg, 126 µmol, 78 % yield).

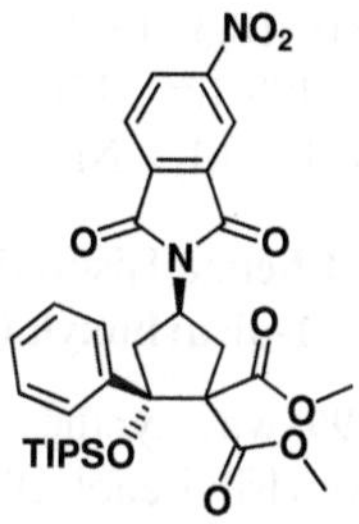

R_f 0.50 (8:2 Pentane/AcOEt). **Mp** 71.7–74.0 °C. **^{1}H NMR** (400 MHz, $CDCl_3$) δ 8.66 (d, 1 H, J = 1.6 Hz, Phth), 8.60 (dd, 1 H, J = 8.1, 1.6 Hz, Phth), 8.04 (d, 1 H, J = 8.1 Hz, Phth), 7.73–7.71 (m, 2 H, Ar), 7.32–7.27 (m, 3 H, Ar), 5.27 (m, 1 H, N–CH), 3.85 (s, 3 H, OMe), 3.77 (t, 1 H, J = 12.2 Hz, CH_2), 3.45 (s, 3 H, OMe), 3.46–3.40 (m, 1 H, CH_2), 2.87 (dd, 1 H, J = 13.8, 8.5 Hz, CH_2), 2.47 (dd, 1 H, J = 12.2, 6.0 Hz, CH_2), 0.95–0.91 (m, 21 H, TIPS). **^{13}C NMR** (101 MHz, $CDCl_3$) δ 171.0, 168.7, 166.3, 166.0, 151.9, 141.5, 136.4, 133.4, 129.5, 128.4, 128.2, 127.3, 124.6, 118.8, 87.7, 70.1, 52.6, 52.4, 48.6, 41.8, 36.3, 18.3, 18.3, 13.8. **IR** 2952 (w), 2869 (w), 1721 (s), 1542 (w), 1379 (w), 1346 (m), 1255 (w), 1132 (m), 1057 (m), 985 (w), 911 (w). **HRMS (ESI)** calcd for $C_{32}H_{40}N_2NaO_9Si^+$ $[M + Na]^+$ 647.2395; found 647.2391.

***Trans*-dimethyl-2-phenyl-4-(4,5,6,7-tetrafluoro-1,3-dioxoisoindolin-2-yl)-2-(tri-isopropylsilyl)oxy) cyclopentane-1,1-dicarboxylate (298)**

Following GP14, compound **298** was synthesized starting from **290** (74.8 mg, 0.200 mmol, 1.00 eq), triisopropylsilyl-enol ether **219a** (83.0 mg, 0.300 mmol, 1.50 eq) and $SnCl_4$ (95 µL 0.21 M, 20 µmol, 10 mol%) as a white solid (82.0 mg, 126 µmol, 63 % yield).

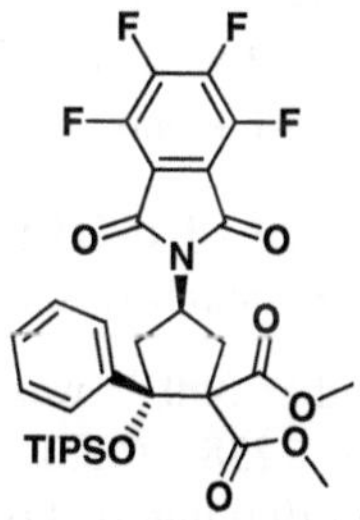

R_f 0.32 (9:1 Pentane/AcOEt). **Mp** 143.5–145.0 °C. **^{1}H NMR** (400 MHz, $CDCl_3$) δ 7.73–7.70 (m, 2 H, Ar), 7.31–7.27 (m, 3 H, Ar), 5.21 (m, 1 H, N–CH), 3.84 (s, 3 H, OMe), 3.74 (t, 1 H, J = 12.2 Hz, CH_2), 3.40 (s, 3 H, OMe), 3.40–3.36 (m, 1 H, CH_2), 2.81 (dd, 1 H, J = 13.8, 8.7 Hz, CH_2), 2.43 (dd, 1 H, J = 12.3, 5.8 Hz, CH_2), 0.96–0.89 (m, 21 H, TIPS). **^{13}C NMR** (101 MHz, $CDCl_3$) δ 170.9, 168.6, 162.5, 145.1 (m), 143.5 (m), 141.4, 128.4, 128.2, 127.3, 113.7 (m), 87.6,

70.1, 52.6, 52.4, 48.8, 41.6, 36.0, 18.3, 18.2, 13.8 [31]. **IR** 2951 (w), 2869 (w), 1723 (s), 1514 (m), 1500 (m), 1408 (m), 1363 (m), 1108 (m), 985 (m). **HRMS (ESI)** calcd for $C_{32}H_{37}F_4NNaO_7Si^+$ $[M + Na]^+$ 674.2168; found 674.2161.

Trans-dimethyl-4-(1,3-dioxo-1H-benzo[f]isoindol-2(3H)-yl)-2-phenyl-2-(triisopropylsilyl)oxy)Cyclopentane-1,1-dicarboxylate (299)

Following GP14, compound **299** was synthesized starting from **257** (56.4 mg, 0.160 mmol, 1.00 eq), triisopropylsilyl-enol ether **219a** (66.2 mg, 0.239 mmol, 1.50 eq) and $SnCl_4$ (19 µL 0.43 M, 8.2 µmol, 5.1 mol%) as a white solid (96.0 mg, 0.152 mmol, 95 % yield).

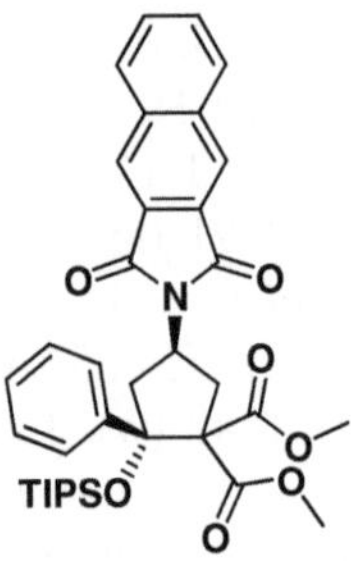

R_f 0.30 (8:2 Hexane/AcOEt). **Mp** 57.5–63.5 °C. **^{1}H NMR** (400 MHz, $CDCl_3$) δ 8.33 (s, 2 H, Ar), 8.06 (dd, 2 H, J = 6.1, 3.3 Hz, Ar), 7.78–7.75 (m, 2 H, Ar), 7.70 (dd, 2 H, J = 6.2, 3.3 Hz, Ar), 7.32–7.28 (m, 3 H, Ar), 5.39–5.30 (m, 1 H, N–CH), 3.85 (s, 3 H, OMe), 3.89–3.82 (m, 3 H, CH_2), 3.47 (s, 3 H, OMe), 3.47–3.40 (m, 3 H, CH_2), 2.96 (dd, 1 H, J = 13.8, 8.6 Hz, CH_2), 2.48 (dd, 1 H, J = 12.3, 5.9 Hz, CH_2), 0.99–0.97 (m, 21 H, TIPS). **^{13}C NMR** (101 MHz, $CDCl_3$) δ 171.1, 168.9, 168.2, 141.9, 135.7, 130.5, 129.3, 128.6, 128.1, 127.8, 127.2, 124.7, 87.8, 70.2, 52.5, 52.3, 48.1, 41.8, 36.3, 18.4, 18.3, 13.9. **IR** 2950 (w), 2868 (w), 1765 (m), 1739 (m), 1712 (s), 1451 (w), 1369 (s), 1258 (w), 1132 (m), 1058 (w), 984 (w), 912 (w). **HRMS** (ESI) calcd for $C_{36}H_{43}NNaO_7Si^+$ $[M + Na]^+$ 652.2701; found 652.2700.

Trans-dimethyl-4-(2,5-dioxo-2,5-dihydro-1H-pyrrol-1-yl)-2-phenyl-2-(triisopropylsilyl)oxy)cyclopentane-1,1-dicarboxylate (300)

Following GP14, compound **300** was synthesized starting from N-maleimide-aminocyclopropane **258** (43.0 mg, 0.170 mmol, 1.00 eq), triisopropylsilyl-enol ether **219a** (79.0 mg, 0.286 mmol, 1.68 eq) and $SnCl_4$ (89 µL 0.21 M, 19 µmol, 11 mol%) as a white solid (38.9 mg, 73.0 µmol, 43 % yield).

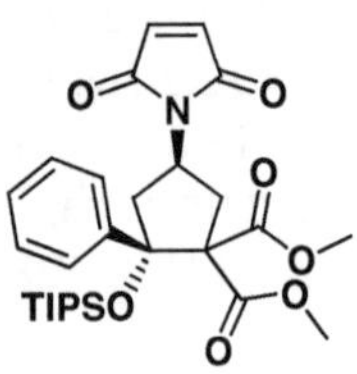

R_f 0.36 (8:2 Pentane/AcOEt). **Mp** 150.0–153.7 °C. **^{1}H NMR** (400 MHz, $CDCl_3$) δ 7.73–7.71 (m, 2 H, Ar), 7.30–7.26 (m, 3 H, Ar), 6.69 (s, 2 H, CH–C=O), 5.05 (m, 1 H, CH–N), 3.83 (s, 3 H, OMe), 3.62 (t, 1 H, J = 12.2 Hz, CH_2), 3.39 (s, 3 H, OMe), 3.32 (ddd, 1 H, J = 13.6, 9.4, 1.0 Hz, CH_2), 2.72 (dd, 1 H, J = 13.7, 8.7 Hz, CH_2), 2.38 (m, 1 H, CH_2), 0.95–0.89 (m, 21 H, *TIPS*). **^{13}C NMR** (101 MHz, $CDCl_3$) δ 171.1, 170.8, 168.8, 141.8, 134.3, 128.4, 128.1, 127.3, 87.6, 70.0, 52.5, 52.2, 47.8, 41.8, 36.4, 18.3, 18.3, 13.8. **IR** 2949 (w), 2868 (w), 1740 (m), 1711 (s), 1383 (w), 1160 (w), 974 (w). **HRMS (ESI)** calcd for $C_{28}H_{39}NNaO_7Si^+$ [M + Na]$^+$ 552.2388; found 552.2390.

***Trans*-dimethyl-4-(2,5-dioxopyrrolidin-1-yl)-2-phenyl-2-(triisopropylsilyl)oxy) cyclopentane-1,1-dicarboxylate (301)**

Following GP14, compound **301** was synthesized starting from N-succinimide-aminocyclopropane **291** (51.0 mg, 0.200 mmol, 1.00 eq), triisopropylsilyl-enol ether **219a** (83.0 mg, 0.300 mmol, 1.50 eq) and $SnCl_4$ (23 μL 0.43 M, 1.0 μmol, 5.0 mol%) as a white solid (23.0 mg, 43.0 μmol, 22 % yield).

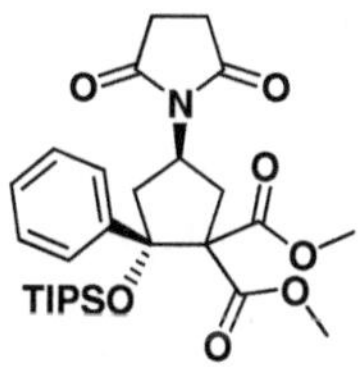

R_f 0.34 (6:4 Hexane/AcOEt). Mp 129.0–136.0 °C. **^{1}H NMR** (400 MHz, $CDCl_3$) δ 7.70–7.67 (m, 2 H, Ar), 7.29–7.24 (m, 3 H, Ar), 5.14–5.05 (m, 1 H, N–CH), 3.81 (s, 3 H, OMe), 3.66 (t, 1 H, J = 12.2 Hz, CH_2), 3.41 (s, 3 H, OMe), 3.30 (ddd, 1 H, J = 13.6, 9.8, 0.7 Hz, CH_2), 2.79 (dd, 1 H, J = 13.7, 8.2 Hz, CH_2), 2.70 (s, 4 H, O=C–CH_2), 2.32 (dd, 1 H, J = 12.3, 6.0 Hz, CH_2), 0.99–0.82 (m, 21 H, TIPS). **^{13}C NMR** (101 MHz, $CDCl_3$) δ 177.3, 170.9, 168.8, 141.7, 128.5, 128.1, 127.2, 87.7, 70.0, 52.5, 52.2, 48.4, 40.9, 35.3, 28.2, 18.3, 18.3, 13.8. **IR** 2949 (w), 2869 (w), 1737 (m), 1704 (s), 1439 (w), 1382 (m), 1257 (w), 1175 (s), 1102 (w), 1030 (w), 972 (w), 915 (w). **HRMS (ESI)** calcd for $C_{28}H_{41}NNaO_7Si^+$ [M + Na]$^+$ 554.2544; found 554.2541.

GP15 for the Synthesis of Monosubstituted Enol Ethers

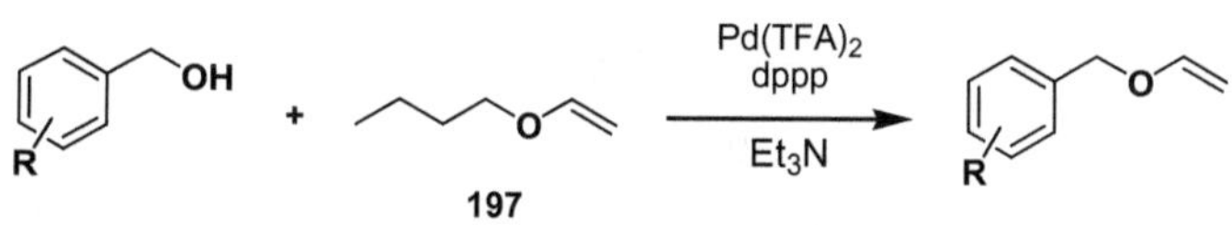

Following a slightly modified procedure [29], palladium(II) trifluoroacetate (0.500 mol%) and 4,7-diphenyl-1,10-phenanthroline (0.500 mol%) were dissolved in 1-(vinyloxy)butane (**197**) (20.0 eq) in an oven-dried 20 mL vial equipped with a stirring bar to obtain a yellow solution. The corresponding alcohol (1.00 eq) and triethylamine (0.075 eq) were then added to the solution. The flask was sealed with a microwave cap and stirred at 75 °C for 24 h. The reaction was cooled to room temperature and filtrated through a plug of activated charcoal and eluted with hexane. The solvent was evaporated under reduced pressure to obtain the crude oils that were purified by a short column chromatography using deactivated silica gel (3 % Et_3N) or basic alumina and hexane as eluent.

(Vinyloxy)methyl)benzene (305b)

Following GP15, compound **305b** was synthesized starting from 1-(vinyloxy) butane (**197**) (12.0 mL, 92.0 mmol, 20.0 eq) and phenylmethanol (500 mg, 4.62 mmol, 1.00 eq) with palladium(II) trifluoroacetate (7.70 mg, 23.0 µmol, 0.500 mol%), 4,7-diphenyl-1,10-phenanthroline (7.70 mg, 23.0 µmol, 0.500 mol %), and triethylamine (35.0 mg, 0.350 mmol, 0.0750 eq). The crude product was purified by a short column chromatography using deactivated silica gel (3 % Et_3N) and hexane as eluent to obtain (vinyloxy)methyl)benzene (**305b**) as a colorless oil (421 mg, 3.14 mmol, 63 % yield).

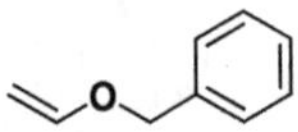

R_f 0.9 (Hexane). **^{1}H NMR** (400 MHz, $CDCl_3$) δ 7.40–7.30 (m, 5 H, Ph), 6.57 (dd, 1 H, J = 14.3, 6.8 Hz, CH_2=CH–O), 4.77 (s, 2 H, CH_2Ph), 4.31 (dd, 1 H, J = 14.3, 1.7 Hz, CH_2=CH–O), 4.09 (m, 1 H, CH_2=CH–O). **^{13}C NMR** (101 MHz, $CDCl_3$) δ 151.7, 136.9, 128.5, 128.0, 127.6, 87.4, 70.1.

The characterization data for **305b** corresponded to the reported values [30].

1-Bromo-4-(vinyloxy)methyl)benzene (305c)

Following GP15, compound **305c** was synthesized starting from 1-(vinyloxy)butane (**197**) (14.0 mL, 107 mmol, 20.0 eq) and (4-bromophenyl)methanol (1.00 g, 5.35 mmol, 1.00 eq) with palladium(II) trifluoroacetate (8.9 mg, 27 µmol, 0.50 mol%), 4,7-diphenyl-1,10-phenanthroline (8.9 mg, 27.0 µmol, 0.500 mol%), and triethylamine (56.0 µmL, 0.401 mmol, 0.075 eq). The crude product was purified by a short column chromatography using basic alumina and hexane as eluent to obtain 1-bromo-4-(vinyloxy)methyl)benzene (**305c**) as a colorless oil (915 mg, 4.29 mmol, 80 % yield).

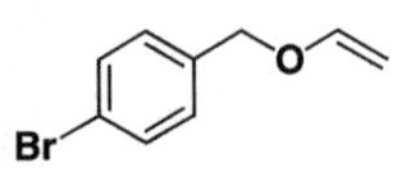

R_f 0.9 (9:1 Hexane/Et_2O). **^{1}H NMR** (400 MHz, $CDCl_3$) δ 7.49 (d, J = 8.4 Hz, 2 H, Ar), 7.23 (d, J = 8.3 Hz, 2 H, Ar), 6.54 (dd, J = 14.3, 6.8 Hz, 1 H, CH_2=CH–O), 4.71 (s, 2 H, CH_2Ar), 4.29 (dd, J = 14.3, 2.3 Hz, 1 H, CH_2=CH–O), 4.10 (dd,

J = 6.8, 2.3 Hz, 1 H, CH_2=CH–O). **^{13}C NMR** (101 MHz, $CDCl_3$) δ 151.5, 136.1, 131.8, 129.2, 122.0, 87.8, 69.4. **IR** 2359 (w), 2316 (w), 1325 (m), 1225 (m), 1091 (m), 993 (s), 895 (m), 847 (m), 684 (s).

1-Nitro-4-(vinyloxy)methyl)benzene (305d)

Following GP15, compound **305d** was synthesized starting from 1-(vinyloxy) butane (**197**) (17.0 mL, 131 mmol, 20.0 eq) and (4-nitrophenyl)methanol (1.00 g, 6.53 mmol, 1.00 eq) with palladium(II) trifluoroacetate (10.9 mg, 33.0 μmol, 0.500 mol%) and 4,7-diphenyl-1,10-phenanthroline (10.9 mg, 33.0 μmol, 0.500 mol%),and triethylamine (68.0 μmL, 0.490 mmol, 0.075 eq). The crude product was purified by a short column chromatography using basic alumina and hexane as eluent to obtain 1-nitro-4-(vinyloxy)methyl)benzene (**305d**) as a colorless oil (973 mg, 5.43 mmol, 83 % yield).

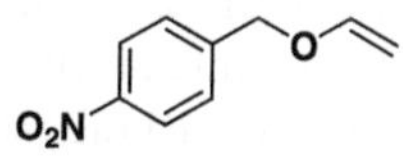

R_f 0.9 (9:1 Hexane/Et_2O). **^{1}H NMR** (400 MHz, $CDCl_3$) δ 8.23 (m, 2 H, Ar), 7.53 (m, 2 H, Ar), 6.57 (dd, 1 H, J = 14.3, 6.8 Hz, CH=C), 4.87 (s, 2 H, OCH_2Ar), 4.30 (dd, 1 H, J = 14.3, 2.5 Hz, C=CH_2), 4.16 (dd, 1 H, J = 6.8, 2.5 Hz, C=CH_2). **^{13}C NMR** (101 MHz, $CDCl_3$) δ 151.1, 147.5, 144.4, 127.6, 123.7, 88.2, 68.5.

The ^{1}H NMR data for **305d** corresponded to the reported values [31].

Synthesis of Esters

2,2,2-Trifluoroethyl benzoate (314)

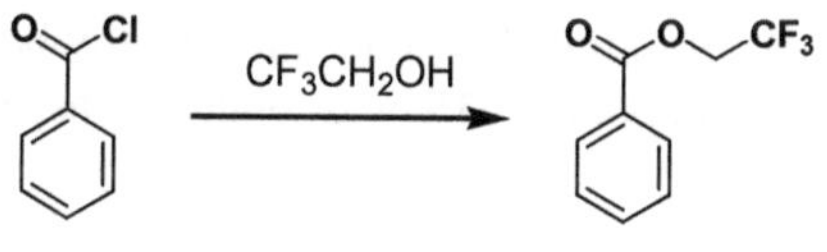

382 **314**

In a 250 mL round bottom flask equipped with a stirring bar, 2,2,2-trifluoroethanol (2.33 mL, 32.3 mmol, 1.00 eq), DMAP (39.5 mg, 0.323 mmol, 0.01 eq) and pyridine (3.14 mL, 38.8 mmol, 1.20 eq) were dissolved in diethyl ether (150 mL) while stirring. A solution of benzoyl chloride **(382)** (5.00 g, 35.6 mmol, 1.10 eq) in diethyl ether (10 mL) was added dropwise to the reaction mixture and the reaction was stirred at room temperature for 12 h. A saturated $NaHCO_3$ solution (100 mL) was added to the crude mixture. The two layers were separated and the organic layer washed with water (50 mL × 3), dried over $MgSO_4$ and the solvent removed under reduced pressure. 2,2,2-trifluoroethyl benzoate was purified by flash column chromatography on silica gel (9:1 to 8:2 Pentane/AcOEt) to obtain (**314**) as a colourless oil (3.50 g, 17.1 mmol, 53 % yield).

R$_f$ 0.8 (9:1 Pentane/Et_2O). **^{1}H NMR** (400 MHz, $CDCl_3$) δ 8.09–8.07 (m, 2 H, Ar), 7.64–7.60 (m, 1 H, Ar), 7.50–7.46 (m, 2 H, Ar), 4.71 (q, 2 H, $J_{F\text{-}H}$ = 8.4 Hz, CH_2–CF_3). **^{13}C NMR** (101 MHz, $CDCl_3$) δ 165.1, 134.0, 130.2, 128.8, 128.5, 123.3 (q, $J_{F\text{-}C}$ = 277 Hz), 60.9 (q, $J_{F\text{-}C}$ = 37 Hz).

The characterization data for **314** correspond to the reported values [32].

GP16 for the Synthesis of Benzyl Esters

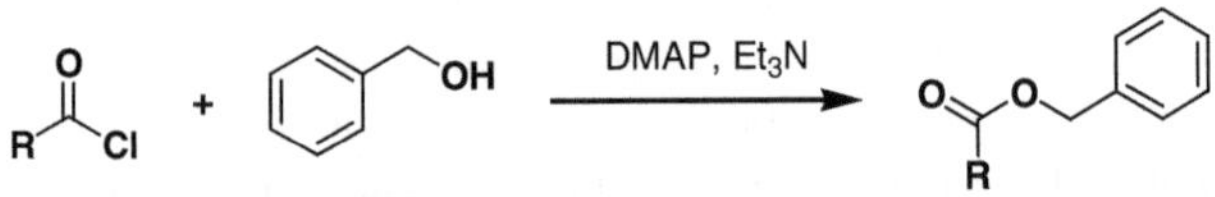

383

In a 250 mL round bottom flask equipped with a stirring bar and a condenser, benzyl alcohol (**383**) (1.00 eq), DMAP (0.01 eq) and triethylamine (1.10 eq) were dissolved in diethyl ether (90 mL) while stirring. A solution of the corresponding acyl chloride (1.20 eq) in diethyl ether (10 mL) was added dropwise to the reaction mixture and the reaction was stirred at reflux for 12 h. A saturated $NaHCO_3$ solution (100 mL) was added to the crude mixture and stirred for 15 min at room temperature. The two layers were separated and the organic layer was washed with water (50 mL × 3), dried over Na_2SO_4 and the solvent removed under reduced pressure. The esters were purified by flash column chromatography on silica gel (9:1–8:2 Pentane/AcOEt).

Benzyl 3-methylbenzoate (315)

Following GP16, compound **315** was synthesized starting from benzyl alcohol (**383**) (1.17 g, 10.8 mmol, 1.00 eq), DMAP (13.0 mg, 0.108 mmol, 0.01 eq), triethylamine (1.66 mL, 11.9 mmol, 1.10 eq) and 3-methylbenzoyl chloride (2.01 g, 13.0 mmol, 1.20 eq). Compound **315** was obtained as a colorless oil (2.37 g, 10.5 mmol, 97 % yield).

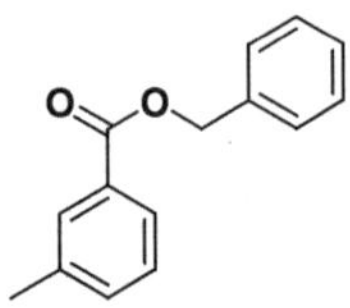

R$_f$ 0.55 (9:1 Hexane/AcOEt). **^{1}H NMR** (400 MHz, $CDCl_3$) δ 7.90–7.88 (m, 2 H, Ph), 7.47–7.45 (m, 2 H, Ph), 7.42–7.31 (m, 5 H, Ph), 5.37 (s, 2 H, CH_2), 2.40 (s, 3 H, Me). **^{13}C NMR** (101 MHz, $CDCl_3$) δ 166.7, 138.3, 136.3, 133.9, 130.3, 130.2, 128.7, 128.4, 128.3, 128.3, 127.0, 66.8, 21.4.

The characterization data for **315** correspond to the reported values [33].

Benzyl thiophene-2-carboxylate (316)

Following GP16, compound **316** was synthesized starting from benzyl alcohol (**383**) (1.08 g, 10.0 mmol, 1.00 eq), DMAP (12.2 mg, 0.100 mmol, 0.01 eq), triethylamine (1.54 mL, 11.0 mmol, 1.10 eq) and thiophene-2-carbonyl chloride (1.76 g, 12.0 mmol, 1.20 eq), compound **316** was obtained as a colorless oil (2.09 g, 9.58 mmol, 96 % yield).

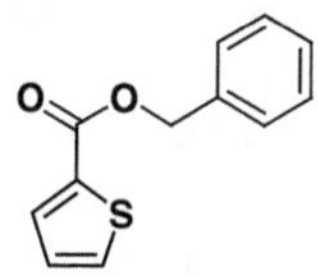

R_f 0.48 (9:1 Hexane/AcOEt). **^{1}H NMR** (400 MHz, $CDCl_3$) δ 7.85–7.84 (m, 1 H, Thiophene), 7.56 (dd, 1 H, J = 5.0, 1.2 Hz, Thiophene), 7.46–7.44 (m, 2 H, Ph), 7.42–7.33 (m, 3 H, Ph), 7.10 (dd, 1 H, J = 4.9, 3.8 Hz, Thiophene), 5.36 (s, 2 H, CH_2–Ph). **^{13}C NMR** (101 MHz, $CDCl_3$) δ 162.1, 135.9, 133.7, 133.7, 132.6, 128.7, 128.4, 128.3, 127.9, 66.8.

The characterization data for **316** correspond to the reported values [34].

Synthesis of Enol ethers
GP17 for the Synthesis of disubstituted Enol Ethers.

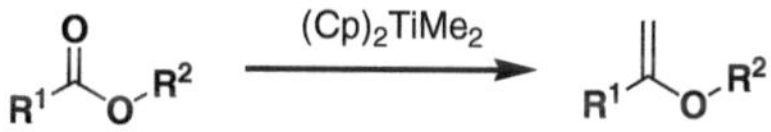

Following a slightly modified procedure [35], a round-bottom flask equipped with a magnetic stirrer was charged with a solution (10–15 % in toluene) of di(cyclopenta-1,3-dien-1-yl)dimethyltitanium (2.20 eq) in toluene, di(cyclopenta-1,3-dien-1-yl)titanium(IV) chloride (0.060 eq) and the corresponding ester (1.00 eq) under inert atmosphere. The red/orange mixture was heated in the dark to 80 °C for 16 h, and then cooled to room temperature. Pentane (50 mL) was added to the mixture and the precipitated solids were removed by filtration through a basic alumina plug (Pentane/diethyl ether 9:1, 3 % Et_3N) to afford a yellow oil. The benzyl enol ethers were purified right before use by flash column chromatography using basic alumina (Pentane, 3 % Et_3N).

(1-(Benzyloxy)vinyl)benzene (305a)

Following GP17, compound **305a** was synthesized starting from di(cyclopenta-1,3-dien-1-yl)dimethyltitanium (19.1 g of a 10.8 % solution in toluene, 9.90 mmol, 2.20 eq), di(cyclopenta-1,3-dien-1-yl)titanium(IV) chloride (67.2 mg,

0.270 mmol, 0.060 eq) and benzyl benzoate (0.955 g, 4.50 mmol, 1.00 eq) to obtain (1-(benzyloxy)vinyl)benzene (**305a**) as a colorless oil (545 mg, 2.59 mmol, 58 % yield).

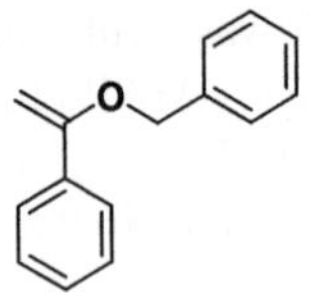

R_f 0.8 (9:1 Hexane/Et_2O). **^{1}H NMR** (400 MHz $CDCl_3$) δ 7.72 (ddd, J = 7.5, 3.3, 1.7 Hz, 2 H, Ar), 7.59–7.29 (m, 8 H, Ar), 5.00 (d, J = 2.1 Hz, 2 H, O–CH_2–Ar), 4.79 (t, J = 2.8 Hz, 1 H, CH=C), 4.36 (t, J = 2.5 Hz, 1 H, CH=C). **^{13}C NMR** (101 MHz, $CDCl_3$) δ 159.8, 137.3, 136.5, 128.6, 128.6, 128.3, 127.9, 127.5, 125.6, 83.3, 69.9. **IR** 2432 (w), 2407 (w), 1361 (m), 1336 (s), 1161 (m), 1064 (s), 994 (s), 862 (s), 782 (m). **HRMS (ESI)** calcd for $C_{15}H_{14}AgO^+$ $[M + Ag]^+$ 317.0090; found 317.0102.

1-(1-Methoxyvinyl)-4-methylbenzene (305e)

Following GP17, compound **305e** was synthesized starting from di (cyclopenta-1,3-dien-1-yl)dimethyltitanium (24.2 g of a 12.6 % solution in toluene, 14.7 mmol, 2.20 eq), di(cyclopenta-1,3-dien-1-yl)titanium(IV) chloride (99.0 mg, 0.400 mmol, 0.060 eq) and 2,2,2-trifluoroethyl benzoate (1.00 g, 6.66 mmol, 1.00 eq) to obtain 1-(1-methoxyvinyl)-4-methylbenzene (**305e**) as a colorless oil (537 mg, 3.63 mmol, 54 % yield).

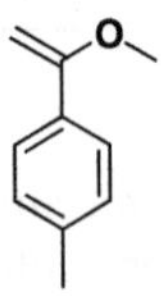

R_f 0.9 (9:1 Hexane/Et_2O). **^{1}H NMR** (400 MHz, $CDCl_3$) δ 7.78–7.45 (m, 2 H, Ar), 7.17 (d, J = 8.0 Hz, 2 H, Ar), 4.64 (dd, J = 2.7, 1.0 Hz, 1 H, C=CH_2), 4.20 (d, J = 2.7 Hz, 1 H, C=CH_2), 3.76 (s, 3 H, OMe), 2.38 (s, 3 H, CH_3Ar). **^{13}C NMR** (101 MHz, $CDCl_3$) δ 161.0, 138.3, 133.7, 128.9, 128.8, 128.8, 125.3, 125.3, 125.3, 125.3, 81.0, 55.2, 21.2. **IR** 2953 (w), 1743 (w), 1706 (w), 1644 (w), 1514 (m), 1303 (s) 1127 (s), 1047 (s), 903 (m), 796 (s). **HRMS (ESI)** calcd for $C_{10}H_{12}AgO^+$ $[M + Ag]^+$ 254.9934; found 254.9898.

(1-(2,2,2-Trifluoroethoxy)vinyl)benzene (305f)

Following GP17, compound **305f** was synthesized starting from di (cyclopenta-1,3-dien-1-yl)dimethyltitanium (18.2 g of a 12.6 % solution in toluene, 11.0 mmol, 2.20 eq), di(cyclopenta-1,3-dien-1-yl)titanium(IV) chloride (75.0 mg, 0.300 mmol, 0.060 eq) and 2,2,2-trifluoroethyl benzoate (1.02 g, 5.00 mmol,

1.00 eq) to obtain (1-(2,2,2-trifluoroethoxy)vinyl)benzene (**305f**) as a colorless oil (621 mg, 3.07 mmol, 61 % yield).

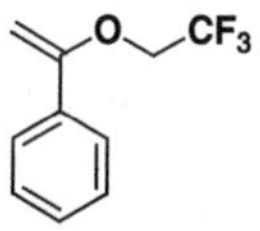

R_f 0.9 (9:1 Hexane/Et_2O). **^{1}H NMR** (400 MHz $CDCl_3$) δ 7.65–7.61 (m, 2 H, Ar), 7.39–7.36 (m, 3 H, Ar), 4.82 (d, J = 3.7 Hz, 1 H, CH=C), 4.31–4.19 (m, 3 H, CH=C, CH_2–CF_3). **^{13}C NMR** (101 MHz, $CDCl_3$) δ 159.1, 135.0, 129.2, 128.4, 125.6, 123.6 (q, J = 277 Hz). 84.3, 65.5 (q, J = 35.8 Hz). **IR** 2374 (w), 1331 (w), 1176 (w), 1047 (s), 966 (s), 818 (m), 801 (m), 656 (s). **HRMS (ESI)** calcd for $C_{10}F_3H_{10}O^+$ $[M + H]^+$ 203.0678; found 203.0678.

The NMR data for **305f** corresponded to the reported values [36].

1-(1-(Benzyloxy)vinyl)-3-methylbenzene 305g

Following GP17, compound **305g** was synthesized starting from di (cyclopenta-1,3-dien-1-yl)dimethyltitanium (17.1 g of a 10.7 % solution in toluene, 8.80 mmol, 2.20 eq), di(cyclopenta-1,3-dien-1-yl)titanium(IV) chloride (60.0 mg, 0.240 mmol, 0.060 eq) and benzyl 3-methylbenzoate (0.905 g, 4.00 mmol, 1.00 eq) to obtain 1-(1-(benzyloxy)vinyl)-3-methylbenzene (**305g**) as a colorless oil (450 mg, 2.01 mmol, 45 % yield).

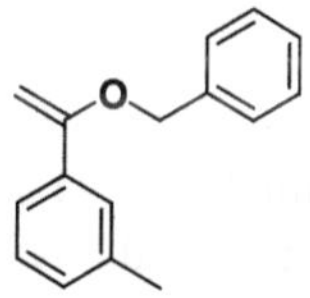

R_f 0.9 (9:1 Hexane/Et_2O). **^{1}H NMR** (400 MHz $CDCl_3$) δ 7.44–7.35 (m, 4 H, Ph), 7.31 (t, J = 7.5 Hz, 2 H, ArMe), 7.28–7.21 (m, 1 H, Ph), 7.15 (ddd, J = 8.3, 6.1, 1.5 Hz, 1 H, ArMe), 7.05 (d, J = 7.5 Hz, 1 H, ArMe), 4.88 (s, 2 H, CH_2–Ph), 4.64 (d, J = 2.8 Hz, 1 H, CH_2=C), 4.22 (d, J = 2.8 Hz, 1 H, CH_2=C), 2.28 (s, 3 H, Me). **^{13}C NMR** (101 MHz, $CDCl_3$) δ 160.1, 137.8, 137.3, 136.5, 129.4, 128.6, 128.2, 127.9, 127.6, 126.3, 122.8, 83.2, 69.9, 21.7. **HRMS (ESI)** calcd for $C_{16}H_{17}O^+$ $[M + H]^+$ 225.1274; found 225.1282.

2-(1-(Benzyloxy)vinyl)thiophene (305h)

Following GP17, compound **305h** was synthesized starting from di (cyclopenta-1,3-dien-1-yl)dimethyltitanium (17.1 g of a 10.7 % solution in toluene, 8.80 mmol, 2.20 eq), di(cyclopenta-1,3-dien-1-yl)titanium(IV) chloride (60.0 mg, 0.240 mmol, 0.060 eq) and benzyl thiophene-2-carboxylate (0.873 g, 4.00 mmol,

1.00 eq) to obtain 2-(1-(benzyloxy)vinyl)thiophene (**305g**) (0.500 g, 4.00 mmol, 58 %) as a colorless oil.

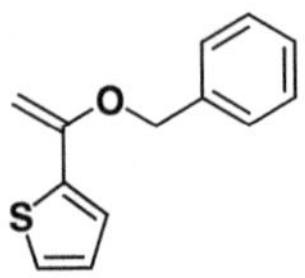

Impurities are present in the NMR sample due to degradation of the product during analysis.

R_f 0.8 (9:1 Hexane/Et_2O). **^{1}H NMR** (400 MHz, $CDCl_3$) δ 7.58–7.35 (m, 6 H, Ar), 7.30 (d, J = 5.0 Hz, 1 H, Ar), 7.07 (dd, J = 5.0, 3.7 Hz, 1 H, Ar), 5.05 (s, 2 H, Benzyl), 4.81 (d, J = 2.9 Hz, 1 H, C=CH), 4.35 (d, J = 3.1 Hz, 1 H, C=CH). **^{13}C NMR** (101 MHz, $CDCl_3$) δ 155.0, 140.4, 136.9, 128.6, 127.9, 127.3, 127.3, 125.2, 124.0, 82.7, 69.8. **HRMS (ESI)** calcd for $C_{13}H_{13}OS^+$ $[M + H]^+$ 217.0682; found 217.0688.

Synthesis of [Cu(BOX)](X)$_2$

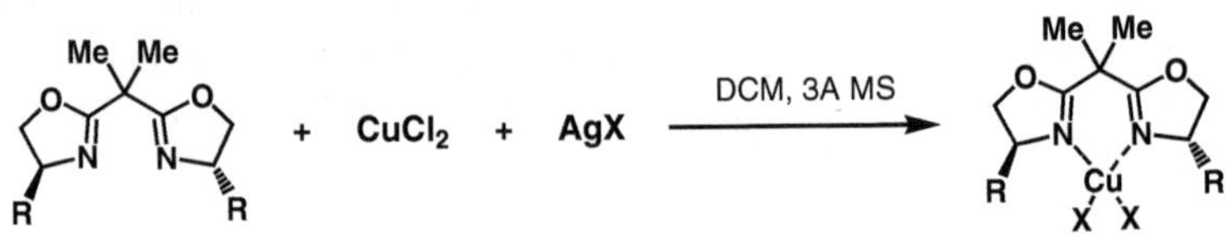

Following a modified procedure [37], an oven-dried Schlenk tube containing a magnetic stirrer was charged with $CuCl_2$ (1.1 mg, 8.0 µmol, 1.0 eq), silver salt (15 µmol, 1.9 eq) and previously activated 3 Å MS in an inert atmosphere (N_2). The flask was sealed with a septum, covered with aluminium foil and removed from the glovebox. Under argon atmosphere,[18] 0.40 mL of a solution of the corresponding BOX ligand (9.6 µmol, 1.2 eq) in dry dichloromethane were added via syringe. The mixture was stirred for 3 h at room temperature and filtrated under Ar into a sealed oven-dried vial using a syringe filter (regenerated cellulose, 0.2 µm), to obtain a bright green solution that was used for the catalysis.[19]

[18]Argon from gas cylinder was used as using central nitrogen supply with Drierite filter gave blue complexes.

[19]Blue complex gave lower er and poorly reproducible results.

GP18 for the racemic [3 + 2] Annulation Reaction

A. Racemic cyclopentylamines or tetrahydrofurylamines were synthesized using 1 equivalent of cyclopropane with 2 equivalents of enol ether or aldehyde in presence of 20 mol% of scandium triflate in dry DCM at 0 °C. Conversion was followed by TLC and when full conversion was reached, the reaction mixture was filtered on a small silica plug. Purification by Preparative TLC afforded material that was submitted to HPLC therefore no yield is given.

B. Racemic cyclopentylamines were synthesized using 1 equivalent of cyclopropane with 2 equivalents of enol ether in presence of 20 mol% of tin tetrachloride in dry DCM at −40 °C. Conversion was followed by TLC and when full conversion was reached, the reaction mixture was filtered on a small silica plug. Purification by Preparative TLC afforded material that was submitted to HPLC therefore no yield is given.

GP19 for the Screening of Conditions for the Catalytic Asymmetric [3 + 2] Annulation Reaction

The corresponding N-protected-aminocyclopropane[20] (40.0 µmol, 1.00 eq) and freshly purified enol ether (50.0 µmol, 1.20 eq) were dissolved in 0.4 mL of dry dichloromethane. The solution was added into a sealed oven-dried vial containing a magnetic stirrer, pre-activated 3 Å MS and 0.4 mL of the solution of the desired complex (0.01 M, 4.00 µmol, 0.1 eq). Dry dichloromethane was used to complete a final volume of 1.0 mL. The mixture was stirred at rt until full conversion was obtained as verified by TLC. The reaction was quenched by addition of 0.3 mL of Et_3N and filtrated through a silica gel plug eluting with 5 mL of a mixture 1:1 Hexane/AcOEt to obtain a yellowish solution. The solvent was evaporated under

[20]Dried by dissolving in benzene then removing the solvent under reduced pressure and drying in high vacuo.

reduced pressure and the crude analyzed by ^{1}**H NMR** and chiral HPLC. The NMR yields indicated were obtained using trimethoxybenzene as internal standard.

Dimethyl-(2*S*,4*S*)-2-(benzyloxy)-4-(1,3-dioxoisoindolin-2-yl)-2-phenylcyclopentane-1,1-dicarboxylate (306)

Chiralcel IA Hexane/^{i}PrOH 95:5, 1 mL/min, λ = 254 nm, tr1 = 18.3 min. tr2 = 21.1 min.

Following GP19, the crude of the reaction using *tert*-butyl-BOX/Cu(ClO_4)$_2$ complex was analyzed: er = 78:22.

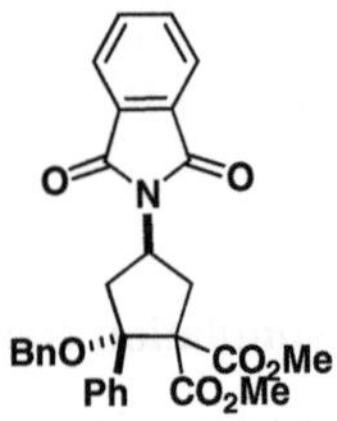

R$_f$ 0.7 (5:5 Pentane/AcOEt). **Mp** 187.0–188.8 °C ^{1}**H NMR** (400 MHz, $CDCl_3$) δ 7.85 (dd, *J* = 5.4, 3.0 Hz, 2 H, Phth), 7.73 (dd, *J* = 5.5, 3.0 Hz, 2 H, Phth), 7.66–7.58 (m, 2 H, Ar), 7.42–7.27 (m, 8 H, Ar), 5.06 (dddd, *J* = 11.6, 10.0, 7.5, 6.2 Hz, 1 H, N–C–H), 4.39 (d, *J* = 11.7 Hz, 1 H, CH_2 Benzyl), 4.08 (d, *J* = 11.7 Hz, 1 H, CH_2 benzyl), 3.82–3.73 (m, 1 H, CH_2), 3.76 (s, 3 H, OMe), 3.65–3.49 (m, 1 H, CH_2), 3.60 (s, 3 H, OMe), 2.88 (dd, *J* = 14.0, 7.5 Hz, 1 H, CH_2), 2.57 (dd, *J* = 13.1, 6.3 Hz, 1 H, CH_2). ^{13}**C NMR** (101 MHz, $CDCl_3$) δ 170.3, 168.5, 168.3, 138.2, 136.5, 134.1, 131.9, 129.2, 128.3, 128.1, 127.3, 127.2, 126.5, 123.3, 89.8, 68.2, 63.5, 52.4, 52.2, 46.4, 36.1, 35.6. **IR** 1737 (s), 1712 (s), 1435 (w), 1379 (m), 1259 (w), 1127 (m). **HRMS (ESI)** calcd for $C_{30}H_{27}NNaO_7^+$ [M + Na]$^+$ 536.1680; found 536.1667.

Dimethyl-(2*S*,4*S*)-2-(benzyloxy)-4-(5-methoxy-1,3-dioxoisoindolin-2-yl)-2-phenylcyclopentane-1,1-dicarboxylate (307)

Chiralcel IA Hexane/^{i}PrOH 80:20, 1 mL/min, λ = 220 nm, tr1 = 15.4 min. tr2 = 56.1 min.

Following GP19, the crude of the reaction using *tert*-butyl-BOX/Cu(ClO_4)$_2$ complex was analyzed: er = 74:26.

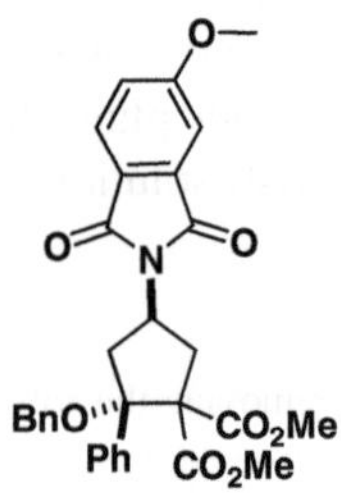

R_f 0.8 (5:5 Pentane/AcOEt). **Mp** 128.3–130.7 °C. **^{1}H NMR** (400 MHz, $CDCl_3$) δ 7.75 (d, J = 8.3 Hz, 1 H, Ar), 7.67–7.58 (m, 2 H, Ar), 7.41–7.27 (m, 9 H, Ar), 7.16 (dd, J = 8.3, 2.3 Hz, 1 H, Ar), 5.10–4.94 (m, 1 H, N–C–H), 4.38 (d, J = 11.7 Hz, 1 H, CH_2 benzyl), 4.06 (d, J = 11.7 Hz, 1 H, CH_2 benzyl), 3.93 (m, 3 H, OMe), 3.83–3.71 (m, 1 H, CH_2), 3.75 (s, 3 H, OMe), 3.59 (s, 3 H, OMe), 3.57–3.47 (m, 1 H, CH_2), 2.86 (dd, J = 14.0, 7.5 Hz, 1 H, CH_2), 2.55 (dd, J = 13.1, 6.3 Hz, 1 H, CH_2). **^{13}C NMR** (101 MHz, $CDCl_3$) δ 170.3, 168.5, 168.1, 168.1, 164.8, 138.2, 136.6, 134.5, 129.2, 128.3, 128.1, 127.3, 127.1, 126.5, 125.0, 123.9, 119.8, 108.0, 89.7, 68.2, 63.4, 56.1, 52.3, 52.2, 46.4, 36.1, 35.6. **IR** 1360 (m), 1336 (s), 1263 (w), 1161 (w), 1127 (w), 1116 (w), 1115 (w), 1065 (s), 995 (m), 967 (m), 956 (m), 863 (m), 690 (m), 689 (m). **HRMS (ESI)** calcd for $C_{31}H_{29}NNaO_8^+$ $[M + Na]^+$ 566.1785; found 566.1788.

Dimethyl-(*2S*,*4S*)-2-(benzyloxy)-4-(5-nitro-1,3-dioxoisoindolin-2-yl)-2-phenylcyclopentane-1,1-dicarboxylate (308)

Chiralcel IA Hexane/^{i}PrOH 85:15, 1 mL/min, λ = 220 nm, tr1 = 33.4 min. tr2 = 36.6 min.

Following GP19, the crude of the reaction using *tert*-butyl-BOX/Cu(ClO_4)$_2$ complex was analyzed: er = 92:8).

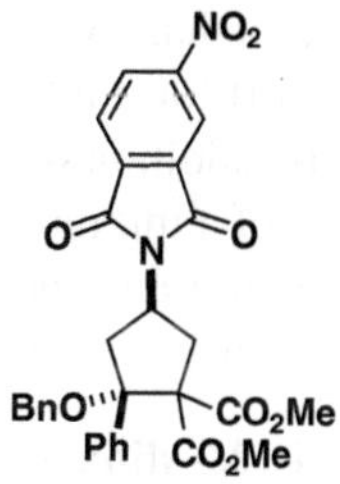

R_f 0.8 (5:5 Pentane/AcOEt). **Mp** 95.5–98.3 °C. **^{1}H NMR** (400 MHz, $CDCl_3$) δ 8.82–8.50 (m, 2 H, Ar), 8.05 (dd, J = 8.1, 0.7 Hz, 1 H, Ar), 7.80–7.57 (m, 2 H, Ar), 7.48–7.13 (m, 8 H, Ar), 5.06 (tdd, J = 7.7, 4.3, 2.2 Hz, 1 H, N–C–H), 4.36 (d, J = 11.8 Hz, 1 H, CH_2 benzyl), 4.10 (d, J = 11.7 Hz, 1 H, CH_2 benzyl), 3.84–3.67 (m, 1 H, CH_2), 3.76 (s, 3 H, OMe), 3.67–3.44 (m, 1 H, CH_2), 3.60 (s, 3 H, OMe), 2.86 (dd, J = 14.0, 7.4 Hz, 1 H, CH_2), 2.60 (dd, J = 13.1, 6.3 Hz, 1 H, CH_2). **^{13}C NMR** (101 MHz, $CDCl_3$) δ 170.2, 168.3, 166.1, 165.8, 151.8, 138.0, 136.3, 136.2, 133.3, 129.4, 129.1, 128.4, 128.3, 127.4, 127.3, 126.5, 124.5, 118.7, 89.8, 68.2, 63.6, 52.4, 52.3, 47.2, 36.0, 35.6. **IR** 1393 (w), 1345 (s), 1201 (m), 1126 (w), 1115 (w), 1069 (m), 1043 (s), 972 (w), 868 (m), 691 (m). **HRMS (ESI)** calcd for $C_{30}H_{26}N_2NaO_9^+$ $[M + Na]^+$ 581.1531; found 581.1540.

GP20 for the Catalytic Asymmetric [3 + 2] Annulation Reaction

Dimethyl 2-(2,5-dioxopyrrolidin-1-yl)cyclopropane-1,1-dicarboxylate (**291**) (51.0 mg, 0.200 mmol, 1.00 eq) and freshly purified enol ether or aldehyde (0.400 mmol, 2.00 eq) were dissolved in 2.0 mL of dry dichloromethane. The solution was added into a sealed oven-dried vial containing a magnetic stirrer, pre-activated 3 Å MS and 2.0 mL of the solution of the copper complex (0.01 M, 0.020 mmol, 0.10 eq). Dry dichloromethane was used to complete a final volume of 5.0 mL. The mixture was stirred at rt until full conversion was observed by TLC. The reaction was quenched by addition of 0.5 mL of Et_3N and filtrated through a silica gel plug eluting with 10 mL of a mixture of 3:7 Hexane/AcOEt. The solvent was evaporated under reduced pressure and the crude analyzed by ^1H NMR. Purification by column chromatography using pentane/AcOEt (6:4–S3:7) afforded the product as a mixture of diastereoisomers. In the case of the reaction with enol ether, it was possible to purify the major diastereoisomer by preparative TLC for characterization and HPLC analysis. For aldehydes, characterization was done directly on the obtained mixture of diastereoisomers.

GP21 for the Catalytic Asymmetric [3 + 2] Annulation Reaction on 1 mmol scale

Dimethyl 2-(2,5-dioxopyrrolidin-1-yl)cyclopropane-1,1-dicarboxylate (**291**) (255 mg, 1.00 mmol, 1.00 eq) and freshly purified enol ether **305b** (268 mg, 2.00 mmol, 2.00 eq) were dissolved in 10.0 mL of dry dichloromethane. The solution was added into a sealed oven-dried vial containing a magnetic stirrer, pre-activated 3 Å MS and 10.0 mL of the solution of the copper complex (0.01 M, 0.100 mmol, 0.10 eq). Dry dichloromethane was used to complete a final volume of 25.0 mL. The mixture was stirred at rt for 2 h and full conversion was observed by TLC. The reaction was quenched by addition of 1 mL of Et_3N and filtrated through a silica gel plug eluting with 50 mL of a mixture of 3:7 Hexane/AcOEt. The solvent was evaporated under reduced pressure and the crude analyzed by ^1H NMR. and purified by column chromatography using pentane/AcOEt (6:4–3:7).

Dimethyl-(2*S*,4*S*)-2-(benzyloxy)-4-(2,5-dioxopyrrolidin-1-yl)-2-phenylcyclopentane-1,1-dicarboxylate (309a)

Following GP20, using (1-(benzyloxy)vinyl)benzene (**305a**) (84.0 mg, 0.400 mmol, 2.00 eq), dimethyl 2-(benzyloxy)-4-(2,5-dioxopyrrolidin-1-yl)-2-phenylcyclopentane-1,1-dicarboxylate (**309a**) (90.3 mg, 0.194 mmol, 97 %) was obtained as a colorless solid.

$$Crude\,analysis : dr = 7{:}1 \text{ between peaks at } 5.01(minor) \text{ and } 4.67(major).$$

$$er_{major} = 96{:}4,\ \text{ChiralcelIA Hexane}/{}^{i}\text{PrOH } 80{:}20,\ 1\ \text{mL/min},\ \lambda = 210\ \text{nm},\ \text{tr1} = 18.2\text{min. tr2} = 24.0\text{min.}$$

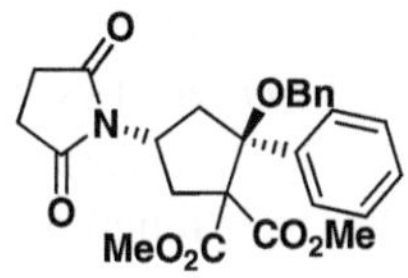

$[\alpha]_D^{25.0}$ − 21.0 (c = 0.43, $CHCl_3$). **R_f** 0.30 (5:5 Hexane/AcOEt). **Mp** 90.1–91.7 °C. **^{1}H NMR** (400 MHz, $CDCl_3$) δ 7.57–7.41 (m, 2 H, Ar), 7.32–7.13 (m, 8 H, Ar), 5.01–4.67 (m, 1 H, N–C–H), 4.33 (d, J = 11.7 Hz, 1 H, CH_2 benzyl), 4.03 (d, J = 11.7 Hz, 1 H, CH_2 benzyl), 3.66 (s, 3 H, OMe), 3.56 (dd, J = 13.1, 11.6 Hz, 1 H, CH_2), 3.49 (s, 3 H, OMe), 3.37 (ddd, J = 14.0, 10.3, 0.9 Hz, 1 H, CH_2), 2.71 (dd, J = 13.9, 7.2 Hz, 1 H, CH_2), 2.63 (s, 4 H, CH_2 succinimide), 2.38 (dd, J = 13.0, 6.4 Hz, 1 H, CH_2). **^{13}C NMR** (101 MHz, $CDCl_3$) δ 177.0, 170.1, 168.3, 138.0, 136.3, 129.1, 128.2, 128.0, 127.1, 127.0, 126.3, 89.6, 68.0, 63.2, 52.2, 52.1, 46.9, 34.9, 34.5, 28.0. **IR** 2255 (w), 1738 (w), 1704 (m), 1382 (w), 1260 (w), 1178 (w), 906 (s). **HRMS (ESI)** calcd for $C_{26}H_{27}NNaO_7^+$ $[M + Na]^+$ 488.1680; found 488.1687.

Dimethyl-(2*S*,4*S*)-4-(2,5-dioxopyrrolidin-1-yl)-2-methoxy-2-(p-tolyl) cyclopentane-1,1-dicarboxylate (309b)

Following GP20, using 1-(1-methoxyvinyl)-4-methylbenzene (**305e**) (59.3 mg, 0.400 mmol, 2.00 eq), dimethyl 4-(2,5-dioxopyrrolidin-1-yl)-2-methoxy-2-(p-tolyl) cyclopentane-1,1-dicarboxylate (**309b**) (77.0 mg, 0.191 mmol, 95 %) was obtained as a colorless solid.

$$\textit{Crude analysis:dr} = 20{:}1 \textit{ between peaks at } 5.17(\textit{minor}) \text{ and } 4.84(\textit{major}).$$

$$er_{major} = 94.5{:}5.5,\ \text{ChiralcelIA Hexane}/{}^i\text{PrOH } 80{:}20,\ 1\ \text{mL/min}, \lambda = 220\ \text{nm}, \text{tr1} = 17.9\ \text{min. tr2} = 22.5\ \text{min.}$$

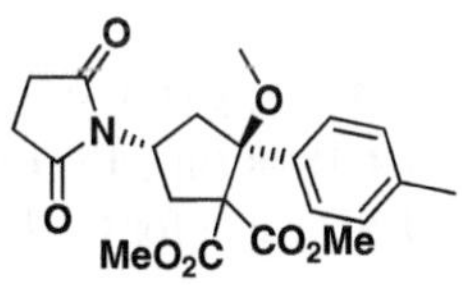

$[\alpha]_D^{25.0}$ 15.7 (c = 0.81, $CHCl_3$). **R_f** 0.3 (3:7 Pentane/AcOEt). **Mp** 97.5–99.0 °C. **^{1}H NMR** (400 MHz, $CDCl_3$) δ 7.36 (d, J = 8.3 Hz, 2 H, Ar), 7.13 (d, J = 8.1 Hz, 2 H, Ar), 4.83 (tt, J = 11.0, 6.6 Hz, 1 H, N–CH), 3.74 (s, 3 H, OMe), 3.58 (s, 3 H, OMe), 3.54–3.42 (m, 1 H, CH_2), 3.36 (dd, J = 14.0, 10.7 Hz, 1 H, CH_2), 2.97 (s, 3 H, Me), 2.80–2.60 (m, 1 H, CH_2), 2.72 (s, 4 H, CH_2 succinimide), 2.40–2.22 (m, 1 H, CH_2), 2.36 (s, 3 H, Me). **^{13}C NMR** (101 MHz, $CDCl_3$) δ 177.1, 170.2, 168.5,

137.6, 133.0, 129.3, 127.8, 89.5, 67.7, 52.2, 52.1, 49.5, 46.5, 34.9, 33.8, 28.0, 21.1. **IR** 1740 (s), 1703 (s), 1436 (w), 1399 (w), 1382 (m), 1382 (m), 1294 (w), 1276 (w), 1276 (w), 1261 (w), 1178 (m). **HRMS (ESI)** calcd for $C_{21}H_{25}NNaO_7^+$ $[M + Na]^+$ 426.1523; found 426.1516.

Dimethyl-(2*S*,4*S*)-4-(2,5-dioxopyrrolidin-1-yl)-2-phenyl-2-(2,2,2-trifluoroethoxy)cyclopentane-1,1-dicarboxylate (309c)

Following GP20, using (1-(2,2,2-trifluoroethoxy)vinyl)benzene (**305f**) (81.0 mg, 0.400 mmol, 2.00 eq), dimethyl 4-(2,5-dioxopyrrolidin-1-yl)-2-phenyl-2-(2,2,2-trifluoroethoxy)cyclopentane-1,1-dicarboxylate (**309c**) (80.4 mg, 0.176 mmol, 88 %) was obtained as a colorless oil.

Crude analysis : $dr = 1.5 : 1$between peaks at 5.06(*minor*) and 4.75(*major*).

anti

$er = 95.5 : 4.5$, ChiralcelIB Hexane/^{i}PrOH 80:20, 1 mL/min, $\lambda = 210$ nm, tr1 = 14.6 min. tr2 = 17.4 min.

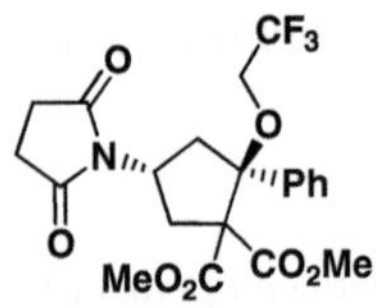

$[\alpha]_D^{25.0}$ 18.9 (c = 0.43, $CHCl_3$). **R_f** 0.2 (4:6 Pentane/AcOEt). **^{1}H NMR** (400 MHz, $CDCl_3$) δ 7.59–7.43 (m, 2 H, Ar), 7.43–7.28 (m, 3 H, Ar), 4.97–4.75 (m, 1 H, N–CH), 3.80 (s, 3 H, OMe), 3.60 (dd, *J* = 13.4, 11.6 Hz, 1 H, CH_2), 3.56–3.44 (m, 1 H, CH_2–CF_3), 3.53 (s, 3 H, OMe), 3.43–3.30 (m, 2 H, CH_2 + CH_2–CF_3), 2.84–2.75 (m, 1 H, CH_2), 2.71 (s, 4 H, CH_2 succinimide), 2.24 (dd, *J* = 13.3, 6.4 Hz, 1 H, CH_2). **^{13}C NMR** (101 MHz, $CDCl_3$) δ 177.0, 169.8, 167.7, 135.0, 129.0, 128.7, 127.6, 123.6 (q, *J* = 278 Hz), 90.4, 67.7, 60.50 (q, *J* = 35 Hz), 52.4, 52.3, 46.6, 34.8, 34.6, 28.0. **IR** 1808 (m), 1771 (s), 1521 (w), 1520 (w), 1459 (w), 1355 (m), 1256 (s). **HRMS (ESI)** calcd for $C_{21}H_{22}F_3NNaO_7^+$ $[M + Na]^+$ 480.1241; found 480.1243.

syn

$er = 96.5:3.5$, ChiralcelIA Hexane/^{i}PrOH90:10, 1 mL/min, $\lambda = 220$ nm, tr1 = 28.6 min. tr2 = 31.1min.

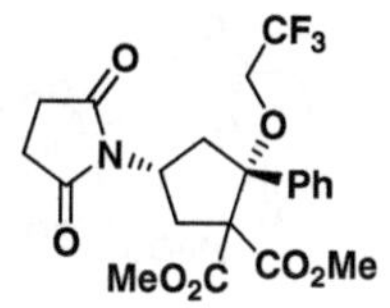

$[\alpha]_D^{25.0}$ − 2.3 (c = 0.39, $CHCl_3$). **R_f** 0.3 (4:6 Pentane/AcOEt). **^{1}H NMR** (400 MHz, $CDCl_3$) δ 7.64–7.52 (m, 2 H, Ar), 7.41–7.29 (m, 3 H, Ar), 5.20–5.06 (m, 1 H, N–CH), 4.06 (dq, J = 10.2, 8.4 Hz, 1 H, CH_2–CF_3), 3.75 (s, 3 H, OMe), 3.61–3.40 (m, 1 H, CH_2 + CH_2–CF_3), 3.50 (s, 3 H, OMe), 3.01 (dd, J = 15.1, 11.0 Hz, 1 H, CH_2), 2.84–2.67 (m, 5 H, CH_2 succinimide + CH_2), 2.34 (dd, J = 12.9, 7.4 Hz, 1 H, CH_2). **^{13}C NMR** (101 MHz, $CDCl_3$) δ 177.1, 169.3, 167.6, 135.2, 129.1, 128.4, 127.6, 124.1 (q, J = 278 Hz), 90.7, 70.0, 60.4 (q, J = 35 Hz), 52.5, 52.1, 47.4, 36.5, 33.7, 28.0. **IR** 1737 (m), 1705 (s), 1447 (w), 1382 (w), 1279 (m), 1256 (w), 1170 (s). **HRMS (ESI)** calcd for $C_{21}H_{22}F_3NNaO_7^+$ $[M + Na]^+$ 480.1241; found 480.1237.

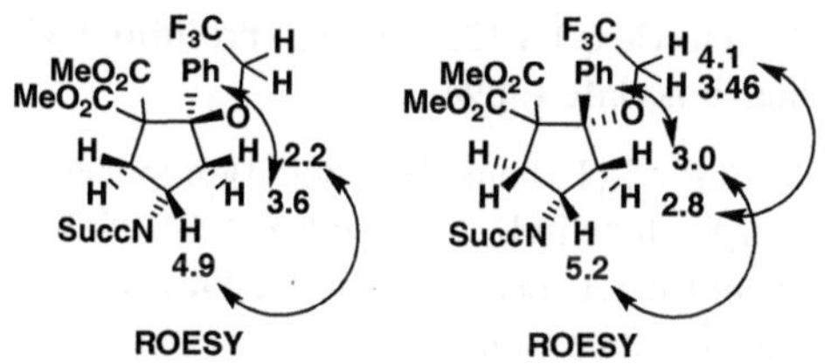

Dimethyl-(2*S*,4*S*)-2-(benzyloxy)-4-(2,5-dioxopyrrolidin-1-yl)-2-(*m*-tolyl) cyclopentane-1,1-dicarboxylate (309d)

Following GP20, using 1-(1-(benzyloxy)vinyl)-3-methylbenzene (**305 g**) (90.0 mg, 0.400 mmol, 2.00 eq), dimethyl 2-(benzyloxy)-4-(2,5-dioxopyrrolidin-1-yl)-2-(m-tolyl)cyclopentane-1,1-dicarboxylate (**309d**) (96.0 mg, 0.199 mmol, 99 %) was obtained as a colorless oil.

Crude analysis : *dr* > 20 : 1.

$er = 95{:}5$, ChiralcelIA Hexane/^{i}PrOH80:20, 1 mL/min, $\lambda = 220$ nm, tr1 = 16.2 min. tr2 = 22.1 min.

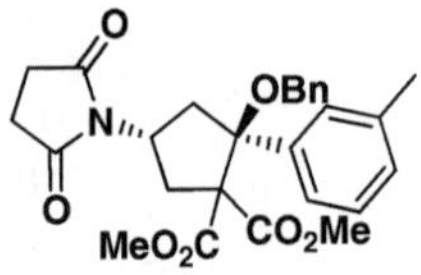

$[\alpha]_D^{25.0}$ − 16.5 (c = 0.44, $CHCl_3$). **R_f** 0.25 (4:6 Pentane/AcOEt). **^{1}H NMR** (400 MHz, $CDCl_3$) δ 7.37–7.29 (m, 4 H, Ar), 7.28–7.23 (m, 3 H, Ar), 7.20 (t, J = 7.7 Hz, 1 H, Ar), 7.11 (d, J = 7.5 Hz, 1 H, Ar), 4.97–4.81 (m, 1 H, N–CH), 4.31 (d, J = 11.7 Hz, 1 H, CH_2 benzyl), 4.03 (d, J = 11.7 Hz, 1 H, CH_2 benzyl), 3.73 (s, 3 H, OMe), 3.65–3.55 (m, 1 H, CH_2), 3.57 (s, 3 H, OMe), 3.44 (dd, J = 13.9, 10.4 Hz, 1 H, CH_2), 2.78 (dd, J = 13.9, 7.0 Hz, 1 H, CH_2), 2.70 (s, 4 H, CH_2 succinimide), 2.45 (dd, J = 13.0, 6.5 Hz, 1 H, CH_2), 2.34 (s, 3 H, Me). **^{13}C NMR** (101 MHz, $CDCl_3$) δ 177.1, 170.2, 168.3, 138.2, 136.5, 136.3, 130.0, 128.8, 128.3, 127.1,[21] 126.4, 126.2, 89.7, 68.1, 63.3, 52.2, 52.1, 46.9, 35.0, 34.6, 28.0, 21.7. **IR** 2924 (w), 1739 (m), 1703 (s), 1435 (w), 1383 (m), 1295 (w), 1259 (w), 1181 (m), 738 (w). **HRMS (ESI)** calcd for $C_{27}H_{29}NNaO_7^+$ $[M + Na]^+$ 502.1836; found 502.1845.

Dimethyl-(*2R,4S*)-2-(benzyloxy)-4-(2,5-dioxopyrrolidin-1-yl)-2-(thiophen-2-yl) cyclopentane-1,1-dicarboxylate (309e)

Following GP20, using 2-(1-(benzyloxy)vinyl)thiophene (**305h**) (87.0 mg, 0.400 mmol, 2.00 eq), dimethyl 2-(benzyloxy)-4-(2,5-dioxopyrrolidin-1-yl)-2-(thiophen-2-yl)cyclopentane-1,1-dicarboxylate (**309e**) (89.1 mg, 0.189 mmol, 94 %) was obtained as a colorless oil.

$$\textit{Crude analysis:dr} = 8{:}1{:}\text{between peaks at } 5.11(\textit{minor}) \text{ and } 4.86(\textit{major}).$$

$$er_{major} = 94{:}6,\ \text{ChiralcelIA Hexane}/^i\text{PrOH } 80{:}20,\ 1\ \text{mL/min},\ \lambda = 210\ \text{nm},\ \text{tr1} = 27.0\text{min. tr2} = 40.2\ \text{min.}$$

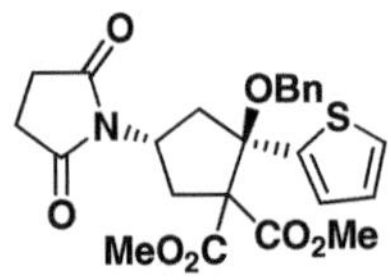

$[\alpha]_D^{25.0}$ − 11.8 (c = 0.44, $CHCl_3$). **R_f** 0.2 (4:6 Pentane/AcOEt). **^{1}H NMR** (400 MHz, $CDCl_3$) δ 7.42–7.14 (m, 7 H, Ar), 6.97 (dd, J = 5.1, 3.6 Hz, 1 H, Thiophene), 4.86 (dddd, J = 11.9, 10.0, 7.9, 6.3 Hz, 1 H, N–CH), 4.33 (d, J = 11.4 Hz, 1 H, CH_2 benzyl),

[21]2 carbon signal overlapping.

4.14 (d, $J = 11.4$ Hz, 1 H, CH_2 benzyl), 3.79 (s, 3 H, OMe), 3.63–3.52 (m, 1 H, CH_2), 3.60 (s, 3 H, OMe), 3.38 (dd, $J = 13.9$, 10.1 Hz, 1 H, CH_2), 2.79 (dd, $J = 13.9$, 7.9 Hz, 1 H, CH_2), 2.70 (s, 4 H, CH_2 succinimide), 2.59 (dd, $J = 12.9$, 6.3 Hz, 1 H, CH_2). **^{13}C NMR** (101 MHz, $CDCl_3$) δ 177.2, 170.2, 168.5, 141.4, 138.0, 128.9, 128.3, 127.4, 126.9, 126.6, 126.0, 87.7, 68.2, 63.9, 52.5, 52.4, 47.1, 36.8, 34.7, 28.2. **IR** 1740 (s), 1704 (s), 1435 (w), 1384 (w), 1270 (w), 1175 (m). **HRMS (ESI)** calcd for $C_{24}H_{25}NNaO_7S^+$ $[M + Na]^+$ 494.1244; found 494.1241.

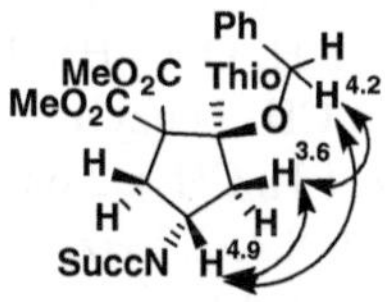

ROESY

Dimethyl-(*2R,4S*)-2-(benzyloxy)-4-(2,5-dioxopyrrolidin-1-yl) cyclopentane-1,1-dicarboxylate (309f)

Following GP20, using (vinyloxy)methyl)benzene (**305b**) (53.7 mg, 0.400 mmol, 2.00 eq), dimethyl 2-(benzyloxy)-4-(2,5-dioxopyrrolidin-1-yl) cyclopentane-1,1-dicarboxylate (**309f**) (74.8 mg, 0.192 mmol, 96 %) was obtained as a colorless solid. Recrystallized from isopropanol.[22]

$$Crude\ analysis{:}dr = 4{:}1 \text{between peaks at } 3.79(major) \text{ and } 3.75(minor).$$

$$er_{major} = 96.5{:}3.5,\ \text{ChiralcelIA Hexane}/^{i}\text{PrOH } 80{:}20,\ 1\ \text{mL/min}, \lambda = 210\ \text{nm},\ \text{tr1} = 27.4\ \text{min. tr2} = 37.8\ \text{min.}$$

Following GP21, dimethyl-2-(benzyloxy)-4-(2,5-dioxopyrrolidin-1-yl) cyclopentane-1,1-dicarboxylate **(309f)** (311 mg, 0.800 mmol, 80 %) was obtained as a colorless solid.

$$Crude\ analysis{:}dr = 4{:}1 \text{between peaks at } 3.03(major) \text{ and } 3.37(minor).$$

$$er_{major} = 95.5{:}4.5,\ \text{ChiralcelIA Hexane}/^{i}\text{PrOH80:20},\ 1\ \text{mL/min},\ \lambda = 210\ \text{nm}.$$

[22]Structure is registered in CCDC under the number CCDC 988525.

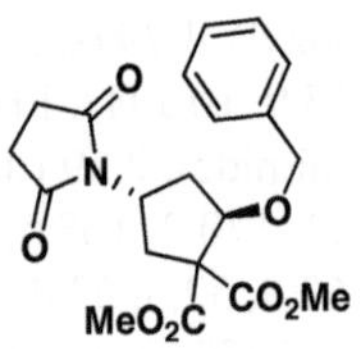

$[\alpha]_D^{25.0}$ – 32.7 (c = 0.43, $CHCl_3$). $\mathbf{R_f}$ 0.3 (5:5 Pentane/AcOEt). **Mp** 106.8–109.5 ° C. **^{1}H NMR** (400 MHz, $CDCl_3$) δ 7.36–7.24 (m, 5 H, Ar), 5.03–4.82 (m, 1 H, N–C–H), 4.75 (dd, *J* = 4.7, 2.7 Hz, 1 H, O–C–H), 4.59 (d, *J* = 11.9 Hz, 1 H, CH_2 benzyl) 4.49 (d, *J* = 11.9 Hz, 1 H, CH_2 benzyl), 3.79 (s, 3 H, OMe), 3.68 (s, 3 H, OMe), 3.03 (dd, *J* = 14.4, 10.6 Hz, 1 H, CH_2), 2.64 (s, 4 H, succinimide), 2.50–2.29 (m, 2 H, CH_2), 2.15 (ddd, *J* = 13.4, 8.3, 2.7 Hz, 1 H, CH_2). **^{13}C NMR** (101 MHz, $CDCl_3$) δ 176.9, 171.0, 169.0, 138.0, 128.3, 127.6, 127.4, 83.0, 71.7, 65.1, 52.9, 52.7, 48.0, 34.0, 33.3, 28.0. **IR** 1737 (m), 1702 (s), 1398 (w), 1397 (w), 1384 (w), 1283 (w), 1262 (w), 1175 (m), 1100 (w). **HRMS (ESI)** calcd for $C_{20}H_{24}NO_7^+$ $[M + H]^+$ 390.1547; found 390.1554.

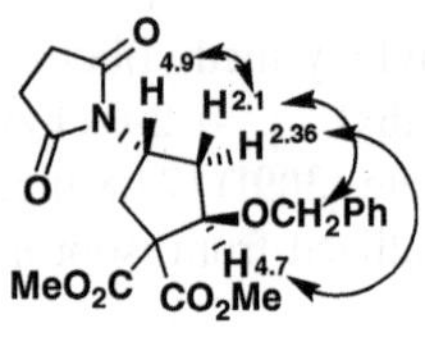

ROESY

Dimethyl-(*2R,4S*)-2-(4-bromobenzyl)oxy)-4-(2,5-dioxopyrrolidin-1-yl) cyclopentane-1,1-dicarboxylate (309g)

Following GP20, using 1-bromo-4-(vinyloxy)methyl)benzene (**305c**) (85.0 mg, 0.400 mmol, 2.00 eq), dimethyl 2-(4-bromobenzyl)oxy)-4-(2,5-dioxopyrrolidin-1-yl)cyclopentane-1,1-dicarboxylate (**309 g**) (68.2 mg, 0. 146 mmol, 73 %) was obtained as a colorless oil.

Crude analysis: *dr* = 5:1 between peaks at 3.36 (*minor*) and 3.00 (*major*).

er_{major} > 94.5:5.5,[23] Chiralcel IB Hexane/^{i}PrOH 80:20, 1 mL/min, λ = 220 nm, tr1 = 26.0 min. tr2 = 29.6 min.

[23]Due to shoulder in the peaks, separation was not complete (cf HPLC spectra).

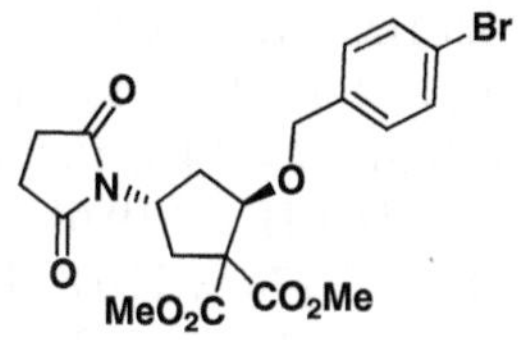

$[\alpha]_D^{25.0}$ − 28.9 (c = 0.46, $CHCl_3$). **R_f** 0.20 (4:6 Hexane/AcOEt). **^{1}H NMR** (400 MHz, $CDCl_3$) δ 7.44 (d, J = 8.3 Hz, 2 H, Ar), 7.13 (d, J = 8.3 Hz, 2 H, Ar), 4.93 (dtd, J = 10.6, 8.7, 6.4 Hz, 1 H, N–C–H), 4.75 (dd, J = 4.9, 3.0 Hz, 1 H, O–C–H), 4.55 (d, J = 11.9 Hz, 1 H, CH_2 benzyl), 4.44 (d, J = 12.1 Hz, 1 H, CH_2 benzyl), 3.79 (s, 3 H, OMe), 3.68 (s, 3 H, OMe), 3.00 (dd, J = 14.4, 10.6 Hz, 1 H, CH_2), 2.65 (s, 4 H, CH_2 succinimide), 2.50–2.24 (m, 2 H, CH_2), 2.13 (ddd, J = 13.5, 8.4, 3.0 Hz, 1 H, CH_2). **^{13}C NMR** (101 MHz, $CDCl_3$) δ 176.8, 171.0, 169.0, 137.0, 131.4, 129.0, 121.4, 83.0, 71.0, 65.0, 52.9, 52.7, 47.8, 34.1, 33.4, 28.0. **IR** 1739 (m), 1703 (s), 1397 (w), 1383 (w), 1283 (w), 1262 (w), 1261 (w), 1176 (w). **HRMS (ESI)** calcd for $C_{20}^{79}BrH_{23}NO_7^+$ $[M + H]^+$ 468.0652; found 468.0661.

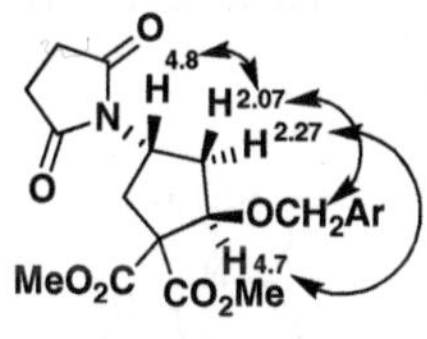

ROESY

Dimethyl-(*2R,4S*)-4-(2,5-dioxopyrrolidin-1-yl)-2-(4-nitrobenzyl)oxy) cyclopentane-1,1-dicarboxylate (309 h)

Following GP20, using 1-nitro-4-(vinyloxy)methyl)benzene (**305d**) (71.7 mg, 0.400 mmol, 2.00 eq), dimethyl 4-(2,5-dioxopyrrolidin-1-yl)-2-(4-nitrobenzyl)oxy) cyclopentane-1,1-dicarboxylate (**309 h**) (71.0 mg, 0.163 mmol, 82 %) was obtained as a colorless solid.

Crude analysis: *dr* = 5:1 between peaks at 4.44 (*minor*) and 4.95 (*major*).

er_{major} = 98:2, Chiralcel IF Hexane/^{i}PrOH 70:30, 1 mL/min, λ = 230 nm, tr1 = 49.9 min. tr2 = 67.0 min.

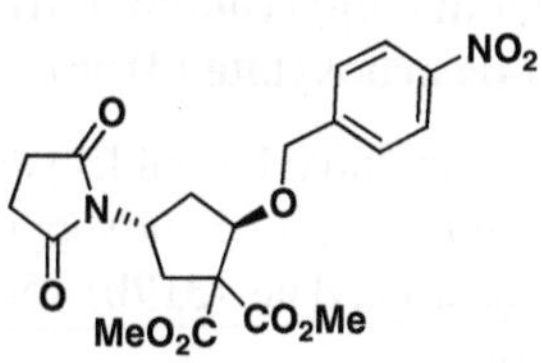

$[\alpha]_D^{25.0}$ − 27.1 (c = 0.43, $CHCl_3$). **R_f** 0.39 (5:5 Hexane/AcOEt). **Mp** 67.9–70.5 ° C. **^{1}H NMR** (400 MHz, $CDCl_3$) δ 8.23–8.11 (m, 2 H, Ar), 7.46–7.39 (m, 2 H, Ar),

4.95 (dtd, J = 10.5, 8.6, 6.5 Hz, 1 H, N–CH), 4.86–4.80 (m, 1 H, CHO), 4.72 (d, J = 13.2 Hz, 1 H, CH_2 benzyl), 4.62 (d, J = 13.2 Hz, 1 H, CH_2 benzyl), 3.80 (s, 3 H, Me), 3.70 (s, 3 H, Me), 3.00 (dd, J = 14.4, 10.5 Hz, 1 H, CH_2), 2.66 (s, 4 H, CH_2 succinimide), 2.45–2.34 (m, 2 H, CH_2), 2.19 (ddd, J = 13.6, 8.7, 3.6 Hz, 1 H, CH_2). **^{13}C NMR** (101 MHz, $CDCl_3$) δ 176.8, 170.8, 168.9, 147.3, 145.5, 127.3, 123.5, 83.4, 70.5, 64.9, 52.9, 52.7, 47.6, 34.0, 33.5, 27.9. **IR** 1737 (m), 1703 (s), 1523 (w), 1348 (m), 1284 (w), 1177 (w). **HRMS (ESI)** calcd for $C_{20}H_{22}N_2NaO_9^+$ $[M + Na]^+$ 457.1218; found 457.1232.

Dimethyl-(*2R,5R*)-5-(2,5-dioxopyrrolidin-1-yl)-2-phenyldihydrofuran-3,3(2H)-dicarboxylate (319a)

Following GP20, using benzaldehyde (**215a**) (42.4 mg, 0.400 mmol, 2.00 eq), dimethyl 5-(2,5-dioxopyrrolidin-1-yl)-2-phenyldihydrofuran-3,3(2 H)-dicarboxylate (**319a**) (59.2 mg, 0.164 mmol, 82 %) was obtained as a colorless oil.

Crude analysis: *dr* = 13:1 between peaks at 5.35 (*minor*) and 5.78 (*major*).

er_{major} = 92:8, Chiralcel IA Hexane/^{i}PrOH 70:30, 1 mL/min, λ = 210 nm, tr1 = 18.9 min. tr2 = 24.3 min.

er_{minor} = 92:8, Chiralcel IA Hexane/^{i}PrOH 70:30, 1 mL/min, λ = 210 nm, tr1 = 11.1 min. tr2 = 13.0 min.

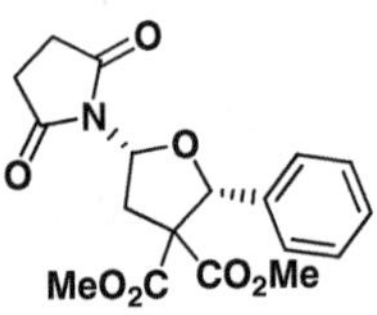

$[\alpha]_D^{25.0}$ 50.9 (c = 0.49, $CHCl_3$). **R_f** 0.4 (4:6 Pentane/AcOEt). **^{1}H NMR** (400 MHz, $CDCl_3$) δ 7.51 (d, J = 7.5 Hz, 2 H, Ar), 7.35–7.23 (m, 3 H, Ar), 5.78 (m, 2 H, N–C–H + Ph–C–H), 4.24–4.05 (m, 1 H, CH_2), 3.83 (s, 3 H, OMe), 3.10 (d, J = 1.7 Hz, 3 H, OMe), 2.75 (d, J = 1.7 Hz, 4 H, CH_2 succinimide), 2.45–2.28 (m, 1 H, CH_2). **^{13}C NMR** (101 MHz, $CDCl_3$) δ 176.2, 170.9, 167.8, 137.5, 128.5, 128.0, 127.6, 82.5, 79.6, 65.0, 53.4, 52.3, 33.1, 28.0. **IR** 1737 (s), 1715 (s), 1436 (w), 1384 (w), 1275 (m), 1233 (w), 1169 (w). **HRMS (ESI)** calcd for $C_{18}H_{20}NO_7^+$ $[M + H]^+$ 362.1234; found 362.1235.

Dimethyl-(*2R,5R*)-5-(2,5-dioxopyrrolidin-1-yl)-2-(4-methoxyphenyl) dihydrofuran-3,3(2 H)-dicarboxylate (319b)

Following GP20, using 4-methoxybenzaldehyde (**215b**) (54.5 mg, 0.400 mmol, 2.00 eq), dimethyl 5-(2,5-dioxopyrrolidin-1-yl)-2-(4-methoxyphenyl) dihydrofuran-3,3(2 H)-dicarboxylate (**319b**) (54.1 mg, 0.138 mmol, 69 %) was obtained as a colorless oil.

Crude analysis: *dr* > 20:1.

er = 96:4, Chiralcel IC Hexane/^{i}PrOH 80:20, 1 mL/min, λ = 220 nm, tr1 = 34.9 min. tr2 = 38.5 min.

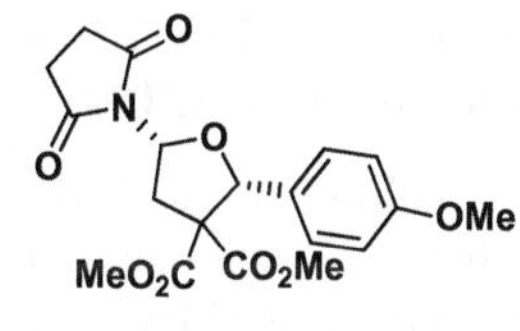

$[\alpha]_D^{25.0}$ 17.8 (c = 0.50, $CHCl_3$). **R_f** 0.30 (4:6 Pentane/AcOEt). **^{1}H NMR** (400 MHz, $CDCl_3$) δ 7.51–7.38 (m, 2 H, Ar), 6.94–6.70 (m, 2 H, Ar), 5.87–5.66 (m, 2 H, N–C–H + Ph–C–H), 4.14 (dd, J = 13.1, 11.0 Hz, 1 H, CH_2), 3.83 (s, 3 H, OMe), 3.78 (s, 3 H, OMe), 3.17 (s, 3 H, OMe), 2.75 (s, 4 H, CH_2 succinimide), 2.35 (dd, J = 13.1, 5.1 Hz, 1 H, CH_2). **^{13}C NMR** (101 MHz, $CDCl_3$) δ 176.2, 170.9, 167.8, 159.7, 129.7, 128.9, 113.3, 82.3, 79.5, 64.9, 55.2, 53.4, 52.4, 33.0, 28.0. **IR** 1732 (s), 1708 (s), 1614 (w), 1516 (w), 1365 (m), 1274 (m), 1250 (s), 1174 (s). **HRMS (ESI)** calcd for $C_{19}H_{22}NO_8^+$ $[M + H]^+$ 392.1340; found 392.1346.

Dimethyl-(*2R,5R*)-5-(2,5-dioxopyrrolidin-1-yl)-2-(3-methoxyphenyl) dihydrofuran-3,3(2 H)-dicarboxylate (319c)

Following GP20, using 3-methoxybenzaldehyde (**215c**) (54.5 mg, 0.400 mmol, 2.00 eq), dimethyl 5-(2,5-dioxopyrrolidin-1-yl)-2-(3-methoxyphenyl) dihydrofuran-3,3(2 H)-dicarboxylate (**319c**) (65.9 mg, 0.168 mmol, 84 %) was obtained as a colorless oil.

Crude analysis: *dr* = 10:1 between peaks at 6.05 (*minor*) and 5.82 (*major*).

er = 93:7, Chiralcel IA Hexane/^{i}PrOH 70:30, 1 mL/min, λ = 210 nm, tr1 = 23.8 min. tr2 = 27.2 min.

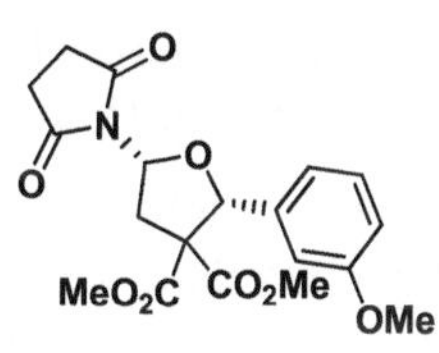

$[\alpha]_D^{25.0}$ 57.8 (c = 0.47, $CHCl_3$). **R_f** 0.4 (4:6 Pentane/AcOEt). **^{1}H NMR** (400 MHz, $CDCl_3$) δ 7.24–7.15 (m, 2 H, Ar), 7.02 (d, J = 7.6 Hz, 1 H, Ar), 6.88–6.73 (m, 1 H, Ar), 5.87–5.68 (m, 2 H, N–C–H + Ph–C–H), 4.15 (dd, J = 13.2, 11.1 Hz, 1 H, CH_2), 3.87–3.85 (m, 6 H, OMe + OMe), 3.16 (s, 3 H, OMe), 2.75 (s, 4 H, CH_2 succinimide), 2.36 (dd, J = 13.1, 5.0 Hz, 1 H, CH_2). **^{13}C NMR** (101 MHz, $CDCl_3$) δ 176.2, 170.8, 167.7, 159.5, 139.2, 128.8, 120.1, 115.2, 112.1, 82.5, 79.6, 65.1, 55.4, 53.5, 52.4, 33.0, 28.0. **IR** 1736 (s), 1715 (s), 1382 (w), 1276 (m), 1233 (w), 1169 (m), 1048 (w). **HRMS (ESI)** calcd for $C_{19}H_{21}NNaO_8^+$ $[M + Na]^+$ 414.1159; found 414.1181.

Dimethyl-(*2R,5R*)-2-(4-chlorophenyl)-5-(2,5-dioxopyrrolidin-1-yl) dihydrofuran-3,3(2 H)-dicarboxylate (319d)

Following GP20, using 4-chlorobenzaldehyde (**215e**) (56.2 mg, 0.400 mmol, 2.00 eq), dimethyl 2-(4-chlorophenyl)-5-(2,5-dioxopyrrolidin-1-yl) dihydrofuran-3,3(2 H)-dicarboxylate (**319d**) (71.0 mg, 0.179 mmol, 90 %) was obtained as a colorless oil.

Crude analysis: *dr* = 14:1 between peaks at 6.36 (*minor*) and 5.80 (*major*).

er = 91:9, Chiralcel IA Hexane/^{i}PrOH 70:30, 1 mL/min, λ = 210 nm, tr1 = 19.5 min. tr2 = 47.2 min.

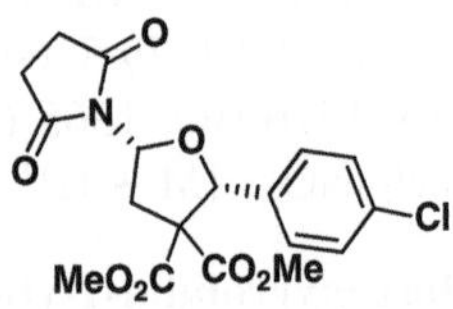

$[\alpha]_D^{25.0}$ 41.8 (c = 0.53, $CHCl_3$). **R_f** 0.3 (4:6 Pentane/AcOEt). **^{1}H NMR** (400 MHz, $CDCl_3$) δ 7.47 (d, *J* = 8.2 Hz, 2 H, Ar), 7.34–7.27 (m, 2 H, Ar), 5.84–5.70 (m, 2 H, N–C–H + Ph–C–H), 4.10 (dd, *J* = 13.2, 10.9 Hz, 1 H, CH_2), 3.84 (d, *J* = 1.2 Hz, 3 H, OMe), 3.18 (d, *J* = 1.3 Hz, 3 H, OMe), 2.76 (s, 4 H, CH_2 succinimide), 2.38 (dd, *J* = 13.2, 5.2 Hz, 1 H, CH_2). **^{13}C NMR** (101 MHz, $CDCl_3$) δ 176.1, 170.7, 167.5, 136.1, 134.3, 129.0, 128.1, 81.8, 79.6, 64.9, 53.5, 52.5, 33.0, 28.0. **IR** 1737 (s), 1718 (s), 1659 (w), 1382 (w), 1278 (m), 1231 (m), 1212 (w), 1168 (m). **HRMS (ESI)** calcd for $C_{18}ClH_{19}NO_7^+$ $[M + H]^+$ 396.0845; found 396.0844.

Dimethyl-(2S,5R)-5-(2,5-dioxopyrrolidin-1-yl)-2-(thiophen-2-yl) dihydrofuran-3,3(2 H)-dicarboxylate (319e)

Following GP20, using thiophene-2-carbaldehyde (**215d**) (44.9 mg, 0.400 mmol, 2.00 eq), dimethyl 5-(2,5-dioxopyrrolidin-1-yl)-2-(thiophen-2-yl)dihydrofuran-3,3 (2 H)-dicarboxylate (**319e**) (71.6 mg, 0.195 mmol, 97 %) was obtained as a colorless oil.

Crude analysis: *dr* > 20:1.

er = 95:5, Chiralcel IA Hexane/^{i}PrOH 70:30, 1 mL/min, λ = 210 nm, tr1 = 24.4 min. tr2 = 31.9 min.

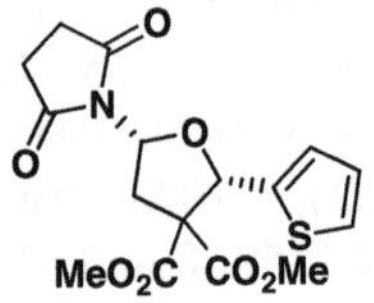

$[\alpha]_D^{25.0}$ 75.3 (c = 0.54, $CHCl_3$). **R_f** 0.4 (4:6 Pentane/AcOEt). **^{1}H NMR** (400 MHz, $CDCl_3$) δ 7.26 (d, *J* = 2.6 Hz, 2 H, Ar), 6.96 (dd, *J* = 5.0, 3.6 Hz, 1 H, Ar), 6.03 (s, 1 H, Ar–C–H), 5.74 (dd, *J* = 11.1, 4.9 Hz, 1 H, N–C–H), 4.22 (dd,

$J = 13.2, 11.1$ Hz, 1 H, CH_2), 3.85 (s, 3 H, OMe), 3.33 (s, 3 H, OMe), 2.74 (s, 4 H, CH_2 succinimide), 2.39 (dd, $J = 13.2, 4.9$ Hz, 1 H, CH_2). **^{13}C NMR** (101 MHz, $CDCl_3$) δ 176.2, 170.4, 167.4, 140.1, 127.0, 126.7, 125.9, 79.4, 78.4, 65.2, 53.6, 52.7, 32.2, 28.0. **IR** 1736 (s), 1716 (s), 1376 (w), 1279 (w), 1169 (w). **HRMS (ESI)** calcd for $C_{16}H_{18}NO_7S^+$ $[M + H]^+$ 368.0799; found 368.0819.

Dimethyl-(*2R,5R*)-5-(2,5-dioxopyrrolidin-1-yl)-2-(E)-styryl)dihydrofuran-3,3(2 H)-dicarboxylate (319f)

Following GP20, using cinnamaldehyde (**215g**) (52.9 mg, 0.400 mmol, 2.00 eq), dimethyl 5-(2,5-dioxopyrrolidin-1-yl)-2-(*E*)-styryl)dihydrofuran-3,3(2 H)-dicarboxylate (**319f**) (74.0 mg, 0.191 mmol, 96 %) was obtained as a colorless oil.

Crude analysis: *dr* = 14:1 between peaks at 5.58 (*minor*) and 5.83 (*major*).

er = 94:6, Chiralcel IA Hexane/^{i}PrOH 70:30, 1 mL/min, λ = 210 nm, tr1 = 30.2 min. tr2 = 56.1 min.

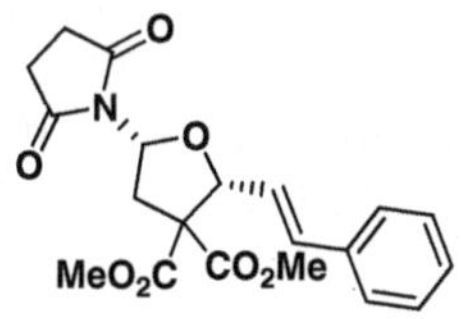

$[\alpha]_D^{25.0}$ 14.8 (c = 0.49, $CHCl_3$). **R_f** 0.4 (4:6 Pentane/AcOEt). **^{1}H NMR** (400 MHz, $CDCl_3$) δ 7.43–7.36 (m, 2 H, Ar), 7.30 (dd, $J = 8.4, 6.5$ Hz, 2 H, Ar), 7.24 (m, 3 H, Ar), 6.76–6.51 (m, 2 H, CH olefin), 5.83 (dd, $J = 10.2, 5.9$ Hz, 1 H, N–C–H), 5.25 (dd, $J = 5.3, 3.1$ Hz, 1 H, O–C–H), 3.95 (dd, $J = 13.3, 10.2$ Hz, 1 H, CH_2), 3.85 (s, 3 H, OMe), 3.61 (s, 3 H, OMe), 2.72 (s, 4 H, CH_2 succinimide), 2.49 (dd, $J = 13.3, 6.0$ Hz, 1 H, CH_2). **^{13}C NMR** (101 MHz, $CDCl_3$) δ 176.3, 170.0, 167.2, 136.2, 134.5, 128.6, 128.1, 126.9, 125.2, 83.0, 80.0, 64.8, 53.5, 53.0, 32.0, 28.0. **IR** 1739 (s), 1717 (s), 1435 (w), 1276 (m), 1231 (m), 1219 (w), 1169 (m). **HRMS (ESI)** calcd for $C_{20}H_{22}NO_7^+$ $[M + H]^+$ 388.1391; found 388.1388.

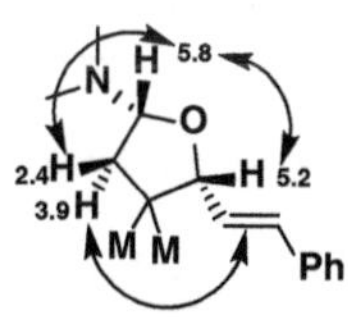

ROESY

Dimethyl-(*2R,5R*)-5-(2,5-dioxopyrrolidin-1-yl)-2-phenethyldihydrofuran-3,3(2 H)-dicarboxylate (319g)

Following GP20, using 3-phenylpropanal (**215j**) (53.7 mg, 0.400 mmol, 2.00 eq), dimethyl 5-(2,5-dioxopyrrolidin-1-yl)-2-phenethyldihydrofuran-3,3(2 H)-dicarboxylate (**319g**) (65.9 mg, 0.169 mmol, 85 %) was obtained as a colorless oil.

Crude analysis: *dr* = 13:1 between peaks at 6.15 (*minor*) and 5.79 (*major*).

er = 91.5:8.5, Chiralcel IA Hexane/^{i}PrOH 70:30, 1 mL/min, λ = 210 nm, tr1 = 10.3 min. tr2 = 18.5 min.

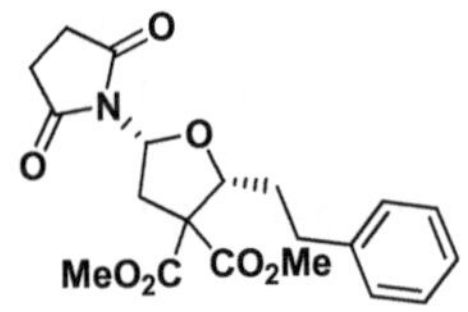

$[\alpha]_D^{25.0}$ 21.6 (c = 0.42, $CHCl_3$). **R_f** 0.4 (4:6 Pentane/AcOEt). **^{1}H NMR** (400 MHz, $CDCl_3$) δ 7.40–7.07 (m, 5 H, Ar), 5.79 (dd, *J* = 10.0, 6.1 Hz, 1 H, N–C–H), 4.65 (dd, *J* = 11.4, 2.7 Hz, 1 H, O–C–H), 3.83 (s, 3 H, OMe), 3.79–3.73 (m, 1 H, CH_2 THF), 3.76 (s, 3 H, OMe), 2.85 (ddd, *J* = 14.7, 10.5, 4.8 Hz, 1 H, CH_2), 2.76 (s, 4 H, CH_2 succinimide), 2.57 (ddd, *J* = 13.6, 10.1, 6.3 Hz, 1 H, CH_2), 2.48 (dd, *J* = 13.2, 6.1 Hz, 1 H, CH_2 THF), 2.37–2.24 (m, 1 H, CH_2), 1.79–1.53 (m, 1 H, CH_2). **^{13}C NMR** (101 MHz, $CDCl_3$) δ 176.2, 170.3, 167.9, 141.8, 128.6, 128.3, 125.8, 80.8, 79.5, 63.6, 53.4, 53.0, 33.4, 32.3, 32.2, 28.0. **IR** 1736 (s), 1712 (s), 1436 (w), 1371 (w), 1274 (m), 1168 (m), 1041 (w). **HRMS (ESI)** calcd for $C_{20}H_{23}NNaO_7^+$ $[M + Na]^+$ 412.1367; found 412.1349.

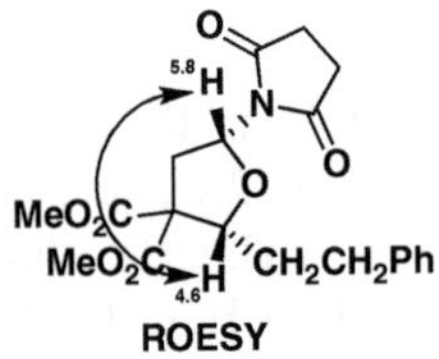

Dimethyl 4-(2,5-dioxopyrrolidin-1-yl)-2-hydroxycyclopentane-1,1-dicarboxylate (321)

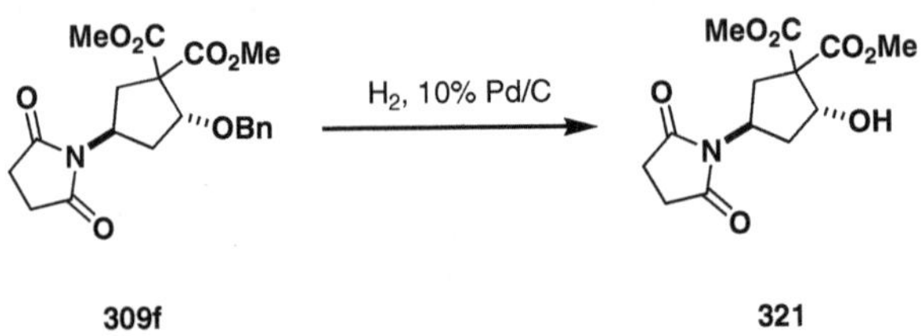

Compound **309f** (25 mg, 0.064 mmol, 1 eq) is added in a vial. Palladium on charcoal 10 % (11 mg, 0.00096 mmol, 0.15 eq) is added under N_2. The atmosphere is flushed with hydrogen and the reaction is stirred at room temperature for 2 h. Celite (*ca* 1 g) is added and the reaction is filtered and rinsed with 10 mL of dichloromethane and 10 mL of methanol. The compound was observed in the crude ^{1}H NMR and by HRMS but could not be purified due to observed decomposition.

HRMS (ESI) calcd for $C_{13}H_{17}NO_7$ $[M + H]^+$ 300.1089; found 300.1079.

5.7 Synthesis of Aminocyclobutanes

Synthesis of methylene malonates
Dimethyl 2-methylenemalonate (323a)

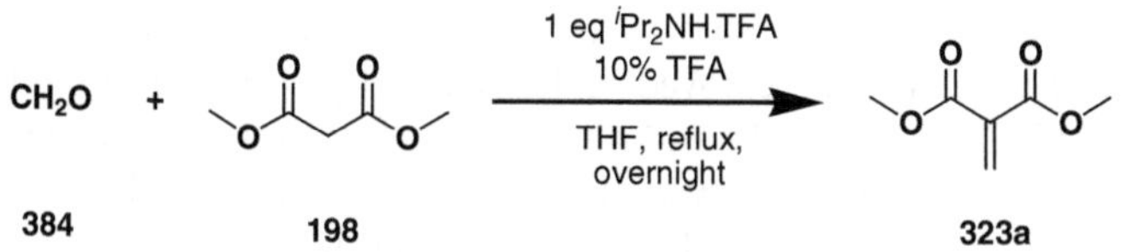

Following a modified procedure [38], dry THF (200 mL), dimethyl malonate (**198**) (18.3 g, 139 mmol, 1 eq), diisopropylamine 2,2,2-trifluoroacetate (29.9 g, 139 mmol, 1 eq), paraformaldehyde (**384**) (8.35 g, 278 mmol, 2 eq) and trifluoroacetic acid (1.07 mL, 13.9 mmol, 0.1 eq) were added to a 500 mL round bottom flask. A condenser was added and the suspension was stirred to reflux for 2 h. Paraformaldehyde (8.35 g, 278 mmol, 2 eq) was added and the reflux was restarted for 6 h. The reaction was cooled to room temperature and THF was removed under reduced pressure (300–50 mbar at 45 °C). The crude mixture was dissolved in diethyl ether (75 mL) and filtered through cotton in a separatory funnel. The organic layer was washed twice with 1 M HCl (50 mL). The aqueous layers were combined and extracted with diethyl ether (25 mL). The organic layers were combined, dried with anhydrous Na_2SO_4, filtered and concentrated under reduced pressure to give a colorless oil (30 g). The crude oil was purified by distillation (all the glassware needed for distillation had been washed with 2 M HCl, rinsed with methanol and dried in the oven at 110 °C prior to use). Dimethyl 2-methylenemalonate (**323a**) (13.6 g, 94.0 mmol, 68 % yield) was collected as a colorless oil between 45 °C at 1.5 mbar and 50 °C at 1 mbar. The product was stored under nitrogen in a freezer and can be kept 2–3 weeks without major degradation. In case of degradation, a short path distillation was enough to obtain pure material.

^{1}H NMR (400 MHz, $CDCl_3$) δ 6.45 (s, 2 H), 4.57 (s, 6 H). **^{13}C NMR** (101 MHz, $CDCl_3$) δ 164.4, 135.4, 134.1, 52.6. **IR** 1792 (w), 1736 (s), 1440 (m), 1340 (m), 1244 (s), 1128 (s). **HRMS (ESI)** calcd for $C_6H_9O_4^+$ $[M + H]^+$ 145.0495; found 145.0502.

Ethyl 2-benzoylacrylate (323c)

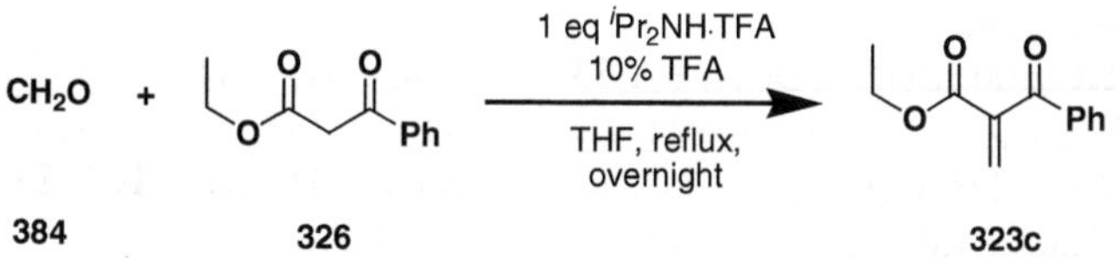

Following a modified procedure [38], dry THF (20 mL), ethyl 3-oxo-3-phenylpropanoate (**326**) (2.50 g, 13.0 mmol, 1 eq), diisopropylamine 2,2,2-trifluoroacetate (2.80 g, 13.0 mmol, 1 eq), paraformaldehyde (**384**) (780 mg, 27.0 mmol, 2 eq) and trifluoroacetic acid (0.100 mL, 1.30 mmol, 0.1 eq) were added to a 50 mL round bottom flask. A condenser was added and the suspension was stirred to reflux for 2 h. Paraformaldehyde (780 mg, 27.0 mmol, 2 eq) was added and the reflux was restarted for 6 h. The reaction was cooled to room temperature and THF was removed under reduced pressure. The crude was dissolved in diethyl ether (20 mL) and filtered through cotton in a separatory funnel. The organic layer was washed twice with 1 M HCl (20 mL). The aqueous layers were combined and extracted with diethyl ether (20 mL). The organic layers were combined, dried over anhydrous Na_2SO_4, filtered and concentrated under reduced pressure to give a yellow oil. The crude oil was purified by column chromatography on Biotage (SNAP cartridge KP-SIL 50 g, 95:5–4:6 Hexane/Ethyl acetate) affording ethyl 2-benzoylacrylate (**323c**) (1.75 g, 8.57 mmol, 66 % yield) as a pale yellow oil.

^{1}H NMR (400 MHz, $CDCl_3$) δ 7.86 (dd, J = 8.4, 1.4 Hz, 2 H, ArH), 7.65–7.56 (m, 1 H, ArH), 7.51–7.42 (m, 2 H, ArH), 6.70 (d, J = 0.8 Hz, 1 H, C=CH_2), 6.07 (d, J = 0.8 Hz, 1 H, C=CH_2), 4.23 (q, J = 7.1 Hz, 2 H, CH_2), 1.20 (t, J = 7.2 Hz, 3 H, CH_3). **HRMS (ESI)** calcd for $C_{12}H_{13}O_3^+$ $[M + H]^+$ 205.0859; found 205.0867.

Data match literature report [39].

Dimethyl 2-(3-phenylpropylidene)malonate (323d)

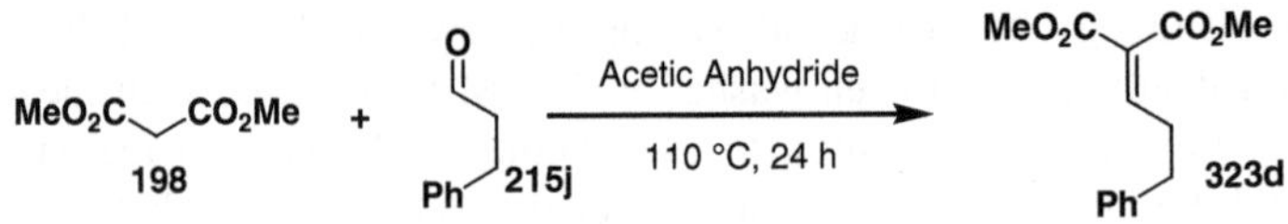

Following a modified reported procedure [40], dimethyl malonate (**198**) (661 mg, 5.00 mmol, 1 eq), acetic anhydride (708 μL, 7.50 mmol, 1.5 eq) and 3-phenylpropanal (**215j**) (1.33 mL, 10.0 mmol, 2 eq) were added in a microwave vial. The vial was sealed and the reaction was stirred 24 h at 110 °C. Evaporation of the acetic anhydride on the rotary evaporator and high vacuum afforded an oil that was purified by column chromatography on Biotage (SNAP cartridge KP-SIL 25 g, 98:2 Hexane/Ethyl acetate to 95:5 Hexane/Ethyl acetate). Dimethyl 2-(3-phenylpropylidene)malonate (**323d**) (690 mg, 2.78 mmol, 56 % yield) was obtained as a yellow oil.

^{1}H NMR (400 MHz, $CDCl_3$) δ 7.33–7.27 (m, 2 H, ArH), 7.24–7.16 (m, 3 H, ArH), 7.07 (dd, J = 7.7, 7.7 Hz, 1 H, C=CH), 3.81–3.75 (m, 6 H, OCH_3), 2.80 (dd, J = 8.7, 6.7 Hz, 2 H, CH_2), 2.63 (dd, J = 7.6, 7.6 Hz, 2 H, CH_2). **HRMS (ESI)** calcd for $C_{14}H_{16}NaO_4^+$ $[M + Na]^+$ 271.0941; found 271.0933.

Data of **323d** match literature report [41].

Dibenzyl 2-ethylidenemalonate (323e)

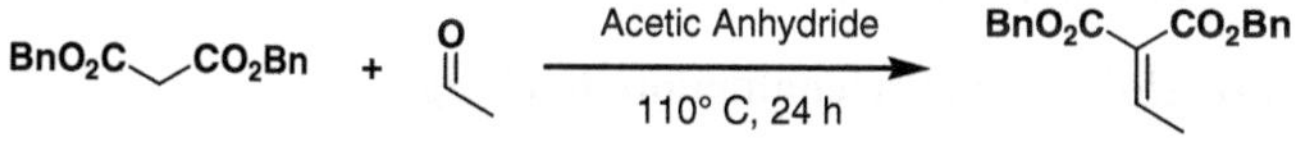

250 385 323e

Following a modified reported procedure [40], dibenzyl malonate (**250**) (2.00 g, 7.03 mmol, 1 eq), acetic anhydride (1.08 g, 10.6 mmol, 1.5 eq) and acetaldehyde (**385**) (1.55 g, 35.2 mmol, 5 eq) were added in a microwave vial. The vial was sealed and the reaction was stirred 24 h at 85 °C. Evaporation of the acetic anhydride on the rotary evaporator and high vacuum afforded an oil that was purified by column chromatography on Biotage (SNAP cartridge KP-SIL 50 g, 95:2–7:3 Hexane/Ethyl acetate). Dibenzyl 2-ethylidenemalonate (**323e**) (1.10 g, 3.54 mmol, 50 % yield + some traces of malonate left) was obtained as a colorless oil.

R_f 0.6 (7:3 *Hexane/Ethyl acetate*). **^{1}H NMR** (400 MHz, $CDCl_3$) δ 7.39–7.30 (m, 10 H, ArH), 7.18 (q, J = 7.3 Hz, 1 H, =CH), 5.28 (s, 2, H, CH_2Ar), 5.22 (s, 2 H, CH_2Ar), 1.96 (d, J = 7.3 Hz, 3 H, Me). **^{13}C NMR** (101 MHz, $CDCl_3$) δ 165.0, 163.6, 146.3, 135.5, 135.3, 129.0, 128.5, 128.5, 128.4, 128.3, 128.2, 128.0, 67.0, 66.8, 15.6. **IR** 3066 (w), 3035 (w), 2954 (w), 1731 (s), 1499 (w), 1262 (s), 1216 (s), 1138 (m), 1050 (m), 1003 (w), 744 (m). **HRMS (ESI)** calcd for $C_{19}H_{19}O_4^+$ $[M + H]^+$ 311.1278; found 311.1281.

Dibenzyl 2-(2,2,2-trifluoroethylidene)malonate (323f)

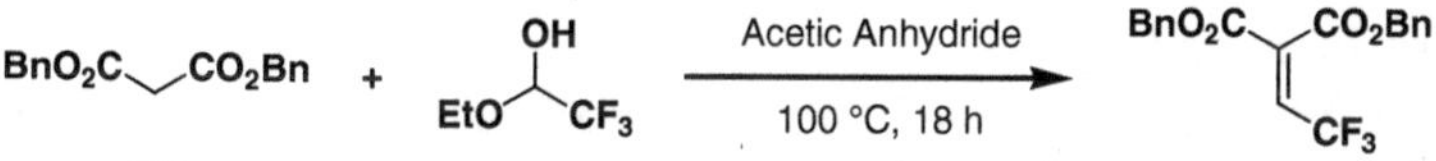

250 386 323f

Dibenzyl malonate (**250**) (300 mg, 1.06 mmol, 1 eq), acetic anhydride (500 µL, 5.28 mmol, 5 eq) and 1-ethoxy-2,2,2-trifluoroethan-1-ol (**386**) (368 µL, 3.17 mmol, 3 eq) were added in a microwave vial. The vial was sealed and the reaction was stirred 18 h at 100 °C. After going back to room temperature the reaction was transferred into a separatory funnel and sat.$NaHCO_3$ (15 mL) and diethyl ether (20 mL) were added. The layers were separated and the organic layer was dried over anhydrous Na_2SO_4. The solvents were evaporated under reduced pressure and the obtained oil was purified by column chromatography on Biotage (SNAP cartridge KP-SIL 25 g, 98:2–95:5 Hexane/Ethyl acetate). Dibenzyl 2-(2,2,2-trifluoroethylidene)malonate (**323f**) (162 mg, 0.401 mmol, 40 % yield, 90 % pure by NMR) was obtained as a colorless oil.

^{1}H NMR (400 MHz, $CDCl_3$) δ 7.34–7.20 (m, 10 H, ArH), 6.73 (q, $J = 7.5$ Hz, 1 H, C=CH), 5.21 (s, 2 H, CH_2), 5.18 (s, 2 H, CH_2). **^{19}F NMR** (376 MHz, $CDCl_3$) δ −62.3.

Data match literature report [42].

Dimethyl 2-(cyclohexylmethylene)malonate (323g)

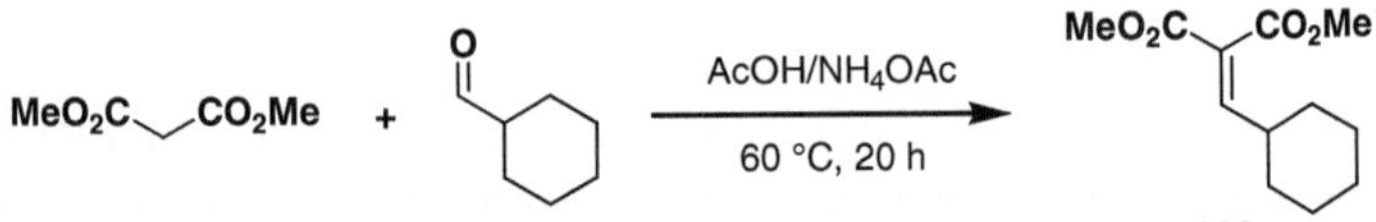

Dimethyl malonate (**198**) (2.14 g, 16.2 mmol, 1 eq), AcOH (30 mL), cyclohexanecarbaldehyde (**215l**) (2.00 g, 17.8 mmol, 1 eq) and ammonium acetate (1.37 g, 17.8 mmol, 1.1 eq) were added to a 50 mL round bottom flask. A condenser was added and the suspension was stirred at 60 °C for 20 h. The reaction was poured in brine (25 mL) and extracted with ethyl acetate (50 mL). The organic layer was washed three times with brine (25 mL), dried over anhydrous Na_2SO_4, filtered and concentrated under reduced pressure. The crude product was purified by column chromatography (SiO_2, Hexane/Ethyl Acetate 95:5) affording dimethyl 2-(cyclohexylmethylene)malonate (**323g**) (2.80 g, 12.4 mmol, 76 % yield) as a yellow oil.

^{1}H NMR (400 MHz, $CDCl_3$) δ 6.85 (d, $J = 10.4$ Hz, 1 H, C=CH), 3.83 (s, 3 H, OCH_3), 3.77 (s, 3 H, OCH_3), 2.37 (dtt, $J = 14.1, 6.6, 3.5$ Hz, 1 H, CH), 1.79–1.61 (m, 5 H, CH_2), 1.36–1.08 (m, 5 H, CH_2). **HRMS (ESI)** calcd for $C_{12}H_{19}O_4^+$ $[M + H]^+$ 227.1278; found 227.1280.

Data of **323g** match literature report [43].

Synthesis of enimides

***tert*-Butyl 5-methyl-2,6-dioxo-3-vinyl-2,3-dihydropyrimidine-1(*6 H*)-carboxylate (327)**

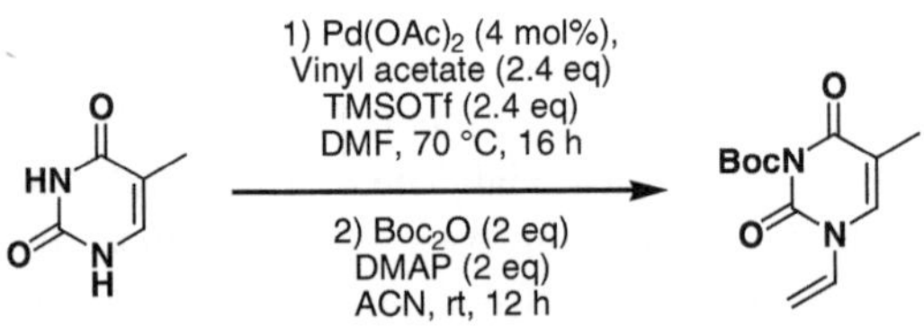

In a sealed vial were added palladium acetate (36.0 mg, 0.160 mmol, 0.04 eq), vinyl acetate (881 μl, 9.52 mmol, 2.4 eq), 5-methylpyrimidine-2,4(*1H,3 H*)-dione (**326**) (500 mg, 3.96 mmol, 1 eq), TMSOTf (1.72 mL, 9.52 mmol, 2.4 eq) and DMF

(10.0 mL). The reaction was stirred for 16 h at 70 °C, then water (25 mL) was added. The reaction was extracted three times with ethyl acetate (30 mL). Then, the organic layers were combined and washed three times with water (30 mL). The organic layer was dried over magnesium sulfate and concentrated under reduced pressure. The crude product was purified by column chromatography, eluting with hexane/ethyl acetate/NEt_3 (7/3/0.01) to obtain 5-methyl-1-vinylpyrimidine-2,4 (*1H,3 H*)-dione (**327**) (269 mg, 1.77 mmol, 45 % yield) as a colorless solid.

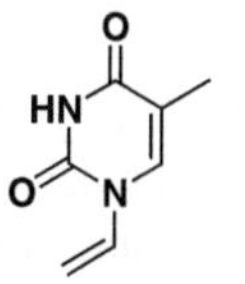

R*f* 0.5 (1:1 Hexane/Ethyl acetate). **Mp** 208.0–209.1 °C. **^{1}H NMR** (400 MHz, $CDCl_3$) δ 9.17 (s, 1 H, NH), 7.34 (s, 1 H, C=C–H), 7.21 (dd, 1 H, *J* = 16.0, 9.1 Hz, –CH=C vinyl), 5.07 (dd, 1 H, *J* = 16.0, 2.1 Hz, $C{=}CH_2$ vinyl, *trans*), 4.91 (dd, 1 H, *J* = 9.1, 2.1 Hz, $C{=}CH_2$ vinyl, *cis*), 1.99 (s, 3 H, Me). **^{13}C NMR** (101 MHz, $CDCl_3$) δ 163.6, 149.3, 134.5, 129.6, 112.1, 100.5, 12.6. **IR** 3180 (w), 3062 (w), 2827 (w), 1645 (s). **HRMS (ESI)** calcd for $C_7H_9N_2O_2^+$ $[M + H]^+$ 153.0659; found 153.0653.

In an oven dried flask, 5-methyl-1-vinylpyrimidine-2,4(*1H,3 H*)-dione (920 mg, 6.05 mmol, 1 eq), di-*tert*-butyl dicarbonate (2.64 g, 12.1 mmol, 2 eq) and dimethylaminopyridine (1.48 g, 12.1 mmol, 2 eq) were stirred in acetonitrile (25.0 mL) for 12 h. Silica was added to the reaction and the solvent was evaporated. The dry residue was loaded on a silica chromatography column and eluted with hexane/ethyl acetate/1 % NEt_3 (95:5–80:20) to provide *tert*-butyl 5-methyl-2,6-dioxo-3-vinyl-2,3-dihydropyrimidine-1(*6 H*)-carboxylate (**327**) (1.15 g, 4.56 mmol, 75 % yield) as a colorless solid.

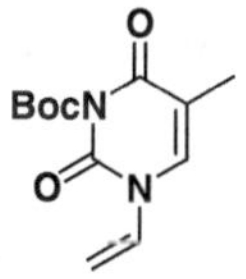

R*f* 0.2 (9:1 Hexane/Ethyl acetate). **Mp** 109.9–111.2 °C. **^{1}H NMR** (400 MHz, $CDCl_3$) δ 7.31 (s, 1 H, C=C–H), 7.15 (dd, 1 H, *J* = 16.0, 9.1 Hz, –CH=C vinyl), 5.09 (dd, 1 H, *J* = 16.0, 2.2 Hz, $C{=}CH_2$ vinyl, *trans*), 4.94 (dd, 1 H, *J* = 9.1, 2.2 Hz, $C{=}CH_2$ vinyl, *cis*), 1.99 (s, 3 H), 1.60 (s, 9 H, Me). **^{13}C NMR** (101 MHz, $CDCl_3$) δ 161.0, 147.6, 147.5, 134.0, 129.6, 111.8, 101.3, 87.1, 27.5, 12.7. **IR** 2982 (w), 2937 (w), 1778 (s), 1721 (s), 1672 (s). **HRMS (ESI)** calcd for $C_{12}H_{16}N_2O_4Na^+$ $[M + Na]^+$ 275.1002; found 275.1008.

(*E*)-2-(Prop-1-en-1-yl)isoindoline-1,3-dione (331)

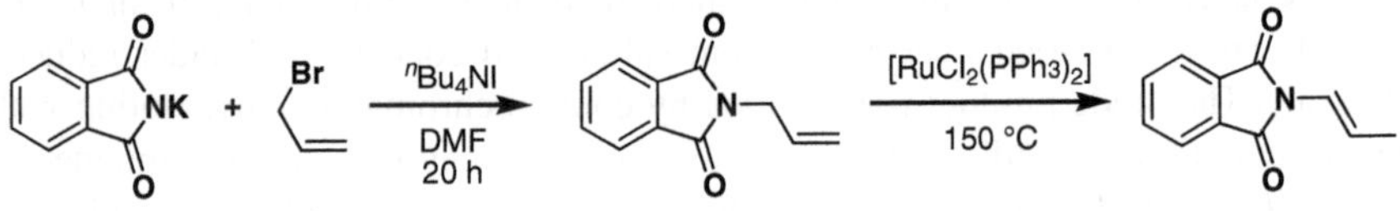

328 329 330 331

Following a modified procedure [44], allyl bromide (**329**) (2.6 mL, 30 mmol, 1.1 eq) was added dropwise at room temperature to a suspension of potassium phthalimide (**328**) (5.0 g, 27 mmol, 1 eq) and Bu_4NI (0.50 g, 1.4 mmol, 0.05 eq) in DMF (10 mL). The mixture was stirred for 20 h at room temperature, and then H_2O (20 mL) was added. The precipitate was isolated by filtration, dried, and recrystallized from isopropanol to give 2-allylisoindoline-1,3-dione (**330**) (3.4 g, 18 mmol, 68 % yield).

2-Allylisoindoline-1,3-dione (**330**) (2.0 g, 11 mmol, 1 eq) was added in a sealed tube under nitrogen atmosphere to $[RuCl_2(PPh_3)_2]$ (0.10 g, 0.11 mmol, 0.01 eq). The solids were heated at 150 °C during 12 h and the reaction was cooled down to room temperature. The black mixture was dissolved in toluene and filtered on a Celite pad. The solvents were evaporated and the brown orange solid was recrystallized in ethanol (20 mL) of (*E*)-2-(prop-1-en-1-yl)isoindoline-1,3-dione (**331**) (1.15 g, 6.10 mmol, 58 % yield) as a yellow solid were collected from the first recrystallization.

^{1}H NMR (400 MHz, $CDCl_3$) δ 7.86 (dd, 2 H, J = 5.2, 3.1 Hz, Phth), 7.72 (dd, 2 H, J = 5.2, 3.0 Hz, Phth), 6.64–6.54 (m, 2 H, CH=CH), 1.85 (d, 3 H, J = 5.1 Hz, CH_3). **HRMS (ESI)** calcd for $C_{11}H_{10}NO_2^+$ $[M + H]^+$ 188.0706; found 188.0713.

^{1}H NMR data of **331** match literature report [44].

(*E*)-2-(Oct-1-en-1-yl)isoindoline-1,3-dione (334)

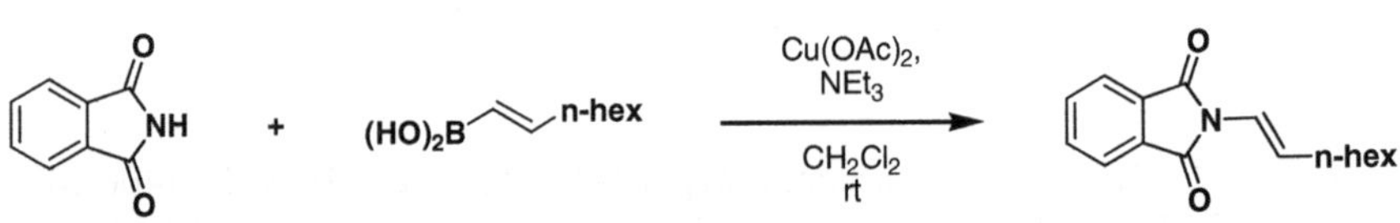

332 333 334

Following a modified procedure [45], phthalimide (**332**) (200 mg, 1.36 mmol, 1 eq), copper(II)acetate monohydrate (300 mg, 1.50 mmol, 1.1 eq), dichloromethane (2 mL) and triethylamine (379 μL, 2.72 mmol, 2 eq) were added under air in a flask. Subsequently, (*E*)-oct-1-en-1-ylboronic acid (**333**) (212 mg, 1.36 mmol, 1 eq) was added as a solid. The reaction was stirred for 12 h. Volatiles were removed under reduced pressure and the crude was directly purified by Biotage

(SNAP cartridge KP-SIL 25 g, 95:5–4:6 Hexane/Ethyl acetate) to give (*E*)-2-(oct-1-en-1-yl)isoindoline-1,3-dione (**334**) (127 mg, 0.490 mmol, 36 % yield) as a pale yellow oil.

R_f 0.6 (7:3 Hexane/Ethyl acetate). **^{1}H NMR** (400 MHz, $CDCl_3$) δ 7.89–7.81 (m, 2 H, Phth), 7.77–7.69 (m, 2 H, Phth), 6.61–6.57 (m, 2 H, CH=CH), 2.21–2.13 (m, 2 H, CH_2), 1.51–1.41 (m, 2 H, CH_2), 1.39–1.24 (m, 6 H, CH_2), 0.89 (t, 3 H, J = 7.0 Hz, CH_3). **^{13}C NMR** (101 MHz, $CDCl_3$) δ 166.8, 134.3, 131.8, 123.5, 123.1, 117.5, 31.7, 31.2, 29.4, 28.9, 22.7, 14.1. **IR** 2956 (w), 2929 (w), 2858 (w), 1779 (w), 1719 (s), 1387 (s). **HRMS (ESI)** calcd for $C_{16}H_{20}NO_2^+$ $[M + H]^+$ 258.1489; found 258.1499.

GP22 for the synthesis of (Z)-enimides

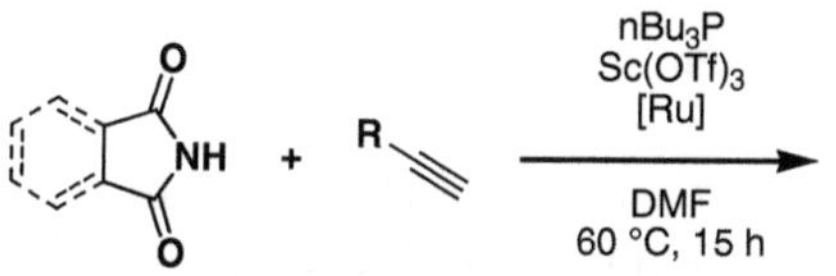

Following a modified procedure [46], in the glovebox, bis-(2-methylallyl) cycloocta-1,5-diene ruthenium (12.8 mg, 0.0400 mmol, 0.02 eq), scandium triflate (39.4 mg, 0.0800 mmol, 0.04 eq) and imide (2.00 mmol, 1 eq) were added in a microwave vial. The vial was sealed with a Teflon septum and subsequently, tri-*n*-butylphosphine (30.0 μL, 0.120 mmol, 0.06 eq), alkyne (4.00 mmol, 2 eq) and freshly distilled DMF (6 mL) were added. The solution was stirred at 60 °C for 15 h and then poured into an aqueous sodium bicarbonate solution (20 mL). The mixture was extracted twice with ethyl acetate (20 mL). The organic fractions were collected, washed with brine, dried over magnesium sulfate, filtered and concentrated under reduced pressure. The crude product was purified by Biotage (SNAP cartridge KP-SIL 25 g, 95:5–4:6 Hexane/Ethyl acetate).

(Z)-2-Styrylisoindoline-1,3-dione (335)

Followong GP22, phthalimide (294 mg, 2.00 mmol, 1 eq) and 1-hexyne (439 μl, 4.00 mmol, 2 eq) were combined to afford (Z)-2-styrylisoindoline-1,3-dione (**335**) (65 mg, 0.26 mmol, 13 % yield) as an orange solid.

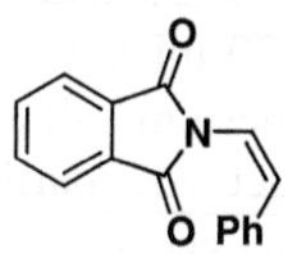

^{1}H NMR (400 MHz, $CDCl_3$) δ 7.91–7.85 (m, 2 H, Phth), 7.79–7.72 (m, 2 H, Phth), 7.28–7.22 (m, 5 H, ArH), 6.71 (d, 1 H, J = 9.1 Hz, CH=CH), 6.34 (d, 1 H,

J = 9.1 Hz, CH=CH). **HRMS (ESI)** calcd for $C_{16}H_{12}NO_2^+$ $[M + H]^+$ 250.0863; found 250.0851.

^{1}H NMR Data of **335** match literature report [46].

(*Z*)-1-(4-Phenylbut-1-en-1-yl)pyrrolidine-2,5-dione (336)

Followong GP22, succinimide (198 mg, 2.00 mmol, 1 eq) and but-3-yn-1-ylbenzene (562 μL, 4.00 mmol, 2 eq) were combined to afford (Z)-1-(4-phenylbut-1-en-1-yl)pyrrolidine-2,5-dione (**336**) (400 mg, 1.75 mmol, 87 % yield) as a brown light solid.

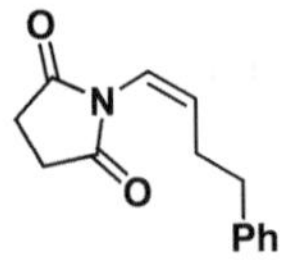

^{1}H NMR (400 MHz, $CDCl_3$) δ 7.31–7.27 (m, 2 H, ArH), 7.22–7.16 (m, 3 H, ArH), 5.92 (d, 1 H, J = 8.6 Hz, CH=CH), 5.76 (m, 1 H, CH=CH), 2.80–2.67 (m, 6 H, CH_2), 2.28 (m, 2 H, CH_2). **HRMS (ESI)** calcd for $C_{14}H_{16}NO_2^+$ $[M + H]^+$ 230.1176; found 230.1175.

^{1}H NMR data match literature report [46].

GP23 for the synthesis of (*E*)-enimides.

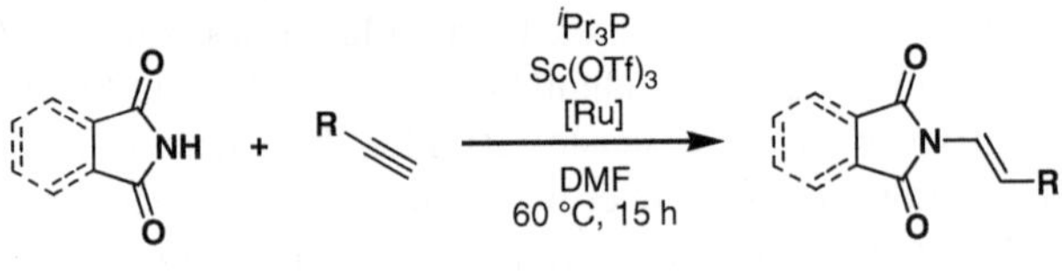

Following a modified procedure [46], bis-(2-methylallyl)cycloocta-1,5-diene ruthenium (31.9 mg, 0.100 mmol, 0.05 eq), scandium triflate (39.4 mg, 0.0800 mmol, 0.04 eq) and imide (2.00 mmol, 1 eq) were added to a microwave vial in the glovebox. The vial was sealed with a Teflon septum and triisopropyl-phosphine (57.0 μL, 0.300 mmol, 0.15 eq), alkyne (4.00 mmol, 2 eq) and freshly distilled DMF (6 mL) were added. The solution was stirred at 60 °C for 15 h and then poured into an aqueous sodium bicarbonate solution (20 mL). The mixture was extracted twice with ethyl acetate (20 mL). The organic fractions were collected, washed with brine, dried over magnesium sulfate, filtered and concentrated under reduced pressure. The crude product was purified by Biotage (SNAP cartridge KP-SIL 25 g, 95:5–4:6 Hexane/Ethyl acetate).

(*E*)-2-(2-Cyclopropylvinyl)isoindoline-1,3-dione (337)

Followong GP23, phthalimide (294 mg, 2.00 mmol, 1 eq) and ethynylcyclopropane (339 μL, 4.00 mmol, 2 eq) were combined to afford (*E*)-2-

(2-cyclopropylvinyl)isoindoline-1,3-dione (**337**) (159 mg, 0.750 mmol, 37 % yield) as a yellow solid.

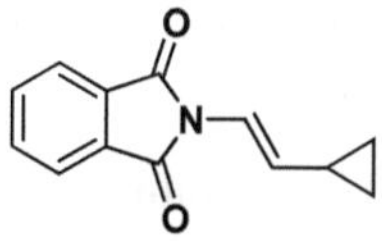

R_f 0.27 (7:3 Hexane/Ethyl acetate). **Mp** 92.2–94.3 °C. **^{1}H NMR** (400 MHz, $CDCl_3$) δ 7.88–7.81 (m, 2 H, Phth), 7.76–7.68 (m, 2 H, Phth), 6.70 (d, 1 H, J = 14.6 Hz, N–CH=), 6.20 (dd, 1 H, J = 14.6, 9.0 Hz, =CH–C), 1.48 (m, 1 H, CH cyclopropane), 0.87–0.75 (m, 2 H, CH_2 cyclopropane), 0.56–0.48 (m, 2 H, CH_2 cyclopropane). **^{13}C NMR** (101 MHz, $CDCl_3$) δ 166.6, 134.2, 131.7, 126.6, 123.4, 115.7, 12.7, 6.9. **IR** 3080 (w), 3016 (w), 1773 (w), 1772 (w), 1713 (s), 1612 (w), 1468 (w), 1393 (s), 1308 (w), 1207 (w), 1098 (w). **HRMS (ESI)** calcd for $C_{13}H_{12}NO_2^+$ $[M + H]^+$ 214.0863; found 214.0854.

(*E*)-1-(Hex-1-en-1-yl)pyrrolidine-2,5-dione (338)

Followong GP23, succinimide (198 mg, 2.00 mmol, 1 eq) and hex-1-yne (463 μL, 4.00 mmol, 2 eq) were combined to afford (*E*)-1-(hex-1-en-1-yl) pyrrolidine-2,5-dione (**338**) (242 mg,1.34 mmol, 67 % yield) as a brown oil.

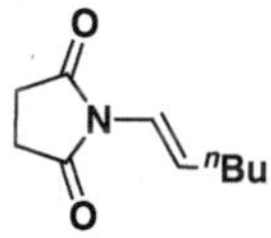

^{1}H NMR (400 MHz, $CDCl_3$) δ 6.61–6.55 (m, 1 H, CH=CH), 6.45–6.40 (m, 1 H, CH=CH), 2.72 (s, 4 H, CH_2), 2.11 (qd, 2 H, J = 7.2, 1.1 Hz, CH_2), 1.46–1.27 (m, 4 H, CH_2), 0.89 (t, 3 H, J = 7.2 Hz, CH_3). **HRMS (ESI)** calcd for $C_{10}H_{16}NO_2^+$ $[M + H]^+$ 182.1176; found 182.1177.

^{1}H NMR data match literature report [46].

(*E*)-1-(4-Phenylbut-1-en-1-yl)pyrrolidine-2,5-dione (339)

Followong GP23, succinimide (198 mg, 2.00 mmol, 1 eq) and but-3-yn-1-ylbenzene (562 μL, 4.00 mmol, 2 eq) were combined to afford (*E*)-1-(4-phenylbut-1-en-1-yl)pyrrolidine-2,5-dione (**339**) (209 mg, 0.910 mmol, 46 % yield) as a brown light solid.

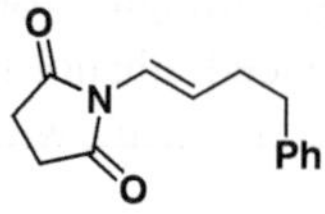

^{1}H NMR (400 MHz, $CDCl_3$) δ 7.34–7.30 (m, 2 H, ArH), 7.25–7.19 (m, 3 H, ArH), 6.75–6.63 (m, 1 H, CH=CH), 6.50 (d, 1 H, J = 14.9 Hz, CH=CH), 2.82–2.72 (m, 6 H, CH_2), 2.51–2.43 (m, 2 H, CH_2). **HRMS (ESI)** calcd for $C_{14}H_{16}NO_2^+$ [M + H]$^+$ 230.1176; found 230.1175.

Data match literature report [46].

(*E*)-2-(5-Chloropent-1-en-1-yl)isoindoline-1,3-dione (340)

Followong GP23, phthalimide (294 mg, 2.00 mmol, 1 eq) and 5-chloropent-1-yne (427 μL, 4.00 mmol, 2 eq) were combined to afford (*E*)-2-(5-chloropent-1-en-1-yl) isoindoline-1,3-dione (**340**) (77 mg, 0.31 mmol, 15 % yield) as a yellow solid.

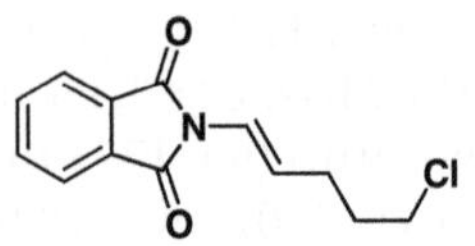

R$_f$ 0.35 (7:3 Hexane/Ethyl acetate). **Mp** 78.8–81.2 °C. **^{1}H NMR** (400 MHz, $CDCl_3$) δ 7.86 (dd, J = 5.5, 3.1 Hz, 2 H, Phth), 7.73 (dd, J = 5.5, 3.1 Hz, 2 H, Phth), 6.68 (dt, J = 14.6, 1.2 Hz, 1 H, N–CH=), 6.56 (dt, J = 14.6, 7.2 Hz, 1 H, =CH–C), 3.59 (dd, J = 6.5 Hz, 6.5 Hz, 2 H, CH_2), 2.39–2.30 (m, 2 H, CH_2), 1.96 (dt, J = 7.9, 6.5 Hz, 2 H, CH_2). **^{13}C NMR** (101 MHz, $CDCl_3$) δ 166.6, 134.4, 131.6, 123.5, 120.4, 118.7, 44.2, 32.1, 28.2. **IR** 2958 (w), 2847 (w), 1780 (w), 1714 (s), 1613 (w), 1436 (w), 1384 (s), 1266 (m), 1113 (w). **HRMS (ESI)** calcd for $C_{13}ClH_{13}NO_2^+$ [M + H]$^+$ 250.0629; found 250.0633.

GP24 for the synthesis of (*E*)-styrylphthalimide:

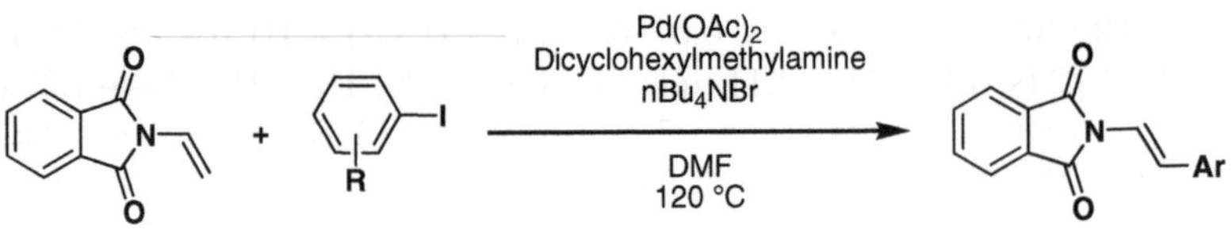

Following a reported procedure [47], a mixture of aryl halide (4.00 mmol, 1 eq), N-vinylphthalimide (693 mg, 4.00 mmol, 1 eq), Cy_2NMe (1.17 g, 6.00 mmol, 1.5 eq), TBAB (1.29 g, 4.00 mmol, 1 eq) and palladium acetate (1.00 mg, 4.00 μmol, 0.001 eq) in DMF (8 mL) was heated at 120 °C (oil bath temperature) in a pressure tube after 5 cycles of vacuum/N_2. When full conversion was observed by TLC, the yellow solution was poured into toluene (40 mL) and rinsed with toluene (10 mL). This solution was filtered through a pad of Celite and concentrated under vacuum. To the yellow oil was added ethanol (20 mL). The product precipitated and was recovered by filtration after rinsing with ethanol.

(*E*)-2-Styrylisoindoline-1,3-dione (341)

Followong GP24, phenyl iodide (816 mg, 4.00 mmol, 1 eq) was used and stirred 1h30 at 120 °C. (*E*)-2-styrylisoindoline-1,3-dione (**341**) (700 mg, 2.77 mmol, 69 % yield) was obtained as a yellow powder.

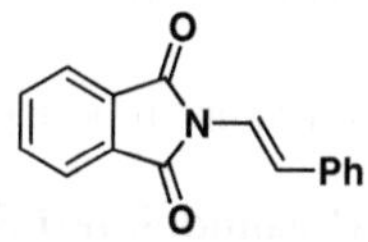

^{1}H NMR (400 MHz, $CDCl_3$) δ 7.91 (dd, 2 H, *J* = 5.5, 3.1 Hz, Phth), 7.77 (dd, 2 H, *J* = 5.4, 3.1 Hz, Phth), 7.66 (d, 1 H, *J* = 15.2 Hz, CH=CH), 7.51–7.45 (m, 2 H, ArH), 7.40–7.32 (m, 3 H, ArH), 7.30–7.24 (m, 1 H, CH=CH). **HRMS (ESI)** calcd for $C_{16}H_{12}NO_2^+$ $[M + H]^+$ 250.0863; found 250.0858.

Data of **341** match literature report [48].

(*E*)-2-(4-Bromostyryl)isoindoline-1,3-dione (342)

Followong GP24, 1-Bromo-4-iodobenzene (1.13 g, 4.00 mmol, 1 eq) was used and stirred 3h30 at 120 °C. (*E*)-2-(4-bromostyryl)isoindoline-1,3-dione (**342**) (330 mg, 1.00 mmol, 25 % yield) was obtained as a yellow powder.

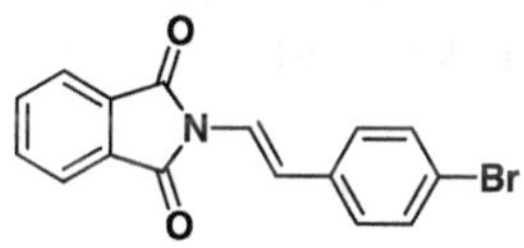

^{1}H NMR (400 MHz, $CDCl_3$) δ 7.95–7.88 (m, 2 H, Phth), 7.81–7.75 (m, 2 H, Phth), 7.60–7.54 (m, 1 H, CH=CH), 7.50–7.45 (m, 2 H, ArH), 7.38–7.31 (m, 3 H, ArH + CH=CH). **^{13}C NMR** (101 MHz, $CDCl_3$) δ 166.3, 135.0, 134.6, 131.8, 131.6, 127.7, 123.7, 121.4, 118.9, 118.1.

Data of **342** match literature report, with a slight shift in the carbon spectra for the area 127.7–121.4 [49].

(*E*)-2-(2-Bromostyryl)isoindoline-1,3-dione (343)

Followong GP24, 1-Bromo-2-iodobenzene (1.13 g, 4.00 mmol, 1 eq) was used and stirred 3h30 at 120 °C. (*E*)-2-(2-bromostyryl)isoindoline-1,3-dione (**343**) (320 mg, 0.981 mmol, 24 % yield) was obtained as a yellow powder after column chromatography Hexane:dichloromethane 95/5–1/1.

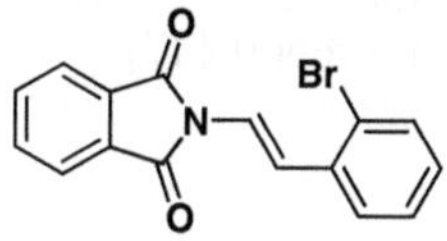

R_f 0.58 (7:3 Hexane/Ethyl acetate).**Mp** 193.2–194.6 °C. **^{1}H NMR** (400 MHz, $CDCl_3$) δ 8.00 (d, 1 H, *J* = 15.0 Hz, CH=), 7.96–7.88 (m, 2 H, Phth), 7.82–7.74 (m, 2 H, Phth), 7.63–7.55 (m, 2 H, Ar), 7.36–7.27 (m, 2 H, Ar + CH=), 7.17–7.10 (m, 1 H, Ar). **^{13}C NMR** (101 MHz, $CDCl_3$) δ 166.3, 136.3, 134.7, 133.1, 131.7, 128.9, 127.7, 126.4, 124.1, 123.8, 119.6, 119.4. **IR** 1724 (s), 1645 (w), 1383 (s), 1100 (w), 1086 (w), 950 (w). **HRMS (ESI)** calcd for $C_{16}^{79}BrH_{11}NO_2^+$ $[M + H]^+$ 327.9968; found 327.9960.

Data of **343** does not match literature report [47].

(*E*)-2-(4-(Trifluoromethyl)styryl)isoindoline-1,3-dione (344)

Followong GP24, 1-Iodo-4-(trifluoromethyl)benzene (1.08 g, 4.00 mmol, 1 eq) was used and stirred 1h30 at 120 °C. (*E*)-2-(4-(trifluoromethyl)styryl) isoindoline-1,3-dione (**344**) (679 mg, 2.14 mmol, 54 % yield) was obtained as a yellow powder.

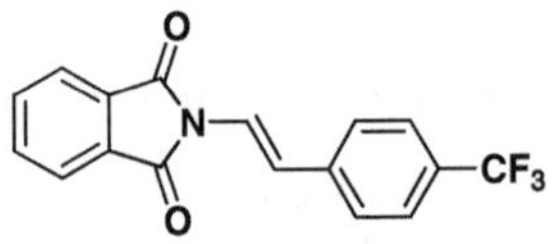

^{1}H NMR (400 MHz, $CDCl_3$) δ 7.96–7.90 (m, 2 H, Phth), 7.83–7.76 (m, 2 H, Phth), 7.70 (d, 1 H, *J* = 15.2 Hz, CH_2), 7.63–7.54 (m, 4 H, ArH), 7.45 (d, 1 H, *J* = 15.2 Hz, CH=CH). **HRMS (ESI)** calcd for $C_{17}F_3H_{11}NO_2^+$ $[M + H]^+$ 318.0736; found 318.0724.

Data of **344** match literature report [48].

(*E*)-2-(4-Methylstyryl)isoindoline-1,3-dione (345)

Followong GP24, 1-Iodo-4-methylbenzene (872 mg, 4.00 mmol, 1 eq) was used and stirred 3h30 at 120 °C. (*E*)-2-(4-methylstyryl)isoindoline-1,3-dione (**345**) (728 mg, 2.77 mmol, 69 % yield) was obtained as a yellow powder.

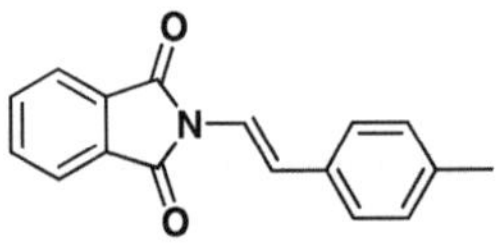

^{1}H NMR (400 MHz, $CDCl_3$) δ 7.93–7.88 (m, 2 H, Phth), 7.79–7.73 (m, 2 H, Phth), 7.63 (d, 1 H, *J* = 15.2 Hz, CH=CH), 7.38 (d, 2 H, *J* = 8.1 Hz, ArH), 7.32 (d, 1 H, *J* = 15.2 Hz, CH=CH), 7.17 (d, 2 H, *J* = 7.9 Hz, ArH), 2.36 (s, 3 H, CH_3). **HRMS (ESI)** calcd for $C_{17}H_{14}NO_2^+$ $[M + H]^+$ 264.1019; found 264.1022.

Data of **345** match literature report [47].

(*E*)-2-(4-Methoxystyryl)isoindoline-1,3-dione (346)

Followong GP24, 1-Iodo-4-methoxybenzene (936 mg, 4.00 mmol, 1 eq) was used and stirred 1h30 at 120 °C. (*E*)-2-(4-methoxystyryl)isoindoline-1,3-dione (**346**) (780 mg, 2.79 mmol, 69 % yield) was obtained as a yellow powder.

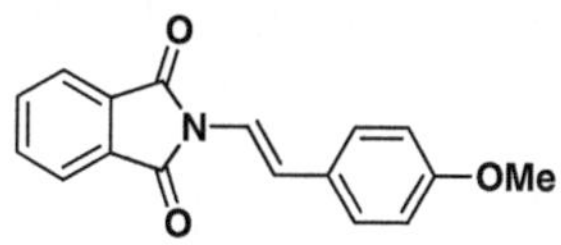

^{1}**H NMR** (400 MHz, $CDCl_3$) δ 7.89 (dd, 2 H, J = 5.5, 3.1 Hz, Phth), 7.75 (dd, 2 H, J = 5.4, 3.0 Hz, Phth), 7.60 (d, 1 H, J = 15.2 Hz, CH=CH), 7.42 (d, 2 H, J = 8.6 Hz, ArH), 7.26–7.22 (m, 1 H, the doublct was covered by $CDCl_3$ peak, CH=CH), 6.90 (d, 2 H, J = 8.8 Hz, ArH), 3.83 (s, 3 H, CH_3). **HRMS (ESI)** calcd for $C_{17}H_{14}NO_3^+$ $[M + H]^+$ 280.0968; found 280.0966.

Data of **346** match literature report [48].

Screening of Lewis acids

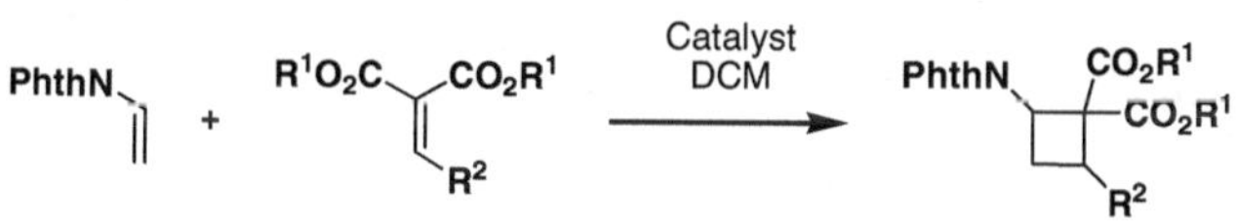

Reaction with $ZnBR_2$:

The reaction vessel was washed with aq HCl, rinsed with MeOH and dried in the oven.

Following a modified procedure [50], $ZnBr_2$ (247 mg, 1.10 mmol, 1.6 eq) was added in a flask and dichloromethane (1.5 mL) was added. Vinyl phthalimide (228 mg, 1.30 mmol, 1.9 eq) was dissolved in a second flask and dichloromethane (1.5 mL) was added. Dimethyl 2-methylenemalonate (100 mg, 0.694 mmol, 1 eq) was added in a third flask and dissolved in dichloromethane (1.5 mL).

The flask containing the $ZnBr_2$ solution was cooled down to −130 °C with heptane/liq N_2. The dimethyl 2-methylenemalonate solution and the vinyl phthalimide solution were then slowly cannulated to the reaction mixture. The solvents were solid under these conditions.

The flask was warmed to −78 °C. When the solution became liquid again, the reaction was stirred for 1 h at −78 °C. Pyridine (0.4 mL) in dichloromethane (1 mL) cooled to −78 °C was added to give an orange solution which was warmed to room temperature. A saturated Rochelle salt solution was added and the layers were separated. The organic layer was dried over $MgSO_4$ and evaporated. The crude mixture was analyzed by ^{1}H NMR and only vinyl phthalimide was observed.

GP25 Reaction with $Yb(OTf)_3$, $Sc(OTf)_3$ and diethyl 2-ethylidenemalonate:

The Lewis acid (0.014 mmol, 0.1 eq) was weighted in the glovebox and dry dichloromethane (200 μL) was added. The reaction was cooled at the indicated temperature. Diethyl 2-ethylidenemalonate (20.0 mg, 0.139 mmol, 1 eq) and 2-vinylisoindoline-1,3-dione (18.5 mg, 0.167 mmol, 1.2 eq) were dissolved in dry dichloromethane (800 μL) and the resulting solution was added dropwise via syringe pump for the indicated time. After the addition was complete, the reaction was filtered over a pad of alumina, eluted with ethyl acetate and concentrated under vacuum. The crude product was analyzed by ^{1}H NMR.

GP26 reaction with $In(OTf)_3$, $Sc(OTf)_3$ and $FeCl_3$-Al_2O_3 and dimethyl 2-methylidenemalonate:

The Lewis acid (0.012 mmol, 0.2 eq) was weighted in the glovebox and dry dichloromethane (0.20 mL) was added. Dimethyl 2-ethylidenemalonate (13 mg, 0.087 mmol, 1 eq) and 2-vinylisoindoline-1,3-dione (10 mg, 0.058 mmol, 1 eq) were dissolved in dry dichloromethane (0.80 mL) and added dropwise. After stirring for the indicated time, the reaction was filtered over a pad of alumina, eluted with ethyl acetate and concentrated under vacuum. The crude product was analyzed by ^{1}H NMR.

Synthesis of cyclobutanes

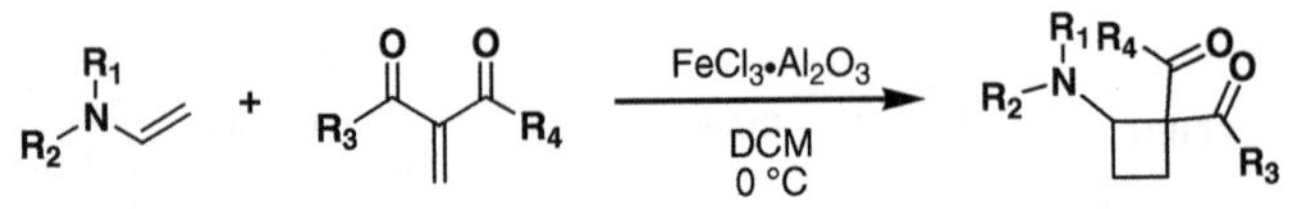

GP27

In the glovebox, iron trichloride supported on alumina [51] (1.00 mmol/g, 20.0 mg, 0.0200 mmol, 0.1 eq) was added to a microwave vial. The vial was sealed with a Teflon septum and taken out of the glovebox. Dry dichloromethane (200 μL) was added and the yellow suspension was cooled to 0 °C. The methylene malonate (0.400 mmol, 2 eq) and vinyl amide (0.200 mmol, 1 eq) were dissolved in dry dichloromethane (800 μL). The solution was then added dropwise to the iron trichloride. The reaction was stirred at 0 °C or room temperature and the conversion was followed by TLC (Hexane/Ethyl Acetate 7:3). When full conversion of the vinyl compound was observed, the reaction was filtered over a pad of alumina, eluted with ethyl acetate and concentrated under vacuum. The crude product was purified by column chromatography using the indicated solvents.

GP28

In the glovebox, iron trichloride supported on alumina (1.00 mmol/g, 20.0 mg, 0.0200 mmol, 0.1 eq) was added to a microwave vial. The vial was sealed with a Teflon septum and took out of the glovebox. Dry dichloromethane (200 μL) was added. The methylene malonate (0.40 mmol, 2 eq) and vinyl amide (0.200 mmol, 1 eq) were dissolved in dry dichloromethane (800 μL). The solution was then added dropwise to the iron trichloride. The reaction was stirred at room temperature and the conversion was followed by TLC (Hexane/Ethyl Acetate 7:3). When full conversion of the vinyl compound was observed, the reaction was filtered over a pad of alumina, eluted with ethyl acetate and concentrated under vacuum. The crude product was purified by column chromatography as indicated.

GP29

In the glovebox, iron trichloride supported on alumina (1.00 mmol/g, 20.0 mg, 0.0200 mmol, 0.1 eq) was added to a microwave vial. The vial was sealed with a Teflon septum and took out of the glovebox. Dry dichloromethane (200 μL) was added and the vinyl amide (0.200 mmol, 1 eq) was dissolved in dry dichloromethane (500 μL) and added to the iron trichloride. The methylene malonate (0.400 mmol, 2 eq or 0.800 mmol, 4 eq) was dissolved in dry dichloromethane (300 μL) and added dropwise over 2 h. When the addition was finished, the reaction was stirred at room temperature until TLC indicated that no more starting material was present. The reaction was filtered over a pad of alumina, eluted with ethyl acetate and concentrated under vacuum. The crude product was purified by column chromatography.

Dimethyl 2-(1,3-dioxoisoindolin-2-yl)cyclobutane-1,1-dicarboxylate (325a)

Following GP27, 2-vinylisoindoline-1,3-dione (**183**) (34.6 mg, 0.200 mmol) and dimethyl 2-methylenemalonate (**323a**) (57.7 mg, 0.400 mmol) were stirred for 20 min at 0 °C. The crude oil was purified by column chromatography on Biotage (SNAP cartridge KP-SIL 10 g, 95:5–5:5 Hexane/Ethyl acetate) affording dimethyl 2-(1,3-dioxoisoindolin-2-yl)cyclobutane-1,1-dicarboxylate (**325a**) (61.1 mg, 0.190 mmol, 96 % yield) as a colorless solid.

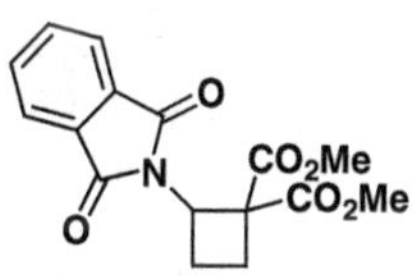

R_f 0.45 (1:1 Hexane/Ethyl acetate). **Mp** 124.1–126.3 °C. **^{1}H NMR** (400 MHz, $CDCl_3$) δ 7.93 (m, 2 H, Phth), 7.80 (m, 2 H, Phth), 5.17 (t, 1 H, J = 10.9 Hz, N–C–H), 3.16 (s, 3 H, CO_2CH_3), 2.98 (s, 3 H, CO_2CH_3), 2.58 (m, 1 H, CH_2), 2.25 (m, 1 H, CH_2), 1.48 (m, 1 H, CH_2), 1.33 (dt, 1 H, J = 13.6, 10.4 Hz, CH_2). **^{13}C NMR** (101 MHz, $CDCl_3$) δ 170.6, 168.7, 168.3, 134.3, 131.9, 123.5, 59.0, 53.2, 53.0, 47.9, 24.7, 21.9. **IR** 2956 (w), 2848 (w), 1782 (w), 1781 (w), 1741 (s), 1721 (s), 1437 (w), 1378 (m), 1266 (m). **HRMS (ESI)** calcd for $C_{16}H_{16}NO_6^+$ $[M + H]^+$ 318.0972; found 318.0978.

Dimethyl 2-(2,5-dioxopyrrolidin-1-yl)cyclobutane-1,1-dicarboxylate (347a)

Following GP27, 1-vinylpyrrolidine-2,5-dione (**284**) (25.0 mg, 0.200 mmol) and dimethyl 2-methylenemalonate (**323a**) (57.7 mg, 0.400 mmol) were stirred for 30 min at 0 °C. The crude oil was purified by column chromatography on Biotage (SNAP cartridge KP-SIL 10 g, 95:5–4:6 Hexane/Ethyl acetate) affording cyclobutane dimethyl 2-(2,5-dioxopyrrolidin-1-yl)cyclobutane-1,1-dicarboxylate (**347a**) (48.8 mg, 0.180 mmol, 91 % yield) as a colorless oil.

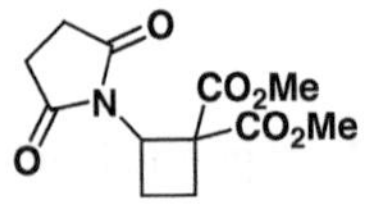

R_f 0.35 (6:4 Hexane/Ethyl acetate). **^{1}H NMR** (400 MHz, $CDCl_3$) δ 5.30 (dd, 1 H, J = 9.2, 9.2 Hz, N–C–H), 3.74 (s, 3 H, CO_2CH_3), 3.70 (s, 3 H, CO_2CH_3), 3.18–3.03 (m, 1 H, CH_2), 2.98–2.86 (m, 1 H, CH_2), 2.77–2.55 (m, 4 H, Succinimide), 2.26–2.11 (m, 2 H, CH_2). **^{13}C NMR** (101 MHz, $CDCl_3$) δ 177.1, 170.5, 168.6, 57.7, 53.1, 52.9, 48.2, 28.0, 24.8, 20.7. **IR** 2956 (w), 2848 (w), 1778 (w), 1736 (s), 1706 (s), 1436 (m), 1378 (s), 1262 (s), 1198 (m), 1116 (s). **HRMS (ESI)** calcd for $C_{12}H_{15}NNaO_6^+$ $[M + Na]^+$ 292.0792; found 292.0799.

Dimethyl 2-(2,5-dioxo-2,5-dihydro-1H-pyrrol-1-yl)cyclobutane-1,1-dicarboxyl ate (348)

Following GP27, 1-vinyl-*1H*-pyrrole-2,5-dione (**249**) (24.6 mg, 0.200 mmol) and dimethyl 2-methylenemalonate (**323a**) (57.7 mg, 0.400 mmol) were stirred for 60 min at 0 °C. The crude oil was purified by column chromatography on Biotage (SNAP cartridge KP-SIL 10 g, 95:5–4:6 Hexane/Ethyl acetate) affording dimethyl 2-(2,5-dioxo-2,5-dihydro-1H-pyrrol-1-yl)cyclobutane-1,1-dicarboxylate (**348**) (28.2 mg, 0.100 mmol, 48 % yield) as a colorless oil.

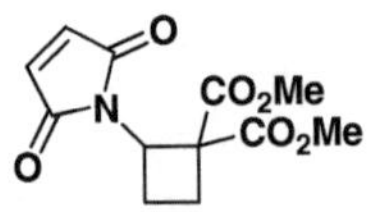

R_f 0.19 (7:3 Hexane/Ethyl acetate). **^{1}H NMR** (400 MHz, $CDCl_3$) δ 6.68 (s, 2 H, Maleimide), 5.27 (dd, 1 H, J = 9.4, 9.4 Hz, N–C–H), 3.74 (s, 3 H, CO_2CH_3), 3.67 (s, 3 H, CO_2CH_3), 3.14 (m, 1 H, CH_2), 2.88 (m, 1 H, CH_2), 2.25 (m, 1 H, CH_2), 2.13 (m, 1 H, CH_2). **^{13}C NMR** (101 MHz, $CDCl_3$) δ 170.5, 170.4, 168.5, 134.2, 58.8, 53.1, 53.0, 47.5, 24.3, 21.7. **IR** 2958 (w), 1736 (s), 1709 (s), 1437 (w), 1405 (w), 1379 (m), 1265 (s), 1107 (m). **HRMS (ESI)** calcd for $C_{12}H_{13}NNaO_6^+$ $[M + Na]^+$ 290.0635; found 290.0629.

Dimethyl 2-(3-(tert-butoxycarbonyl)-5-methyl-2,4-dioxo-3,4-dihydropyrimidin-1 (2 *H*)-yl)cyclobutane-1,1-dicarboxylate (349)

Following GP27, (*tert*-butyl 5-methyl-2,6-dioxo-3-vinyl-2,3-dihydropyrimidine-1(6 *H*)-carboxylate (**327**) (50.5 mg, 0.200 mmol) and dimethyl 2-methylenemalonate (**323a**) (57.7 mg, 0.400 mmol) were stirred for 3 h at room temperature. The crude oil was purified by column chromatography on Biotage (SNAP cartridge KP-SIL 10 g, 95:5–4:6 Hexane/Ethyl acetate) affording dimethyl 2-(3-(tert-butoxycarbonyl)-

5-methyl-2,4-dioxo-3,4-dihydropyrimidin-1(2 H)-yl)cyclobutane-1,1-dicarboxylate (**349**) (60.0 mg, 0.150 mmol, 76 % yield) as a colorless oil.

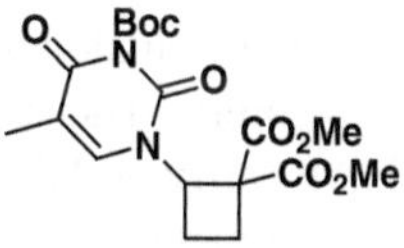

R*f* 0.18 (6:4 Hexane/Ethyl acetate). **^{1}H NMR** (400 MHz, $CDCl_3$) δ 7.05 (s, 1 H, C=CH), 5.22 (dd, 1 H, *J* = 9.5, 9.5 Hz, N–C–H), 3.76 (s, 3 H, CO_2CH_3), 3.69 (s, 3 H, CO_2CH_3), 2.96–2.84 (m, 1 H, CH_2), 2.81–2.72 (m, 1 H, CH_2), 2.38–2.27 (m, 1 H, CH_2), 2.18–2.07 (m, 1 H, CH_2), 1.92 (s, 3 H, Me), 1.59 (s, 9 H, Boc). **^{13}C NMR** (101 MHz, $CDCl_3$) δ 170.1, 168.5, 161.3, 149.1, 147.8, 138.0, 110.0, 86.7, 59.1, 56.2, 53.2, 53.1, 27.5, 23.6, 22.9, 12.5. **IR** 2984 (w), 2957 (w), 1784 (m), 1734 (s), 1713 (m), 1667 (s), 1437 (m), 1371 (m), 1263 (s), 1238 (s), 1147 (s), 1108 (m). **HRMS (ESI)** calcd for $C_{18}H_{25}N_2O_8^+$ $[M + H]^+$ 397.1605; found 397.1605.

Dimethyl 2-(1,3-dioxoisoindolin-2-yl)-3-methylcyclobutane-1,1-dicarboxylate (325c)

Following GP27, (*E*)-2-(prop-1-en-1-yl)isoindoline-1,3-dione (**331**) (37.4 mg, 0.200 mmol) and dimethyl 2-methylenemalonate (**323a**) (57.7 mg, 0.400 mmol) were stirred for 20 min at 0 °C. The crude oil was purified by column chromatography on Biotage (SNAP cartridge KP-SIL 10 g, 95:5–4:6 Hexane/Ethyl acetate) affording dimethyl 2-(1,3-dioxoisoindolin-2-yl)-3-methylcyclobutane-1,1-dicarboxylate (**325c**) (55.2 mg 0.170 mmol, 83 % yield) as a colorless solid. The product was recrystallized in ethanol.[24]

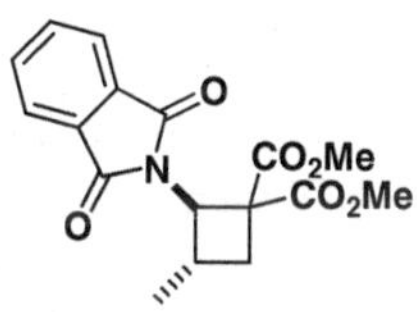

R*f* 0.26 (7:3 Hexane/Ethyl acetate). **Mp** 112.1–114.3 °C. **^{1}H NMR** (400 MHz, $CDCl_3$) δ 7.83 (m, 2 H, Phth), 7.72 (m, 2 H, Phth), 4.96 (d, 1 H, *J* = 10.3 Hz, N–C–H), 3.71 (m, 4 H, CO_2CH_3 and C–H–CH_3), 3.60 (s, 3 H, CO_2CH_3), 3.00 (dd, 1 H, *J* = 10.9, 9.3 Hz, CH_2), 1.74 (dd, 1 H, *J* = 11.5, 9.5 Hz, CH_2), 1.16 (d, 3 H, *J* = 6.6 Hz, CH_3). **^{13}C NMR** (101 MHz, $CDCl_3$) δ 170.4, 168.6, 168.3, 134.2, 131.8, 123.4, 56.3, 54.9, 53.0, 52.9, 32.1, 29.9, 19.3. **IR** 3006 (w), 2955 (w), 2869 (w), 1780 (w), 1735 (s), 1716 (s), 1437 (w), 1379 (s), 1261 (s). **HRMS (ESI)** calcd for $C_{17}H_{18}NO_6^+$ $[M + H]^+$ 332.1129; found 332.1124.

---

[24]The crystal structure has been deposited at the Cambridge Crystallographic Data Centre and allocated the deposition number CCDC 933180.

Dimethyl 2-(1,3-dioxoisoindolin-2-yl)-3-hexylcyclobutane-1,1-dicarboxylate (325d)

Following GP27, (*E*)-2-(oct-1-en-1-yl)isoindoline-1,3-dione (**334**) (51.5 mg, 0.200 mmol) and dimethyl 2-methylenemalonate (57.7 mg, 0.400 mmol) were stirred for 60 min at room temperature. The crude oil was purified by column chromatography on Biotage (SNAP cartridge KP-SIL 10 g, 95:5–4:6 Hexane/Ethyl acetate) affording dimethyl 2-(1,3-dioxoisoindolin-2-yl)-3-hexylcyclobutane-1,1-dicarboxylate **(325d)** (68.2 mg, 0.170 mmol, 85 % yield) as a colorless oil.

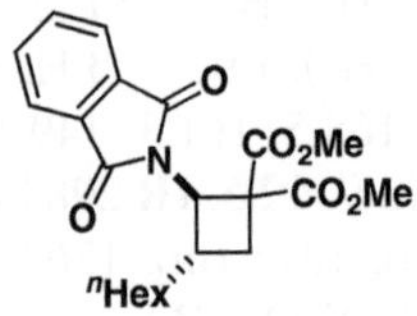

R_f 0.3 (*7:3 Hexane/Ethyl acetate*). **^{1}H NMR** (400 MHz, $CDCl_3$) δ 7.86–7.80 (m, 2 H, Phth), 7.74–7.69 (m, 2 H, Phth), 5.03 (d, 1 H, J = 10.2 Hz, N–C–H), 3.73 (s, 3 H, CO_2CH_3), 3.65-3.54 (m, 4 H, CO_2CH_3 + CH_2 cyclobutane), 2.99 (ddd, 1 H, J = 11.4, 9.3, 0.5 Hz, CH_2 cyclobutane), 1.73 (dd, 1 H, J = 11.5, 9.3 Hz, CH_2 cyclobutane), 1.60–1.40 (m, 2 H, CH_2 hexyl), 1.25–1.10 (m, 8 H, CH_2 hexyl), 0.84–0.74 (m, 3 H, CH_3 hexyl). **^{13}C NMR** (101 MHz, $CDCl_3$) δ 170.4, 168.6, 168.2, 134.2, 131.7, 123.4, 56.1, 53.6, 53.0, 53.0, 34.8, 34.7, 31.7, 30.7, 29.1, 26.6, 22.5, 14.0. **IR** 2955 (w), 2925 (w), 2855 (w), 1781 (w), 1738 (s), 1716 (s). **HRMS (ESI)** calcd for $C_{22}H_{28}NO_6^+$ $[M + H]^+$ 402.1911; found 402.1914.

Dimethyl-3-cyclopropyl-2-(1,3-dioxoisoindolin-2-yl) cyclobutane-1,1-dicarboxylate (325e)

Following GP27, (*E*)-2-(2-cyclopropylvinyl)isoindoline-1,3-dione (**337**) (42.6 mg, 0.200 mmol) and dimethyl 2-methylenemalonate (**323a**) (57.7 mg, 0.400 mmol) were stirred for 60 min at room temperature. The crude oil was purified by column chromatography on Biotage (SNAP cartridge KP-SIL 10 g, 95:5–4:6 Hexane/Ethyl acetate) affording dimethyl-3-cyclopropyl-2-(1,3-dioxoisoindolin-2-yl) cyclobutane-1,1-dicarboxylate (**325e**) (53.1 mg, 0.150 mmol, 74 % yield) as a colorless oil.

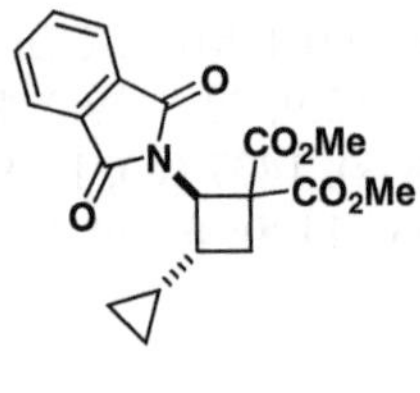

R_f 0.20 (7:3 Hexane/Ethyl acetate). **^{1}H NMR** (400 MHz, $CDCl_3$) δ 7.84 (dd, $J = 5.4, 3.1$ Hz, 2 H, Phth), 7.72 (dd, $J = 5.5, 3.0$ Hz, 2 H, Phth), 5.10 (d, $J = 10.4$ Hz, 1 H, N–C–H), 3.75 (s, 3 H, CO_2CH_3), 3.59 (s, 3 H, CO_2CH_3), 3.36–3.20 (m, 1 H, CH_2 cyclobutane), 2.92 (dd, $J = 11.3, 9.4$ Hz, 1 H, CH_2 cyclobutane), 1.83 (dd, $J = 11.5, 9.6$ Hz, 1 H, CH_2 cyclobutane), 0.85 (qt, $J = 8.1, 4.9$ Hz, 1 H, CH cyclopropane), 0.55-0.27 (m, 2 H, CH_2 cyclopropane), 0.23–0.04 (m, 2 H, CH_2 cyclopropane). **^{13}C NMR** (101 MHz, $CDCl_3$) δ 170.4, 168.7, 168.4, 134.3, 131.9, 123.5, 56.0, 53.1, 53.1, 53.0, 38.4, 29.8, 13.6, 2.7, 2.6. **IR** 3081 (w), 3005 (w), 2957 (w), 1779 (w), 1737 (s), 1715 (s), 1436 (m), 1377 (s), 1265 (s), 1199 (m), 1144 (m), 1049 (w). **HRMS (ESI)** calcd for $C_{19}H_{20}NO_6^+$ $[M + H]^+$ 358.1285; found 358.1283.

Dimethyl 3-(3-chloropropyl)-2-(1,3-dioxoisoindolin-2-yl)cyclobutane-1,1-dicarboxylate (325f)

Following GP28, (*E*)-2-(5-chloropent-1-en-1-yl)isoindoline-1,3-dione (**340**) (49.9 mg, 0.200 mmol) and dimethyl 2-methylenemalonate (**323a**) (57.7 mg, 0.400 mmol) were stirred for 60 min at room temperature. The crude oil was purified by column chromatography on Biotage (SNAP cartridge KP-SIL 10 g, 95:5–4:6 Hexane/Ethyl acetate) affording dimethyl 3-(3-chloropropyl)-2-(1,3-dioxoisoindolin-2-yl)cyclobutane-1,1-dicarboxylate (**325f**) (75.1 mg, 0.190 mmol, 95 % yield) as a colorless oil.

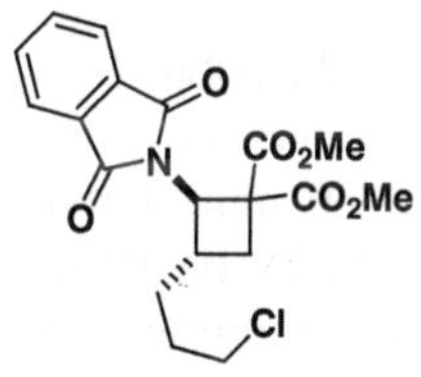

R_f 0.25 (7:3 Hexane/Ethyl acetate). **^{1}H NMR** (400 MHz, $CDCl_3$) δ 7.89–7.81 (m, 2 H, Phth), 7.78–7.69 (m, 2 H, Phth), 5.06 (d, 1 H, $J = 10.1$ Hz, N–C–H), 3.75 (s, 3 H, CO_2CH_3), 3.71–3.57 (m, 4 H, CO_2CH_3 + Ar–C–H), 3.52–3.43 (m, 2 H, CH_2–Cl), 3.03 (dd, 1 H, $J = 11.3, 9.5$ Hz, CH_2 cyclobutane), 1.78 (dd, 1 H, $J = 11.5, 9.3$ Hz, CH_2 cyclobutane), 1.74–1.65 (m, 4 H, CH_2 alkyl chain). **^{13}C NMR** (101 MHz, $CDCl_3$) δ 170.2, 168.4, 168.2, 134.3, 131.7, 123.4, 56.0, 53.5, 53.1, 53.0, 44.6, 34.1, 33.0, 30.6, 29.8. **IR** 2954 (w), 1781 (w), 1736 (s), 1714 (s), 1436 (w), 1378 (s), 1263 (s), 1262 (s), 1200 (m), 1072 (m). **HRMS (ESI)** calcd for $C_{19}ClH_{21}NO_6^+$ $[M + H]^+$ 394.1052; found 394.1053.

Dimethyl-3-butyl-2-(2,5-dioxopyrrolidin-1-yl)cyclobutane-1,1-dicarboxylate (347b)

Following GP28, (*E*)-1-(hex-1-en-1-yl)pyrrolidine-2,5-dione (**338**) (36.2 mg, 0.200 mmol) and dimethyl 2-methylenemalonate (**323a**) (57.7 mg, 0.400 mmol) were stirred for 60 min at room temperature. The crude oil was purified by column chromatography on Biotage (SNAP cartridge KP-SIL 10 g, 95:5–4:6 Hexane/Ethyl acetate) affording dimethyl-3-butyl-2-(2,5-dioxopyrrolidin-1-yl)cyclobutane-1,1-dicarboxylate (**347b**) (52.4 mg, 0.160 mmol, 81 % yield) as a colorless oil.

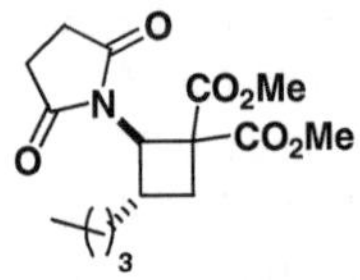

R_f 0.34 (1:1 Hexane/Ethyl acetate). **^{1}H NMR** (400 MHz, $CDCl_3$) δ 4.89 (d, $J = 9.9$ Hz, 1 H, N-C-H), 3.73 (s, 3 H, CO_2CH_3), 3.70 (s, 3 H, CO_2CH_3), 3.51 (m, 1 H, CH_2 cyclobutane), 2.96 (dd, $J = 11.7, 9.2$ Hz, 1 H, CH_2 cyclobutane), 2.75–2.58 (m, 4 H, CH_2 succinimide), 1.72 (dd, $J = 11.6, 9.1$ Hz, 1 H, CH_2 cyclobutane), 1.52–1.37 (m, 2 H, CH_2 butyl), 1.24 (m, 2 H, CH_2 butyl), 1.20–1.10 (m, 2 H, CH_2 butyl), 0.84 (t, $J = 7.1$ Hz, 3 H, CH_3 butyl). **^{13}C NMR** (101 MHz, $CDCl_3$) δ 177.2, 170.5, 168.6, 55.1, 54.0, 53.1, 53.0, 34.5, 33.5, 30.8, 28.7, 28.0, 22.5, 14.0. **IR** 2957 (w), 2956 (w), 2856 (w), 1784 (w), 1736 (s), 1707 (s), 1436 (m), 1375 (m), 1262 (s), 1181 (m), 1132 (s), 1041 (w). **HRMS (ESI)** calcd for $C_{16}H_{23}NNaO_6^+$ $[M + Na]^+$ 348.1418; found 348.1411.

Dimethyl 2-(2,5-dioxopyrrolidin-1-yl)-3-phenethylcyclobutane-1,1-dicarboxylate (347c)

Following GP28, (*E*)-1-(4-phenylbut-1-en-1-yl)pyrrolidine-2,5-dione (**339**) (45.9 mg, 0.200 mmol) and dimethyl 2-methylenemalonate (**323a**) (57.7 mg, 0.400 mmol) were stirred for 60 min at room temperature. The crude oil was purified by column chromatography on Biotage (SNAP cartridge KP-SIL 10 g, 95:5–4:6 Hexane/Ethyl acetate) affording dimethyl 2-(2,5-dioxopyrrolidin-1-yl)-3-phenethylcyclobutane-1,1-dicarboxylate (**347c**) (61.7 mg, 0.170 mmol, 83 % yield, 8:1 dr determined by integration of the peaks at 4.90 (*maj*), and 5.09 (*min*) in the crude ^{1}H NMR) as a colorless oil.

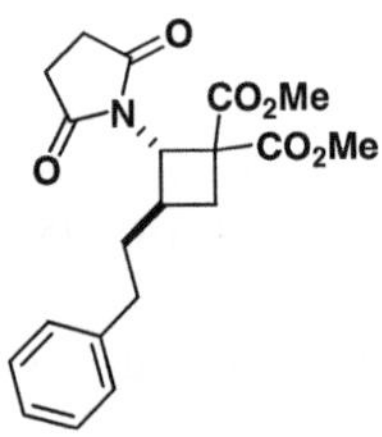

R_f 0.20 (1:1 Hexane/Ethyl acetate). **^{1}H NMR** (400 MHz, $CDCl_3$) on a 8(*maj.*):1 (*min.*) mixture δ 7.27–7.22 (m, 18 H, Ar *maj* + Ar *min* with chloroform peak), 7.19–7.05 (m, 26 H, Ar *maj* + Ar *min*), 6.86–6.80 (m, 1 H, Ar *min*), 5.09 (d, $J = 10.5$ Hz, 1 H, N–C–H *min*), 4.90 (d, $J = 9.8$ Hz, 8 H, N–C–H *maj*), 3.72 (m, 30H, CO_2CH_3 *maj* + CO_2CH_3 *min*), 3.68 (s, 24 H, CO_2CH_3 *maj*), 3.59 (dtd, $J = 16.8, 9.4, 7.3$ Hz, 8 H, CH_2 cyclobutane *maj*), 3.46 (dd, $J = 11.0, 4.3$ Hz, 1 H, CH_2 cyclobutane *min*), 2.95

(dd, J = 11.4, 9.6 Hz, 8 H, CH_2 cyclobutane *maj*), 2.67–2.45 (m, 52 H, CH_2 succinimide *maj* + CH_2 succinimide *min* + CH_2 chain *maj*), 2.95 (dd, J = 11.4, 9.6 Hz, 1 H, CH_2 cyclobutane *min*), 2.13–1.99 (m, 2 H, CH_2 chain *min*), 1.85 (ddq, J = 17.0, 8.7, 6.8 Hz, 18 H CH_2 chain *maj* + CH_2 chain *min*), 1.73 (dd, J = 11.7, 9.1 Hz, 8 H, CH_2 cyclobutane *maj*), 1.54 (td, J = 12.9, 4.5 Hz, 1 H, CH_2 cyclobutane *min*). **^{13}C NMR** (101 MHz, $CDCl_3$) δ 177.0, 170.3, 168.4, 141.3, 128.3, 128.1, 125.9, 55.0, 54.0, 53.0, 52.9, 36.2, 33.3, 33.1, 30.7, 27.8. Only major isomer. **IR** 2955 (w), 1784 (w), 1783 (w), 1739 (s), 1708 (s), 1673 (w), 1438 (w), 1437 (w), 1384 (m), 1374 (m), 1315 (w), 1267 (s), 1197 (w), 1151 (m), 1150 (m). **HRMS (ESI)** calcd for $C_{20}H_{23}NNaO_6^+$ $[M + Na]^+$ 396.1418; found 396.1427.

Dimethyl 2-(1,3-dioxoisoindolin-2-yl)-3-phenylcyclobutane-1,1-dicarboxylate (325 g)

Following GP29, (*E*)-2-styrylisoindoline-1,3-dione (**341**) (49.9 mg, 0.200 mmol) and dimethyl 2-methylenemalonate (**323a**) (57.7 mg, 0.400 mmol) were stirred for 1 h at room temperature after the end of the slow addition. The crude oil was purified by column chromatography on Biotage (SNAP cartridge KP-SIL 10 g, 95:5–4:6 Hexane/Ethyl acetate) affording dimethyl 2-(1,3-dioxoisoindolin-2-yl)-3-phenylcyclobutane-1,1-dicarboxylate (**325 g**) (70.6 mg, 0.179 mmol, 90 % yield) as a colorless oil.

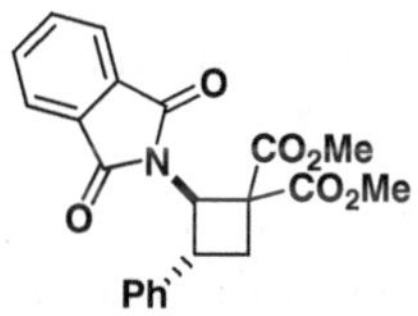

R_f 0.28 (7:3 Hexane/Ethyl acetate). **^{1}H NMR** (400 MHz, $CDCl_3$) δ 7.63 (dd, 2 H, J = 3.7, 2.1 Hz, Phth), 7.55 (dd, 2 H, J = 3.8, 2.1 Hz, Phth), 7.29-7.24 (m, 4 H, Ph), 7.23–7.19 (m, 1 H, Ph), 5.52 (dd, 1 H, J = 7.4, 0.4 Hz, N–C–H), 4.86–4.96 (m, 1 H, H–C–Ph), 3.76 (s, 3 H, CO_2CH_3), 3.65 (s, 3 H, CO_2CH_3), 3.25 (ddd, 1 H, J = 7.7, 6.5, 0.4 Hz, CH_2), 2.21 (dd, 1 H, J = 7.8, 6.9 Hz, CH_2). **^{13}C NMR** (101 MHz, $CDCl_3$) δ 158.8, 157.4, 157.2, 134.0, 128.8, 126.6, 124.1, 122.9, 122.5, 119.7, 63.1, 61.4, 60.7, 60.7, 48.4, 42.9. **IR** 2199 (w), 1833 (m), 1790 (w), 1155 (s), 1127 (s), 940 (m), 890 (s), 842 (s). **HRMS (ESI)** calcd for $C_{22}H_{20}NO_6^+$ $[M + H]^+$ 394.1285; found 394.1271.

Dimethyl-3-(4-bromophenyl)-2-(1,3-dioxoisoindolin-2-yl)cyclobutane-1,1-dicarboxylate (325h)

Following GP29, (*E*)-2-(4-bromostyryl)isoindoline-1,3-dione (**342**) (65.6 mg, 0.200 mmol) and dimethyl 2-methylenemalonate (115 mg, 0.800 mmol) were stirred for 3 h at room temperature after the end of the slow addition. The crude oil was purified by column chromatography on Biotage (SNAP cartridge KP-SIL 10 g, 95:5–4:6 Hexane/Ethyl acetate) affording dimethyl-3-(4-bromophenyl)-2-

(1,3-dioxoisoindolin-2-yl)cyclobutane-1,1-dicarboxylate (**325h**) (47.0 mg, 0.100 mmol, 50 % yield) as a colorless oil as well as (*E*)-2-(4-bromostyryl) isoindoline-1,3-dione (**342**) (15.4 mg, 0.0469 mmol, 23, 65 % yield **325 h** b.r.s.m.).

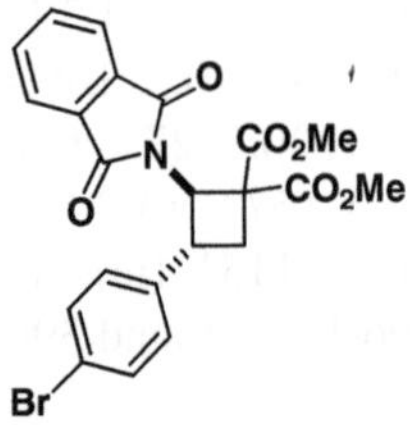

R*f* 0.26 (7:3 Hexane/Ethyl acetate). **^{1}H NMR** (400 MHz, $CDCl_3$) δ 7.87–7.81 (m, 2 H, Phth), 7.76–7.70 (m, 2 H, Phth), 7.45–7.40 (m, 2 H, Ar), 7.18–7.13 (m, 2 H, Ar), 5.47 (dd, *J* = 10.9, 0.7 Hz, 1 H, N–C–H), 4.88 (q, *J* = 10.1 Hz, 1 H, Ar–C–H), 3.78 (s, 3 H, CO_2CH_3), 3.67 (s, 3 H, CO_2CH_3), 3.26 (ddd, *J* = 11.6, 9.5, 0.8 Hz, 1 H, CH_2), 2.19 (dd, *J* = 11.5, 10.0 Hz, 1 H, CH_2). **^{13}C NMR** (101 MHz, $CDCl_3$) δ 170.0, 168.3, 168.2, 139.5, 134.3, 131.8, 131.6, 128.6, 123.5, 121.1, 55.9, 53.9, 53.20, 53.15, 38.0, 31.8. **IR** 2955 (w), 1782 (w), 1738 (s), 1721 (s), 1492 (w), 1437 (w), 1378 (s), 1267 (m), 1201 (m), 1049 (w). **HRMS (ESI)** calcd for $C_{22}^{79}BrH_{19}NO_6^+$ $[M + H]^+$ 472.0390; found 472.0385.

Dimethyl-3-(2-bromophenyl)-2-(1,3-dioxoisoindolin-2-yl) cyclobutane-1,1-dicarboxylate (325i)

Following GP29, (*E*)-2-(2-bromostyryl)isoindoline-1,3-dione (**343**) (65.6 mg, 0.200 mmol) and dimethyl 2-methylenemalonate (**323a**) (57.7 mg, 0.400 mmol) were stirred for 5 min at room temperature after the end of the slow addition. The crude oil was purified by column chromatography on Biotage (SNAP cartridge KP-SIL 10 g, 95:5–4:6 Hexane/Ethyl acetate) affording dimethyl-3-(2-bromophenyl)-2-(1,3-dioxoisoindolin-2-yl)cyclobutane-1,1-dicarboxylate (**325i**) (87.4 mg, 0.185 mmol, 93 % yield) as a colorless oil.

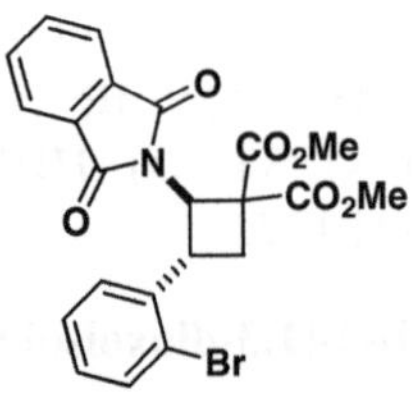

R*f* 0.32 (7:3 Hexane/Ethyl acetate). **^{1}H NMR** (400 MHz, $CDCl_3$) δ 7.88–7.79 (m, 2 H, Phth), 7.77–7.66 (m, 2 H, Phth), 7.50 (dd, *J* = 8.0, 1.2 Hz, 1 H, Ar), 7.38 (ddd, *J* = 7.9, 1.6, 0.6 Hz, 1 H, Ar), 7.32–7.27 (m, 1 H, Ar), 7.08 (dddd, *J* = 7.9, 7.3, 1.7, 0.5 Hz, 1 H, Ar), 5.74 (dd, *J* = 10.8, 0.8 Hz, 1 H, N–C–H), 5.21–5.06 (m, 1 H, Ar–C–H), 3.77 (s, 3 H, CO_2CH_3), 3.69 (s, 3 H, CO_2CH_3), 3.53 (ddd, *J* = 11.7, 9.6,

0.9 Hz, 1 H, CH_2), 2.06 (dd, J = 11.6, 9.7 Hz, 1 H, CH_2). **^{13}C NMR** (101 MHz, $CDCl_3$) δ 170.3, 168.3, 168.2, 139.9, 134.4, 133.1, 131.7, 128.7, 127.9, 127.3, 124.0, 123.6, 55.9, 53.30, 53.28, 51.5, 38.8, 31.9. **IR** 3062 (w), 2955 (w), 2848 (w), 1783 (w), 1738 (s), 1722 (s), 1471 (w), 1437 (m), 1379 (s), 1267 (s). **HRMS (ESI)** calcd for $C_{22}^{79}BrH_{19}NO_6^+$ [M + H]$^+$ 472.0390; found 472.0405.

Dimethyl-2-(1,3-dioxoisoindolin-2-yl)-3-(4-(trifluoromethyl)phenyl) cyclobutane-1,1-dicarboxylate (325j)

Following GP29, (*E*)-2-(4-(trifluoromethyl)styryl)isoindoline-1,3-dione (**344**) (63.5 mg, 0.200 mmol) and dimethyl 2-methylenemalonate (**323a**) (115 mg, 0.800 mmol) were stirred for 5 min at room temperature after the end of the slow addition. The crude oil was purified by column chromatography on Biotage (SNAP cartridge KP-SIL 10 g, 95:5 to 4:6 Hexane/Ethyl acetate) affording dimethyl-2-(1,3-dioxoisoindolin-2-yl)-3-(4-(trifluoromethyl)phenyl)cyclobutane-1,1-dicarboxy late (**325j**) (35.0 mg, 0.0760 mmol, 38 % yield, with a polymeric impurity) as a colorless oil as well as (*E*)-2-(4-(trifluoromethyl)styryl)isoindoline-1,3-dione (**344**) (10.3 mg, 0.0324 mmol, 16, 45 % yield of **325j** b.r.s.m.).

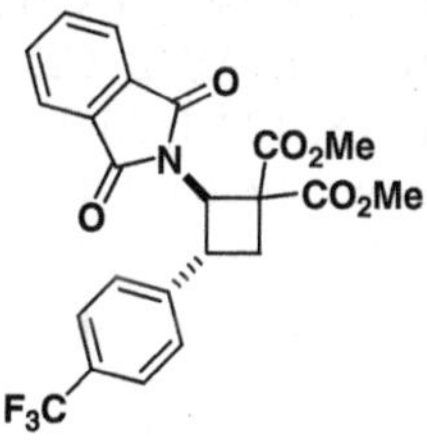

R_f 0.25 (7:3 Hexane/Ethyl acetate). **^{1}H NMR** (400 MHz, $CDCl_3$) δ 7.85 (dd, J = 5.5, 3.1 Hz, 2 H), 7.73 (dd, J = 5.5, 3.1 Hz, 2 H, Phth), 7.56 (d, J = 8.1 Hz, 2 H, Ar), 7.39 (d, J = 8.0 Hz, 2 H, Ar), 5.53 (d, J = 10.8 Hz, 1 H, N–C–H), 4.99 (q, J = 10.1 Hz, 1 H, Ar–C–H), 3.79 (s, 3 H, OMe), 3.68 (s, 3 H, OMe), 3.37–3.21 (m, 1 H, CH_2 cyclobutane), 2.25 (dd, J = 11.6, 10.0 Hz, 1 H, CH_2 cyclobutane). **^{13}C NMR** (101 MHz, $CDCl_3$) δ 169.9, 168.1, 168.0, 144.5, 134.3, 131.5, 129.5 (q, $J_{C\text{-}F}$ = 32.4 Hz), 127.0, 125.7 (q, J = 3.3 Hz), 124.1 (q, J = 272.2 Hz), 123.5, 55.8, 53.6, 53.14, 53.08, 38.1, 31.5. **IR** 2957 (w), 1782 (w), 1737 (s), 1720 (s), 1620 (w), 1438 (w), 1378 (s), 1266 (s), 1200 (m), 1164 (s), 1124 (s), 1049 (m). **HRMS (ESI)** calcd for $C_{23}H_{18}F_3NNaO_6^+$ [M + Na]$^+$ 484.0978; found 484.0996.

Dimethyl-2-(1,3-dioxoisoindolin-2-yl)-3-(p-tolyl)cyclobutane-1,1-dicarboxylate (325 k) and dimethyl-3-(1,3-dioxoisoindolin-2-yl)-2-(p-tolyl)cyclobutane-1,1-dicarboxylate (351a)

Following GP27, (*E*)-2-(4-methylstyryl)isoindoline-1,3-dione (**345**) (52.7 mg, 0.200 mmol) and dimethyl 2-methylenemalonate (**323a**) (57.7 mg, 0.400 mmol) were stirred for 2h30 at room temperature. The crude oil was purified by column chromatography on Biotage (SNAP cartridge KP-SIL 10 g, 95:5–4:6 Hexane/Ethyl acetate) affording cyclobutane **325k** (56.1 mg, 0.140 mmol, 69 % yield) as a colorless oil and cyclobutane **351a** (22.5 mg, 0.0600 mmol, 28 % yield) as a colorless oil. The structure of **325k** was confirmed by 2D-NMR experiments. The structure of **351a** was assigned in analogy with the one of product **351b**.

325k

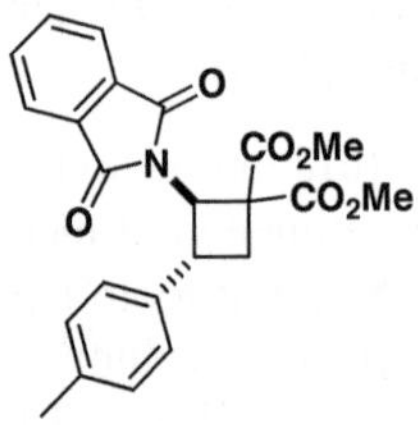

R*f* 0.27 (7:3 Hexane/Ethyl acetate). **^{1}H NMR** (400 MHz, $CDCl_3$) δ 7.86–7.78 (m, 2 H, Phth), 7.74–7.67 (m, 2 H, Phth), 7.22–7.16 (m, 2 H, Ar), 7.15–7.07 (m, 2 H, Ar), 5.51 (dd, *J* = 10.9, 0.8 Hz, 1 H, N–C–H), 4.88 (dd, *J* = 10.1, 10.1, Hz, 1 H, Ar–C–H), 3.78 (s, 3 H, CO_2CH_3), 3.67 (s, 3 H, CO_2CH_3), 3.25 (ddd, *J* = 11.5, 9.5, 0.9 Hz, 1 H, CH_2 cyclobutane), 2.30 (s, 3 H, CH_3), 2.20 (dd, *J* = 11.5, 10.1 Hz, 1 H, CH_2 cyclobutane). **^{13}C NMR** (101 MHz, $CDCl_3$) δ 170.2, 168.5, 168.2, 137.5, 136.9, 134.2, 131.7, 129.4, 126.8, 123.4, 55.9, 54.1, 53.11, 53.08, 38.2, 32.1, 21.1. **IR** 3023 (w), 2955 (w), 1781 (w), 1736 (s), 1718 (s), 1518 (w), 1437 (w), 1378 (s), 1265 (s), 1199 (m), 1048 (m), 881 (w). **HRMS (ESI)** calcd for $C_{23}H_{21}NNaO_6^+$ $[M + Na]^+$ 430.1261; found 430.1257.

351a

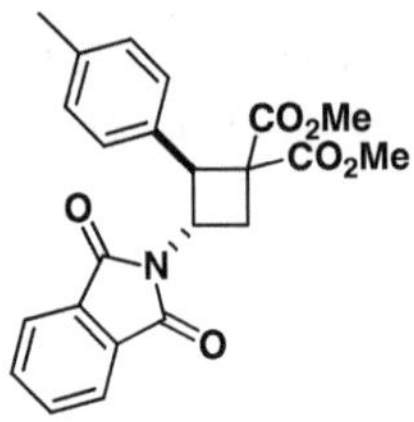

R*f* 0.32 (7:3 Hexane/Ethyl acetate). **^{1}H NMR** (400 MHz, $CDCl_3$) δ 7.85–7.77 (m, 2 H, Phth), 7.74–7.66 (m, 2 H, Phth), 7.32–7.27 (m, 2 H, Ar), 7.13–7.07 (m, 2 H, Ar), 5.44 (dt, *J* = 10.2, 9.1 Hz, 1 H, N–C–H), 5.18 (d, *J* = 10.3 Hz, 1 H, Ar–C–H), 3.87 (s, 3 H, CO_2CH_3), 3.35 (s, 3 H, CO_2CH_3), 3.29 (dd, *J* = 11.4, 9.2 Hz, 1 H, CH_2 cyclobutane), 3.00 (ddd, *J* = 11.4, 9.0, 0.8 Hz, 1 H, CH_2 cyclobutane), 2.28 (s, 3 H, CH_3). **^{13}C NMR** (101 MHz, $CDCl_3$) δ 171.2, 169.5, 168.3, 137.1, 134.2, 133.0, 131.7, 129.0, 127.4, 123.3, 54.4, 52.9, 52.3, 49.6, 42.2, 30.8, 21.1. **IR** 2954 (w), 2926 (w), 2863 (w), 1776 (w), 1734 (s), 1716 (s), 1437 (w), 1384 (m), 1277 (m), 1203 (m), 1128 (m). **HRMS (ESI)** calcd for $C_{23}H_{22}NO_6^+$ $[M + H]^+$ 408.1442; found 408.1451.

Dimethyl-3-(1,3-dioxoisoindolin-2-yl)-2-(4-methoxyphenyl) cyclobutane-1,1-dicarboxylate (351b)

Following GP27, (*E*)-2-(4-methoxystyryl)isoindoline-1,3-dione (**345**) (55.9 mg, 0.200 mmol) and dimethyl 2-methylenemalonate (57.7 mg, 0.400 mmol) were stirred for 60 min at room temperature. The crude oil was purified by column chromatography on Biotage (SNAP cartridge KP-SIL 10 g, 95:5–4:6 Hexane/Ethyl acetate) affording dimethyl-3-(1,3-dioxoisoindolin-2-yl)-2-(4-methoxyphenyl) cyclobutane-1,1-dicarboxylate (**351b**) (69.0 mg, 0.160 mmol, 81 % yield) as a colorless oil.

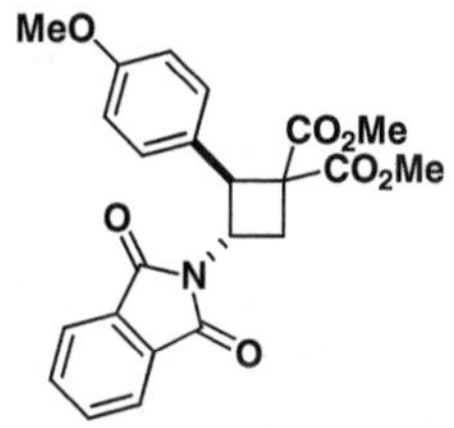

R_f 0.15 (7:3 Hexane/Ethyl acetate). **^{1}H NMR** (400 MHz, $CDCl_3$) δ 7.84–7.78 (m, 2 H, Phth), 7.73–7.67 (m, 2 H, Phth), 7.36–7.30 (m, 2 H, Ar), 6.87–6.80 (m, 2 H, Ar), 5.46–5.38 (m, 1 H, Ar–C–H), 5.15 (d, 1 H, J = 10.3 Hz, N–C–H), 3.87 (s, 3 H, CO_2CH_3), 3.76 (s, 3 H, Ar–OMe), 3.36 (s, 3 H, CO_2CH_3), 3.28 (dd, 1 H, J = 11.3, 9.2 Hz, CH_2), 2.99 (dd, 1 H, J = 11.3, 8.9 Hz, CH_2). **^{13}C NMR** (101 MHz, $CDCl_3$) δ 171.3, 169.5, 168.3, 159.0, 134.2, 131.7, 128.8, 128.1, 123.4, 113.8, 55.2, 54.4, 52.9, 52.4, 49.4, 42.4, 30.6. **IR** 3030 (w), 2955 (w), 1782 (w), 1739 (s), 1721 (s), 1378 (s), 1267 (s). **HRMS** (ESI) calcd for $C_{23}H_{22}NO_7^+$ $[M + H]^+$ 424.1391; found 424.1387.

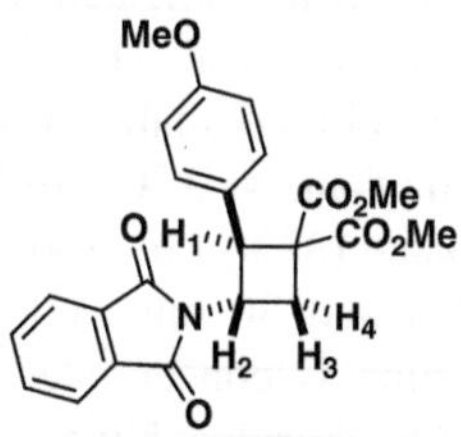

ROESY : H1-H4
H2-H3

Di*iso*propyl-2-(1,3-dioxoisoindolin-2-yl)cyclobutane-1,1-dicarboxylate (325l)

In the glovebox, scandium trifluoromethanesulfonate (49.2 mg, 0.100 mmol, 0.1 eq) was added to a microwave vial. The vial was sealed with a Teflon septum and took out of the glovebox. Dry dichloromethane (1.00 mL) was added. The solution was cooled to 0 °C and vinyl phthalimide (**183**) (190 mg, 1.10 mmol, 1.1 eq) was added in 2.00 mL dichloromethane. A 2.00 mL dichloromethane solution of di*iso*propyl 2-methylenemalonate (**323j**)[25] (1.03 g, 18 % purity: 200 mg, 1.00 mmol, 1 eq) is added dropwise over 30 min. When the addition is finished, 0.5 mL pyridine is added, the reaction is concentrated and purified by column chromatography on Biotage (SNAP cartridge KP-SIL 10 g, 95:5–5:5 Hexane/Ethyl acetate) affording di*iso*propyl-2-(1,3-dioxoisoindolin-2-yl) cyclobutane-1,1-dicarboxylate (**325l**) (245 mg, 0.656 mmol, 66 % yield) as a colorless solid.

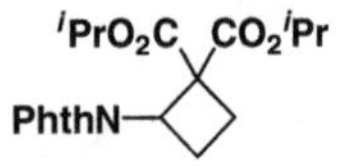

R_f 0.5 (6:4 Hexane/Ethyl acetate) **^{1}H NMR** (400 MHz, $CDCl_3$) δ 7.83 (dd, J = 5.5, 3.0 Hz, 2 H, Ar Phth), 7.71 (dd, J = 5.5, 3.1 Hz, 2 H, Ar Phth), 5.46 (td, J = 9.4, 0.9 Hz, 1 H, N–C–H), 5.06 (sep, J = 6.3 Hz, 1 H, CH isopropyl), 4.93 (sep, J = 6.3 Hz, 1 H, CH isopropyl), 3.33–3.13 (m, 1 H, CH_2), 2.98 (dddd, J = 11.7, 10.5, 4.1, 1.0 Hz, 1 H, CH_2), 2.32 (dtd, J = 11.4, 9.1, 4.1 Hz, 1 H, CH_2), 2.12 (dt, J = 11.8, 8.8 Hz, 1 H, CH_2), 1.28–1.19 (m, 6 H, 2x CH_3 isopropyl), 1.10 (d, J = 6.2 Hz, 3 H, CH_3 isopropyl), 0.80 (d, J = 6.3 Hz, 3 H, CH_3 isopropyl). **^{13}C NMR** (101 MHz, $CDCl_3$) δ 169.7, 168.1, 167.7, 134.1, 131.9, 123.3, 69.4, 69.1, 59.1, 47.4, 24.4, 21.6, 21.5, 21.4, 21.4, 21.1. **IR** 2919 (w), 1642 (w), 1582 (s), 1213 (s), 1095 (m), 921 (m). **HRMS (ESI)** calcd for $C_{20}H_{24}NO_6^+$ $[M + H]^+$ 374.1598; found 374.1587.

Diallyl 2-(1,3-dioxoisoindolin-2-yl)cyclobutane-1,1-dicarboxylate (325 m)

In the glovebox, scandium trifluoromethanesulfonate (125 mg, 0.255 mmol, 0.05 eq) was added to a microwave vial. The vial was sealed with a Teflon septum and took out of the glovebox. Vinyl phthalimide (**183**) (883 mg, 5.10 mmol, 1 eq) is added in 8.00 mL dichloromethane at 0 °C. An 8.00 mL dichloromethane solution of diallyl 2-methylenemalonate (**323i**) (See footnote 25) (1.00 g, 5.10 mmol, 1 eq) is added dropwise over 2 h and 30 min. When the addition is finished, 2 mL pyridine is added, the reaction is concentrated and purified by column chromatography on Biotage (SNAP cartridge KP-SIL 10 g, 95:5 to 5:5 Hexane/Ethyl acetate) affording diallyl 2-(1,3-dioxoisoindolin-2-yl)cyclobutane-1,1-dicarboxylate (**325m**) (927 mg, 2.51 mmol, 49 % yield) as a colorless solid.

[25] The methylenation crude was used directly without isolation of methylenemalonate compound.

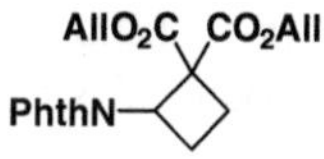

^{1}H NMR (400 MHz, $CDCl_3$) δ 7.75 (dt, *J* = 5.5, 2.9 Hz, 2 H, Ar Phth), 7.66 (td, *J* = 5.3, 4.6, 2.4 Hz, 2 H, Ar Phth), 5.80 (dddt, *J* = 16.0, 13.3, 7.0, 4.3 Hz, 1 H, CH=CH_2 allyl), 5.52 (tdd, *J* = 10.3, 6.7, 5.2 Hz, 1 H, CH=CH_2 allyl), 5.47–5.39 (m, 1 H, N–C–H), 5.27–5.18 (m, 1 H, CH=CH_2 allyl), 5.14 (dt, *J* = 10.5, 1.6 Hz, 1 H, CH=CH_2 allyl), 4.98 (dq, *J* = 17.2, 1.5 Hz, 1 H, CH=CH_2 allyl), 4.85 (ddd, *J* = 10.4, 2.6, 1.4 Hz, 1 H, CH=CH_2 allyl), 4.58 (ddq, *J* = 6.1, 4.7, 1.4 Hz, 2 H, CH_2–CH=CH_2 allyl), 4.48–4.29 (m, 2 H, CH_2–CH=CH_2), 3.29–3.09 (m, 1 H, CH_2), 3.05–2.86 (m, 1 H, CH_2), 2.26 (dtd, *J* = 11.0, 9.0, 3.7 Hz, 1 H, CH_2), 2.12 (dt, *J* = 11.6, 8.9 Hz, 1 H, CH_2). **^{13}C NMR** (101 MHz, $CDCl_3$) δ 169.5, 168.0, 167.6, 134.1, 131.7, 131.4, 131.1, 123.3, 118.6, 118.4, 66.5, 66.2, 59.0, 47.5, 24.4, 21.6. **IR** 2896 (w), 2895 (w), 1641 (w), 1575 (s), 1211 (s), 1077 (m), 947 (m). **HRMS (ESI)** calcd for $C_{20}H_{20}NO_6^+$ $[M + H]^+$ 370.1285; found 370.1284.

Ethyl 1-benzoyl-2-(1,3-dioxoisoindolin-2-yl)cyclobutane-1-carboxylate (325n)

Following GP27, 2-vinylisoindoline-1,3-dione (**183**) (34.6 mg, 0.200 mmol) and dimethyl 2-methylenemalonate (**323c**) (82.0 mg, 0.400 mmol) were stirred for 4 h at 0 °C. The crude oil was purified by column chromatography on Biotage (SNAP cartridge KP-SIL 10 g, 95:5–4:6 Hexane/Ethyl acetate) affording ethyl 1-benzoyl-2-(1,3-dioxoisoindolin-2-yl)cyclobutane-1-carboxylate (**325n**) (46.8 mg, 0.120 mmol, 62 % yield, 2.5:1 dr determined by integration of the peaks at 3.31–3.23 (*maj*), and 3.04-2.93(min) in the crude ^{1}H NMR) as a colorless oil.

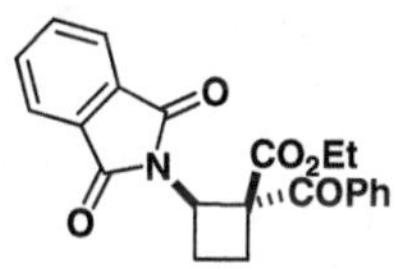

R_f 0.35 (7:3 Hexane/Ethyl acetate). **^{1}H NMR** (400 MHz, $CDCl_3$) on a 3(*maj.*):1 (*min.*) mixture δ 7.91–7.82 (m, 12 H, Phth and Ar *maj.*), 7.75–7.68 (m, 6 H, Phth *maj.*), 7.64–7.56 (m, 6 H, Phth and Ar *min.*), 7.56–7.50 (m, 4 H, Ar *maj.*), 7.44–7.39 (m, 6 H, Ar *maj.*), 7.20–7.10 (m, 3 H, Ar *min.*), 5.85–5.80 (m, 1 H, N–C–H *min.*), 5.76 (dd, 3 H, *J* = 9.5, 9.5 Hz, N–C–H *maj.*), 4.19 (qd, 2 H, *J* = 7.1, 1.4 Hz, COOEt *min.*), 4.00–3.85 (m, 6 H, COOEt *maj.*), 3.51–3.37 (m, 4 H, CH_2 *maj.* and *min.*), 3.31–3.23 (m, 3 H, CH_2 *maj.*), 3.04–2.93 (m, 1 H, CH_2 *min.*), 2.56–2.44 (m, 1 H, CH_2 *min.*), 2.33–2.23 (m, 4 H, CH_2 *maj.* and *min.*), 2.22–2.14 (m, 3 H, CH_2 *maj.*), 1.11 (t, 3 H, *J* = 7.1 Hz, COOEt *min.*), 0.77 (t, 10 H, *J* = 7.2 Hz, COOEt *maj.*). **^{13}C NMR** (101 MHz, $CDCl_3$) *maj.* δ 193.5, 169.1, 168.2, 134.2, 133.9, 133.5, 131.8, 129.0, 128.7, 123.4, 65.6, 62.1, 46.9, 25.3, 22.0, 13.5. **^{13}C NMR** (101 MHz, $CDCl_3$) *min.* δ 192.7, 171.7, 167.9, 135.4, 133.9, 132.6, 131.3, 128.5, 128.2, 123.0, 62.2, 62.0, 48.2, 25.5, 21.8, 13.9. **IR** 1780 (w), 1734 (s), 1719 (s),

1436 (w), 1377 (s), 1263 (s), 1200 (w). **HRMS (ESI)** calcd for $C_{22}H_{20}NO_5^+$ $[M + H]^+$ 378.1336; found 378.1338.

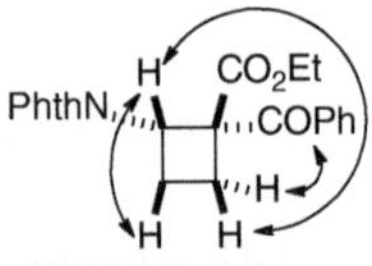

ROESY for minor

Diethyl 2-(1,3-dioxoisoindolin-2-yl)-4-methylcyclobutane-1,1-dicarboxylate (325b)

Following GP28, 2-vinylisoindoline-1,3-dione (**183**) (34.6 mg, 0.200 mmol) and diethyl 2-ethylidenemalonate (**323b**) (74.5 mg, 0.400 mmol) were stirred for 18 h at room temperature. The crude oil was purified by column chromatography on Biotage (SNAP cartridge KP-SIL 10 g, 95:5 to 4:6 Hexane/Ethyl acetate) affording diethyl 2-(1,3-dioxoisoindolin-2-yl)-4-methylcyclobutane-1,1-dicarboxylate (**325b**) (45.2 mg, 0.130 mmol, 63 % yield, 1.2:1 dr determined by integration of the peaks at 4.91 (*maj*), and 5.58 (*min*) in the crude ^{1}H NMR) as a colorless oil.

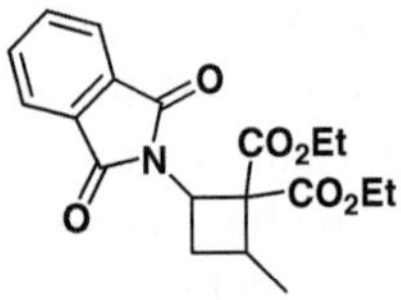

$\mathbf{R_f}$ 0.25 (7:3 Hexane/Ethyl acetate). $\mathbf{^1H}$ **NMR** (400 MHz, $CDCl_3$) on a 1.6 (*maj.*):1(*min.*) mixture δ 7.84–7.79 (m, 5.2 H, Phth *maj* + *min*), 7.72–7.67 (m, 5.2 H, Phth *maj* + *min*), 5.58 (ddd, J = 10.0, 8.3, 1.2 Hz, 1 H, N–C–H *min*), 4.91 (dd, J = 10.9, 8.7 Hz, 1.6 H, N–C–H *maj*), 4.27–4.18 (m, 5.2 H, CO_2CH_2 *maj* + *min*), 4.17–3.91 (m, 5.2 H, CO_2CH_2 *maj* + *min*), 3.65-3.54 (m, 1 H, CH cyclobutane *min*), 3.27–3.12 (m, 2.6 H, CH_2 cyclobutane *maj* + *min*), 2.78 (ddq, J = 10.7, 8.1, 7.0 Hz, 1.6 H, CH cyclobutane *maj*), 2.46 (dt, J = 10.8, 8.5 Hz, 1.6 H, CH_2 cyclobutane *maj*), 1.98 (ddd, J = 11.7, 10.0, 5.4 Hz, 1 H, CH_2 cyclobutane *min*), 1.32 (d, J = 7.0 Hz, 4.6 H, CH_3 maj), 1.28–1.21 (m, 7.8 H, $CO_2CH_2CH_3$ *maj* + *min*), 1.14 (d, J = 7.3 Hz, 3 H, CH_3 min), 1.05 (t, J = 7.1 Hz, 4.6 H, $CO_2CH_2CH_3$ *maj*), 0.90 (t, J = 7.1 Hz, 3 H, $CO_2CH_2CH_3$ *min*). $\mathbf{^{13}C}$ **NMR** (101 MHz, $CDCl_3$) δ 170.2, 168.5, 168.4, 168.3, 168.1, 167.5, 134.1, 134.0, 132.0, 131.8, 123.3, 123.2, 62.4, 62.0, 61.7,[26] 61.6, 60.9, 47.5, 45.1, 33.8, 31.1, 29.6, 28.7, 16.5, 16.2, 14.2, 14.0, 13.9, 13.6. **IR** 2929 (w), 2851 (w), 1780 (w), 1732 (s), 1713 (s), 1614 (w), 1468 (w), 1377 (s), 1256 (s), 1213 (m), 1072 (m). **HRMS (ESI)** calcd for $C_{19}H_{21}NNaO_6^+$ $[M + Na]^+$ 382.1261; found 382.1248.

[26]Two peaks under this signal as determined by HMBC.

Dimethyl 2-(1,3-dioxoisoindolin-2-yl)-4-phenethylcyclobutane-1,1-dicarboxylate (325o)

Following GP28, 2-vinylisoindoline-1,3-dione (**183**) (34.6 mg, 0.200 mmol) and dimethyl 2-(3-phenylpropylidene)malonate (**323d**) (99.0 mg, 0.400 mmol) were stirred for 3h30 at room temperature. The crude oil was purified by column chromatography on Biotage (SNAP cartridge KP-SIL 10 g, 95:5–4:6 Hexane/Ethyl acetate) affording dimethyl 2-(1,3-dioxoisoindolin-2-yl)-4-phenethylcyclobutane-1,1-dicarboxylate (**325o**) (60.0 mg, 0.140 mmol, 71 % yield, 1.3:1 dr determined by integration of the peaks at 4.93 (*maj*), and 5.58 (*min*) in the crude ^{1}H NMR) as a colorless oil.

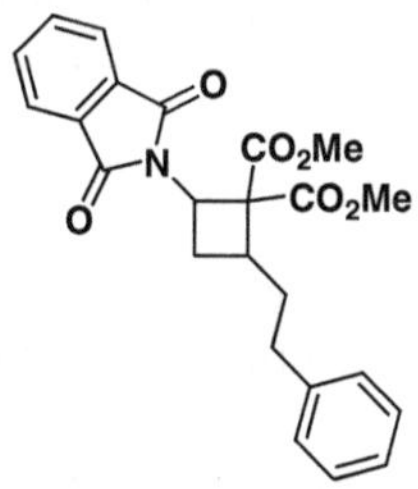

R_f 0.43 (7:3 Hexane/Ethyl acetate). **^{1}H NMR** (400 MHz, $CDCl_3$) on a 1.5 (*maj.*):1(*min.*) mixture δ 7.87–7.79 (m, 5 H, Phth *maj* + Phth *min*), 7.75–7.67 (m, 5 H, Phth *maj* + Phth *min*), 7.32–7.25 (m, 5 H, Ar *maj* + Ar *min* with chloroform peak), 7.19 (dd, *J* = 7.5, 4.9 Hz, 7.5 H, Ar *maj* + Ar *min*), 5.58 (ddd, *J* = 10.2, 7.7, 1.1 Hz, 1 H, N–C–H *min*), 4.93 (dd, *J* = 11.1, 8.6 Hz, 1.5 H, N–C–H *maj*), 3.78 (s, 4.5 H, CO_2CH_3 *maj*), 3.74 (s, 3 H, CO_2CH_3 *min*), 3.60 (s, 4.5 H, CO_2CH_3 *maj*), 3.54 (s, 3 H, CO_2CH_3 *min*), 3.21 (q, *J* = 10.9 Hz, 1.5 H, CH_2 cyclobutane *maj*), 3.09 (ddd, *J* = 12.0, 10.1, 7.8 Hz, 1 H, CH_2 cyclobutane *min*), 2.75–2.54 (m, 7.5 H, 1 × CH_2 cyclobutane *maj*, 2 × CH_2 chain *maj*, 1 × CH_2 cyclobutane *min*, 2 × CH_2 chain *min*), 2.45 (dt, *J* = 10.6, 8.2 Hz, 1.5 H, CH_2 cyclobutane *maj*), 2.24–2.06 (m, 2.5 H, CH_2 chain *maj* + CH_2 cyclobutane *min*), 2.01–1.89 (m, 1.5 H, CH_2 chain *maj*), 1.89-1.80 (m, 1 H, CH_2 chain *min*), 1.76—1.63 (m, 1 H, CH_2 chain *min*). **^{13}C NMR** (101 MHz, $CDCl_3$) δ 170.5, 169.0, 168.9, 168.3, 168.0, 167.9, 141.7, 141.6, 134.2, 134.0, 131.9, 131.7,128.45, 128.44, 128.40, 126.0, 125.9, 123.4, 123.2, 62.1, 61.5, 52.9, 52.8, 52.7, 52.2, 47.7, 45.4, 38.6, 36.6, 33.5, 33.4, 33.1, 32.9, 28.5, 27.3.[27] **IR** 3028 (w), 2953 (w), 1780 (w), 1737 (s), 1715 (s), 1605 (w), 1496 (w), 1455 (w), 1436 (w), 1379 (s), 1262 (m), 1199 (m), 1159 (w). **HRMS (ESI)** calcd for $C_{24}H_{23}NNaO_6^+$ $[M + Na]^+$ 444.1418; found 444.1418.

Dimethyl 2-cyclohexyl-4-(1,3-dioxoisoindolin-2-yl)cyclobutane-1,1-dicarboxylate (325p)

Following GP28, 2-vinylisoindoline-1,3-dione (**183**) (34.6 mg, 0.200 mmol) and dimethyl 2-(cyclohexylmethylene)malonate (**323g**) (91.0 mg, 0.400 mmol) were

[27] A peak is not resolved in the 128.45–128.40 massif.

stirred for 18 h at room temperature. The crude oil was purified by column chromatography on Biotage (SNAP cartridge KP-SIL 10 g, 95:5–4:6 Hexane/Ethyl acetate) affording dimethyl 2-cyclohexyl-4-(1,3-dioxoisoindolin-2-yl) cyclobutane-1,1-dicarboxylate (**325p**) (47.0 mg, 0.120 mmol, 59 % yield, 1.2:1 dr determined by integration of the peaks at 4.96 (*min*), and 5.38 (*maj*) in the crude ^{1}H NMR) as a colorless oil.

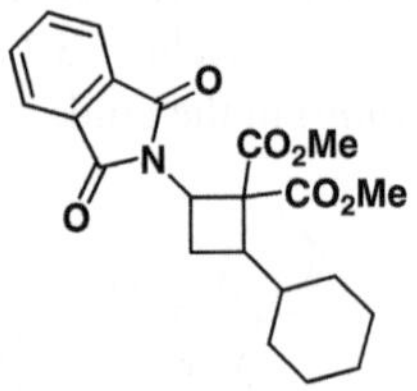

R*f* 0.30 (7:3 Hexane/Ethyl acetate). **^{1}H NMR** (400 MHz, $CDCl_3$) on a 1.2 (*maj.*):1 (*min.*) mixture δ 7.87–7.79 (m, 4.4 H, Phth *maj* + *min*), 7.75–7.68 (m, 4.4 H, Phth *maj* + *min*), 5.38 (ddd, *J* = 10.5, 5.1, 1.1 Hz, 1.2 H, N–C–H *maj*), 4.93 (dd, *J* = 11.4, 8.0 Hz, 1 H, N–C–H *min*), 3.79-3.76 (m, 6.6 H, CO_2CH_3 *maj* + *min*), 3.59 (s, 3 H, CO_2CH_3 *min*), 3.56–3.48 (m, 4.8 H, CO_2CH_3 *maj* + CH cyclobutane *maj*), 3.22 (td, *J* = 11.1, 9.9 Hz, 1 H, CH_2 cyclobutane *min*), 2.72 (ddd, *J* = 12.4, 10.4, 5.1 Hz, 1.2 H, CH_2 cyclobutane *maj*), 2.45–2.27 (m, 3.2 H, CH_2 cyclobutane *maj* + *min* + CH cyclobutane *min*), 1.97–1.58 (m, 12.6 H, CH cyclohexyl *maj* + *min*), 1.40-0.98 (m, 8 H, CH_2 cyclohexyl *maj* + *min*), 0.96-0.73 (m, 3.6 H, CH_2 cyclohexyl *maj*). **^{13}C NMR** (101 MHz, $CDCl_3$) δ 171.0, 170.3, 170.2, 169.8, 168.5, 168.2, 134.3, 134.1, 132.0, 131.9, 123.5, 123.4, 62.5, 61.1, 53.0, 52.8, 52.6, 52.2, 48.0, 46.1, 45.2, 44.0, 39.5, 39.4, 31.4, 31.3, 30.0, 29.9, 29.6, 27.4, 26.7, 26.5, 26.1, 26.0, 25.9, 25.8. **IR** 2923 (w), 2851 (w), 1780 (w), 1732 (s), 1732 (s), 1714 (s), 1613 (w), 1436 (w), 1376 (s), 1265 (s), 1204 (m), 1157 (w), 1061 (m). **HRMS (ESI)** calcd for $C_{22}H_{25}NNaO_6^+$ $[M + Na]^+$ 422.1574; found 422.1590.

Dibenzyl-2-(1,3-dioxoisoindolin-2-yl)-4-methylcyclobutane-1,1-dicarboxylate (325q)

Following GP28, 2-vinylisoindoline-1,3-dione (**183**) (34.6 mg, 0.200 mmol) and dibenzyl 2-ethylidenemalonate (**323e**) (124 mg, 0.400 mmol) were stirred for 18 h at room temperature. The crude oil was purified by column chromatography on Biotage (SNAP cartridge KP-SIL 10 g, 95:5–4:6 Hexane/Ethyl acetate) affording dibenzyl-2-(1,3-dioxoisoindolin-2-yl)-4-methylcyclobutane-1,1-dicarboxylate (**325q**) (60.5 mg, 0.125 mmol, 62 % yield, 3:1 dr determined by integration of the peaks at 3.64 (*maj*), and 2.49 (*min*) in the crude ^{1}H NMR) as a colorless oil.

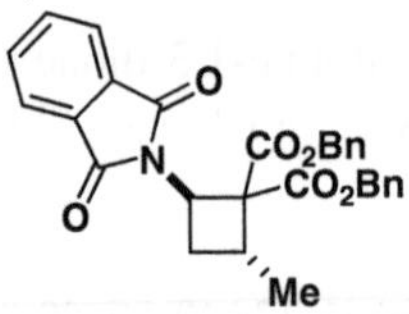

R*f* 0.22 (7:3 Hexane/Ethyl acetate). **^{1}H NMR** (400 MHz, $CDCl_3$) on a 2 (*maj*):1 (*min*) diatereomeric mixture δ 7.77–7.68 (m, 6 H, Phth), 7.65 (ddd, *J* = 5.8, 3.2, 1.9 Hz, 6 H, Phth), 7.34–7.27 (m, 15 H, Ar), 7.19–7.06 (m, 6 H, Ar), 7.06–6.98 (m, 6 H, Ar), 6.97–6.90 (m, 3 H, Ar), 5.62 (ddd, *J* = 9.8, 8.4, 1.2 Hz, 2 H, N–C–H *maj*), 5.27 – 5.12 (m, 6 H, CH_2–Ar *maj* + *min*), 5.06–4.88 (m, 7 H, N–C–H *min*, CH_2–Ar *maj* + *min*), 3.64 (dddd, *J* = 10.1, 7.4, 5.2, 1.3 Hz, 2 H, CH-Me cyclobutane *maj*), 3.28–3.13 (m, 3 H, CH_2 cyclobutane *maj* + *min*), 2.86–2.74 (m, 1 H, CH–Me cyclobutane *min*), 2.49 (dt, *J* = 10.7, 8.3 Hz, 1 H, CH_2 cyclobutane *min*), 1.99 (ddd, *J* = 11.6, 9.9, 5.3 Hz, 2 H, CH_2 cyclobutane *maj*), 1.32 (d, *J* = 7.0 Hz, 3 H, Me cyclobutane *min*), 1.08 (d, *J* = 7.4 Hz, 6 H, Me cyclobutane *maj*). **^{13}C NMR** (101 MHz, $CDCl_3$) δ 169.8, 168.3, 168.1, 168.1, 168.0, 167.2, 135.5, 135.2, 134.9, 134.8, 133.9, 133.8, 131.8, 131.6, 128.5, 128.5, 128.4, 128.3, 128.3, 128.2, 128.2, 128.1, 128.1, 128.0, 128.0, 126.0, 123.3, 123.2, 67.5, 67.4, 67.3, 67.0, 62.4, 62.1, 47.5, 45.1, 34.2, 31.3, 29.8, 28.8, 16.4, 16.2. **IR** 2470 (w), 2447 (w), 2386 (w), 1453 (w), 1418 (s), 1407 (s), 1199 (w), 1138 (s), 1045 (m), 1008 (m), 956 (w). **HRMS (ESI)** calcd for $C_{29}H_{26}NO_6^+$ $[M + H]^+$ 484.1755; found 484.1756.

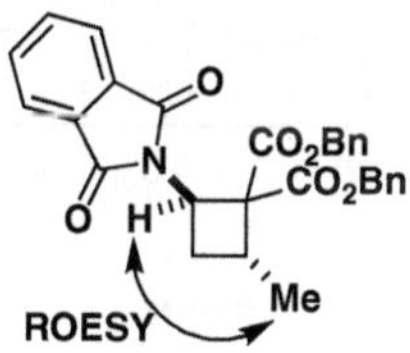

Dibenzyl-2-(1,3-dioxoisoindolin-2-yl)-4-(trifluoromethyl)cyclobutane-1,1-dicarboxylate (325r)

Following GP27, 2-vinylisoindoline-1,3-dione (**183**) (34.6 mg, 0.200 mmol) and dibenzyl 2-(2,2,2-trifluoroethylidene)malonate (**323f**) (146 mg, 0.400 mmol) were stirred for 18 h at room temperature. The crude oil was purified by column chromatography on Biotage (SNAP cartridge KP-SIL 10 g, 95:5–4:6 Hexane/Ethyl acetate) affording dibenzyl-2-(1,3-dioxoisoindolin-2-yl)-4-(trifluoromethyl) cyclobutane-1,1-dicarboxylate (**325r**) (82.2 mg, 0.153 mmol, 76 % yield) as a colorless oil.

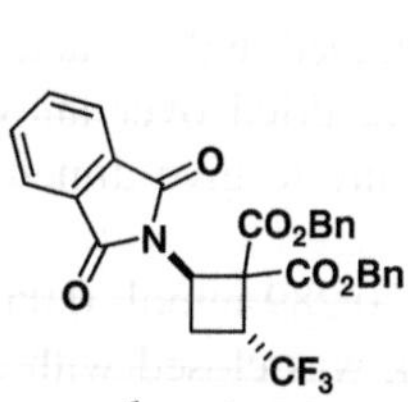

R*f* 0.18 (7:3 Hexane/Ethyl acetate). **^{1}H NMR** (400 MHz, $CDCl_3$) δ 7.75–7.64 (m, 4 H, Phth), 7.33–7.26 (m, 5 H, Ar), 7.11–7.05 (m, 1 H, Ar), 7.02–6.97 (m, 2 H, Ar), 6.92–6.88 (m, 2 H, Ar), 5.87–5.67 (m, 1 H, N–C–H), 5.30–5.01 (m, 2 H, CH_2Ar), 5.00–4.77 (m, 2 H, CH_2Ar), 4.47–4.26 (m, 1 H, CH_2 cyclobutane), 3.23 (ddd, *J* = 12.8, 10.8, 7.7 Hz, 1 H, CH_2 cyclobutane), 2.60 (ddd, *J* = 12.7, 10.4,

6.5 Hz, 1 H, CH_2 cyclobutane). **^{13}C NMR** (101 MHz, $CDCl_3$) δ 167.7, 166.2, 165.9, 134.6, 134.3, 134.1, 131.3, 128.5, 128.5, 128.5, 128.3, 128.29, 128.26, 125.6 (q, $J = 278$ Hz), 123.6, 68.4, 68.3, 59.3, 45.3, 39.6 (q, $J = 31.4$ Hz), 22.6. **^{19}F NMR** (376 MHz, $CDCl_3$) δ −62.3. **IR** 3035 (w), 2963 (w), 1781 (m), 1741 (s), 1736 (s), 1456 (w), 1380 (s), 1276 (s), 1142 (s), 1088 (m). **HRMS (ESI)** calcd for $C_{29}H_{22}F_3NNaO_6^+$ $[M + Na]^+$ 560.1291; found 560.1277.

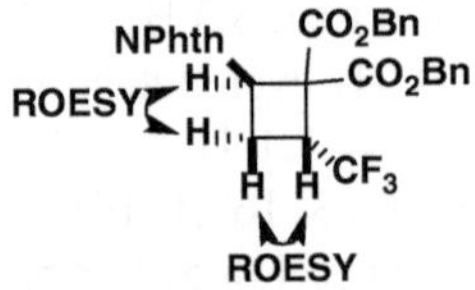

Sequential synthesis of aminocyclobutanes
Dimethyl 2-(1,3-dioxoisoindolin-2-yl)cyclobutane-1,1-dicarboxylate (325a)

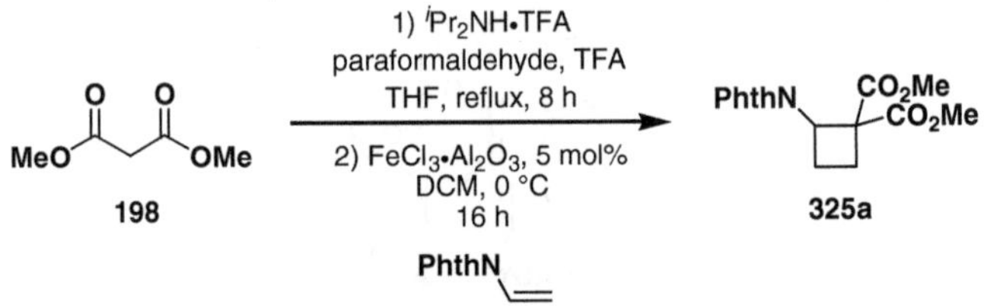

Dimethyl malonate (**198**) (1.32 mL, 11.6 mmol, 2 eq), diisopropylamine 2,2,2-trifluoroacetate (2.49 g, 11.6 mmol, 2 eq), paraformaldehyde (**384**) (0.695 mg, 23.1 mmol, 4 eq) and trifluoroacetic acid (89.0 μL, 1.16 mmol, 0.2 eq) were added to tetrahydrofuran (20 mL). A condenser was added and the suspension was stirred at reflux for 2 h. Paraformaldehyde (**384**) (0.695 mg, 23.1 mmol, 4 eq) was added and the reflux was continued for 6 h. The reaction was cooled to room temperature and the tetrahydrofuran was removed under reduced pressure (300–50 mbar at 45 ° C). The crude was dissolved in diethyl ether (25 mL) and filtered through cotton in a separatory funnel. The organic layer was washed twice with 1 M HCl (25 mL). The aqueous layers were combined and extracted with diethyl ether (25 mL). The organic layers were combined, dried over anhydrous Na_2SO_4, filtered and concentrated under reduced pressure to give dimethyl crude 2-methylenemalonate as colorless oil.

The iron catalyst (289 mg, 0.289 mmol, 0.05 eq) was weighted in an oven-dry flask in a glovebox. The flask was closed with a silicon septum, taken out of the glovebox and put under positive pressure of nitrogen and dichloromethane (5 mL) was added. 2-vinylisoindoline-1,3-dione (**183**) (1.00 g, 5.77 mmol, 1 eq) was dissolved in dichloromethane (5 mL)and added to the yellow suspension. Finally, the crude dimethyl 2-methylenemalonate was dissolved in dichloromethane (5 mL) and added to the reaction in one portion. The reaction was stirred at room temperature for 16 h and then filtered over a basic alumina plug, eluting with ethyl acetate. The solvents were evaporated and the brown solid was purified by column chromatography on Biotage (SNAP cartridge KP-SIL 50 g, 95:5–4:6 Hexane/Ethyl acetate) affording dimethyl 2-(1,3-dioxoisoindolin-2-yl) cyclobutane-1,1-dicarboxylate (**325a**) (1.55 g, 4.89 mmol, 85 % yield) as a colorless solid.

1-(*Tert*-butyl) 1-methyl 2-(1,3-dioxoisoindolin-2-yl)cyclobutane-1,1-dicarboxy late (325s)

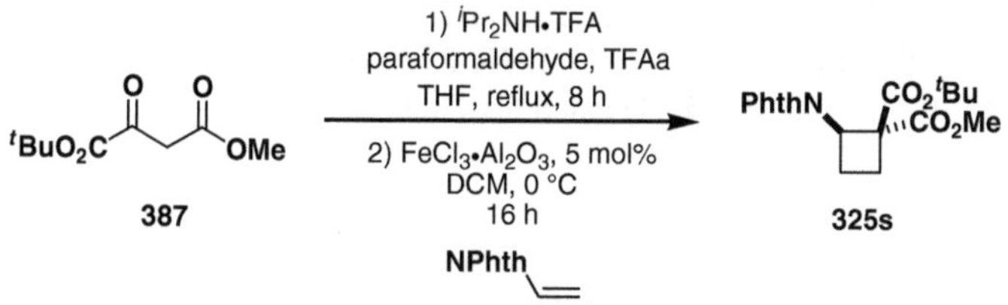

Tert-butyl methyl malonate (**387**) (1.95 mL, 11.6 mmol, 2 eq), diisopropylamine 2,2,2-trifluoroacetate (2.49 g, 11.6 mmol, 2 eq), paraformaldehyde (**384**) (0.695 mg, 23.1 mmol, 4 eq) and trifluoroacetic acid (89.0 μL, 1.16 mmol, 0.2 eq) were added to tetrahydrofuran (20 mL). A condenser was added and the suspension was stirred at reflux for 2 h. Paraformaldehyde (**384**) (0.695 mg, 23.1 mmol, 4 eq) was added and the reflux was continued for 6 h. The reaction mixture was cooled to room temperature and the tetrahydrofuran was removed under reduced pressure. The crude product was dissolved in diethyl ether (25 mL) and filtered through cotton in a separatory funnel. The organic layer was washed twice with 1 M HCl (25 mL). The aqueous layers were combined and extracted with diethyl ether (25 mL). The organic layers were combined, dried over anhydrous Na_2SO_4, filtered and concentrated under reduced pressure to give crude methylenemalonate (2.9 g) as colorless oil.

The iron catalyst (289 mg, 0.289 mmol, 0.05 eq) was weighted in an oven-dry flask in a glovebox. The flask was closed with a silicon septum, taken out of the glovebox and put under a positive pressure of nitrogen. Dichloromethane (5 mL) was added. The reaction was cooled to 0 °C and 2-vinylisoindoline-1,3-dione (**183**) (1.00 g, 5.77 mmol, 1 eq) was dissolved in dichloromethane (5 mL) and added to the yellow suspension dropwise. Finally, the crude methylenemalonate was dissolved in dichloromethane (5 mL) and added to the reaction mixture dropwise. The reaction mixture was stirred at 0 °C for 3h30 and then filtered over a basic alumina plug, eluting with ethyl acetate. The solvents were evaporated and the brown solid was purified by column chromatography on Biotage (SNAP cartridge KP-SIL 50 g, 95:5–4:6 Hexane/Ethyl acetate) affording 1-(Tert-butyl) 1-methyl 2-(1,3-dioxoisoindolin-2-yl)cyclobutane-1,1-dicarboxylate (**325 s**) (1.97 g,

5.48 mmol, 95 % yield, 1.7:1 dr determined by integration of the peaks at 2.45-2.33 (*maj*), and 2.31–2.21 (*min*) in the crude ^{1}H NMR) as a colorless oil.

R$_f$ 0.22 (7:3 Hexane/Ethyl acetate). **^{1}H NMR** (400 MHz, $CDCl_3$) on a 1.5 (*maj*):1(*min*) diatereomeric mixture δ 7.95–7.78 (m, 5 H, Phth), 7.78–7.64 (m, 5 H, Phth), 5.58–5.26 (m, 2.5 H, N–C–H *major* + *minor*), 3.75 (s, 4.5 H, OMe *major*), 3.57 (s, 3 H, OMe *minor*), 3.36–3.10 (m, 2.5 H, CH_2 cyclobutane *major* + *minor*), 3.08–2.83 (m, 2.5 H, CH_2 cyclobutane *major* + *minor*), 2.45–2.33 (m, 1.5 H, CH_2 cyclobutane *major*), 2.31–2.21 (m, 1 H, CH_2 cyclobutane *minor*) 2.20–2.10 (m, 2.5 H, CH_2 cyclobutane *major* + *minor*), 1.44 (s, 9 H, *^{t}Bu minor*), 1.16 (s, 13.5 H, tBu *major*). **^{13}C NMR** (101 MHz, $CDCl_3$) δ 171.2, 169.0, 169.0, 168.3, 168.2, 167.2, 134.3, 134.2, 132.1, 131.9, 123.4, 123.4, 82.5, 82.2, 60.1, 59.5, 53.0, 52.8, 47.7, 47.6, 27.9, 27.5, 24.8, 24.3, 21.7, 21.3. **IR** 2979 (w), 1732 (s), 1717 (s), 1372 (s), 1266 (s), 1129 (m). **HRMS (ESI)** calcd for $C_{19}H_{22}NO_6^+$ $[M + H]^+$ 360.1442; found 360.1437.

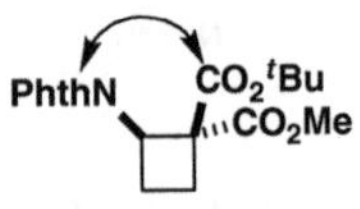

ROESY: major

Dibenzyl 2-(1,3-dioxoisoindolin-2-yl)cyclobutane-1,1-dicarboxylate (325t)

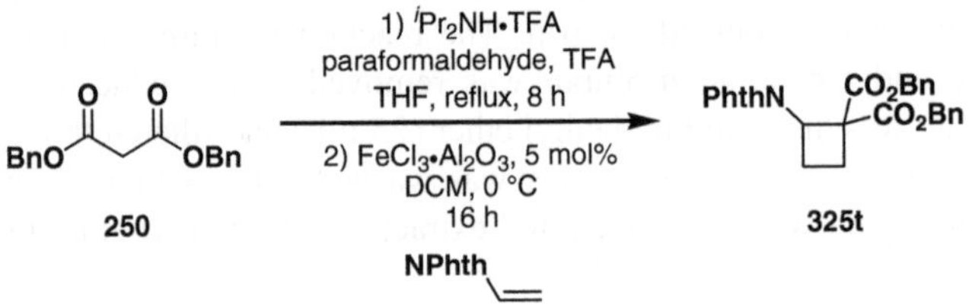

Dibenzyl malonate (**250**) (2.89 mL, 11.6 mmol, 2 eq), diisopropylamine 2,2,2-trifluoroacetate (2.49 g, 11.6 mmol, 2 eq), paraformaldehyde (**384**) (0.695 mg, 23.1 mmol, 4 eq) and trifluoroacetic acid (89.0 μL, 1.16 mmol, 0.2 eq) were added to tetrahydrofuran (20 mL). A condenser was added and the suspension was stirred at reflux for 2 h. Paraformaldehyde (**384**) (0.695 mg, 23.1 mmol, 4 eq) was added and the reflux was continued for 6 h. The reaction was cooled to room temperature and the tetrahydrofuran was removed under reduced pressure. The crude was re-taken in diethyl ether (25 mL) and filtered through cotton in a separatory funnel. The organic layer was washed twice with 1 M HCl (25 mL). The aqueous layers were combined and extracted with diethyl ether (25 mL). The organic layers were combined, dried over anhydrous Na_2SO_4, filtered and concentrated under reduced pressure to give dibenzyl 2-methylenemalonate crude as colorless oil.

The iron catalyst (289 mg, 0.289 mmol, 0.05 eq) was weighted in an oven-dry flask in a glovebox. The flask was closed with a silicon septum, taken out of the glovebox and put under a positive pressure of nitrogen. Dichloromethane (5 mL) was added. The reaction was cooled to 0 °C and 2-vinylisoindoline-1,3-dione (**183**) (1.00 g, 5.77 mmol, 1 eq) was dissolved in dichloromethane (5 mL) and added to the yellow suspension dropwise. Finally, the crude dibenzyl 2-methylenemalonate was dissolved in dichloromethane (5 mL) and added to the reaction mixture dropwise. The reaction mixture was stirred at room temperature for 2 h and then filtered over a basic alumina plug, eluting with ethyl acetate. The solvents were evaporated and the brown solid was purified by column chromatography on Biotage (SNAP cartridge KP-SIL 50 g, 95:5–4:6 Hexane/Ethyl acetate) affording dibenzyl 2-(1,3-dioxoisoindolin-2-yl)cyclobutane-1,1-dicarboxylate (**325t**) (2.60 g, 5.54 mmol, 96 % yield) as a colorless oil that solidify upon storage at 4 °C.

R_f 0.45 (6:4 Hexane/Ethyl acetate). **Mp** 132.1–133.8 °C. **^{1}H NMR** (400 MHz, $CDCl_3$) δ 7.74–7.68 (m, 2 H, Phth), 7.68–7.61 (m, 2 H, Phth), 7.34–7.26 (m, 5 H, Ar), 7.14–7.07 (m, 1 H, Ar), 7.06–7.00 (m, 2 H, Ar), 6.97–6.91 (m, 2 H, Ar), 5.50 (td, *J* = 9.5, 0.9 Hz, 1 H, N–C–H), 5.17 (q, *J* = 12.3 Hz, 2 H, CH_2Bn), 5.02–4.87 (m, 2 H, CH_2Bn), 3.34–3.16 (m, 1 H, CH_2 cyclobutane), 3.02 (dddd, *J* = 11.7, 10.5, 3.8, 1.0 Hz, 1 H, CH_2 cyclobutane), 2.32 (dtd, *J* = 11.2, 9.1, 3.7 Hz, 1 H, CH_2 cyclobutane), 2.18 (dt, *J* = 11.6, 8.8 Hz, 1 H, CH_2 cyclobutane). **^{13}C NMR** (101 MHz, $CDCl_3$) δ 169.7, 168.1, 167.9, 135.3, 134.7, 134.0, 131.6, 128.5, 128.3, 128.2, 128.1, 128.1, 126.0, 123.3, 67.7, 67.5, 59.2, 47.6, 24.4, 21.8. **IR** 3034 (w), 2360 (w), 2339 (w), 1780 (w), 1738 (s), 1721 (s), 1379 (s), 1378 (s), 1261 (m), 1260 (m). **HRMS (ESI)** calcd for $C_{28}H_{24}NO_6^+$ $[M + H]^+$ 470.1598; found 470.1607.

Dimethyl 2-(1,3-dioxoisoindolin-2-yl)-3-methylcyclobutane-1,1-dicarboxylate (325q)

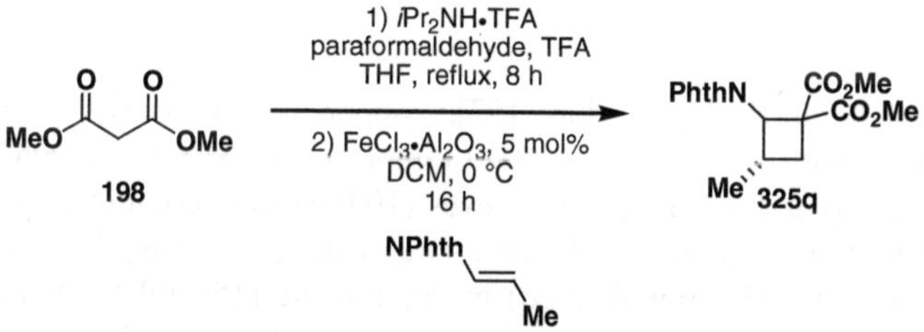

Dimethyl malonate (**250**) (1.22 mL, 10.7 mmol, 2 eq), diisopropylamine 2,2,2-trifluoroacetate (2.30 g, 10.7 mmol, 2 eq), paraformaldehyde (**384**) (0.640 mg, 21.4 mmol, 4 eq) and trifluoroacetic acid (82.0 μl, 1.07 mmol, 0.2 eq) were added to tetrahydrofuran (20 mL). A condenser was added and the suspension was stirred at reflux for 2 h. Paraformaldehyde (**384**) (0.640 mg, 21.4 mmol) was added and the reflux was continued for 6 h. The reaction was cooled to room temperature and the tetrahydrofuran was removed under reduced pressure. The crude was dissolved in diethyl ether (25 mL) and filtered through cotton in a separatory funnel. The organic layer was washed twice with 1 M HCl (25 mL). The aqueous layers were combined and extracted with diethyl ether (25 mL). The organic layers were combined, dried over anhydrous Na_2SO_4, filtered and concentrated under reduced pressure to give dimethyl 2-methylenemalonate crude as colorless oil.

The iron catalyst (267 mg, 0.267 mmol, 0.05 eq) was weighted in an oven-dry flask in a glovebox. The flask was closed with a silicon septum, taken out of the glovebox and put under a positive pressure of nitrogen. Dichloromethane (5 mL) was added. The reaction was cooled to 0 °C and (*E*)-2-(prop-1-en-1-yl) isoindoline-1,3-dione (**331**) (1.00 g, 5.34 mmol, 1 eq) was dissolved in dichloromethane (5 mL) and added to the yellow suspension dropwise. Finally, the crude dimethyl 2-methylenemalonate was dissolved in dichloromethane (5 mL) and added to the reaction mixture dropwise. The reaction mixture was stirred at room temperature for 4 h and then filtered over a basic alumina plug, eluting with ethyl acetate. The solvents were evaporated and the brown solid was purified by column chromatography on Biotage (SNAP cartridge KP-SIL 50 g, 95:5–4:6 Hexane/Ethyl acetate) affording dimethyl 2-(1,3-dioxoisoindolin-2-yl)-3-methylcyclobutane-1,1-dicarboxylate (**325q**) (1.53 g, 4.62 mmol, 86 % yield) as a colorless solid.

Synthesis of labeled reagents
2-(Trimethylsilyl)ethynyl)isoindoline-1,3-dione (353)

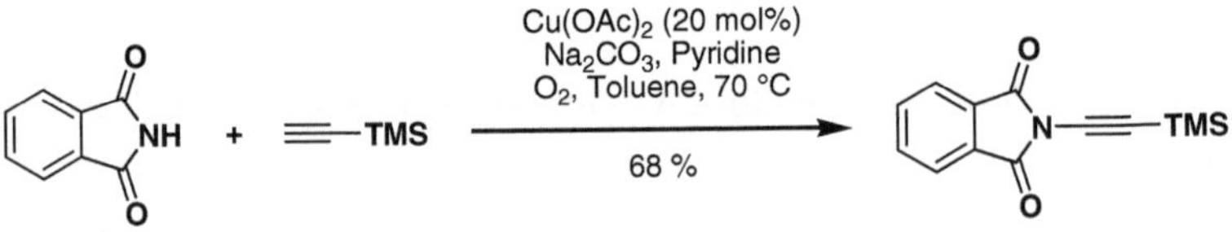

Following a reported procedure [52], copper acetate (0.617 g, 3.40 mmol, 0.2 eq), phthalimide (**332**) (12.5 g, 85.0 mmol, 5 eq), sodium carbonate (3.60 g, 34.0 mmol, 2 eq) and 4 Å molecular sieves (10.0 g) were combined in a 1 L three neck round bottom flask equipped with a large magnetic stirring bar. A solution of pyridine (2.75 mL, 34.0 mmol, 2 eq) in dry toluene (150 mL) was added to the reaction flask. The mixture was stirred vigorously and the reaction atmosphere was flushed using oxygen from a balloon. Finally a large balloon of oxygen was connected to the flask and the reaction was stirred in a preheated oil bath at 70 °C. After 2 h of stirring at 70 °C, a solution of ethynyltrimethylsilane (**352**) (2.42 mL, 17.0 mmol, 1 eq) in dry toluene (20 mL) was added to the flask in 2 h using syringe pump. After the end of addition, the reaction was stirred for 15 additional hours at 70 °C. The reaction mixture was filtered warm through a glass frit and the filtrate was concentrated under reduced pressure. The residue was suspended in diethyl ether (50 mL) and washed with sat. NH_4Cl (50 mL) The layers were separated and the organic layer was dried over anhydrous Na_2SO_4, filtered and concentrated to dryness to give an off-white solid. Purification of the crude solid by column chromatography (SiO_2, hexane/ethyl acetate, 95:5 to 80:20,) afforded 2-(trimethylsilyl)ethynyl)isoindoline-1,3-dione (**353**) (2.80 g, 11.5 mmol, 68 %) as a white fluffy solid.

^{1}H NMR (400 MHz, $CDCl_3$) δ 7.93 (dd, J = 5.5, 3.1 Hz, 2 H), 7.81 (dd, J = 5.5, 3.1 Hz, 2 H), 0.29 (s, 9 H).

Data of **353** match literature report [52].

(Z)-2-(Vinyl-2-d)isoindoline-1,3-dione (355)

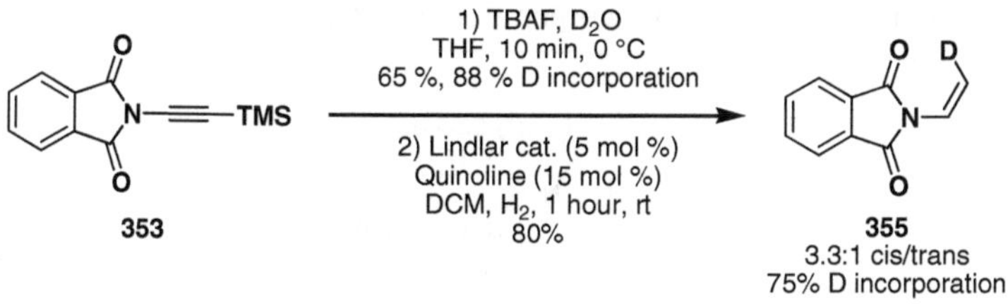

2-(Trimethylsilyl)ethynyl)isoindoline-1,3-dione (**353**) (300 mg, 1.23 mmol, 1 eq) was dissolved in thetrahydrofuran (1 mL), then D_2O (0.5 mL) was added and the reaction was stirred at 0 °C for 5 min. TBAF (1 M in thetrahydrofuran, 1.48 mL, 1.48 mmol, 1.2 eq) was added in another flask containing tetrahydrofuran (2 mL) and D_2O (0.5 mL). The TBAF solution was then added dropwise to the ynimide solution. After 10 min at 0 °C (some ice has formed during the process), ethyl acetate (10 mL) was added, followed by sat. NH_4Cl (10 mL). The layers were stirred vigorously and the organic layer was separated, dried over anhydrous Na_2SO_4, filtered and concentrated to dryness to give an orange solid. The crude product was purified by column chromatography (SiO_2, 7:3, hexane/ethyl acetate) affording deuterated ynamide **354** (138 mg, 0.802 mmol, 65, 88 % deuterium incorporation determined by the integrations of the peak at 7.88–7.79 and 3.34) as colorless solid.

^{1}H NMR (400 MHz, $CDCl_3$) δ 8.02–7.88 (m, 16 H), 7.88–7.79 (m, 16 H), 3.34 (s, 1 H). **^{13}C NMR** (126 MHz, $CDCl_3$) δ 165.1, 135.4, 131.0, 124.5, 67.9[28].

Lindlar catalyst (25 mg, 0.012 mmol, 0.05 eq) and quinoline (5.0, 0.039 mmol, 0.16 eq) were added in dichloromethane (1.0 mL) in a flask under nitrogen atmosphere. The mixture was stirred for 5 min and 2-(ethynyl-d) isoindoline-1,3-dione (40 mg, 0.23 mmol, 1 eq) was added as a solution in dichloromethane (0.5 mL). The reaction atmosphere was flushed with hydrogen and a hydrogen balloon was connected to the flask. The reaction was stirred for 1 h at room temperature and then filtered on a Celite pad, eluting with dichloromethane. The solvent was removed under reduced pressure and the crude product was purified by column chromatography on Biotage (SNAP cartridge KP-SIL 10 g,

[28]The deuterated carbon was not resolved.

95:5–4:6 Hexane/Ethyl acetate) affording (Z)-2-(vinyl-2-d)isoindoline-1,3-dione (**355**) (32.5 mg, 0.190 mmol, 80 % yield with 20 % of the saturated compound, 3.3:1 Z/E ratio and 75 % deuterium incorporation) as a colorless solid.

$\mathbf{R_f}$ 0.60 (7:3 Hexane/Ethyl acetate). **^{1}H NMR** (400 MHz, $CDCl_3$) δ 7.99–7.63 (m, 4.7 H, Phth), 6.97–6.77 (m, 1 H, H_1 *no-D* + *Z* + *E*), 6.09 (dd, *J* = 16.4, 5.2 Hz, 0.43 H, H_2 *no-D* + *E*), 5.05 (dd, *J* = 9.9, 5.4 Hz, 0.83 H, H_3 *no-D* + *Z*), 3.81–3.68 (m, 0.35 H, CH_2 saturated), 1.35–1.19 (m, 0.50 H, CH_3 saturated). **^{13}C NMR** (101 MHz, $CDCl_3$) δ 166.7, 134.6, 131.8, 124.0, 123.8, 104.3 (t, *J* = 20 Hz). **IR** 1776 (w), 1724 (s), 1617 (w), 1468 (w), 1383 (s). **HRMS (ESI)** calcd for $C_{10}H_7[^2H]NO_2^+$ $[M + H]^+$ 175.0611; found 175.0620.

Determination of deuterium incorporation and Z/E ratio:

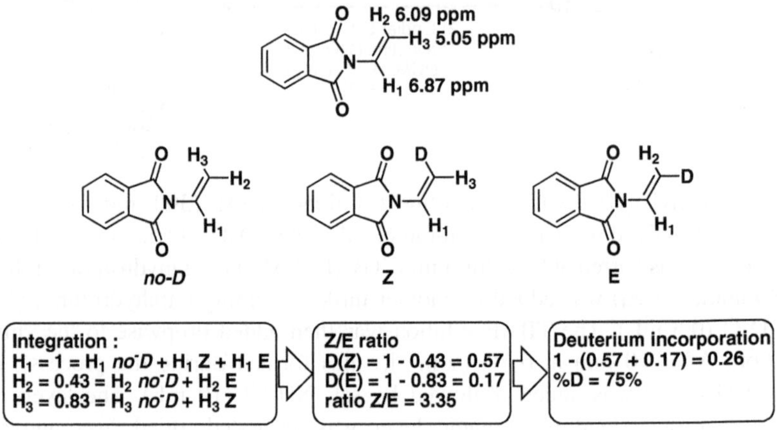

Dimethyl 2-(1,3-dioxoisoindolin-2-yl)cyclobutane-1,1-dicarboxylate-3-d (356)

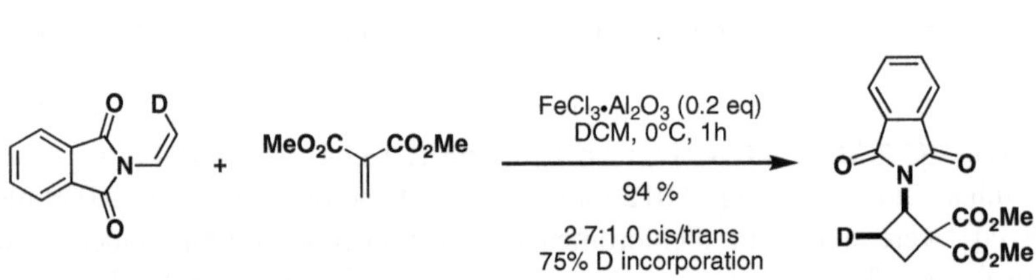

355 **323a** **356**

In the glovebox, Iron trichloride supported on alumina (1.00 mmol/g, 11.0 mg, 0.011 mmol, 0.1 eq) was added to a microwave vial. The vial was sealed with a Teflon septum and taken out of the glovebox. Dry dichloromethane (200 μL) was added and the yellow suspension was cooled to 0 °C. Dimethyl 2-methylenemalonate (**323a**) (16.4 mg, 0.114 mmol, 2 eq) and (Z)-2-(vinyl-2-D) isoindoline-1,3-dione (**355**) (12.4 mg, 0.0570 mmol, 1 eq) were dissolved in dry dichloromethane (800 μL). The solution was then added dropwise to the iron trichloride solution. The reaction was stirred at 0 °C for 1 h. The reaction mixture was filtered over a pad of alumina, eluted with ethyl acetate and concentrated under vacuum. The crude product was purified by column chromatography on Biotage (SNAP cartridge KP-SIL 10 g, 95:5–4:6 Hexane/Ethyl acetate) affording dimethyl 2-(1,3-dioxoisoindolin-2-yl)cyclobutane-1,1-dicarboxylate-3-d (**356**) (17.0 mg, 0.0530 mmol, 94 % yield, 2.7:1.0 diastereomeric ratio and 75 % deuterium incorporation determined by the same method than previously, integrating the peaks at 5.51–5.39 (no-D + cis + trans) 3.32–3.15 (trans + no-D) and 2.36–2.23 (cis + no-D) as a colorless solid. The relative stereochemistry was determined by ROESY experiments.

R_f 0.45 (1:1 Hexane/Ethyl acetate). **^{1}H NMR** (500 MHz, $CDCl_3$) δ 7.91–7.77 (m, 2 H, Phth), 7.75–7.64 (m, 2 H, Phth), 5.51–5.39 (m, 1 H, H_1 *cis + trans + no-D*), 3.75 (s, 3 H, OMe), 3.60 (s, 3 H, OMe), 3.32–3.15 (m, 0.46 H, H_2 *trans + no-D*), 3.05–2.85 (m, 1 H, CH_2 cyclobutane), 2.36–2.23 (m, 0.83 H, H_3 *cis + no-D*), 2.23–2.09 (m, 1 H, CH_2 cyclobutane). **^{13}C NMR** (101 MHz, $CDCl_3$) δ 170.4, 168.5, 168.1, 134.1, 131.7, 123.3, 58.74, 58.69, 53.0, 52.8, 47.6, 47.50, 47.48, 24.4, 24.3, 21.7, 21.6, 21.4, 21.2. **IR** 2959 (w), 2923 (w), 2852 (w), 1779 (w), 1738 (s), 1715 (s), 1436 (w), 1377 (s), 1263 (s). **HRMS (ESI)** calcd for $C_{16}H_{15}[^2H]NO_6^+$ $[M + H]^+$ 319.1034; found 319.1025.

Determination of deuterium incorporation and cis/trans ratio:

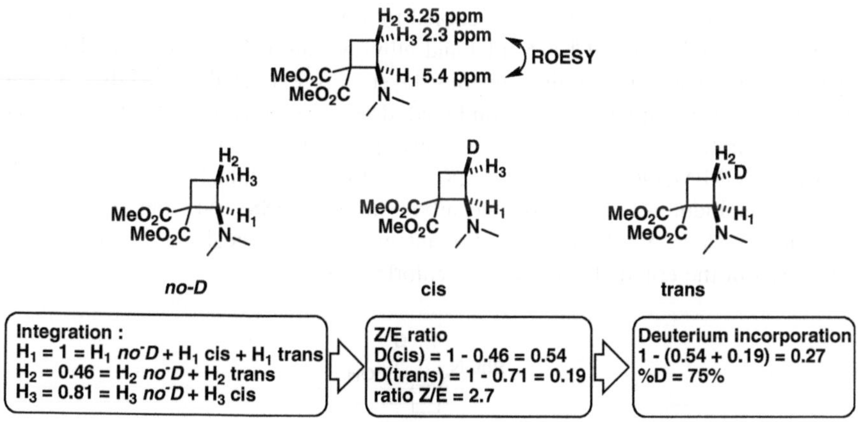

Synthesis of dipeptide
Ethyl (-2-(1,3-dioxoisoindolin-2-yl)cyclobutane-1-carbonyl)glycinate (357)

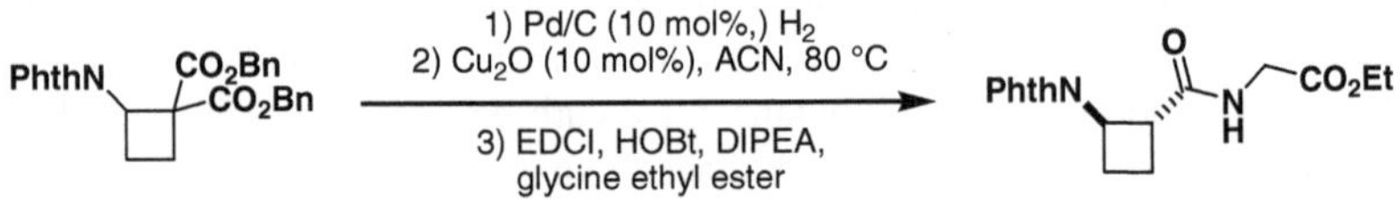

325t 357

Under nitrogen, dibenzyl 2-(1,3-dioxoisoindolin-2-yl) cyclobutane-1,1-dicarboxylate (**325t**) (1.00 g, 2.13 mmol, 1 eq) was dissolved in technical ethanol (15 mL). Palladium on charcoal 5 % (0.453 g, 0.213 mmol, 0.1 eq) was added and the reaction atmosphere was purged with hydrogen. The reaction was stirred at room temperature for 5 h. Celite (ca 5 g) was added to the reaction and the suspension was filtered through a pad of Celite. The cake was rinsed abundantly with hot ethanol. The solvents were removed on a rotary evaporator and the solid crude diacid was directly used for the next step.

In a glovebox, copper (I) oxide (30.5 mg, 0.213 mmol, 0.1 eq) was added in a vial which was closed with a silicon septum and removed from the glovebox. The crude diacid was quickly added as a solid, the vial was sealed and three cycles of vacuum/N_2 were performed. Dry acetonitrile (2 mL) was added and the reaction was stirred in an oil bath at 80 °C for 3 h when no more starting material was detected by NMR. The reaction was cooled to room temperature and poured into a separatory funnel. 1 M HCl (15 mL) and ethyl acetate (15 mL) were added. The layers were separated and the aqueous layer was extracted with ethyl acetate (15 mL). The organic layers were combined, dried over anhydrous Na_2SO_4, filtered and concentrated to dryness. The crude was purified on column chromatography (SiO_2, DCM/MeOH/AcOH, 98:2:0.01 to 90:10:0.01) affording 2-(1,3-dioxoisoindolin-2-yl)cyclobutanecarboxylic acid (**388**) (447 mg, 1.83 mmol, 86 % on two steps, 5:1 dr determined by integration of the peaks at 5.02 (*maj*), and 4.91 (*min*) in the crude ^{1}H NMR) as a colorless oil.

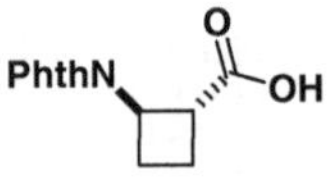

R_f 0.23 (90:10:0.01 DCM/MeOH/AcOH). **Mp** 103.8–105.9 °C. **^{1}H NMR** (400 MHz, $CDCl_3$) on a 5 (*maj*):1 (*min*) diastereomeric mixture δ 7.81 (m, 12 H, Phth), 7.71 (m, 12 H, Phth), 5.02 (q, *J* = 9.1 Hz, 5 H, N–C–H, *major*), 4.91 (q, *J* = 9.2 Hz, 1 H, N–C–H, *minor*), 4.12 (q, *J* = 9.4 Hz, 1 H, CH_2 cyclobutane, *minor*), 3.65–3.54 (m, 5 H, CH_2 cyclobutane, *major*), 3.21–3.08 (m, 5 H, CH_2

cyclobutane, *major*), 2.79 (m, 1 H, CH_2 cyclobutane, *minor*), 2.64 (m, 5 H, CH_2 cyclobutane, *major*), 2.42 (m, 5 H, CH_2 cyclobutane, *major*), 2.31–1.97 (m, 8 H, CH_2 cyclobutane, *major* + *minor*). **^{13}C NMR** (101 MHz, $CDCl_3$) δ 178.2, 177.2, 168.5, 168.2, 134.1, 134.0, 131.8, 131.7, 123.34, 123.29, 46.3, 45.5, 43.2, 42.5, 24.6, 24.0, 19.7, 18.9.

2-(1,3-Dioxoisoindolin-2-yl)cyclobutanecarboxylic acid (**388**) (200 mg, 0.816 mmol, 1 eq), N-ethyl-N-isopropylpropan-2-amine (316 mg, 2.45 mmol, 3 eq) and Hydroxybenzotriazole (187 mg, 1.22 mmol, 1.5 eq), were added in a flask, dichloromethane (2 mL) was added and the reaction was cooled to 0 °C. Then, 3-((ethylimino)methylene)amino)-N,N-dimethylpropan-1-amine hydrochloride (235 mg, 1.22 mmol, 1.5 eq) and finally ethyl 2-aminoacetate hydrochloride (114 mg, 0.816 mmol, 1 eq) were added and the reaction was warmed to room temperature and stirred for 24 h. The reaction was then concentrated to dryness and purified on column chromatography (SiO_2, DCM/MeOH/AcOH, 98:2:0.01 to 90:10:0.01) affording ethyl 2-(2-(1,3-dioxoisoindolin-2-yl)cyclobutanecarboxamido)acetate (**357**) (244 mg, 0.740 mmol, 91 %, 5:1 dr determined by integration of the peaks at 3.33–3.14 (*maj*), and 4.26–4.13 (*min*) in the isolated product ^{1}H NMR) as a colorless oil.

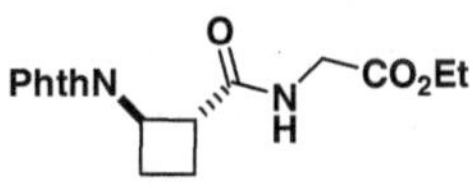

R_f 0.35 (9/1/0.01 DCM/MeOH/AcOH). **^{1}H NMR** (400 MHz, $CDCl_3$) on a 5 (*maj*):1(*min*) diatereomeric mixture δ 7.88–7.75 (m, 12 H, Phth *major* + *minor*), 7.74–7.62 (m, 12H, Phth *major* + *minor*), 6.24 (s, 1 H, NH *minor*), 5.97 (s, 5 H, NH, *major*), 5.01–4.83 (m, 6 H, N–C–H *major* + *minor*), 4.26–4.13 (m, 2 H, CH_2 O–CH_2 *minor*), 4.11–3.96 (m, 13 H, 2 H O–CH_2 *major* + 2 H CH_2 glycine *minor* + 1H CH_2 cyclobutane *minor*), 3.95–3.79 (m, 10 H, CH_2 glycine *major*), 3.66–3.49 (m, 5 H, CH_2 cyclobutane *major*), 3.33–3.14 (m, 5 H, CH_2 cyclobutane *major*), 2.80–2.57 (m, 6 H, CH_2 cyclobutane *major* + *minor*), 2.57–2.44 (m, 5 H, CH_2 cyclobutane *major*), 2.30–2.02 (m, 8 H, 1 H CH_2 cyclobutane *major* + 2 H CH_2 cyclobutane *minor*), 1.26 (t, J = 7.2 Hz, 3 H, CH_3 *minor*), 1.19 (t, J = 7.1 Hz, 15 H, CH_3 *major*). **^{13}C NMR** (101 MHz, $CDCl_3$) δ 172.2, 171.3, 169.8, 169.6, 168.6, 168.3, 134.1, 133.7, 132.0, 131.7, 123.2, 123.1, 61.4, 61.2, 46.9, 46.8, 44.5, 44.0, 41.3, 41.3, 24.3, 24.0, 20.6, 19.0, 14.03, 13.98. **IR** 3595 (w), 3377 (w), 2987 (w), 1777 (w), 1748 (w), 1713 (s), 1662 (w), 1538 (w), 1381 (s), 1201 (m), 1027 (w). **HRMS (ESI)** calcd for $C_{17}H_{19}N_2O_5^+$ $[M + H]^+$ 331.1288; found 331.1290.

5.8 [4 + 2] Annulation of Aminocyclobutanes

GP29 for the synthesis of TBS enol ethers:

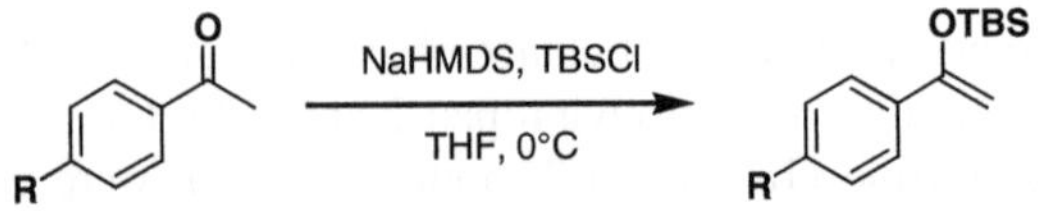

Ketone (1 eq) is dissolved in dry THF (20 mL) under nitrogen atmosphere and cooled down to 0 °C. NaHMDS (1.2 eq, 1.9 M in THF) is added dropwise. The orange solution is stirred at room temperature for 1 h and then cooled to 0 °C. The chlorosilane (1.2 eq) is dissolved in THF (3 mL) and added dropwise. The reaction is stirred at room temperature for 2 h and then concentrated under reduced pressure. The crude oil is filtered through a silica plug, eluting with 250 mL hexanes, concentrated under reduced pressure and purified by bulb to bulb distillation.

Methyl 4-(1-(*tert*-butyldimethylsilyl)oxy)vinyl)benzoate (360a)

Following GP29, methyl 4-acetylbenzoate (1.0 g, 5.6 mmol, 1 eq), NaHMDS (3.5 mL, 6.7 mmol, 1.2 eq and *tert*-butylchlorodimethylsilane (1.0 g, 6.7 mmol, 1.2 eq) were reacted. Methyl 4-(1-(*tert*-butyldimethylsilyl)oxy)vinyl)benzoate (**360a**) (0.50 g, 1.7 mmol, 31 %) was obtained as a colorless oil.

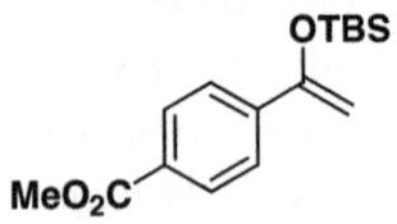

^{1}H NMR (400 MHz, DMSO-d_6) δ 8.07–7.86 (m, 2 H, Ar), 7.86–7.68 (m, 2 H, Ar), 5.20 (d, J = 2.2 Hz, 1 H, CH_2 vinyl), 4.59 (d, J = 2.2 Hz, 1 H, CH_2 vinyl), 3.86 (s, 3 H, OMe), 0.98 (s, 9 H, OTBS), 0.22 (s, 6 H, OTBS). **^{13}C NMR** (101 MHz, DMSO) δ 165.8, 153.9, 141.4, 129.2, 125.0, 93.8, 52.1, 30.7, 25.6, 18.0, −4.8.[29] **IR** 2888 (w), 2866 (w), 2821 (w), 2820 (w), 2787 (w), 1583 (s), 1464 (w), 1146 (m), 1106 (s), 937 (m), 926 (m), 828 (m). **HRMS (ESI)** calcd for $C_{16}H_{25}O_3Si^+$ $[M + H]^+$ 293.1567; found 293.1560.

***Tert*-butyl(1-(4-methoxyphenyl)vinyl)oxy)dimethylsilane (360b)**

Following GP29, 1-(4-methoxyphenyl)ethanone (2.0 g, 13 mmol, 1 eq), NaHMDS (8.4 mL, 16 mmol, 1.2 eq and *tert*-butylchlorodimethylsilane (2.4 g, 16 mmol, 1.2 eq) were reacted. *Tert*-butyl(1-(4-methoxyphenyl)vinyl)oxy)dimethylsilane (**360b**) (2.5 g, 9.5 mmol, 71 %) was obtained as a colorless oil.

[29]One peak corresponding to the Me of TBDMS has been cut due to small acquisition window.

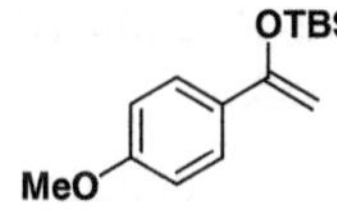

^{1}H NMR (400 MHz, DMSO-d_6) δ 7.54 (d, J = 8.9 Hz, 2 H, Ar), 6.93 (d, J = 8.9 Hz, 2 H, Ar), 4.89 (d, J = 1.8 Hz, 1 H, CH_2 vinyl), 4.32 (d, J = 1.8 Hz, 1 H, CH_2 vinyl), 3.77 (s, 3 H, OMe), 0.98 (s, 9 H, OTBS), 0.20 (s, 6 H, OTBS). **^{13}C NMR** (101 MHz, DMSO) δ 159.4, 154.7, 129.5, 126.2, 113.5, 89.4, 55.1, 25.7, 18.0, −4.7.

NMR data of **360b** corresponded to the reported spectra [53].

Tert-butyl(1-(4-fluorophenyl)vinyl)oxy)dimethylsilane (360c)

Following GP29, 1-(4-fluorophenyl)ethanone (2.0 g, 15 mmol, 1 eq), NaHMDS (9.1 mL, 17 mmol, 1.2 eq and *tert*-butylchlorodimethylsilane (2.6 g, 17 mmol, 1.2 eq) were reacted. *Tert*-butyl(1-(4-fluorophenyl)vinyl)oxy)dimethylsilane (**360c**) (0.80 g, 3.2 mmol, 22 %) was obtained as a colorless oil.

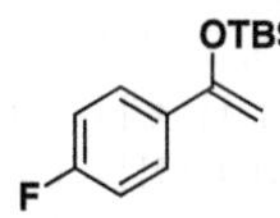

^{1}H NMR (400 MHz, DMSO-d_6) δ 7.65 (dd, J = 9.0, 5.5 Hz, 2 H, Ar), 7.21 (t, J = 8.9 Hz, 2 H, Ar), 5.00 (d, J = 2.1 Hz, 1 H, CH_2 vinyl), 4.43 (d, J = 2.0 Hz, 1 H, CH_2 vinyl), 0.98 (d, J = 0.8 Hz, 9 H, OTBS), 0.21 (d, J = 0.9 Hz, 6 H, OTBS). **^{13}C NMR** (101 MHz, DMSO) δ 162.1 (d, J = 245.5 Hz), 153.9, 133.5 (d, J = 3.2 Hz), 126.9 (d, J = 8.4 Hz), 115.1 (d, J = 21.6 Hz), 91.1, 25.6, 17.9, −4.8. **IR** 2891 (w), 2866 (w), 2821 (w), 2788 (w), 1564 (w), 1470 (w), 1456 (m), 1456 (m), 1351 (m), 1311 (w), 1144 (s), 1144 (s), 1122 (s), 1085 (m), 1060 (s), 1060 (s), 823 (s). **HRMS (ESI)** calcd for $C_{14}FH_{21}KOSi^{+}$ $[M + K]^{+}$ 291.0977; found 291.1180.

GP30 for [4 + 2] annulations

4 Å MS pellets (ca 20 mg) were added in an oven dried 5 mL round bottom flask. The flask was closed with a silicon septum and three cycles of vacuum/N_2 were performed. Aminocyclobutane (0.200 mmol, 1 eq) and silyl enol ether (0.300 mmol, 1.5 eq) were dissolved in dichloromethane (2 mL) and added to the reaction flask. The solution was cooled to −40 °C using an acetonitrile/N_2 bath. A solution of tin tetrachloride (0.43 mol/L, 93.0 µL, 0.0400 mmol, 0.2 eq) was added dropwise and the reaction was stirred at −40 °C until full conversion was observed by NMR. The reaction was then quenched by adding triethylamine (0.1 mL) and the solvent was removed under reduced pressure. The reaction was purified by column chromatography on Biotage (SNAP cartridge KP-SIL 25 g, 95:5–4:6 Hexane/Ethyl acetate) affording the aminocyclohexanes.

Dimethyl-4-(1,3-dioxoisoindolin-2-yl)-2-phenyl-2-(triisopropylsilyl)oxy) cyclohexane-1,1-dicarboxylate (358a)

Following GP30 using dimethyl 2-(1,3-dioxoisoindolin-2-yl) cyclobutane-1,1-dicarboxylate (**325a**) (63.5 mg, 0.200 mmol, 1 eq) and triisopropyl (1-phenylvinyl)oxy)silane (**219a**) (83.0 mg, 0.300 mmol, 1.5 eq), dimethyl-4-(1,3-dioxoisoindolin-2-yl)-2-phenyl-2-(triisopropylsilyl)oxy) cyclohexane-1,1-dicarboxylate (**358a**) (98.0 mg, 0.165 mmol, 82 %) was obtained as a colorless solid.

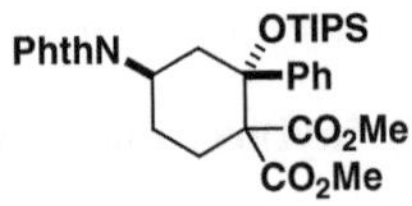

R*f* 0.6 (1:1 Hexane/Ethyl acetate). **^{1}H NMR** (400 MHz, $CDCl_3$) δ 7.84 (dd, *J* = 5.4, 3.0 Hz, 2 H, Ar Phth), 7.72 (dd, *J* = 5.4, 3.0 Hz, 2 H, Ar Phth), 7.52–7.44 (m, 2 H, Ar), 7.26–7.21 (m, 3 H, Ar), 4.86 (ddd, *J* = 12.5, 7.9, 4.7 Hz, 1 H, N-C-H), 4.02 (dd, *J* = 13.6, 12.7 Hz, 1 H, CH_2), 3.66 (s, 3 H, OMe), 3.62 (s, 3 H, OMe), 2.99 (td, *J* = 14.3, 3.8 Hz, 1 H, CH_2), 2.31 (dt, *J* = 13.9, 3.5 Hz, 1 H, CH_2), 2.27–2.06 (m, 2 H, CH_2), 1.88–1.77 (m, 1 H, CH_2), 1.19–1.08 (m, 3 H, OTIPS), 1.06–0.95 (m, 18 H, OTIPS). **^{13}C NMR** (101 MHz, $CDCl_3$) δ 169.9, 169.7, 168.3, 143.4, 133.9, 132.0, 128.7, 127.6, 126.2, 123.1, 80.4, 64.4, 52.2, 51.7, 46.3, 36.8, 29.3, 25.0, 18.4, 18.4, 14.2. **IR** 2883 (w), 2797 (w), 2259 (w), 1594 (s), 1570 (s), 1208 (w), 1070 (w), 852 (m), 726 (w). **HRMS (ESI).**[30]

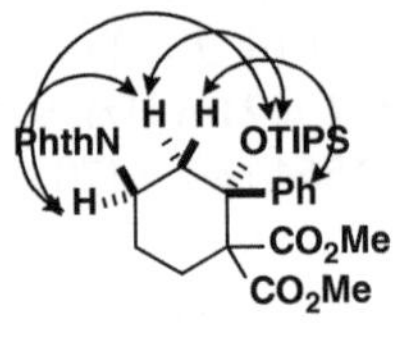

ROESY

Dibenzyl-4-(1,3-dioxoisoindolin-2-yl)-2-phenyl-2-(triisopropylsilyl)oxy) cyclohexane-1,1-dicarboxylate (358b)

Following GP30 using dibenzyl 2-(1,3-dioxoisoindolin-2-yl) cyclobutane-1,1-dicarboxylate (**325u**) (94.0 mg, 0.200 mmol, 1 eq) and tri*iso*propyl (1-phenylvinyl)oxy)silane (**219a**) (83.0 mg, 0.300 mmol, 1.5 eq), dibenzyl-4-(1,3-dioxoisoindolin-2-yl)-2-phenyl-2-(tri*iso*propylsilyl)oxy) cyclohexane-1,1-dicarboxylate (**358b**) (96.0 mg, 0.129 mmol, 64 %) was obtained as a colorless solid.

[30]HRMS could not be detected for this compound.

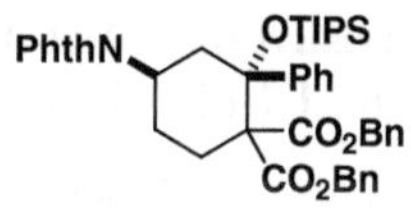

^{1}H NMR (400 MHz, $CDCl_3$) δ 7.85 (dd, J = 5.4, 3.0 Hz, 2 H, Ar Phth), 7.72 (dd, J = 5.4, 3.1 Hz, 2 H, Ar Phth), 7.54–7.40 (m, 2 H, Ar), 7.25–6.91 (m, 13 H, Ar), 5.18–4.84 (m, 4 H, 2 × CH_2 benzyl), 4.93–4.81 (m, 1 H, N–C–H), 4.07 (t, J = 13.1 Hz, 1 H, CH_2), 3.02 (td, J = 14.2, 3.8 Hz, 1 H, CH_2), 2.36 (dt, J = 13.9, 3.4 Hz, 1 H, CH_2), 2.31–2.01 (m, 2 H, CH_2), 1.93–1.71 (m, 1 H, CH_2), 1.14 (ddt, J = 13.4, 8.7, 7.2 Hz, 3 H, OTIPS), 1.04 (d, J = 7.2 Hz, 9 H), 0.97 (d, J = 7.2 Hz, 9 H). **^{13}C NMR** (101 MHz, $CDCl_3$) δ 169.2, 169.1, 168.3, 143.3, 135.34, 135.2, 134.0, 132.0, 128.7, 128.3, 128.3, 128.3, 128.1, 128.0, 128.0, 127.4, 126.3, 123.12, 80.5, 66.8, 66.5, 64.56, 46.3, 36.9, 29.5, 25.0, 18.5, 18.5, 14.3. **IR** 2947 (w), 2868 (w), 2362 (w), 2339 (w), 2338 (w), 1755 (w), 1711 (s), 1462 (w), 1370 (m), 1230 (m), 1031 (s). **HRMS (ESI)** calcd for $C_{45}H_{51}NNaO_7Si^+$ $[M + Na]^+$ 768.3327; found 768.3323.

Dimethyl-2-(*tert*-butyldimethylsilyl)oxy)-4-(1,3-dioxoisoindolin-2-yl)-2-phenylcyclohexane-1,1-dicarboxylate (358c)

Following GP30 using dimethyl 2-(1,3-dioxoisoindolin-2-yl) cyclobutane-1,1-dicarboxylate (**325a**) (63.5 mg, 0.200 mmol, 1 eq) and *tert*-butyldimethyl(1-phenylvinyl)oxy)silane (**219b**) (70.4 mg, 0.300 mmol, 1.5 eq), dimethyl-2-(*tert*-butyldimethylsilyl)oxy)-4-(1,3-dioxoisoindolin-2-yl)-2-phenylcyclohexane-1,1-dicarboxylate (**358c**) (105 mg, 0.190 mmol, 95 % yield) was obtained as a colorless solid.

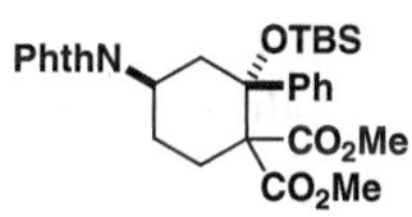

Mp 188.8–190.0 °C. **^{1}H NMR** (400 MHz, $CDCl_3$) δ 7.83 (dd, J = 5.5, 3.0 Hz, 2 H, Phth), 7.72 (dd, J = 5.4, 3.1 Hz, 2 H, Phth), 7.44–7.38 (m, 2 H, Ar), 7.26–7.21 (m, 3 H, Ar), 4.86 (tt, J = 12.5, 4.7 Hz, 1 H, N–C–H), 3.96 (dd, J = 13.6, 12.6 Hz, 1 H, CH_2), 3.66 (s, 3 H, OMe), 3.64 (s, 3 H, OMe), 2.91 (td, J = 14.2, 3.7 Hz, 1 H, CH_2 cyclohexane), 2.31 (dt, J = 13.8, 3.5 Hz, 1 H, CH_2), 2.23–2.01 (m, 2 H, CH_2 cyclohexane), 1.88–1.74 (m, 1 H, CH_2 cyclohexane), 1.06 (s, 9 H, Si–^{t}Bu), 0.14 (s, 3 H, Si–Me), −0.55 (s, 3 H, Si–Me). **^{13}C NMR** (101 MHz, $CDCl_3$) δ 170.1, 169.9, 168.4, 143.0, 134.1, 132.1, 128.8, 127.8, 126.5, 123.3, 79.7, 64.6, 52.3, 52.0, 46.6, 36.7, 29.4, 26.4, 25.3, 19.3, −1.3, −2.5. **IR** 2953 (w), 2890 (w), 2856 (w), 2361 (w), 2339 (w), 1735 (s), 1714 (s), 1373 (m), 1274 (m), 1245 (m), 1039 (m), 1038 (m). **HRMS (ESI)** calcd for $C_{30}H_{37}NNaO_7Si^+$ $[M + Na]^+$ 574.2231; found 574.2228.

Dimethyl-2-(*tert*-butyldimethylsilyl)oxy)-4-(1,3-dioxoisoindolin-2-yl)-2-(4-methoxyphenyl)cyclohexane-1,1-dicarboxylate (358d)

Following GP30 using dimethyl 2-(1,3-dioxoisoindolin-2-yl) cyclobutane-1,1-dicarboxylate (**325a**) (65.5 mg, 0.200 mmol, 1 eq) and *tert*-butyl (1-(4-methoxyphenyl)vinyl)oxy)dimethylsilane (**360b**) (79.0 mg, 0.300 mmol, 1.5 eq), dimethyl-2-(*tert*-butyldimethylsilyl)oxy)-4-(1,3-dioxoisoindolin-2-yl)-2-(4-methoxyphenyl)cyclohexane-1,1-dicarboxylate (**358d**) (105.2 mg, 0.181 mmol, 90 %) was obtained as a colorless solid.

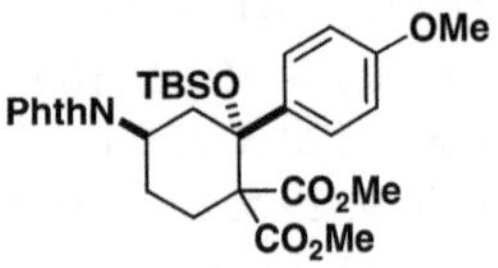

^{1}H NMR (400 MHz, $CDCl_3$) δ 7.83 (dd, J = 5.5, 3.0 Hz, 2 H, Ar Phth), 7.71 (dd, J = 5.4, 3.1 Hz, 2 H, Ar Phth), 7.34–7.29 (m, 2 H, Ar), 6.91–6.64 (m, 2 H, Ar), 5.05–4.69 (m, 1 H, N–C–H), 3.92 (dd, J = 13.5, 12.6 Hz, 1 H, CH_2), 3.79 (s, 3 H, OMe), 3.67 (s, 3 H, OMe), 3.64 (s, 3 H, OMe), 2.89 (td, J = 14.2, 3.7 Hz, 1 H, CH_2), 2.30 (dt, J = 13.8, 3.5 Hz, 1 H, CH_2), 2.22–1.96 (m, 2 H, CH_2), 1.90–1.69 (m, 1 H, CH_2), 1.05 (s, 9 H, OTBS), 0.14 (s, 3 H, OTBS), −0.50 (s, 3 H, OTBS). **^{13}C NMR** (101 MHz, $CDCl_3$) δ 171.5, 171.2, 169.7, 160.12, 136.5, 135.4, 133.4, 131.2, 124.6, 113.0, 80.7, 65.8, 56.5, 53.6, 53.2, 47.8, 38.2, 30.7, 27.7, 26.6, 20.6, −0.0, −1.1. **IR** 2953 (w), 2933 (w), 2856 (w), 1731 (s), 1712 (s), 1612 (w), 1373 (m), 1373 (m), 1254 (m), 1037 (s), 840 (m). **HRMS (ESI)** calcd for $C_{31}H_{40}NO_8Si^+$ $[M + H]^+$ 582.2518; found 582.2500.

Dimethyl-2-(*tert*-butyldimethylsilyl)oxy)-4-(1,3-dioxoisoindolin-2-yl)-2-(4-fluorophenyl)cyclohexane-1,1-dicarboxylate (358e)

Following GP30 using dimethyl 2-(1,3-dioxoisoindolin-2-yl) cyclobutane-1,1-dicarboxylate (**325a**) (63.5 mg, 0.200 mmol, 1 eq) and *tert*-butyl (1-(4-fluorophenyl)vinyl)oxy)dimethylsilane (**360c**) (76.0 mg, 0.300 mmol, 1.5 eq), dimethyl-2-(*tert*-butyldimethylsilyl)oxy)-4-(1,3-dioxoisoindolin-2-yl)-2-(4-fluorophenyl)cyclohexane-1,1-dicarboxylate (**358e**) (113 mg, 0.198 mmol, 99 %) was obtained as a colorless solid.

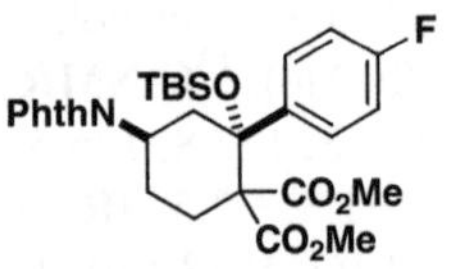

^{1}H NMR (400 MHz, $CDCl_3$) δ 7.83 (dd, J = 5.5, 3.0 Hz, 2 H, Ar Phth), 7.72 (dd, J = 5.4, 3.0 Hz, 2 H, Ar Phth), 7.48–7.34 (m, 2 H, Ar), 6.93 (dd, J = 9.3, 8.2 Hz, 2 H, Ar), 4.84 (tt, J = 12.6, 4.6 Hz, 1 H, N–C–H), 3.97 (dd, J = 13.6, 12.6 Hz, 1 H,

CH_2), 3.67 (s, 3 H, OMe), 3.64 (s, 3 H, OMe), 2.88 (td, J = 14.2, 3.7 Hz, 1 H, CH_2), 2.31 (dt, J = 13.9, 3.4 Hz, 1 H, CH_2), 2.22–1.94 (m, 2 H, CH_2), 1.91–1.74 (m, 1 H, CH_2), 1.05 (s, 9 H, OTBS), 0.17 (s, 3 H, OTBS), −0.51 (s, 3 H, OTBS). **^{13}C NMR** (101 MHz, $CDCl_3$) δ 169.8, 169.7, 168.3, 134.0, 132.0, 130.5 (d, J = 8.0 Hz), 123.2, 113.1 (d, J = 21.1 Hz), 79.2, 64.4, 52.2, 51.8, 46.3, 36.7, 29.2, 26.2, 25.1, 19.1, −1.4, −2.5.[31] **IR** 2953 (w), 2952 (w), 2857 (w), 2856 (w), 1735 (s), 1715 (s), 1375 (m), 1245 (m), 1244 (m), 1039 (m). **HRMS (ESI)** calcd for $C_{30}H_{36}FNNaO_7Si^+$ $[M + Na]^+$ 592.2137; found 592.2142.

Dimethyl-2-(*tert*-butyldimethylsilyl)oxy)-4-(1,3-dioxoisoindolin-2-yl)-2-(4-(methoxycarbonyl)phenyl)cyclohexane-1,1-dicarboxylate (358f)

Following GP30 using dimethyl 2-(1,3-dioxoisoindolin-2-yl) cyclobutane-1,1-dicarboxylate (**325a**) (63.5 mg, 0.200 mmol, 1 eq) and methyl 4-(1-(tert-butyldimethylsilyl)oxy)vinyl)benzoate (**360a**) (88.0 mg, 0.300 mmol, 1.5 eq). A mixture of two diastereomers of dimethyl-2-(*tert*-butyldimethylsilyl) oxy)-4-(1,3-dioxoisoindolin-2-yl)-2-(4-methoxycarbonyl)phenyl) cyclohexane-1,1-dicarboxylate (**358f**) (99.6 mg, 0.163 mmol, 82 %) was obtained as a colorless solid.

dr = 2:1, Diastereoisomeric ratio was calculated in ^{1}H crude NMR integrating peaks at 0.17 and 0.12. Analysis of isolated fractions of the chromatography column allowed the characterization of major *cis* diastereomer.

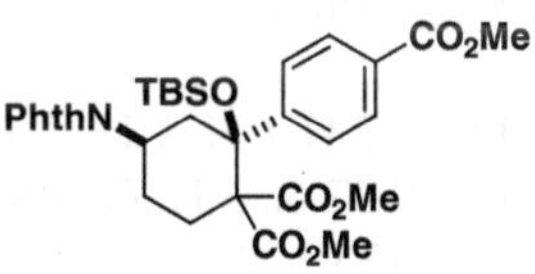

^{1}H NMR (400 MHz, $CDCl_3$) δ 7.97–7.91 (m, 2 H, Ar), 7.84 (dd, J = 5.4, 3.0 Hz, 2 H, Ar Phth), 7.72 (dd, J = 5.4, 3.0 Hz, 2 H, Ar Phth), 7.55–7.48 (m, 2 H, Ar), 4.85 (tt, J = 12.5, 4.7 Hz, 1 H, N–C–H), 3.99 (dd, J = 13.6, 12.6 Hz, 1 H, CH_2), 3.90 (s, 3 H, OMe), 3.66 (s, 3 H, OMe), 3.63 (s, 3 H, OMe), 2.90 (td, J = 14.2, 3.7 Hz, 1 H, CH_2), 2.32 (dt, J = 13.8, 3.4 Hz, 1 H, CH_2), 2.21–2.01 (m, 2 H, CH_2), 1.83 (dd, J = 13.1, 3.5 Hz, 1 H, CH_2), 1.05 (s, 9 H, OTBS), 0.17 (s, 3 H, OTBS), −0.56 (s, 3 H, OTBS). **^{13}C NMR** (101 MHz, $CDCl_3$) δ 169.6, 169.6, 168.3, 166.9, 148.1, 134.0, 131.9, 129.2, 128.8, 127.6, 123.2, 79.4, 64.4, 52.2, 52.0, 51.9, 46.2, 36.5, 29.2, 26.2, 25.1, 19.1, −1.4, −2.5. **IR** 2955 (w), 2955 (w), 2902 (w), 2902 (w), 2901 (w), 2901 (w), 2860 (w), 1716 (s), 1375 (m), 1375 (m), 1281 (m), 1281 (m), 1246 (w), 1040 (m), 1040 (m). **HRMS (ESI)** calcd for $C_{32}H_{39}NNaO_9Si^+$ $[M + Na]^+$ 632.2286; found 632.2282.

[31] Carbon with fluoride not resolved.

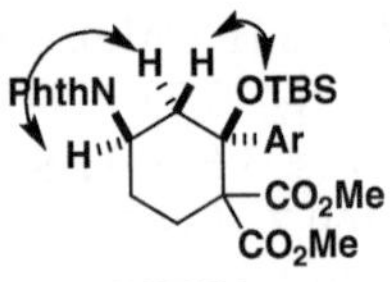

ROESY

Dibenzyl-2-(*tert*-butyldimethylsilyl)oxy)-4-(1,3-dioxoisoindolin-2-yl)-2-phenylcyclohexane-1,1-dicarboxylate (358g)

Following GP30 using dibenzyl 2-(1,3-dioxoisoindolin-2-yl) cyclobutane-1,1-dicarboxylate (**325u**) (94.0 mg, 0.200 mmol, 1 eq) and *tert*-butyldimethyl(1-phenylvinyl)oxy)silane (**219a**) (70.4 mg, 0.300 mmol, 1.5 eq), dibenzyl-2-(*tert*-butyldimethylsilyl)oxy)-4-(1,3-dioxoisoindolin-2-yl)-2-phenylcyclohexane-1,1-dicarboxylate (**358 g**) (101 mg, 0.143 mmol, 72 %) was obtained as a colorless solid.

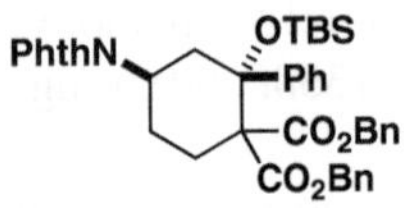

^{1}H NMR (400 MHz, $CDCl_3$) δ 7.84 (dd, *J* = 5.5, 3.0 Hz, 2 H, Ar Phth), 7.72 (dd, *J* = 5.4, 3.0 Hz, 2 H, Ar Phth), 7.43–7.36 (m, 2 H, Ar), 7.24–7.03 (m, 13 H, Ar), 5.20–4.82 (m, *4 H*, 2 × CH_2 benzyl), 4.92–4.81 (m, 1 H, N–C–H), 4.00 (t, *J* = 13.1 Hz, 1 H, CH_2), 2.93 (td, *J* = 14.1, 3.7 Hz, 1 H, CH_2), 2.36 (dt, *J* = 13.8, 3.4 Hz, 1 H, CH_2), 2.26–2.00 (m, 2 H, CH_2), 1.90–1.69 (m, 1 H, CH_2), 1.05 (s, 9 H, OTBS), 0.13 (s, 3 H, OTBS), −0.57 (s, 3 H, OTBS). **^{13}C NMR** (101 MHz, $CDCl_3$) δ 169.3, 169.0, 168.3, 142.7, 135.3, 135.1, 133.9, 132.0, 128.6, 128.3, 128.3, 128.2, 128.0, 128.0, 127.4, 126.4, 123.1, 79.5, 66.8, 66.6, 64.5, 46.3, 36.6, 29.4, 26.2, 25.1, 19.1.[32] **IR** 2954 (w), 2954 (w), 2953 (w), 2953 (w), 2935 (w), 2935 (w), 2858 (w), 1714 (s), 1373 (m), 1273 (w), 1237 (w), 1034 (m). **HRMS (ESI)**. calcd for $C_{42}H_{46}NO_7Si^+$ $[M + H]^+$ 704.3038; found 704.3019.

Di*iso*propyl-2-(tert-butyldimethylsilyl)oxy)-4-(1,3-dioxoisoindolin-2-yl)-2-phenylcyclohexane-1,1-dicarboxylate (358h)

Following GP30 using di*iso*propyl-2-(1,3-dioxoisoindolin-2-yl) cyclobutane-1,1-dicarboxylate (**325l**) (74.7 mg, 0.200 mmol, 1 eq) and *tert*-butyldimethyl(1-phenylvinyl)oxy)silane (**219a**) (70.4 mg, 0.300 mmol, 1.5 eq), di*iso*propyl-2-(*tert*-butyldimethylsilyl)oxy)-4-(1,3-dioxoisoindolin-2-yl)-2-phenylcyclohexane-1,1-dicarboxylate (66.4 mg, 0.109 mmol, 55, 68 % brsm) was obtained as a colorless solid in addition with di*iso*propyl-2-(1,3-dioxoisoindolin-2-yl)cyclobutane-1,1-dicarboxylate (**358h**) (66.4 mg, 0.109 mmol, 55 %).

[32]Two peaks of the TBS have been cut during analysis. One aromatic peak overlapped.

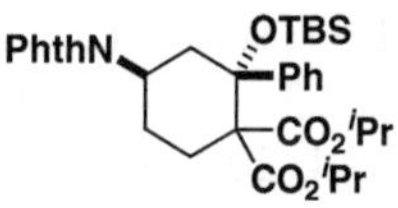

^{1}H NMR (400 MHz, $CDCl_3$) δ 7.72 (td, J = 5.1, 2.7 Hz, 2 H, Ar Phth), 7.63–7.49 (m, 2 H, Ar Phth), 7.40–7.31 (m, 2 H, Ar), 7.13–7.05 (m, 3 H, Ar), 4.91 (dsept J = 10.3, 6.2 Hz, 2 H, CH isopropyl), 4.83–4.62 (m, 1 H, N–C–H), 3.76 (t, J = 13.1 Hz, 1 H, CH_2), 2.75 (td, J = 14.1, 3.7 Hz, 1 H, CH_2), 2.19 (dt, J = 13.6, 3.4 Hz, 1 H, CH_2), 2.08 (qd, J = 13.1, 3.8 Hz, 1 H, CH_2), 2.00–1.90 (m, 1 H, CH_2), 1.76–1.62 (m, 1 H, CH_2), 1.08–0.96 (m, 12H, CH_3 isopropyl), 0.95 (s, 9 H, OTBS), 0.00 (s, 3 H, OTBS), −0.67 (s, 3 H, OTBS). **^{13}C NMR**[33] (101 MHz, $CDCl_3$) δ 170.8, 170.2, 169.8, 144.7, 135.4, 133.5, 130.4, 128.7, 127.8, 124.6, 80.8, 70.5, 69.6, 65.6, 47.9, 38.3, 31.0, 27.8, 26.6, 23.0, 22.9, 22.7, 20.6, −0.0, −1.2. **IR** 2914 (w), 2865 (w), 2865 (w), 2824 (w), 2823 (w), 2785 (w), 2785 (w), 2785 (w), 1634 (w), 1575 (s), 1210 (m), 1073 (m), 931 (m), 847 (m). **HRMS (ESI)**.calcd for $C_{34}H_{45}NNaO_7Si^+$ $[M + Na]^+$ 630.2857; found 630.2873.

Dimethyl-2-(*tert*-butyldimethylsilyl)oxy)-4-(1,3-dioxoisoindolin-2-yl)-3-methyl-2-phenylcyclohexane-1,1-dicarboxylate (358i)

Following GP30 using dimethyl 2-(1,3-dioxoisoindolin-2-yl) cyclobutane-1,1-dicarboxylate (**325a**) (63.5 mg, 0.200 mmol, 1 eq) and (*Z*)-*tert*-butyldimethyl(1-phenylprop-1-en-1-yl)oxy)silane (74.6 mg, 0.300 mmol, 1.5 eq), a mixture of two diastereomers of dimethyl-2-(*tert*-butyldimethylsilyl)oxy)-4-(1,3-dioxoisoindolin-2-yl)-3-methyl-2-phenylcyclohexane-1,1-dicarboxylate (**358i**) (102 mg, 0.181 mmol, 90 %) was obtained as a colorless solid.

dr = 2:1, Diastereoisomeric ratio was calculated in ^{1}H crude NMR integrating peaks at 4.75 and 4.33. Analysis of isolated fractions of the chromatography column allowed the characterization of each diastereomer.

Major *cis*-diastereoisomer

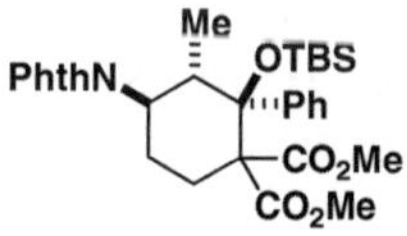

^{1}H NMR (400 MHz, $CDCl_3$) δ 7.99–7.86 (m, 2 H, Ar Phth), 7.86–7.73 (m, 2 H, Ar Phth), 7.70–7.59 (m, 2 H, Ar), 7.32–7.24 (m, 3 H, Ar), 4.62 (td, J = 12.0, 4.7 Hz, 1 H, N–C–H), 4.38 (dq, J = 11.5, 6.8 Hz, 1 H, CH–Me), 3.71 (s, 3 H, OMe), 3.53 (s, 3 H, OMe), 3.03–2.80 (m, 1 H, CH_2), 2.42–2.25 (m, 2 H, CH_2), 1.93–1.80 (m, 1 H, CH_2), 1.17 (s, 9 H, OTBS), 0.92 (d, J = 6.8 Hz, 3 H, CH_3), 0.42

[33] 2 peaks overlap.

(s, 3 H, OTBS), 0.00 (s, 3 H, OTBS). **^{13}C NMR**[34] (101 MHz, $CDCl_3$) δ 169.8, 169.7, 168.6, 140.7, 134.1, 132.3, 131.6, 129.7, 127.2, 123.4, 85.5, 65.8, 52.2, 52.2, 51.5, 36.3, 29.4, 26.6, 25.4, 20.4, 13.6, −0.1, −1.6. **IR** 2951 (w), 2950 (w), 2857 (w), 1736 (s), 1713 (s), 1372 (m), 1371 (m), 1276 (m), 1240 (m), 1134 (m), 1051 (m), 836 (w). **HRMS (ESI)** calcd for $C_{31}H_{40}NO_7Si^+$ $[M + H]^+$ 566.2569; found 566.2563.

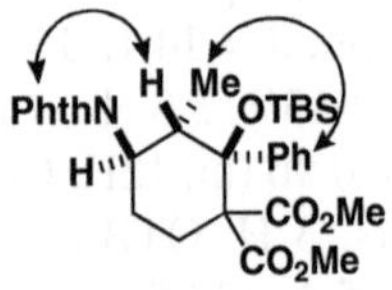

ROESY

Minor *trans*-diastereoisomer

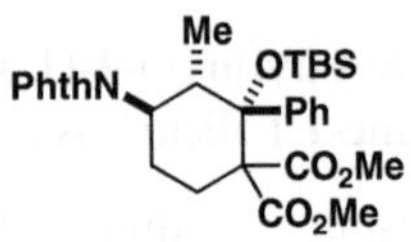

^{1}H NMR (400 MHz, $CDCl_3$) δ 8.11–8.01 (m, 2 H, Ar Phth), 7.81 (ddd, J = 7.7, 1.3, 0.7 Hz, 1 H, Ar), 7.63 (ddd, J = 7.0, 1.6, 0.7 Hz, 1 H, Ar), 7.56–7.37 (m, 5 H, Ar + Ar Phth), 4.76 (dd, J = 9.4, 7.2 Hz, 1 H, CH–Me), 4.15 (ddd, J = 12.3, 9.4, 2.9 Hz, 1 H, N–C–H), 3.87 (s, 3 H, OMe), 3.40 (s, 3 H, OMe), 2.88 (td, J = 14.2, 3.9 Hz, 1 H, CH_2), 2.42 (dt, J = 14.2, 3.4 Hz, 1 H, CH_2), 2.00 (dd, J = 13.3, 3.3 Hz, 1 H, CH_2), 1.63–1.46 (m, 1 H, CH_2), 1.18 (d, J = 7.2 Hz, 3 H, CH_3), 0.94 (s, 9 H, OTBS), −0.03 (s, 3 H, OTBS), −0.34 (s, 3 H, OTBS). **^{13}C NMR** the sample was degraded and a mixture of the product and a side product was present (probably the open product as there is a peak at 200 ppm). **IR** 2955 (w), 2859 (w), 1740 (m), 1703 (s), 1702 (s), 1329 (w), 1272 (m), 1221 (w), 1093 (m), 1072 (m), 863 (w), 842 (w). **HRMS (ESI)**. calcd for $C_{31}H_{40}NO_7Si^+$ $[M + H]^+$ 566.2569; found 566.2568.

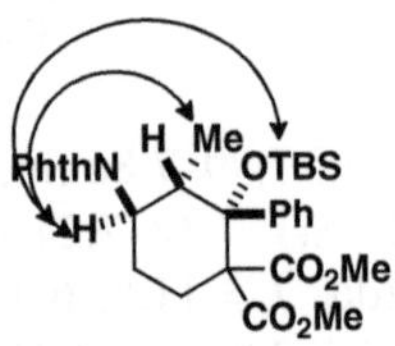

ROESY

[34]Peaks of impurities are present in the aromatic area.

Dimethyl-4-amino-2-(tert-butyldimethylsilyl)oxy)-2-phenylcyclohexane-1,1-dic arboxylate (361)

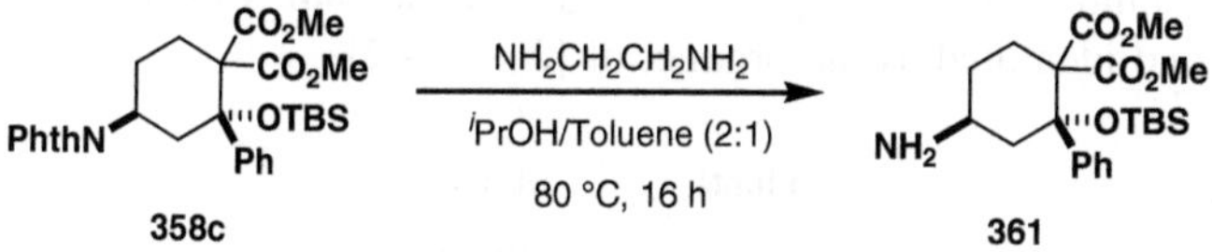

Dimethyl 2-(tert-butyldimethylsilyl)oxy)-4-(1,3-dioxoisoindolin-2-yl)-2-phenylcyclohexane-1,1-dicarboxylate (**358c**) (50 mg, 0.091 mmol, 1 eq) in iso-propanol (1.0 mL) and toluene (0.5 mL) was added in an oven dried 5 mL round bottom flask followed by diaminoethane (27.2 mg, 0.45 mmol, 5 eq). The vial was sealed and the solution was heated to 80 °C for 16 h. The solvent was removed under reduced pressure and the crude was purified by column chromatography (SiO_2, 9:1 DCM/Methanol) affording dimethyl 4-amino-2-(tert-butyldimethylsilyl) oxy)-2-phenylcyclohexane-1,1-dicarboxylate (**361**) (33 mg, 0.079 mmol, 87 % yield, >95 % pure) as a colorless oil.

$\mathbf{R_f}$ 0.20 (9:1 DCM/Methanol). **^{1}H NMR** (400 MHz, $CDCl_3$) δ 7.44–7.38 (m, 2 H, Ar), 7.25–7.22 (m, 3 H, Ar), 3.60 (s, 3 H, COOMe), 3.55 (s, 3 H, COOMe), 3.23 (ddt, J = 11.7, 8.7, 4.2 Hz, 1 H, N-CH), 2.81–2.61 (m, 2 H, CH_2), 2.24–2.06 (m, 2 H, CH_2), 1.99–1.85 (m, 1 H, CH_2), 1.43–1.22 (br, 7 H, NH_2 and H_2O),[35] 1.02–0.84 (m, 10 H, Si–^{t}Bu and CH_2), 0.05 (s, 3 H, SiMe), −0.60 (s, 3 H, SiMe). **^{13}C NMR** (101 MHz, $CDCl_3$) δ 170.2, 169.9, 143.2, 128.6, 127.4, 126.1, 79.5, 64.4, 51.8, 51.5, 46.0, 44.6, 32.5, 29.4, 26.0, 18.9, −1.4, −3.1. **IR** 3056 (w), 2953 (w), 2888 (w), 2860 (w), 1732 (m), 1447 (w), 1436 (w), 1266 (s), 1155 (w), 1136 (w), 1044 (m), 1031 (m). **HRMS (ESI)** calcd for $C_{22}H_{36}NO_5Si^+$ $[M + H]^+$ 422.2357; found 422.2353.

Dimethyl 2,4-bis(1,3-dioxoisoindolin-2-yl)cyclohexane-1,1-dicarboxylate (362)

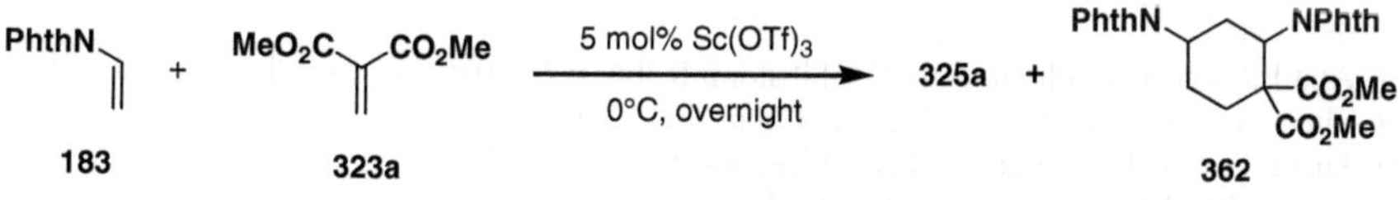

Scandium triflate (0.16 g, 0.33 mmol, 5 mol%) was added in a flask under a nitrogen atmosphere and 5 mL dichloromethane were added. The suspension was cooled to 0 °C with an ice bath and vinyl phthalimide (1.6 g, 9.5 mmol, 1.4 eq) dissolved in 10 mL of dichloromethane was added. Vinylidene malonate (1.0 g, 6.9 mmol, 1 eq) was dissolved in 10 mL dichloromethane and added dropwise over

[35] NH_2 peak over-integrated due to water in $CDCl_3$.

30 min. After the addition was complete, the reaction was concentrated and left in the fridge overnight at 4 °C. Column chromatography on Biotage (SNAP cartridge KP-SIL 25 g, 95:5–4:6 Hexane/Ethyl acetate) afforded the desired product **325a** (1.3 g, 4.1 mmol, 59 %) along with a side product **362** which was isolated for analysis[36] and identified as the product of [2 + 2 + 2].

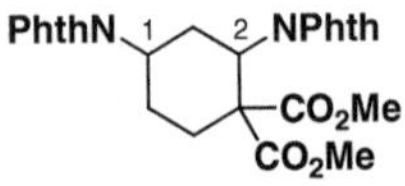

^{1}H NMR (400 MHz, $CDCl_3$) δ 7.87 (dd, J = 5.4, 3.1 Hz, 2 H, Ar phth), 7.80 (dd, J = 5.5, 3.1 Hz, 2 H, Ar phth), 7.75 (dd, J = 5.4, 3.1 Hz, 2 H, Ar phth), 7.70 (dd, J = 5.4, 3.1 Hz, 2 H, Ar phth), 5.65 (dt, J = 7.1, 1.6 Hz, 1 H, N–C^2–H), 5.42 (ddd, J = 12.5, 7.5, 5.2 Hz, 1 H, N–C^1–H), 3.93 (s, 3 H, OMe), 3.50 (s, 3 H, OMe), 3.17 (ddd, J = 14.6, 12.9, 7.1 Hz, 1 H, CH_2), 3.00 (td, J = 14.2, 3.7 Hz, 1 H, CH_2), 2.58–2.16 (m, 2 H, CH_2), 2.05–1.71 (m, 2 H, CH_2). **^{13}C NMR** (101 MHz, $CDCl_3$) δ 170.1, 169.7, 168.3, 168.2, 134.3, 133.9, 131.9, 131.4, 123.5, 123.1, 57.9, 53.3, 52.7, 48.3, 45.5, 30.7, 27.9, 24.5. **IR** 2893 (w), 2817 (w), 1568 (s), 1206 (m), 1206 (m), 1172 (m), 1157 (m), 1084 (m), 1084 (m), 1037 (m), 885 (m), 716 (s). **HRMS (ESI)** (See footnote 30).

References

1. am Ende DJ, DeVries KM, Clifford PJ, Brenek SJ (1998) Org Process Res Dev 2:382
2. Brice JL, Meerdink JE, Stahl SS (1845) Org Lett 2004:6
3. Yar M, Fritz SP, Gates PJ, McGarrigle EM, Aggarwal VK (2012) Eur J Org Chem 2012:160
4. Wyatt P, Hudson A, Charmant J, Orpen AG, Phetmung H (2006) Org Biomol Chem 4:2218
5. Kitamura M, Tashiro N, Miyagawa S, Okauchi T (1037) Synthesis 2011:7
6. González-Bobes F, Fenster MDB, Kiau S, Kolla L, Kolotuchin S, Soumeillant M (2008) Adv Synth Catal 350:813
7. Miller JA, Hennessy EJ, Marshall WJ, Scialdone MA, Nguyen ST (2003) J Org Chem 68:7884
8. Abu-Elfotoh A-M, Phomkeona K, Shibatomi K, Iwasa S (2010) Angew Chem Int Ed 49:8439
9. Zhao J-F, Tan B-H, Loh T-P (2011) Chem Sci 2:349
10. Zhang L, Sun J, Kozmin SA (2006) Tetrahedron 62:11371
11. Yu JQ, Wu HC, Corey EJ (2005) Org Lett 7:1415
12. Larsen SD, May P, Romines K, Schnute ME, Tanis SP (2004) Patent: US2004/138449 A1
13. Magnus P, Barth L (1995) Tetrahedron 51:11075
14. Kanie O, Crawley SC, Palcic MM, Hindsgaul O (1993) Carbohydr Res 243:139
15. Krapcho AP, Jahngen EGE Jr, Lovey AJ, Short FW (1091) Tetrahedron Lett 1974:15
16. Takacs JM, Xu Z, Jiang X-T, Leonov AP, Theriot GC (2002) Org Lett 4:3843
17. Kacprzak K (2003) Synth Commun 33:1499
18. Baret N, Dulcere J-P, Rodriguez J, Pons J-M, Faure R (2000) Eur J Org Chem 2000:1507

[36] No yield was recorded.

19. Tolnai GL, Ganss S, Brand JP, Waser J (2012) Org Lett 15:112
20. Brand JP, Charpentier J, Waser J (2009) Angew Chem Int Ed 48:9346
21. Kanie O, Crawley SC, Palcic MM, Hindsgaul O (1993) Carbohydr Res 243:139
22. Kaburagi Y, Tokuyama H, Fukuyama T (2004) J Am Chem Soc 126:10246
23. Mazzocchi PH, Wilson P, Khachik F, Klingler L, Minamikawa S (1983) J Org Chem 48:2981
24. Hinde NJ, Hall CD (1998) J Chem Soc Perkin Trans 2:1249
25. Kacprzak K (2003) Synth Commun 33:1499
26. Bayer E, Geckeler K (1979) Angew Chem Int Ed 18:533
27. Sibi MP, Liu M (2000) Org Lett 2:3393
28. de Nanteuil F, Waser J (2011) Angew Chem Int Ed 50:12075
29. Bosch M, Schalf M (2003) J Org Chem 68:5225
30. Harada N-A, Nishikata T, Nagashima H (2012) Tetrahedron 68:3243
31. Matysiak S, Fitznar H-P, Schnell R, Pfleiderer W (1998) Helv Chim Acta 81:1545
32. Nummert V, Piirsalu M, Mäemets V, Vahur S, Koppel IA (2009) J Phys Org Chem 22:1155
33. Liu H, Shi G, Pan S, Jiang Y, Zhang Y (2013) Org Lett 15:4098
34. Ji M, Wang X, Lim YN, Kang Y-W, Jang H-Y (2013) Eur J Org Chem 35:7881
35. Payack JF, Hughes DL, Cai D, Cottrell IF, Verhoeven TR (2002) Org Synth 79:19
36. Vuluga D, Legros J, Crousse B, Bonnet-Delpon D (2009) Eur J Org Chem 2009:3513
37. Evans DA, Peterson GS, Johnson JS, Barnes DM, Campos KR, Woerpel KA (1998) J Org Chem 63:4541
38. Bugarin A, Jones KD, Connell BT (2010) Chem Commun 46:1715
39. De Fusco C, Fuoco T, Croce G, Lattanzi A (2012) Org Lett 14:4078
40. Jabin I, Revial G, Monnier-Benoit N, Netchitailo P (2001) J Org Chem 66:256
41. Nickerson DM, Mattson AE (2012) Chem Eur J 18:8310
42. Wen L, Shen Q, Lu L (2010) Org Lett 12:4655
43. Nickerson DM, Mattson AE (2012) Chem Eur J 18:8310
44. Stojanovic A, Renaud P, Schenk K (2004) Helv Chim Acta 81:268
45. Lam PYS, Vincent G, Bonne D, Clark CG (2003) Tetrahedron Lett 44:4927
46. Gooßen LJ, Blanchot M, Brinkmann C, Gooßen K, Karch R, Rivas-Nass A (2006) J Org Chem 71:9506
47. Alacid E, Nájera C (2008) Adv Synth Catal 350:1316
48. Susanto W, Chu C-Y, Ang WJ, Chou T-C, Lo L-C, Lam Y (2012) J Org Chem 77:2729
49. Pawluć P, Franczyk A, Walkowiak J, Hreczycho G, Kubicki M, Marciniec B (2012) Tetrahedron 68:3545
50. Parsons AT, Johnson JS (2009) J Am Chem Soc 131:14202
51. Benfatti F, Nanteuil FD, Waser J (2011) Org Lett 14:386
52. Alford JS, Davies HML (2012) Org Lett 14:6020
53. Ventura DL, Li Z, Coleman MG, Davies HML (2009) Tetrahedron 65:3052

MIX
Papier aus verantwortungsvollen Quellen
Paper from responsible sources
FSC® C105338
FSC
www.fsc.org

If you have any concerns about our products,
you can contact us on
ProductSafety@springernature.com

In case Publisher is established outside the EU,
the EU authorized representative is:
Springer Nature Customer Service Center GmbH
Europaplatz 3, 69115 Heidelberg, Germany

Printed by Libri Plureos GmbH
in Hamburg, Germany